T0180837

Studies in Computational Intelligence

Volume 547

Series editor

Janusz Kacprzyk, Polish Academy of Sciences, Warsaw, Poland
e-mail: kacprzyk@ibspan.waw.pl

For further volumes:
http://www.springer.com/series/7092

About this Series

The series "Studies in Computational Intelligence" (SCI) publishes new developments and advances in the various areas of computational intelligence—quickly and with a high quality. The intent is to cover the theory, applications, and design methods of computational intelligence, as embedded in the fields of engineering, computer science, physics and life sciences, as well as the methodologies behind them. The series contains monographs, lecture notes, and edited volumes in computational intelligence spanning the areas of neural networks, connectionist systems, genetic algorithms, evolutionary computation, artificial intelligence, cellular automata, self-organizing systems, soft computing, fuzzy systems, and hybrid intelligent systems. Of particular value to both the contributors and the readership are the short publication timeframe and the world-wide distribution, which enable both wide and rapid dissemination of research output.

Oscar Castillo · Patricia Melin
Witold Pedrycz · Janusz Kacprzyk

Editors

Recent Advances on Hybrid Approaches for Designing Intelligent Systems

 Springer

Editors
Oscar Castillo
Patricia Melin
Division of Graduate Studies
 and Research
Tijuana Institute of Technology
Tijuana
Mexico

Janusz Kacprzyk
Systems Research Institute
Polish Academy of Sciences
Warsaw
Poland

Witold Pedrycz
Department of Electrical and Computer
 Engineering
University of Alberta
Edmonton, AB
Canada

ISSN 1860-949X   ISSN 1860-9503 (electronic)
ISBN 978-3-319-38274-6   ISBN 978-3-319-05170-3 (eBook)
DOI 10.1007/978-3-319-05170-3
Springer Cham Heidelberg New York Dordrecht London

Printed on acid-free paper

Springer is part of Springer Science+Business Media (www.springer.com)

Preface

We describe, in this book, recent advances on hybrid intelligent systems using soft computing techniques for diverse areas of application, such as intelligent control and robotics, pattern recognition, time series prediction, and optimization complex problems. Soft Computing (SC) consists of several intelligent computing paradigms, including fuzzy logic, neural networks, and bio-inspired optimization algorithms, which can be used to produce powerful hybrid intelligent systems. The book is organized into five main parts, which contain a group of papers around a similar subject. The first part consists of chapters with the main theme of type-2 fuzzy logic, which basically consists of chapters that propose new models and applications for type-2 fuzzy systems. The second part contains papers with the main theme of bio-inspired optimization algorithms, which are basically chapters using nature-inspired techniques to achieve optimization of complex optimization problems in diverse areas of application. The third part contains chapters that deal with new models and applications of neural networks in real world problems. The fourth part contains chapters with the theme of intelligent optimization methods, which basically consider the proposal of new methods of optimization to solve complex real world optimization problems. The fifth part contains chapters with the theme of evolutionary methods and intelligent computing, which are chapters considering soft computing methods for applications related to diverse areas, such as natural language processing, recommending systems and optimization.

In the part of type-2 fuzzy logic, there are nine chapters that describe different contributions that propose new models and concepts, which can be the considered as the basis for achieving applications for real-world problems that can have a better management of uncertainty. In the part of bio-inspired algorithms, there are 11 chapters that describe different contributions on proposing new bio-inspired algorithms and their application to solve complex optimization problems. The bio-inspired methods include variations of ant colony optimization and particle swarm optimization, as well as new nature inspired paradigms. In the part of neural networks, there are 10 chapters that describe different contributions of new algorithms and models for neural networks and their application to diverse complex problems in pattern recognition and time series prediction. In the part of intelligent optimization applications, there are 10 contributions that describe the development of new models and algorithms relevant to complex optimization problems, as well as the application of these intelligent optimization techniques in real-world applications. In the part of evolutionary methods and intelligent computing there are 10 contributions on models and algorithms based on computational intelligent techniques, including novel evolutionary approaches, that are presented, as well as their applications to different real-world problems, such as in recommending systems and natural language processing.

In conclusion, the edited book comprises chapters on diverse aspects of bio-inspired models, soft computing and hybrid intelligent systems for control, mobile robotics, pattern recognition, time series prediction, and other complex real world problems. There are theoretical aspects as well as application chapters.

January 8, 2014   Oscar Castillo
                                                                               Patricia Melin
                                                                               Witold Pedrycz
                                                                               Janusz Kacprzyk

Contents

Part I
Type-2 Fuzzy Logic

Part I
Type-2 Fuzzy Logic

Genetic Algorithm Optimization for Type-2 Non-singleton Fuzzy Logic Controllers

Ricardo Martínez-Soto, Oscar Castillo and Juan R. Castro

Abstract In this chapter we study the automatic design of type-2 non-singleton fuzzy logic controller. To test the controller we use an autonomous mobile robot for the trajectory tracking control. We take the basis of the interval type-2 fuzzy logic controller of previous work for the extension to the type-2 non-singleton fuzzy logic controller. A genetic algorithm is used to obtain an automatic design of the type-2 non-singleton fuzzy logic controller (NSFLC). Simulation results are obtained with Simulink showing the behavior of the mobile robot whit this type of controller.

1 Introduction

Optimization is a term used to refer to a branch of computational science concerned with finding the "best" solution to a problem. Optimization algorithms are search methods, where the goal is to find a solution to an optimization problem, such that a given quantity is optimized, possibly subject to a set of constraints [6, 12, 22]. Some optimization methods are based on populations of solutions [21]. Unlike the classic methods of improvement for trajectory tracking, in this case each iteration of the algorithm has a set of solutions. These methods are based on generating, selecting, combining and replacing a set of solutions. Since they

R. Martínez-Soto (✉) · O. Castillo
Division of Graduate Studies and Research, Tijuana Institute of Technology,
Tijuana, Mexico
e-mail: mc.ricardo.martinez@hotmail.com; molerick@hotmail.com

O. Castillo
e-mail: ocastillo@hafsamx.org

J. R. Castro
School of Engineering UABC University, Tijuana, Mexico
e-mail: jrcastror@uabc.edu.mx

O. Castillo et al. (eds.), *Recent Advances on Hybrid Approaches for Designing Intelligent Systems*, Studies in Computational Intelligence 547,
DOI: 10.1007/978-3-319-05170-3_1, © Springer International Publishing Switzerland 2014

maintain and they manipulate a set, instead of a unique solution throughout the entire search process, they use more computer time than other metaheuristic methods. This fact can be aggravated because the "convergence" of the population requires a great number of iterations. For this reason a concerted effort has been dedicated to obtaining methods that are more aggressive and manage to obtain solutions of quality in a nearer horizon.

Mobile robots have attracted considerable interest in the robotics and control research community, because they have non-holonomic properties caused by non-integrable differential constrains [13, 23]. The motion of non-holonomic mechanical systems [11] is constrained by its own kinematics, so the control laws are not derivable in a straightforward manner (Brockett's condition [5]).

Furthermore, most reported designs rely on intelligent control approaches as fuzzy logic control (FLC) [3, 19, 26, 29, 30, 35, 36] and neural networks [11, 33]. However the majority of the publications mentioned above, have concentrated on kinematics models for mobile robots, which are controlled by the velocity input, while less attention has been paid to the control problems of non-holonomic dynamics systems, where forces and torques are true inputs: Bloch [4], Driankov et al. [10] and Chwa [8], used a sliding mode control to the tracking control problem.

This chapter is concerned with the fuzzy logic controller design specially the type-2 non-singleton fuzzy systems, where we use a genetic algorithm for the automatic design. In this case, we used an autonomous mobile robot to test the fuzzy controller for trajectory tracking control. We expect that the type-2 non-singleton fuzzy logic controller win, improve the simulation results when tested under perturbations because in theory it has more ability to handle uncertainties. In this study we only present the first part of the entire research that consist in the design of the type-2 non-singleton fuzzy logic controller using the genetic algorithms.

This chapter is organized as follows: Sect. 2 presents the theoretical basis and problem statement. Section 2.3.1 introduces the fuzzy logic control design where a GA is used to select the parameters of the membership functions of the type-2 non-singleton fuzzy logic controller to apply on the autonomous mobile robot. Robustness properties of the closed-loop system are achieved with a fuzzy logic control system using a Takagi–Sugeno model where the angular velocity error and the linear velocity error, are considered as the linguistic variables. Section 3 provides simulation results of the trajectory tracking of the autonomous mobile robot using the controller described in Sect. 2.3.1. Finally, Sect. 4 presents the conclusions.

2 Theoretical Basis and Problem Statement

This section describes the theoretical basis of the chapter as well as the problem definition, and some basis about Type-2 Fuzzy Systems and genetic algorithms are first presented.

2.1 Type-2 Non-singleton Type-2 Fuzzy Logic Systems

It is known that Type-1 Fuzzy Logic Systems (FLS) are unable to directly handle rule uncertainties, because they use a type-1 fuzzy sets that are certain [37]. On the other hand, type-2 FLS are very useful in circumstances where it is difficult to determine an exact membership function, and there are measurement uncertainties [28]. Type-2 fuzzy sets enable modeling and minimizing the effects of uncertainties in rule-based FLS [16, 27].

As a justification for the use of type-2 fuzzy sets, in [24] at least four sources of uncertainties not considered in type-1 FLS are mentioned:

1. The meaning of the words that are used in the antecedents and consequents of rules can be uncertain (words mean different things to different people).
2. Consequents may have histogram of values associated with them, especially when knowledge is extracted from a group of experts who do not all agree.
3. Measurements that activate a type-1 FLS may be noisy and therefore uncertain.
4. The data used to tune the parameters of a type-1 FLS may also be noisy.

All of these uncertainties translate into uncertainties about fuzzy set membership functions. Type-2 fuzzy sets are able to model such uncertainties because their membership functions are themselves fuzzy.

A type-2 FLS is again characterized by IF–THEN rules, but its antecedent or consequent sets are now of type-2. Type-2 FLS can be used when the circumstances are too uncertain to determine exact membership grades such as when the training data is corrupted by noise. Similar to type-1 FLS, a type-2 FLS includes a fuzzifier, a rule base, fuzzy inference engine, and an output processor, as we can see in Fig. 1. The output processor includes a type-reducer and defuzzifier; it generates a type-1 fuzzy set output (from the type- reducer) or a crisp number (from the defuzzifier) [20, 25].

Now a type-2 FLS whose inputs are modeled as type-2 fuzzy numbers is referred to as a type-2- non-singleton FLS [32]. A type-2 non-singleton FLS is described by the same diagram as is type-2 FLS, Fig. 1. The rules of a type-2 non-singleton FLS are the same as those for a type-2 FLS. What is different is the fuzzifier, which treats the inputs as type-2 fuzzy sets, and the effect of this change is on the inference block. The output of the inference block will again be a type-2 fuzzy set; so, the type-reducers and defuzzifier that we used for Type-2 FLS apply as well to a type-2 non-singleton FLS.

2.1.1 Rules

The structure of the rules in a type-1 FLS and a type-2 FLS is the same, but in the latter the antecedents and the consequents will be represented by type-2 fuzzy sets. So for a type-2 FLS with p inputs $x_1 \in X_1, ..., x_p \in X_p$ and one output $y \in Y$, which is a multiple input single output (MISO) system, if we assume there are M rules, the

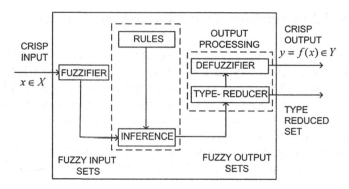

Fig. 1 Type-2 fuzzy logic system

lth rule in the type-2 FLS can be written as follows [31]:R^l : IF x_1 is $\tilde{F}_1^l$ and ... and x_p is $\tilde{F}_p^l$, THEN y is $\tilde{G}^l l = 1, \ldots, M(1)$.

2.1.2 Interval Type-2 Non-singleton TSK Interval Type-2 Fuzzy Inference System

The IT2FIS Takagi-Sugeno-Kang system is designed with **n** inputs, **m** outputs and **r** rules [34]. The kth rule with interval type-2 fuzzy antecedents $\tilde{\tilde{A}}_{k,j} \in \{\mu_{i,l_{k,i}}\}$, interval type-1 fuzzy set are used for the consequents sets, $\tilde{Y}_{j,k}(\tilde{X}) = \sum_{i=1}^{n} \tilde{\theta}_{i,j}^k \cdot \tilde{X}_i + \tilde{\theta}_{0,j}^k$ and real facts are inferred as a direct reasoning:

$$R^k : IF\ x_1\ is\ \tilde{\tilde{A}}_{k,1}\ and\ldots and\ x_n\ is\ \tilde{\tilde{A}}_{k,n}\ THEN\ y_1\ is\ f_{1,k}\ and\ldots and\ y_m\ is\ f_{m,k}$$

$$\frac{H : IF\ x_1\ is\ \tilde{\tilde{A}}_1'\ and\ldots and\ x_n\ is\ \tilde{\tilde{A}}_n'}{C : y_1\ is\ \hat{y}_1\ and\ldots and\ y_m\ is\ \hat{y}_m}$$

The type-2 non-singleton fuzzy inference system defuzzifier output is:

$$\hat{y}_j(\tilde{X}) = \frac{\hat{y}_j^l(x^-) + \hat{y}_j^r(x^+)}{2} \tag{1}$$

2.2 Genetic Algorithms (GA)

Genetic Algorithms (GAs) are adaptive heuristic search algorithms based on the evolutionary ideas of natural selection and genetic processes [21]. The basic principles of GAs were first proposed by John Holland in 1975, inspired by the mechanism of natural selection where stronger individuals are likely the winners in a competing environment [1, 2, 9]. GA assumes that the potential solution of any problem is an individual and can be represented by a set of parameters. These parameters are regarded as the genes of a chromosome and can be structured by a string of values in binary form. A positive value, generally known as a fitness value, is used to reflect the degree of "goodness" of the chromosome for the problem which would be highly related with its objective value. The GA works as follows:

1. Start with a randomly generated population of n chromosomes (candidate solutions to a problem).
2. Calculate the fitness of each chromosome in the population.
3. Repeat the following steps until *n* offspring have been created:

   a. Select a pair of parent chromosomes from the current population, the probability of selection being an increasing function of fitness. Selection is done with replacement, meaning that the same chromosome can be selected more than once to become a parent.
   b. With probability (crossover rate), perform crossover to the pair at a randomly chosen point to a form two offspring.
   c. Mutate the two offspring at each locus with probability (mutation rate), and place the resulting chromosomes in the new population.

4. Replace the current population with the new population.
5. Go to step 2.

The simple procedure just described above is the basis for most applications of GAs.

2.3 Problem Statement

To test the optimized FLCs obtained with the bio-inspired methods; we used a mobile robot model that we were used in previous works [22, 23]. The model considered is a unicycle mobile robot (Fig. 2), it consists of two driving wheels mounted of the same axis and a front free wheel.

A unicycle mobile robot is an autonomous, wheeled vehicle capable of performing missions in fixed or uncertain environments [18]. The robot body is symmetrical around the perpendicular axis and the center of mass is at the geometric center of the body. It has two driving wheels that are fixed to the axis that passes through center of mass "C" and one passive wheel prevents the robot from tipping over as it moves on a plane.

In what follows, it is assumed that the motion of the passive wheel can be ignored in the dynamics of the mobile robot represented by the following set of equations [13]:

Fig. 2 Wheeled mobile robot

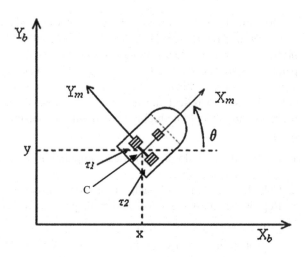

$$M(q)\dot{v} + C(q,\dot{q})v + Dv = \tau + P(t) \tag{2}$$

$$\dot{q} = \underbrace{\begin{bmatrix} \cos\theta & 0 \\ \sin\theta & 0 \\ 0 & 1 \end{bmatrix}}_{J(q)} \underbrace{\begin{bmatrix} v \\ w \end{bmatrix}}_{\upsilon} \tag{3}$$

where $q = (x, y, \theta)^T$ is the vector of the configuration coordinates; $\upsilon = (v, w)^T$ is the vector of velocities; $\tau = (\tau_1, \tau_2)$ is the vector of torques applied to the wheels of the robot where τ_1 and τ_2 denote the torques of the right and left wheel, respectively (Fig. 2); $P \in R^2$ is the uniformly bounded disturbance vector; $M(q) \in R^{2\times 2}$ is the positive-definite inertia matrix; $C(q,\dot{q})\vartheta$ is the vector of centripetal and Coriolis forces; and $D \in R^{2\times 2}$ is a diagonal positive-definite damping matrix. Equation (3) represents the kinematics of the system, where (x, y) is the position in the X–Y (world) reference frame; θ is the angle between the heading direction and the x-axis; v and w are the linear and angular velocities, respectively. Furthermore, the system (2)–(3) has the following non-holonomic constraint:

$$\dot{y}\cos\theta - \dot{x}\sin\theta = 0 \tag{4}$$

which corresponds to a no-slip wheel condition preventing the robot from moving sideways [17]. The system (3) fails to meet Brockett's necessary condition for feedback stabilization [5], which implies that no continuous static state-feedback controller exists that, stabilizes the closed-loop system around the equilibrium point.

The control objective is to design a fuzzy logic controller for τ that ensures

$$\lim_{t\to\infty} \|q_d(t) - q(t)\| = 0, \tag{5}$$

for any continuously, differentiable, bounded desired trajectory $q_d \in R^3$ while attenuating external disturbances.

2.3.1 Fuzzy Logic Control Design

In this section a no-singleton type-2 fuzzy logic controller is designed to force the real velocities of the mobile robot (2) and (3) to match those required in equations (4) of Theorem 3 and as a consequence to satisfy the control objective (5).

We design a Takagi–Sugeno type-2 non-singleton fuzzy logic controller for the autonomous mobile robot, using linguistic variables in the input and mathematical functions in the output. The linear (ϑ_d) and the angular (w_d) velocity errors were taken as input variables and the right (τ_1) and left (τ_2) torques as the outputs. The membership functions used in the input are trapezoidal for the Negative (N) and Positive (P), and triangular for the Zero (Z) linguistics terms. The interval used for this fuzzy controller is $[-50\ 50]$.

The rule set of the FLC contain 9 rules, which govern the input–output relationship of the FLC and this adopts the Takagi–Sugeno style inference engine [7, 14, 15], and we use a single point in the outputs (constant values), obtained using weighted average defuzzification procedure. To find the best fuzzy controller, we used genetic algorithms to find the parameters of the membership functions (Table 1).

Basing on the optimal type-1 FLC, we set the parameters of the membership functions of a type-2 non-singleton fuzzy logic controller (FLC) using a genetic algorithm. We set the parameters of the GA using a chromosome with 12 genes of real values that represent the two inputs, linear velocity error and angular velocity error; the two outputs, left torque and right torque, are constants values that are no include in the genetic algorithm; we used different values in the genetic operator's; mutation and single point crossover. Table 2 shows the parameters of the membership functions, the minimal and the maximum values in the search range for the GA to find the best fuzzy controller system for the mobile robot.

3 Simulation Results

In this section we present the simulations results using the type-2 non-singleton fuzzy logic controller for trajectory tracking of an autonomous mobile robot. Table 3 present the results showing in the 17th row the best results taking account the average error of the positions errors where bold means the best value.

Figure 3 shows the membership functions of input 1 and input 2 of the optimal type-2 non-singleton FLC and Fig. 4 shows the evolution of the individuals of the genetic algorithm giving the best FLC to control the robot mobile.

Table 1 Fuzzy rules of the FLC

	Negative	Zero	Positive
Negative	N	N	Z
Zero	N	Z	P
Positive	Z	P	P

Table 2 Parameters of the membership functions

MF Type	Point	Min value	Max value
Trapezoidal	a	−50	−50
	b	−50	−50
	c	−15	−5.05
	d	−1.5	−0.5
Triangular	a	−5	−1.75
	b	0	0
	c	1.75	5
Trapezoidal	a	0.5	1.5
	b	5.05	15
	c	50	50
	d	50	50

Table 3 Results of the type-2 non-singleton FLC obtained by GA

No	Indiv.	Gen.	% Remp.	Cross.	Mut.	Average error
1	28	18	0.9	0.8	0.3	0.06729112
2	38	15	0.9	0.8	0.2	0.06729835
3	15	25	0.9	0.9	0.2	0.06733408
4	20	50	0.9	0.8	0.2	0.06727392
5	30	20	0.9	0.8	0.3	0.06728935
6	18	30	0.9	0.8	0.2	0.06730611
7	25	50	0.9	0.7	0.2	0.06728692
8	45	30	0.9	0.7	0.3	0.06727011
9	50	15	0.9	0.8	0.3	0.06729371
10	10	40	0.9	0.8	0.3	0.06729535
11	50	20	0.9	0.8	0.2	0.06728903
12	25	80	0.9	0.8	0.3	0.06727028
13	10	15	0.9	0.7	0.2	0.06735353
14	20	35	0.9	0.7	0.3	0.06728564
15	30	30	0.9	0.8	0.4	0.06727038
16	15	40	0.9	0.8	0.4	0.06728775
17	**25**	**45**	**0.8**	**0.8**	**0.3**	**0.06727002**
18	25	20	0.9	0.8	0.2	0.06729877
19	10	30	0.9	0.8	0.3	0.06729535

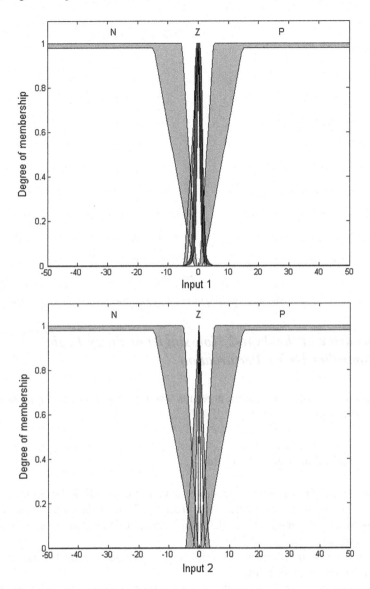

Fig. 3 Input 1 and input 2 membership functions of the optimized type-2 non-singleton FLC

Figure 5a shows the stability of the position errors e_x, e_y and e_{theta} and Fig. 5b shows the control result of the autonomous robot mobile using the optimized type-2 non-singleton FLC obtained with the GA.

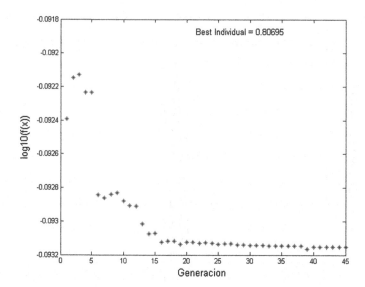

Fig. 4 Evolution of the individuals of the genetic algorithm for the optimized controller

3.1 Behavior of the Type-2 Non-singleton Fuzzy Logic Controller Under Perturbation

In this section we present simulation results when we apply a certain kind of noise during the trajectory tracking.

3.1.1 Perturbation: Pulse Generator

In the first case we use a pulse generator that simulates a little hit on the autonomous mobile robot during the trajectory tracking. Table 4 shows the values of the intensity of the hit (amplitude) of the pulse generator and the average error of the position errors.

Figure 6 shows the simulation results when we apply a hit of 0.03 (amplitude) to the autonomous mobile robot.

We can observe in the plot of the positions errors, that the controller maintains the stability. On the other hand when we hit the mobile robot with a value of 0.3, the robot loses the control, Fig. 7 shows the behavior.

We can observe that even though the error on the orientation variable increases, the controller tries to achieve stability during trajectory tracking.

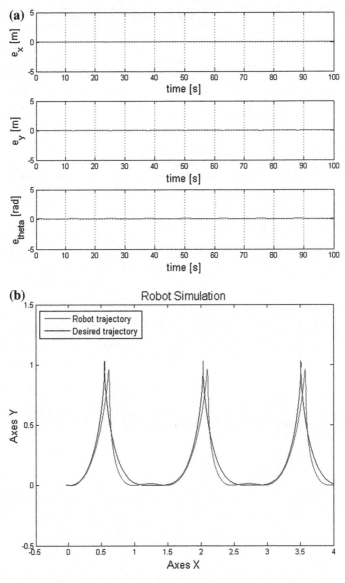

Fig. 5 a Position errors of *x*, *y* and *theta*; **b** closed-loop response of the autonomous mobile robot with the optimized FLC

3.1.2 Perturbation: Equation

In this kind of perturbation we use the following equation:

$$P = \varepsilon \text{Cos} wt$$

Table 4 Noise values using a pulse generator for the autonomous mobile robot

No.	Pulse generator (Amplitude)	Average error
1	0.03	0.067560
2	0.06	0.067074
3	0.09	0.068600
4	0.10	0.069654
5	0.13	0.074555
6	0.16	0.645227
7	0.19	0.733167
8	0.22	2.245156
9	0.26	2.305332
10	0.30	2.313345

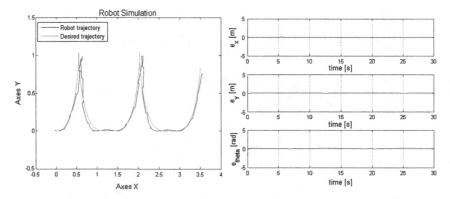

Fig. 6 Close loop response and the position errors x, y and *theta*

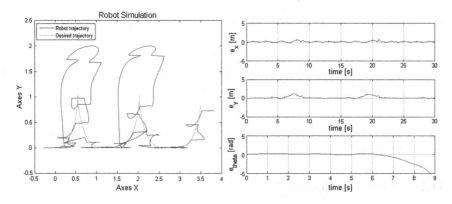

Fig. 7 Shows the close loop response of the autonomous mobile robot and the position errors

Table 5 Perturbation values using the equation applied to the autonomous mobile robot	No.	$P = \varepsilon\,\mathrm{Cos}\,wt$ (Amplitude)	Average error
	1	0.03	0.06646071
	2	0.06	0.06475203
	3	0.09	0.07073969
	4	0.10	0.07886378
	5	0.13	2.26142135
	6	0.16	2.31754076
	7	0.19	2.35259117
	8	0.22	2.37966707
	9	0.26	2.40458794
	10	0.30	2.42460445

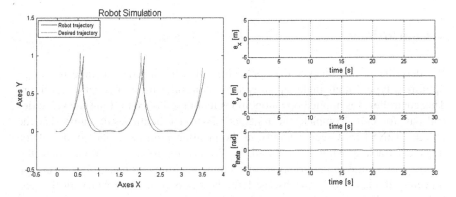

Fig. 8 Close loop response of the autonomous mobile robot and the position errors

where ε take values from 0.03 to 0.3, $w = 1$ and t is a time interval from 1 to 30 s. We use this equation to simulate a little hit to the mobile robot during the trajectory tracking as in previous perturbation (pulse generator).

Table 5 presents the different values of intensity for the hit of the mobile robot and the average error obtained by the position errors of the type-2 non-singleton fuzzy logic controller.

Figure 8 shows the trajectory of the mobile robot and the position errors when applied a hit value of 0.03.

We can observe that the controller maintain the stability of the position errors and the trajectory tracking of the mobile robot. However when we apply a hit value of 0.3 the controller try to follow the trajectory tracking showed in the Fig. 9.

We can observe that the perturbation affects the trajectory tracking increasing the average error obtained by position errors, however the controller tries to maintain the stability of the mobile robot following the trajectory the best possible.

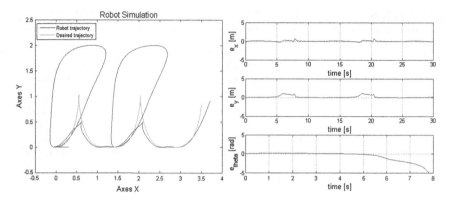

Fig. 9 Close loop response of the mobile robot and the position errors

4 Conclusions

We describe in this chapter the application of a bio-inspired method to design optimized type-2 non-singleton fuzzy logic controller using a genetic algorithm. This controller was applied on autonomous mobile robot controlling the velocities and pair forces to complete the trajectory goal. The main result shows that the optimal T2NST2FLC gets stability in the position errors and god follow of the trajectory tracking without perturbations. We test the robustness of the controller applying two kinds of perturbations that simulate a little hit on the robot during the trajectory tracking. We can observe how the behavior of the robot with the perturbation called pulse generator, when the perturbation increases the behavior of the mobile robot seems destructive movements, however the controller maintain the trajectory. On the other hand the perturbations called equation applied hit with minor disturbances, but the controller maintain the course.

We have achieved satisfactory results with the GA to design type-2 non-singleton fuzzy logic controllers for autonomous mobile robot; the next step is to solve the problem using other bio-inspired methods to design type-2 non-singleton fuzzy logic controller to improve the trajectory control of the autonomous mobile robot and made a comparison with interval type-2 fuzzy logic controllers.

References

1. Alcalá, R., Alcalá-Fdez, J., Herrera, F.: A proposal for the genetic lateral tuning of linguistic fuzzy systems and its interaction with rule selection. IEEE Trans. Fuzzy Syst. **15**(4), 616–635 (2007)
2. Alcalá, R., Gacto, M.J., Herrera, F., Alcalá-Fdez, J.: A multi-objective genetic algorithm for tuning and rule selection to obtain accurate and compact linguistic fuzzy rule-based systems. Int. J. Uncertainty Fuzziness Knowl. based Syst. **15**(5), 539–557 (2007)

3. Bentalba, S., El Hajjaji, A., Rachid, A.: Fuzzy control of a mobile robot: a new approach. In: Proceedings of the IEEE International Conference on Control Applications, pp. 69–72. Hartford, CT, Oct 1997
4. Bloch, A.M.: Nonholonomic Mechanics and Control. Springer, New York (2003)
5. Brockett, R.W.: Asymptotic stability and feedback stabilization. In: Millman, R.S., Sussman, H.J. (eds.) Differential Geometric Control Theory, pp. 181–191. Birkhauser, Boston (1983)
6. Casillas, J., Cordon, O., del Jesús, M.J., Herrera, F.: Genetic tuning of fuzzy rule deep structures preserving interpretability and its interaction with fuzzy rule set reduction. IEEE Trans. Fuzzy Syst. **13**(1), 13–29 (2005)
7. Chi, Z., Yan, H., Pham, T.: Fuzzy Algorithms: With Applications to Image Processing and Pattern recognition. World Scientific, Singapore (1996)
8. Chwa, D.: Sliding -mode tracking control of nonholonomic wheeled mobile robots in polar coordinates. IEEE Trans. Control Syst. Technol. **12**(4), 633–644 (2004)
9. Cordon, O., Gomide, F., Herrera, F., Hoffmann, F., Magdalena, L.: Ten years of genetic fuzzy systems: current framework and new trends. Fuzzy Sets Syst. **141**(1), 5–31 (2004)
10. Driankov, D., Hellendoorn, H., Reinfrank, M.: An Introduction to Fuzzy Control. Springer, Berlin (1993)
11. Fierro, R., Lewis, F.L.: Control of a nonholonomic mobile robot using neural networks. IEEE Trans. Neural Networks **9**(4), 589–600 (1998)
12. Fogel, D.B.: An introduction to simulated evolutionary optimization. IEEE Trans. Neural Networks **5**(1), 3–14 (1994)
13. Fukao, T., Nakagawa, H., Adachi, N.: Adaptive tracking control of a nonholonomic mobile robot. IEEE Trans. Rob. Autom. **16**(5), 609–615 (2000)
14. Hagras, H.A.: A hierarchical type-2 fuzzy logic control architecture for autonomous mobile robots. IEEE Trans. Fuzzy Syst. **12**(4), 524–539 (2004)
15. Ishikawa, S.: A method of indoor mobile robot navigation by fuzzy control. In: Proceedings of International Conference on Intelligent Robots Systems, pp. 1013–1018. Osaka, Japan (1991)
16. Karnik, N.N., Mendel, J.: Centroid of a type-2 fuzzy set. Inf. Sci. **132**(1–4), 195–220 (2001)
17. Kristic, M., Kanellakopoulos, I., Kokotovic, P.: Nonlinear and adaptive control design, Wiley-Interscience (1995)
18. Lee, T.-C., Song, K.-T., Lee, C.-H., Teng, C–.C.: Tracking control of unicycle-modeled mobile robot using a saturation feedback controller. IEEE Trans. Control Syst. Technol. **9**, 305–318 (2001)
19. Lee, T.H., Leung, F.H.F., Tam, P.K.S.: Position control for wheeled mobile robot using a fuzzy controller, pp. 525–528. IEEE (1999)
20. Liang, Q., Mendel, J.: Interval type-2 fuzzy logic systems: theory and design. IEEE Trans. Fuzzy Syst. **8**(5), 535–550 (2000)
21. Man, K.F., Tang, K.S., Kwong, S.: Genetic Algorithms, pp. 5–10. Springer, Concepts Des. (2000)
22. Martinez, R., Castillo, O., Aguilar, L.T., Rodríguez, A.: Optimization of type-2 fuzzy logic controllers for mobile robots using evolutionary methods. In: Proceedings of the 2009 IEEE International Conference on Systems, Man, and Cybernetics San Antonio, pp. 4909–4914. IEEE, TX, Oct (2009)
23. Martinez, R., Castillo, O., Aguilar, L.T.: Intelligent control for a perturbed autonomous wheeled mobile robot using type-2 fuzzy logic and genetic algorithms. J. Autom. Mob. Rob. Intell. Syst. **2** (2008)
24. Mendel, J., John, R.: Type-2 fuzzy sets made simple. IEEE Trans. Fuzzy Syst. **10**(April), 117–127 (2002)
25. Mendel, J.M., George, C.: Type-2 fuzzy logic systems. IEEE Trans. Fuzzy Syst. **7**, 643–658 (1999)
26. Mendel, J.: On a 50% savings in the computation of the centroid of a symmetrical interval type-2 fuzzy set. Inf. Sci. **172**(3–4), 417–430 (2005)

27. Mendel, J., John, R.I.B.: Type-2 fuzzy sets made simple, IEEE Trans. Fuzzy Syst. **10**(2), (2002)
28. Mendel, J.: Uncertain Rule-based Fuzzy Logic Systems. Prentice Hall (2001)
29. Passino, K.M., Yurkovich, S.: Fuzzy Control. Addison Wesley Longman, USA (1998)
30. Pawlowski, S., Dutkiewicz, P., Kozlowski, K., Wroblewski, W.: Fuzzy logic implementation in mobile robot control. In: Second Workshop on Robot Motion and Control, pp. 65–70. IEEE, Oct 2001
31. Sahab, N., Hagras, H.: A type-2 nonsingleton type-2 fuzzy logic system to handle linguis-tic and numerical uncertainties in realworld environments. IEEE Int. Symp. Adv. Type-2 Fuzzy Logic Syst. (2011)
32. Sahab, N., Hagras, H.: Adaptive non-singleton type-2 fuzzy logic systems: a way forward for handling numerical uncertainties in real world applications. Int. J. Comput. Commun. Control **5**(3), 503–529 (2011) (ISSN 1841-9836)
33. Song, K.T., Sheen, L.H.: Heuristic fuzzy-neural network and its application to reactive navigation of a mobile robot. Fuzzy Sets Syst. **110**(3), 331–340 (2000)
34. Takagi, T., Sugeno, M.: Fuzzy identification of systems and its application to modeling and control. IEEE Trans. Syst. Man Cybern. **15**(1) (1985)
35. Tsai, C.-C., Lin, H.-H., Lin, C.-C.: Trajectory tracking control of a laser-guided wheeled mobile robot. In: Proceedings of the IEEE International Conference on Control Applications, vol. 2, pp. 1055–1059. IEEE, Taipei, Sep 2004
36. Ulyanov, S.V., Watanabe, S., Yamafuji, K., Litvintseva, L.V., Rizzotto, G.G.: Soft computing for the intelligent robust control of a robotic unicycle with a new physical measure for mechanical controllability, Soft Computing, vol. 2, pp. 73–88. Springer, Berlin 1998
37. Zadeh, L.A.: Outline of a new approach to the analysis of complex systems and decision processes. IEEE Trans. Syst. Man Cybern. **3**(1), 28–44 (1973)

Hierarchical Genetic Algorithms for Type-2 Fuzzy System Optimization Applied to Pattern Recognition and Fuzzy Control

Daniela Sánchez and Patricia Melin

Abstract In this chapter a new method of hierarchical genetic algorithm for fuzzy inference systems optimization is proposed. This method was used in two applications, the first was to perform the combination of responses of modular neural networks for human recognition based on face, iris, ear and voice, and the second one for fuzzy control of temperature in the shower benchmark problem. The results obtained by non-optimized type-2 fuzzy inference system can be improved using the proposed hierarchical genetic algorithm as can be verified by the simulations.

1 Introduction

The prudent combination of different intelligent techniques can produce powerful hybrid intelligent systems [15, 30]. These kind of systems have become in part important of investigation, because different new techniques have emerged to help themselves [18, 20, 24, 25]. Some of these techniques are fuzzy logic, neural networks, genetic algorithms, ant colony optimization and particle swarm optimization [9–13]. There are many works where have been used this kind of systems, these works have combined different techniques and they have obtained good results [4, 9, 16, 17, 22, 23]. In this chapter different intelligent techniques are combined such as neural networks, type-2 fuzzy logic and genetic algorithms.

D. Sánchez · P. Melin (✉)
Tijuana Institute of Technology, Tijuana, Mexico
e-mail: pmelin@tectijuana.mx

D. Sánchez
e-mail: danielasanchez.itt@hotmail.com

O. Castillo et al. (eds.), *Recent Advances on Hybrid Approaches for Designing Intelligent Systems*, Studies in Computational Intelligence 547, DOI: 10.1007/978-3-319-05170-3_2, © Springer International Publishing Switzerland 2014

The proposed method was applied to human recognition based on face, iris, ear and voice, and fuzzy control of temperature in the Shower benchmark problem.

This chapter is organized as follows: The basic concepts used in this work are presented in Sect. 2. Section 3 contains the general architecture of the proposed method. Section 4 presents experimental results for human recognition and fuzzy control and in Sect. 5, the conclusions of this work are presented.

2 Basic Concepts

In this section the basic concepts used in this research work are presented.

2.1 Modular Neural Networks

A neural network (NN) is said to be modular if the computation performed by the network can be decomposed into two or more modules. The modular neural networks are comprised of modules which can be categorized on the basis of both distinct structure and functionality which are integrated together via an integrating unit. With functional categorization, each module is a neural network which carries out a distinct identifiable subtask [1, 14]. There is evidence that shows that the use of modular neural networks implies a significant learning improvement comparatively to a single neural network [15].

2.2 Type-2 Fuzzy Logic

Fuzzy logic is an area of soft computing that enables a computer system to reason with uncertainty [2]. The concept of a type-2 fuzzy set was introduced by Zadeh (1975) as an extension of the concept of an ordinary fuzzy set (henceforth called a "type-1 fuzzy set"). When we cannot determine the membership of an element in a set as 0 or 1, fuzzy sets of type-1 are used. Similarly, when the situation is so fuzzy that we have trouble determining the membership grade even as a crisp number in [0,1], fuzzy sets of type-2 are used. Uncertainty in the primary memberships of a type-2 fuzzy set, $\tilde{A}$, consists of a bounded region that we call the "footprint of uncertainty" (FOU). Mathematically, it is the union of all primary membership functions [4, 19]. The basics of fuzzy logic do not change from type-1 to type-2 fuzzy sets, and in general will not change for type-n. A higher type number just indicates a higher degree of fuzziness [3].

2.3 Hierarchical Genetic Algorithms

A Genetic algorithm (GA) is an optimization and search technique based on the principles of genetics and natural selection [8, 21, 26].

Hierarchical genetic algorithm (HGA) is a kind of genetic algorithm [27], but the main difference between HGA and GAs is the structure of the chromosome. The basic idea under hierarchical genetic algorithm is that for complex systems which cannot be easily represented or resolved, this type of GA can be a better choice. The complicated chromosomes may provide a good new way to solve the problem [28, 29].

3 Proposed Method

The proposed hierarchical genetic algorithm performs the optimization of type-2 fuzzy inference system. The proposed method can be used for any application that uses a fuzzy inference system. In this chapter two applications were used, first the human recognition based on face, iris, ear and voice, and the second, the fuzzy control of temperature in the shower benchmark problem.

The main idea of the proposed method is to perform the optimization of type-2 inference systems that allow us to have better results than fuzzy inference systems non-optimized. In this work some parameters of these type-2 fuzzy inference systems such as the type of system, type of membership functions (Trapezoidal or Gaussian), percentage of rules, the number of membership functions and their parameters. The number of inputs or outputs can be easily establish depending of the application, for this reason the number of inputs or outputs that the fuzzy inference systems will have, depends on the problem or applications.

Figure 1 shows an example of fuzzy integrators. The chromosome of the proposed hierarchical genetic algorithm is shown in Fig. 2.

3.1 Application to Pattern Recognition

In this case, the number of inputs will depend of the number of biometric measures that are being used, here, the responses of each biometric measure are combined by the fuzzy inference systems, for this reason, the fuzzy integrator must have the best architecture in case of a error or failure of some module. In this work, 4 biometric measures are used (face, iris, ear and voice). In Fig. 3, an example of fuzzy integrator for this application is shown. It can notice that each input corresponds to one biometric measure, and the output corresponds to the final answer. The 4 inputs and the inputs have 3 trapezoidal membership functions, an example of these variables are shown in Fig. 4.

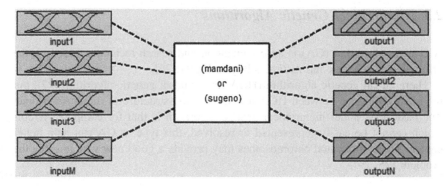

Fig. 1 Example of the type-2 fuzzy inference system

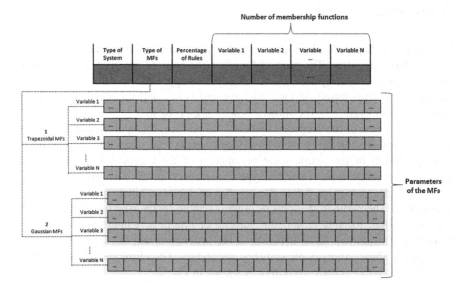

Fig. 2 The chromosome of the hierarchical genetic algorithm for the type-2 fuzzy inference systems

3.1.1 Databases

The databases that were used are described below.

Face

The database of human iris from the Institute of Automation of the Chinese Academy of Sciences was used [6]. It contains of 5 images per person, and it consists of 500 persons, but in this work only 77 persons are used. The image dimensions are 640 × 480, BMP format. Only the first 77 persons were used. Figure 5 shows examples of the human iris images from CASIA database.

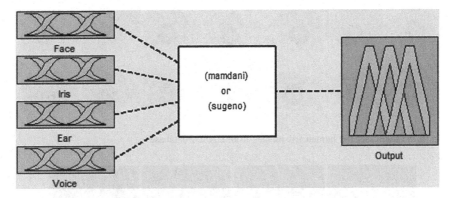

Fig. 3 The architecture of proposed method for the modular neural network

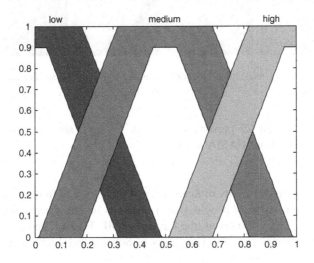

Fig. 4 Example of variable

Fig. 5 Examples of the face images from CASIA database

Ear

The database of human iris from the Institute of Automation of the Chinese Academy of Sciences was used [7]. It contains of 14 images (7 for each eye) per person, and it consists of 99 persons. The image dimensions are 320×280, JPEG

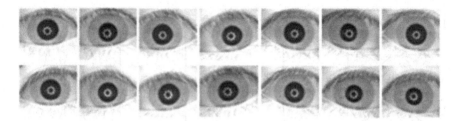

Fig. 6 Examples of the human iris images from CASIA database

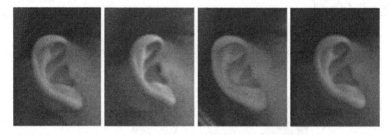

Fig. 7 Examples of ear recognition laboratory from the University of Science and Technology Beijing (USTB)

format. Only the first 77 persons were used. Figure 6 shows examples of the human iris images from CASIA database.

Iris

The database of the University of Science and Technology of Beijing was used [5]. The database consists of 77 people, which contain 4 images per person, the image dimensions are 300 × 400 pixels, the format is BMP. Figure 7 shows examples of the human ear images from the University of Science and Technology Beijing.

Voice

In the case of voice, the database was made from students of Tijuana Institute of Technology, and it consist of 10 voice samples (of 77 persons), WAV format. The word that they said in Spanish was "ACCESAR". To preprocess the voice the Mel Frequency Cepstral Coefficients were used.

3.2 Application to Fuzzy Control

In this case, the fuzzy control of temperature in the shower benchmark problem was used. The variables of this problem are shown in Fig. 8. This problem has 2 inputs (Temp and Flow) and 2 outputs (Cold and Hot). In Fig. 9, an example of fuzzy integrator for this application is shown.

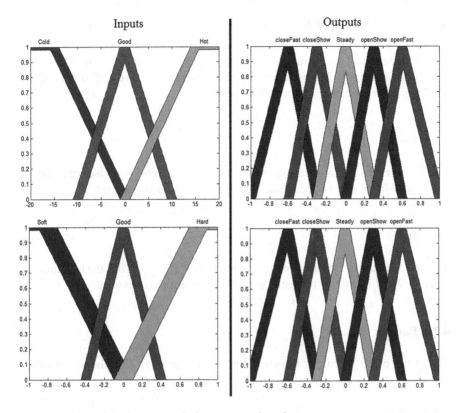

Fig. 8 Variables of the fuzzy control of temperature in a shower

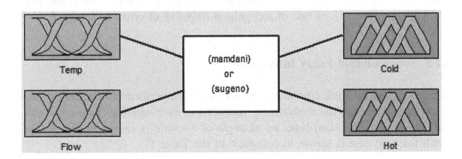

Fig. 9 Example of the fuzzy inference systems

4 Experimental Results

The results obtained (in applications, human recognition and fuzzy control) are presented in this section.

4.1 Human Recognition

The architecture of each training and the percentage of data for training were established randomly. Thirty trainings for each biometrics were performed and the results of each biometric are presented below.

4.1.1 Face

The first 5 results for face are shown in Table 1. The training #1 is the best training where a 87.01 of rate of recognition (0.12987 of error) is obtained.

4.1.2 Iris

The first 5 results for iris are shown in Table 2. The training #10 is the best training where a 98.27 of rate of recognition (0.01732 of error) is obtained.

4.1.3 Ear

The first 5 results for ear are shown in Table 3. The training #5 is the best training.

4.1.4 Voice

The first 5 results for voice are shown in Table 4. The training #7 is the best training where a 93.18 of rate of recognition (0.06818 of error) is obtained.

4.1.5 Non-optimized Fuzzy Integration

Different trainings were combined and 3 cases were performed. The fuzzy integrator with 4 trapezoidal membership functions in each variable (inputs and output) is used, of Mamdani type, an example of variable is shown in Fig. 4. The result for each case is shown in column 6 of the Table 5.

4.1.6 Optimized Fuzzy Integration

Fifteen evolutions for each case were performed and the results are shown in Table 6.

In Fig. 10, the best fuzzy integrator for the case #2 is shown, it is a fuzzy integrator of Sugeno type, and with Gaussian Membership Functions.

Table 1 The first 5 results for face

Training	Images for training (%)	Persons per module	Recognition rate
FT1	77	Module #1(1–25)	68.83 %
		Module #2(26–32)	0.31169
		Module #3(33–47)	
		Module #4(48–57)	
		Module #5(58–77)	
FT2	69	Module #1(1–12)	85.71 %
		Module #2(13–22)	0.14286
		Module #3(23–35)	
		Module #4(36–37)	
		Module #5(38–42)	
		Module #6(43–44)	
		Module #7(45–58)	
		Module #8(59–61)	
		Module #9(62–66)	
		Module #10(67–77)	
FT3	19	Module #1(1–6)	52.92 %
		Module #2(7–17)	0.47078
		Module #3(18–19)	
		Module #4(20–29)	
		Module #5(30–35)	
		Module #6(36–47)	
		Module #7(48–56)	
		Module #8(57–67)	
		Module #9(68–77)	
FT4	11	Module #1(1–21)	45.78 %
		Module #2(22–34)	0.54221
		Module #3(35–53)	
		Module #4(54–55)	
		Module #5(56–73)	
		Module #6(74–77)	
FT5	49	Module #1(1–15)	60.17 %
		Module #2(16–24)	0.39827
		Module #3(25–36)	
		Module #4(37–51)	
		Module #5(52–53)	
		Module #6(54–66)	
		Module #7(67–77)	

4.2 Fuzzy Control

The Simulation plant for the fuzzy control of temperature for the shower in Matlab is shown in Fig. 11.

Table 2 The first 5 results for face

Training	Images for training (%)	Persons per module	Recognition rate
IT1	19	Module #1(1–37)	79.10 %
		Module #2(38–64)	0.20897
		Module #3(65–69)	
		Module #4(70–77)	
IT2	73	Module #1(1–24)	96.10 %
		Module #2(25–77)	0.03896
IT3	71	Module #1(1–77)	90.91 %
			0.09091
IT4	16	Module #1(1–21)	63.20 %
		Module #2(22–42)	0.36797
		Module #3(43–66)	
		Module #4(67–77)	
IT5	80	Module #1(1–10)	98.27 %
		Module #2(11–18)	0.01732
		Module #3(19–28)	
		Module #4(29–39)	
		Module #5(40–52)	
		Module #6(53–64)	
		Module #7(65–77)	

Using the non-optimized fuzzy integrator presented in Fig. 9, an error of 0.1589 is obtained. Fifteen evolutions were perform and the best result is shown in Table 7. In Fig. 12, the best fuzzy inference system obtained is shown.

Fifteen evolutions for each case were performed and the results are shown in Table 7.

In Fig. 12, the best fuzzy integrator obtained is shown; it is a fuzzy integrator of Mamdani type, and with Gaussian Membership Functions. In both applications the fuzzy inference systems obtained have Gaussian membership functions.

5 Conclusions

In this work a new hierarchical genetic algorithm was presented, the main idea of this HGA is to perform the optimization of type-2 fuzzy systems. For this reason the optimization of parameters was performed (type of system, type of membership functions (Trapezoidal or Gaussian), percentage of rules, the number of member-ship functions and their parameters). As the results show, better results can be obtained when a optimization is performed. This HGA was tested in two different applications (combination of responses of modular neural networks for human recognition and fuzzy control of the temperature of a shower) but this HGA

Table 3 The first 5 results for face

Training	Images for training (%)	Persons per module	Recognition rate
ET1	74	Module #1(1–5)	94.81 %
		Module #2(6–15)	0.05195
		Module #3(16–26)	
		Module #4(27–30)	
		Module #5(31–43)	
		Module #6(44–49)	
		Module #7(50–58)	
		Module #8(59–71)	
		Module #9(72–77)	
ET2	66	Module #1(1–21)	77.92 %
		Module #2(22–34)	0.22078
		Module #3(35–45)	
		Module #4(46–64)	
		Module #5(65–77)	
ET3	51	Module #1(1–3)	79.22 %
		Module #2(4–11)	0.20779
		Module #3(12–17)	
		Module #4(18–33)	
		Module #5(34–50)	
		Module #6(51–55)	
		Module #7(56–61)	
		Module #8(62–75)	
		Module #9(76–77)	
ET4	81	Module #1(1–12)	97.40 %
		Module #2(13–14)	0.02597
		Module #3(15–23)	
		Module #4(24–43)	
		Module #5(44–58)	
		Module #6(59–77)	
ET5	56	Module #1(1–27)	82.47 %
		Module #2(28–52)	0.17532
		Module #3(53–55)	
		Module #4(56–77)	

can be used in other applications because can be easily adaptable. In human recognition, the main objective was the minimization of the error of recognition and, in the case of fuzzy control, the minimization of the error of simulation. In the future the combination of membership functions in a same variable will be implemented, waiting of obtaining better results in any application that uses type-2 fuzzy inference systems.

Table 4 The first 5 results for face

Training	Images for training (%)	Persons per module	Recognition rate
VT1	38	Module #1(1–5)	92.86 %
		Module #2(6–12)	0.07143
		Module #3(13–23)	
		Module #4(24–32)	
		Module #5(33–35)	
		Module #6(36–42)	
		Module #7(43–50)	
		Module #8(51–58)	
		Module #9(59–67)	
		Module #10(68–77)	
VT2	57	Module #1(1–23)	91.88 %
		Module #2(24–31)	0.08117
		Module #3(32–52)	
		Module #4(53–77)	
VT3	58	Module #1(1–32)	91.23 %
		Module #2(33–53)	0.08766
		Module #3(54–77)	
VT4	37	Module #1(1–38)	86.36 %
		Module #2(39–77)	0.13636
VT5	64	Module #1(1–13)	93.18 %
		Module #2(14–26)	0.06818
		Module #3(27–32)	
		Module #4(33–37)	
		Module #5(38–50)	
		Module #6(51–52)	
		Module #7(53–66)	
		Module #8(67–68)	
		Module #9(69–77)	

Table 5 Non-optimized results

Case	Face	Iris	Ear	Voice
1	FT4	IT5	ET5	VT2
	45.78 %	98.27 %	82.47 %	91.88 %
2	FT1	IT4	ET2	VT1
	87.01 %	63.20 %	77.92 %	91.77 %
3	FT2	IT2	ET4	VT5
	85.71 %	96.10 %	97.40 %	93.18 %

Table 6 Non-optimized results

Case	Non-optimized	Best	Average
1	85.06 %	99.02 %	97.73 %
	0.1494	0.0097	0.0227
2	79.11 %	93.72 %	92.76 %
	0.2089	0.0628	0.0724
3	92.21 %	100 %	99.77 %
	0.0779	0	0.0023

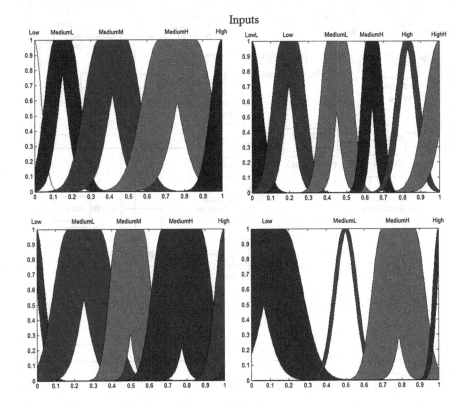

Fig. 10 The best fuzzy integrator for the case #2

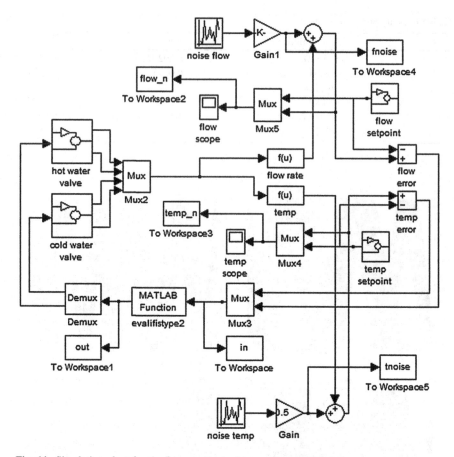

Fig. 11 Simulation plant for the fuzzy control of temperature in the shower

Table 7 Comparison between non-optimized and optimized results

Non-optimized	Best	Average
0.1589	0.00075	0.0040

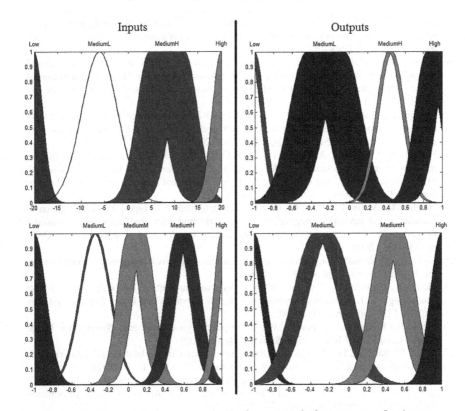

Fig. 12 The best fuzzy inference system for the fuzzy control of temperature of a shower

References

1. Azamm, F.: Biologically inspired modular neural networks. Ph.D. thesis, Virginia Polytechnic Institute and State University, Blacksburg, Virginia, May 2000
2. Castillo, O., Melin, P.: Soft Computing for Control of Non-linear Dynamical Systems. Springer, Heidelberg (2001)
3. Castillo, O., Melin, P.: Type-2 Fuzzy Logic Theory and Applications, pp. 29–43. Springer, Berlin (2008)
4. Castro, J.R., Castillo, O., Melin, P., Mendoza O., Rodríguez Díaz A.: An interval type-2 fuzzy neural network for chaotic time series prediction with cross-validation and akaike test. In: Soft Computing for Intelligent Control and Mobile Robotics, pp. 269–285. Springer, Berlin (2011)
5. Database Ear Recognition Laboratory from the University of Science and Technology Beijing (USTB). http://www.ustb.edu.cn/resb/en/index.htm asp. Accessed 21 Sept 2009
6. Database of Face. Institute of Automation of Chinese Academy of Sciences (CASIA). http://biometrics.idealtest.org/dbDetailForUser.do?id=9. Accessed 11 Nov 2012
7. Database of Human Iris. Institute of Automation of Chinese Academy of Sciences (CASIA). http://www.cbsr.ia.ac.cn/english/IrisDatabase.asp. Accessed 21 Sept 2009
8. Haupt, R., Haupt, S.: Practical Genetic Algorithms, 2nd edn, pp. 42–43. Wiley-Interscience, New York (2004)

9. Hidalgo, D., Castillo, O., Melin, P.: An optimization method for designing type-2 fuzzy inference systems based on the footprint of uncertainty using genetic algorithms. Expert Syst. Appl. **39**(4), 4590–4598 (2012)
10. Hidalgo, D., Castillo, O., Melin, P.: Optimization with genetic algorithms of modular neural networks using interval type-2 fuzzy logic for response integration: The case of multimodal biometry. In: International Joint Conference on Neural Networks (IJCNN), pp. 738–745 (2008)
11. Hidalgo, D., Castillo, O., Melin, P.: Type-1 and type-2 fuzzy inference systems as integration methods in modular neural networks for multimodal biometry and its optimization with genetic algorithms. Inf. Sci. **179**(13), 2123–2145 (2009)
12. Hidalgo, D., Castillo, O., Melin, P.: Type-1 and type-2 fuzzy inference systems as integration methods in modular neural networks for multimodal biometry and its optimization with genetic algorithms. In: Soft Computing for Hybrid Intelligent Systems, pp. 89–114, 1st edn. Springer, Berlin (2008)
13. Hidalgo, D., Melin, P., Licea, G., Castillo, O.: Optimization of type-2 fuzzy integration in modular neural networks using an evolutionary method with applications in multimodal biometry. In: Mexican International Conference on Artificial Intelligence (MICAI), pp. 454–465 (2009)
14. Khan, A., Bandopadhyaya, T., Sharma, S.: Classification of stocks using self organizing map. Int. J. Soft Comput. Appl. **4**, 19–24 (2009)
15. Melin, P., Castillo, O.: Hybrid Intelligent Systems for Pattern Recognition Using Soft Computing: An Evolutionary Approach for Neural Networks and Fuzzy Systems, 1st edn, pp. 119–122. Springer, Berlin (2005)
16. Melin, P., Kacprzyk, J., Pedrycz, W.: Bioinspired Hybrid Intelligent Systems for Image Analysis and Pattern Recognition. Springer, Berlin (2009)
17. Melin, P., Sánchez, D., Castillo, O.: Genetic optimization of modular neural networks with fuzzy response integration for human recognition. Inf. Sci. **197**, 1–19 (2012)
18. Melin, P., Sánchez, D., Cervantes, L.: Hierarchical genetic algorithms for optimal type-2 fuzzy sys-tem design. In: Annual Meeting of the North American Fuzzy Information Processing Society (NAFIPS), pp. 324–329 (2011)
19. Mendel, J.: Uncertain Rule-Based Fuzzy Logic Systems: Introduction and New Directions. Prentice-Hall, Upper Saddle River (2001)
20. Mendoza, O., Melin, P., Licea, G.: A hybrid approach for image recognition combining type-2 fuzzy logic, modular neural networks and the Sugeno integral. Inf. Sci. **179**(13), 2078–2101 (2009)
21. Mitchell, M.: An Introduction to Genetic Algorithms, 3rd edn. A Bradford Book, Cambridge (1998)
22. Oh, S., Pedrycz, W., Roh, S.: Hybrid fuzzy set-based polynomial neural networks and their development with the aid of genetic optimization and in-formation granulation. Appl. Soft Comput. **9**(3), 1068–1089 (2009)
23. Pal, S.K., Mitra, S., Mitra, P.: Rough fuzzy MLP: modular evolution, rule generation and evolution. IEEE Trans. Knowl. Data Eng. **15**(1), 14–25 (2003)
24. Sánchez, D., Melin P.: Optimization of modular neural networks and type-2 fuzzy integrators using hierarchical genetic algorithms for human recognition. In: IFSA World Congress and AFSS International Conference, World Congress (IFSA-AFSS), pp. OS-414-1–OS-414-8 (2011)
25. Sánchez, D., Melin, P., Castillo O.: Hierarchical genetic algorithms for optimal type-2 fuzzy system design in modular neural networks. In: International Joint Conference on Neural Networks (IJCNN), pp. 1267–1274 (2011)
26. Segovia, J., Szczepaniak, P.S., Niedzwiedzinski, M.: E-Commerce and Intelligent Methods, 1st edn, p. 181. Physica-Verlag, Berlin (2002)
27. Tang, K.S., Man, K.F., Kwong, S., Liu, Z.F.: Minimal fuzzy memberships and rule using hierarchical genetic algorithms. IEEE Trans. Ind. Electron **45**(1), 162–169 (1998)

28. Wang, C., Soh, Y.C., Wang, H.: A hierarchical genetic algorithm for path planning in a static environment with obstacles. In: Canadian Conference on Electrical and Computer Engineering, 2002. IEEE CCECE 2002, vol. 3, pp. 1652–1657 (2002)
29. Worapradya, K., Pratishthananda, S.: Fuzzy supervisory PI controller using hierarchical genetic algorithms. In: 5th Asian Control Conference, 2004, vol. 3, pp. 1523–1528 (2004)
30. Zhang, Z., Zhang, C.: An agent-based hybrid intelligent system for financial investment planning. In: 7th Pacific Rim International Conference on Artificial Intelligence (PRICAI), pp. 355–364 (2002)

Designing Type-2 Fuzzy Systems Using the Interval Type-2 Fuzzy C-Means Algorithm

Elid Rubio and Oscar Castillo

Abstract In this work, the Interval Type-2 Fuzzy C-Mean (IT2FCM) algorithm was used for the design of Type-2 Fuzzy Inference Systems using centroids and fuzzy membership matrices for the lower and upper bound of the interval obtained by the IT2FCM algorithm in each data clustering realized by this algorithm, with these elements obtained by IT2FCM algorithm we design the Mamdani, and Sugeno Fuzzy Inference systems for classification of data sets and time series prediction.

1 Introduction

Due to need to find interesting patterns or groups of data with similar characteristics in a given data set the Clustering algorithms [1–6] have been proposed to satisfy this need. Currently there are various fuzzy clustering algorithms. The Fuzzy C-Means (FCM) algorithm [1, 5] has been the base to developing other clustering algorithms, which have been widely used satisfactory in pattern recognition [1], data mining [7], classification [8], image segmentation [9, 10], data analysis and modeling [2]. The popularity of the fuzzy clustering algorithms is due to the fact that allow a datum belong to different data clusters into a given data set.

However this method is not able to handle uncertainty found in a given dataset during the process of data clustering; because of this the FCM was extended to IT2FCM using Type-2 Fuzzy Logic Techniques [11, 12]. This extension of the FCM algorithm has been applied to the creation of membership functions [13–16],

E. Rubio · O. Castillo (✉)
Tijuana Institute of Technology, Tijuana, Mexico
e-mail: ocastillo@tectijuana.edu.mx

E. Rubio
e-mail: elid.rubio@hotmail.com

O. Castillo et al. (eds.), *Recent Advances on Hybrid Approaches for Designing*
Intelligent Systems, Studies in Computational Intelligence 547,
DOI: 10.1007/978-3-319-05170-3_3, © Springer International Publishing Switzerland 2014

classification [17]. In this work the creation of fuzzy systems are presented using the IT2FCM algorithm using centroids matrices and fuzzy partition for the lower and upper limits of the range, with these matrices obtained using IT2FCM are created functions of membership for each variable input and output of the fuzzy system and its rules of inference.

2 Overview of Interval Type-2 Fuzzy Sets

Type-2 Fuzzy Sets is an extension of the Type-1 Fuzzy Sets proposed by Zadeh in 1975, this extension was designed with the aim of mathematically represent the vagueness and uncertainty of linguistic problems and this way overcome limitations of Type-1 Fuzzy Sets and thereby provide formal tools to work with intrinsic imprecision in different type of problems. Type-2 Fuzzy Sets are able to describe uncertainty and vagueness in information, usually are used to solve problems where the available information is uncertain. These Fuzzy Sets include a secondary membership function to model the uncertainty of Type-1 Fuzzy Sets [11, 12].

A Type-2 Fuzzy set in the universal set X, is denoted as $\tilde{A}$, and can be characterized by a type-2 fuzzy membership function $\mu_{\tilde{A}} = (x, u)$ as:

$$\tilde{A} = \int_{x \in X} \mu_{\tilde{A}}(x)/x = \int_{x \in X} \left[\int_{u \in J_x} f_x(u)/u \right] /x, J_x \subseteq [0, 1] \tag{1}$$

where J_x is the primary membership function of x which is the domain of the secondary membership function $f_x(u)$.

The shaded region shown in Fig. 1a is normally called footprint of uncertainty (FOU). The FOU of $\tilde{A}$ is the union of all primary membership that are within the lower and upper limit of the interval of membership functions and can be expressed as:

$$FOU(\tilde{A}) = \bigcup_{\forall x \in X} J_x = \{(x, u)|u \in J_x \subseteq [0, 1]\} \tag{2}$$

The lower membership function (LMF) and upper membership function (UMF) are denoted by $\underline{\mu}_{\tilde{A}}(x)$ and $\overline{\mu}_{\tilde{A}}(x)$ are associated with the lower and upper bound of $FOU(\tilde{A})$ respectively, i.e. The UMF and LMF of $\tilde{A}$ are two type-1 membership functions that bound the FOU as shown in Fig. 1a. By definition they can be represented as:

$$\underline{\mu}_{\tilde{A}}(x) = \underline{FOU(\tilde{A})} \forall x \in X \tag{3}$$

$$\overline{\mu}_{\tilde{A}}(x) = \overline{FOU(\tilde{A})} \forall x \in X \tag{4}$$

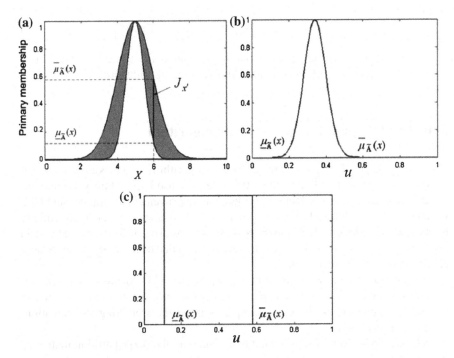

Fig. 1 **a** Type-2 Membership Function, **b** Secondary membership function and, **c** Interval secondary membership function

The secondary membership function is a vertical slice of $\mu_{\tilde{A}} = (x, u)$ as shown in Fig. 1b. The Type-2 Fuzzy Sets are capable modeling uncertainty, where type-1 fuzzy sets cannot. The computation operations required by type-2 fuzzy systems are considerably and, undesirably large, this is due these operations involve numerous embedded type-2 fuzzy sets which consider all possible combinations of the secondary membership values [11, 12]. However with the aim of reduce the computational complexity was proposed Interval Type-2 Fuzzy Sets, where the secondary membership functions are interval sets expressed as:

$$\tilde{A} = \int_{x \in X} \left[\int_{u \in J_x} 1/u \right] / x \tag{5}$$

Figure 1c shows the membership function of an Interval Type-2 Fuzzy Sets. The secondary memberships are all uniformly weighted for each primary membership of x. Therefore J_x can be expressed as:

$$J_x = \left\{ (x, u) | u \in \left[\underline{\mu}_{\tilde{A}}(x), \overline{\mu}_{\tilde{A}}(x) \right] \right\} \tag{6}$$

Moreover, $FOU(\tilde{A})$ in (2) can be expressed as:

$$FOU(\tilde{A}) = \bigcup_{\forall x \in X} \left\{ (x, u) | u \in \left[\underline{\mu_{\tilde{A}}}(x), \overline{\mu}_{\tilde{A}}(x) \right] \right\} \tag{7}$$

As a result of whole this, the computational complexity using Interval Type-2 Fuzzy Sets is reduced only to calculate simple interval arithmetic.

3 Interval Type-2 Fuzzy C-Means algorithm

The Interval Type-2 Fuzzy C-Means (IT2FCM) algorithm [13–15] is an extension of the Fuzzy C-Means (FCM) algorithm, this extension use Type-2 Fuzzy Techniques in combination with the C-Means algorithm, and improvement the traditional FCM, which uses Type-1 Fuzzy Logic [18, 19].This method is able to handle uncertainty found in a given dataset during the process of data clustering and thereby makes data clustering less susceptible to noise to achieve the goal that data can be clustered more appropriately and more accurately.

The weighting (fuzzification) exponent m in the IT2FCM algorithm is represented by an interval rather than a precise numerical value, i.e. $m = [m_1, m_2]$, where m_1 and m_2 represent the lower and upper limit of the weighting (fuzzification) exponent respectively.

Because the m value is represented by an interval, the fuzzy partition matrix μ_{ij} must be calculated to the interval $[m_1, m_2]$, per this reason μ_{ij} would be given by a membership interval $[\underline{\mu}_i(x_j), \overline{\mu}_i(x_j) \ \overline{\mu}_i(x_j)]$ where $\underline{\mu}_i(x_j)$ and $\overline{\mu}_i(x_j)$ represents the lower and upper limit of the belonging interval of datum x_j to a clustering v_i, updating the lower an upper limit of the range of the fuzzy membership matrices can be expressed as:

$$\underline{\mu}_i(x_j) = \min \left\{ \left[\sum_{k=1}^{c} \left(\frac{d_{ij}^2}{d_{ij}^2} \right)^{\frac{2}{m_1-1}} \right]^{-1}, \left[\sum_{k=1}^{c} \left(\frac{d_{ij}^2}{d_{ij}^2} \right)^{\frac{2}{m_2-1}} \right]^{-1} \right\} \tag{8}$$

$$\overline{\mu}_i(x_j) = \max \left\{ \left[\sum_{k=1}^{c} \left(\frac{d_{ij}^2}{d_{ij}^2} \right)^{\frac{2}{m_1-1}} \right]^{-1}, \left[\sum_{k=1}^{c} \left(\frac{d_{ij}^2}{d_{ij}^2} \right)^{\frac{2}{m_2-1}} \right]^{-1} \right\} \tag{9}$$

The procedure for updating the cluster prototypes in the IT2 FCM algorithm should take into account the degree of belonging interval to calculate the centroids of the fuzzy membership matrix for the lower and upper limit these centroids will be given by the following equations:

$$\underline{v}_i = \frac{\sum_{j=1}^{n} \left(\underline{\mu}_i(x_j) \right)^{m_1} x_j}{\sum_{j=1}^{n} \left(\underline{\mu}_i(x_j) \right)^{m_1}} \tag{10}$$

$$\bar{v}_i = \frac{\sum\limits_{j=1}^{n} \left(\bar{\mu}_i(x_j)\right)^{m_1} x_j}{\sum\limits_{j=1}^{n} \left(\bar{\mu}_i(x_j)\right)^{m_1}} \tag{11}$$

The resulting interval of the coordinates of positions of the centroids of the clusters. Type-reduction and defuzzification use type-2 fuzzy operations. The centroids matrix and the fuzzy partition matrix are obtained by the type-reduction as shown in the following equations:

$$v_j = \frac{\underline{v}_j + \bar{v}_j}{2} \tag{12}$$

$$\mu_i(x_j) = \frac{\underline{\mu}_i(x_j) + \bar{\mu}_i(x_j)}{2} \tag{13}$$

Based on all this, the IT2 FCM algorithm consists of the following steps:

1. Establish c, m_1, m_2.
2. Initialize fuzzy partition matrices $\underline{\mu}_i(x_j)$ and $\bar{\mu}_i(x_j)$, such that with restriction in.

$$\sum\limits_{i=1}^{c} \underline{\mu}_i(x_j) = 1 \tag{14}$$

$$\sum\limits_{i=1}^{c} \bar{\mu}_i(x_j) = 1 \tag{15}$$

3. Calculate the centroids for the lower and upper fuzzy partition matrix using the Eqs. (10) and (11) respectively.
4. Calculating the update of the fuzzy partition matrices for lower and upper bound of the interval using the Eqs. (8) and (9) respectively.
5. Type reduction of the fuzzy partition matrix and centroid, if the problem requires using the Eqs. (12) and (13) respectively.
6. Repeat steps 3 to 5 until $|J_{\tilde{m}}(t) - J_{\tilde{m}}(t-1)| < \varepsilon$.

This extension on the FCM algorithm is intended to realize that this algorithm is capable of handling uncertainty and is less susceptible noise.

4 Designing Type-2 Fuzzy Inference Systems Using Interval Type-2 Fuzzy C-Means Algorithm

The creation of Type-2 Inference Systems is performed taking into account the lower and upper centroid matrices and the lower and upper fuzzy membership matrices generated by the Interval Type-2 Fuzzy C-Means.

But for creating a Fuzzy Inference System is necessary to make inputs and outputs variables and rules of inference, the number of input and output variables will be given for the number of dimensions or characteristics of the data input and data output respectively. The number of membership functions for each input and output variable, is given by the number of clusters specified to IT2FCM algorithm for the input and output data clustering.

In creating the membership functions of the input and output variables, the centroid matrices and the fuzzy membership matrices of the lower and upper bounds of the interval are used, in this particular case type-2 Gaussian membership functions are created this is because that the parameter for this membership functions are center and standard deviation.

The centers for the Gaussian membership functions for the lower and upper bound of the interval are provided by the IT2FCM algorithm. The standard deviation for the lower and upper Gaussian membership functions can be found using the following equations:

$$\underline{\sigma}_i = \frac{1}{n} \sum_{j=1}^{n} \sqrt{-\frac{\left(x_j - \underline{v}_i\right)^2}{2 \ln \underline{\mu}_i(x_j)}} \tag{16}$$

$$\overline{\sigma}_i = \frac{1}{n} \sum_{j=1}^{n} \sqrt{-\frac{\left(x_j - \overline{v}_i\right)^2}{2 \ln \overline{\mu}_i(x_j)}} \tag{17}$$

As shown in Fig. 2, the calculation of the standard deviation is performed using the matrices of centroids, the fuzzy membership matrices and the dataset to which applied the IT2FCM algorithm to find the above matrices. Once the standard deviation is calculated for each centroid found in the matrices of the lower and upper centroids, one proceeds to create the functions for the lower and upper limit of the interval as shown in Fig 2, with the following equations:

$$\underline{gauss} = \min(gauss_1, gauss_2) \tag{18}$$

$$\overline{gauss} = \begin{cases} 1 & if \min(\underline{v}_i, \overline{v}_i) < x < \max(\underline{v}_i, \overline{v}_i) \\ \max(gauss_1, gauss_2) & otherwise \end{cases} \tag{19}$$

where

$$gauss_1(x, \underline{\sigma}_i, \underline{v}_i) = e^{-\frac{1}{2}\left(\frac{x - \underline{v}_i}{\underline{\sigma}_i}\right)^2} \tag{20}$$

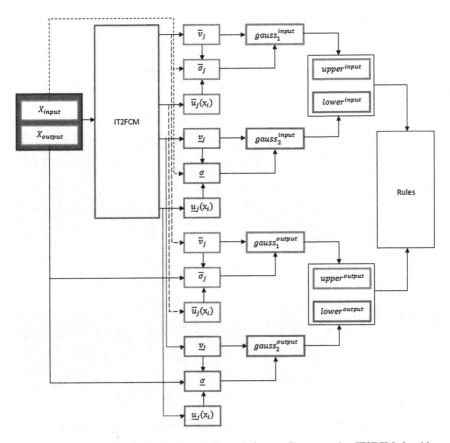

Fig. 2 Block diagram for designing Type-2 Fuzzy Inference Systems using IT2FCM algorithm

$$gauss_2(x, \overline{\sigma}_i, \overline{v}_i) = e^{-\frac{1}{2}\left(\frac{x-\overline{v}_i}{\overline{\sigma}_i}\right)^2} \tag{21}$$

Once the input and output variables are established with their respective membership functions are performed the creation of rules as shown in Fig. 2. In this particular case the amount of rules created will depend on the number of functions of membership, that is, if we have a fuzzy system with an input variable and an output variable with three functions of membership in each variable rules would formed as follows Table 1.

4.1 Type-2 Fuzzy Systems Designed for Time Series Prediction

In this section we present the structure of the Mamdani and Sugeno T2 FIS designed by IT2FCM algorithm for Mackey-Glass time series prediction and the

Table 1 Example of rule creation with IT2FCM

| if input₁ is mf₁ then output₁ is mf₁ |
| if $input_1$ is mf_1 then $output_1$ is mf_1 |
| if $input_1$ is mf_2 then $output_1$ is mf_2 |
| $\vdots$ |
| if $input_1$ is mf_n then $output_1$ is mf_n |

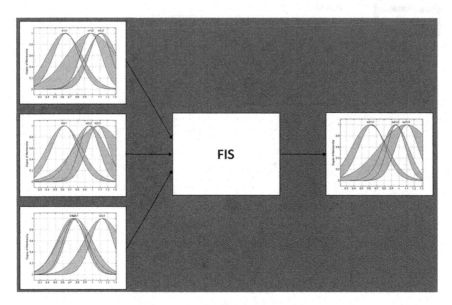

Fig. 3 T2 FIS Mamdani designed by IT2FCM for Mackey-Glass time series forecast

result obtained by T2 FIS designed. In Figs. 3 and 4, are shown the structure of Mamdani and Sugeno respectively of the T2FIS designed for the prediction of time series Mackey-Glass using IT2 FCM algorithm.

From the Mackey-Glass time series 800 pairs of data points were extracted [20, 21, 22–24, 25–27].The work of the T2FIS consist in predict $x(t)$ from to a data set created with three column vectors with 6 periods of delay in each vector of the time series, that is, $x(t-18)$, $x(t-12)$, and $x(t-6)$. Therefore the format of the training data is:

$$[x(t-18), x(t-12), x(t-6) ; x(t)] \tag{22}$$

where $t = 19$ to 818 and $x(t)$ is the desired prediction of the time series. The first 400 pairs of data are used to create the T2 FIS, while the other 400 pairs of data are used to test the T2 FIS designing. In Figs. 5, 6, one can observe the result of the prediction of the time series Mackey-Glass using Type-2 Fuzzy Systems designed by IT2FCM algorithm.

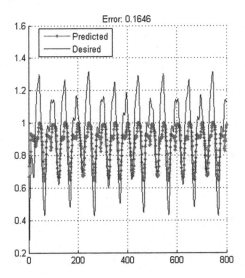

Fig. 4 Results of the Mackey-Glass time series prediction using the Mamdani Type-2 Fuzzy Inference System designed by IT2FCM

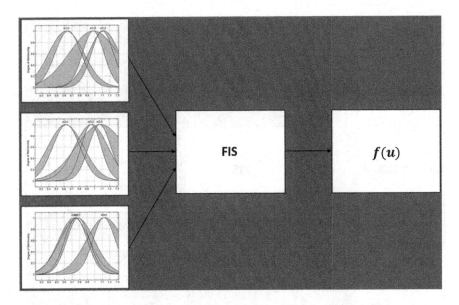

Fig. 5 T2 FIS Sugeno designed by IT2FCM for time series forecast Mackey-Glass

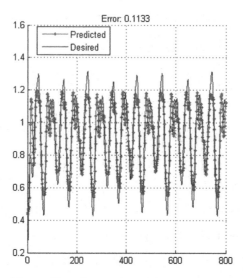

Fig. 6 Result of the Mackey-Glass time series prediction using the Sugeno type-2 fuzzy inference system designed by IT2FCM

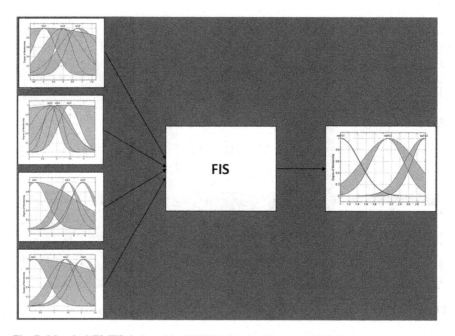

Fig. 7 Mamdani T2 FIS designed by IT2FCM for classification of Iris data set

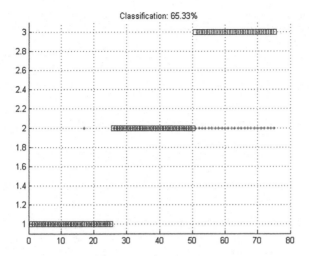

Fig. 8 Classification results for Iris data set using Mamdani T2 FIS designed by IT2FCM

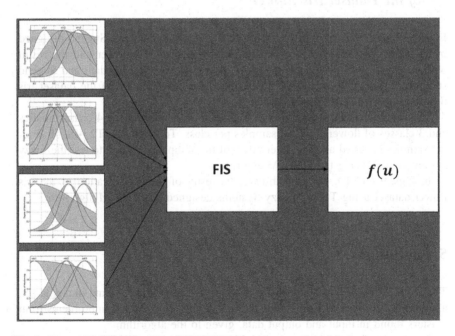

Fig. 9 Sugeno T2 FIS designed by IT2FCM for classification of the Iris data set

Fig. 10 Classification results
for Iris data set using the
SugenoT2 FIS designed by
IT2FCM

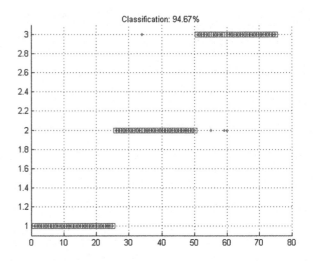

4.2 Type-2 Fuzzy Systems Designed for Classification of the Dataset Iris Flower

In this section we present the structure of T2 FIS Mamdani and Sugeno designed by IT2FCM algorithm for classification of the dataset iris flower and the result obtained by T2 FIS designed. In Figs. 7, 8, are shown the Mamdani and Sugeno structure respectively of the T2FIS designed for classification of the dataset iris flower using IT2 FCM algorithm.

The iris flower dataset consists of 150 samples in each of the 4 characteristics and 3 classes of flowers, i.e., 50 samples per class. To create the T2 FIS, 50 % of the samples are used and the other 50 % of the samples are used to test the fuzzy system created using the IT2FCM algorithm.

In Figs. 9 and 10, one can observe the result of the classification of the iris flower dataset using Type-2 Fuzzy Systems designed by IT2FCM algorithm.

5 Conclusions

The IT2FCM algorithm allows us to design FIS from clusters found in a data set, allowing us to create FIS for classification or time series prediction problems from clusters found in input and output data, given to the algorithm.

The results obtained by the Sugeno FIS designed by the IT2FCM algorithm for the problem of classification of iris flower was good achieving a 94.67 % of correct classification, but the result obtained by Mamdani FIS designed by IT2FCM algorithm for the above problem was not as good as expected getting a 65.33 % of correct classification.

The results obtained by the Sugeno FIS designed by the IT2FCM algorithm for the problem of time series prediction was 0.1133 of error prediction, but the results obtained by the Mamdani FIS designed by the IT2FCM algorithm for the above problem was 0.1646 of error prediction.

References

1. Bezdek, J.: Pattern Recognition with Fuzzy Objective Function Algorithms. Plenum, Berlin (1981)
2. Chang, X., Li, W., Farrell, J.: A C-means clustering based fuzzy modeling method. Fuzzy Systems, 2000. In: The Ninth IEEE International Conference on FUZZ IEEE 2000, vol. 2, 2000, pp. 937–940
3. Gustafson D. E., Kessel W. C.: Fuzzy clustering with a fuzzy covariance matrix. In: Proceeding IEEE Conference Decision Control, San Diego, CA, pp. 761–766 (1979)
4. Krishnapuram, R., Keller, J.: A possibilistic approach to clustering. IEEE Trans. Fuzzy Sys. 1(2), 98–110 (1993)
5. Kruse, R., Döring, C., Lesot, M. J.: fundamentals of fuzzy clustering. In: Advances in Fuzzy Clustering and its Applications, The Atrium, Southern Gate, Chichester John Wiley & Sons Ltd, West Sussex PO19 8SQ, England, pp. 3–30 (2007)
6. Pal, N.R., Pal, K., Keller, J.M., Bezdek, J.C.: A Possibilistic Fuzzy c-Means Clustering Algorithm. IEEE Trans. Fuzzy Sys 13(4), 517–530 (2005)
7. Hirota, K., Pedrycz, W.: Fuzzy computing for data mining. In: Proceeding of the IEEE, vol. 87(9), 1575–1600 (1999)
8. Iyer, N.S., Kendel, A., Schneider, M.: Feature-based fuzzy classification for interpretation of mamograms. Fuzzy Sets Syst. 114, 271–280 (2000)
9. Philips, W.E., Velthuinzen, R.P., Phuphanich, S., Hall, L.O., Clark, L.P., Sibiger, M.L.: Aplication of fuzzy c-means segmentation technique for tissue differentiation in MR images of hemorrhagic glioblastoma multifrome. Magn. Reson. Imaging 13(2), 277–290 (1995)
10. Yang, M.-S., Hu, Y.-J., Lin, K.C.-R., Lin, C.C.-L.: Segmentation techniques for tissue differentiation in MRI of ophthalmology using fuzzy clustering algorithms. Magn. Reson. Imaging 20, 173–179 (2002)
11. Karnik, N., Mendel, M.: Operations on type-2 set. Fuzzy Set Syst. 122, 327–348 (2001)
12. Mendel, J.: Uncertain Rule-Based Fuzzy Logic Systems: Introduction and New Directions, pp. 213–231. Prentice-Hall Inc, New Jersey (2001)
13. Choi, B., Rhee, F.: Interval type-2 fuzzy membership function generation methods for pattern recognition. Inf. Sci. 179(13), 2102–2122 (2009)
14. Hwang, C., Rhee, F.: Uncertain fuzzy clustering: interval type-2 fuzzy approach to C-means. IEEE Trans. Fuzzy Syst. 15(1), 107–120 (2007)
15. Rubio, E., Castillo O.: Interval type-2 fuzzy clustering for membership function generation. In: IEEE workshop on hybrid intelligent models and applications (HIMA),13(18), 16–19 April (2013)
16. Rubio E., Castillo O.: Optimization of the interval type-2 fuzzy C-means using particle swarm optimization. In: Proceedings of NABIC 2013, Fargo, USA, pp. 10–15
17. Ceylan, R., Özbay, Y., Karlik, B.: A novel approach for classification of ECG arrhythmias: Type-2 fuzzy clustering neural network. Expert Syst. Appl. 36(3), 6721–6726 (2009). Part 2
18. Yen, J., Langari, R.: Fuzzy Logic: Intelligence, Control, and Information, Upper Saddle River. Prentice Hall, New Jersey (1999)
19. Zadeh, L.A.: The concept of a linguistic variable and its application to approximate reasoning-I. Inform. Sci. 8(3), 199–249 (1975)

20. Jang, J.S.R.: ANFIS: Adaptive-network-based fuzzy inference systems. IEEE Trans. Syst. Man Cybern. **23**, 665–685 (1992)
21. Melin, P., Soto, J., Castillo, O., Soria, J.: A new approach for time series prediction using ensembles of ANFIS models. Experts Syst. Appl. **39**(3), 3494–3506 (2012). Elsevier
22. Pulido, M., Mancilla, A., Melin P.: An ensemble neural network architecture with fuzzy response integration for complex time series prediction. Evol. Des. Intell. Syst. Model. Simul. Control, pp. 85–110 (2009)
23. Pulido, M., Mancilla, A., Melin, P.: Ensemble neural networks with fuzzy logic integration for complex time series prediction. Int. J. Intell. Eng. Inform. **1**(1), 89–103 (2010)
24. Pulido, M., Mancilla, A., Melin, P.: Ensemble neural networks with fuzzy integration for complex time series prediction. Bio-inspired Hybrid Intell. Syst. Image Anal. Pattern Recognit, PP. 143–155 (2009)
25. Soto, J., Castillo, O, Soria J., A new approach for time series prediction using ensembles of ANFIS models. Soft Computing for Intelligent Control and Mobile Robotics, pp. 318–483
26. Soto, J., Melin, P., Castillo, O.: Time series prediction using ensembles of neuro-fuzzy models with interval type-2 and type-1 fuzzy integrators. In: Proceedings of International Joint Conference on Neural Nerworks, Dallas, Texas, USA, August 4–9, pp. 189–194 (2013)
27. Wang, C., Zhang, J.P.: Time series prediction based on ensemble ANFIS. In: Proceedings of the fourth international conference on machine learning and cybernetics, Guangzhou, pp. 18–21 August (2005)

Neural Network with Fuzzy Weights Using Type-1 and Type-2 Fuzzy Learning with Gaussian Membership Functions

Fernando Gaxiola, Patricia Melin and Fevrier Valdez

Abstract In this chapter type-1 and type-2 fuzzy inferences systems are used to obtain the type-1 or type-2 fuzzy weights in the connection between the layers of a neural network. We used two type-1 or type-2 fuzzy systems that work in the backpropagation learning method with the type-1 or type-2 fuzzy weight adjustment. The mathematical analysis of the proposed learning method architecture and the adaptation of type-1 or type-2 fuzzy weights are presented. The proposed method is based on recent methods that handle weight adaptation and especially fuzzy weights. In this work neural networks with type-1 fuzzy weights or type-2 fuzzy weights are presented. The proposed approach is applied to the case of Mackey–Glass time series prediction.

1 Introduction

Neural networks have been applied in several areas of research, like in the time series prediction area, which is the study case for this chapter, especially in the Mackey–Glass time series.

The approach presented in this chapter works with type-1 and type-2 fuzzy weights in the neurons of the hidden and output layers of the neural network used for prediction of the Mackey–Glass time series. These type-1 and interval type-2 fuzzy weights are updated using the backpropagation learning algorithm. We used

F. Gaxiola · P. Melin (✉) · F. Valdez
Tijuana Institute of Technology, Tijuana, Mexico
e-mail: pmelin@tectijuana.mx

F. Gaxiola
e-mail: fergaor_29@hotmail.com

F. Valdez
e-mail: fevrier@tectijuana.edu.mx

O. Castillo et al. (eds.), *Recent Advances on Hybrid Approaches for Designing Intelligent Systems*, Studies in Computational Intelligence 547,
DOI: 10.1007/978-3-319-05170-3_4, © Springer International Publishing Switzerland 2014

two type-1 inference systems and two type-2 inference systems with Gaussian membership functions.

The proposed approach is applied to time series prediction for the Mackey–Glass series. The objective is obtaining the minimal prediction error for the data of the time series.

We used a supervised neural network, because this type of network is the most commonly used in the area of timer series prediction.

The weights of a neural network are an important part in the training phase, because these affect the performance of the learning process of the neural network.

This chapter is focused in the managing of weights, because on the practice of neural networks, when performing the training of neural networks for the same problem is initialized with different weights or the adjustment are in a different way each time it is executed, but at the final is possible to reach a similar result.

The next section presents a background about modifications of the backpropagation algorithm and different management strategies of weights in neural networks, and basic concepts of neural networks. Section 3 explains the proposed method and the problem description. Section 4 describes the neural network with type-1 fuzzy weights proposed in this chapter. Section 5 describes the neural network with type-2 fuzzy weights proposed in this chapter. Section 6 presents the simulation results for the proposed method. Finally, in Sect. 7 some conclusions are presented.

2 Background and Basic Concepts

In this section a brief review of basic concepts is presented.

2.1 Neural Network

An artificial neural network (ANN) is a distributed computing scheme based on the structure of the nervous system of humans. The architecture of a neural network is formed by connecting multiple elementary processors, this being an adaptive system that has an algorithm to adjust their weights (free parameters) to achieve the performance requirements of the problem based on representative samples [8, 26].

The most important property of artificial neural networks is their ability to learn from a training set of patterns, i.e. they are able to find a model that fits the data [9, 34].

The artificial neuron consists of several parts (see Fig. 1). On one side are the inputs, weights, the summation, and finally the transfer function. The input values are multiplied by the weights and added: $\sum x_i w_{ij}$. This function is completed with the addition of a threshold amount i. This threshold has the same effect as an input with value -1. It serves so that the sum can be shifted left or right of the origin.

Fig. 1 Schematics of an modular artificial neural network

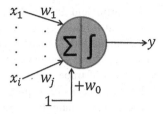

After addition, we have the function f applied to the sum resulting in the final value of the output, also called y_i [28], obtaining the following equation:

$$y_i = f\left(\sum_{i=1}^{n} x_i w_{ij}\right) \qquad (1)$$

where f may be a nonlinear function with binary output $+ -1$, a linear function $f(z) = z$, or as sigmoidal logistic function:

$$f(z) = \frac{1}{1 + e^{-z}} \qquad (2)$$

2.2 Overview of Related Works

The backpropagation algorithm and its variations are the most useful basic training methods in the area of neural networks. However, these algorithms are usually too slow for practical applications.

When applying the basic backpropagation algorithm to practical problems, the training time can be very high. In the literature we can find that several methods have been proposed to accelerate the convergence of the algorithm [2, 18, 28, 37].

There exists many works about adjustment or managing of weights but only the most important and relevant for this research will be mentioned here [4, 10, 31, 36]:

Momentum method. Rumelhart, Hinton and Williams suggested adding in the increased weights expression a momentum term β, to filter the oscillations that can be formed a higher learning rate that lead to great change in the weights [19, 32] .

Adaptive learning rate. focuses on improving the performance of the algorithm by allowing the learning rate changes during the training process (increase or decrease) [19].

Conjugate Gradient algorithm. A search of weight adjustment along conjugate directions. Versions of conjugate gradient algorithm differ in the way in which a constant βk is calculated.

- Fletcher-Reeves update [12].
- Polak-Ribiere updated [12].
- Powell-Beale Restart [3, 33].
- Scaled Conjugate Gradient [29].

Kamarthi and Pittner [24], focused in obtaining a weight prediction of the network at a future epoch using extrapolation.

Ishibuchi et al. [21], proposed a fuzzy network, where the weights are given as trapezoidal fuzzy numbers, denoted as four trapezoidal fuzzy numbers for the four parameters of trapezoidal membership functions.

Ishibuchi et al. [22], proposed a fuzzy neural network architecture with symmetrical fuzzy triangular numbers for the fuzzy weights and biases, denoted by the lower, middle and upper limit of the fuzzy triangular numbers.

Feuring [11], based on the work by Ishibuchi, where triangular fuzzy weights are used, developed a learning algorithm in which the backpropagation algorithm is used to compute the new lower and upper limits of weights. The modal value of the new fuzzy weight is calculated as the average of the new computed limits.

Castro et al. [6], use interval type-2 fuzzy neurons for the antecedents and interval of type-1 fuzzy neurons for the consequents of the rules. This approach handles the weights as numerical values to determine the input of the fuzzy neurons, as the scalar product of the weights for the input vector.

Gaxiola et al. [13–17], proposed at neural network with type-2 fuzzy weights using triangular membership functions.

Recent works on type-2 fuzzy logic have been developed in time series prediction, like that of Castro et al. [6], and other researchers [1, 7].

3 Proposed Method and Problem Description

The objective of this work is to use type-1 and interval type-2 fuzzy sets to generalize the backpropagation algorithm to allow the neural network to handle data with uncertainty. The Mackey–Glass time series (for $\tau = 17$) is utilized for testing the proposed approach.

The updating of the weights will be done differently to the traditional updating of the weights performed with the backpropagation algorithm (Fig. 2).

The proposed method performs the updating of the weights working with interval type-2 fuzzy weights, this method uses two type-1 and two type-2 inference systems with Gaussian membership functions for obtaining the type-1 and interval type-2 fuzzy weights using in the neural network, and obtaining the outputs taking into account the possible change in the way we work internally in the neuron, and the adaptation of the weights given in this way (Figs. 3, 4) [30].

We developed a method for adjusting weights to achieve the desired result, searching for the optimal way to work with type-1 fuzzy weights or type-2 fuzzy weights [23].

We used the sigmoid activation function for the hidden neurons and the linear activation function for the output neurons, and we utilized this activation functions because these functions have obtained good results in similar approaches.

Fig. 2 Scheme of current management of numerical weights (type-0) for the inputs of each neuron

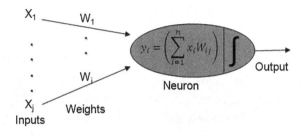

Fig. 3 Schematics of each neuron with the proposed management of weights using type-1 fuzzy sets

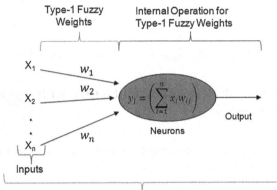

Fig. 4 Schematics of each neuron with the proposed management of weights using interval type-2 fuzzy sets

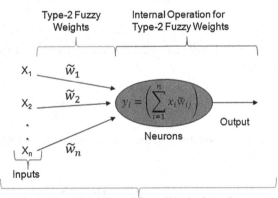

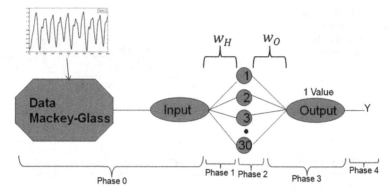

Fig. 5 Proposed neural network architecture with type-1 fuzzy weights

4 Neural Network Architecture with Type-1 Fuzzy Weights

The proposed neural network architecture with type-1 fuzzy weights (see Fig. 5) is described as follows:

Layer 0: Inputs.

$$x = [x_1, x_2, \ldots, x_n]$$ (3)

Layer 1: type-1 fuzzy weights for the hidden layer.

$$w_{ij}$$ (4)

Layer 2: Equations of the calculations in the hidden neurons using type-1 fuzzy weights.

$$Net = \sum_{i=1}^{n} x_i w_i$$ (5)

Layer 3: Equations of the calculations in the outputs neurons using type-1 fuzzy weights.

$$Out = \sum_{i=1}^{n} y_i w_i$$ (6)

Layer 4: Obtain the output of the neural network.

We considered neural network architecture with 1 neuron in the output layer and 30 neurons in the hidden layer.

This neural network used two type-1 fuzzy inference systems, one in the connections between the input neurons and the hidden neurons, and the other in the connections between the hidden neurons and the output neuron. In the hidden layer

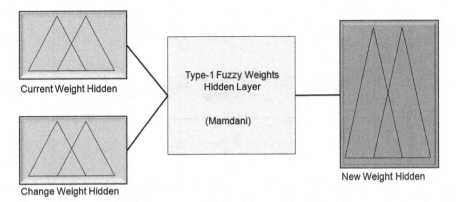

Fig. 6 Structure of the two type-1 fuzzy inference systems that were used to obtain the type-1 fuzzy weights in the hidden and output layer

and output layer of the network we are updating the weights using the two type-1 fuzzy inference system that obtains the new weights in each epoch of the network on base at the backpropagation algorithm.

The two type-2 fuzzy inference systems have the same structure and consist of two inputs (the current weight in the actual epoch and the change of the weight for the next epoch) and one output (the new weight for the next epoch) (see Fig. 6).

We used two Gaussian membership functions with their corresponding range for delimiting the inputs and outputs of the two type-1 fuzzy inference systems (see Figs. 7, 8).

We obtain the two type-1 fuzzy inference systems empirically.

The two type-1 fuzzy inference systems used the same six rules, the four combinations of the two Gaussian membership function and two rules added for null change of the weight (see Fig. 9).

5 Neural Network Architecture with Type-2 Fuzzy Weights

The proposed neural network architecture with interval type-2 fuzzy weights (see Fig. 10) is described as follows:

Layer 0: Inputs.

$$x = [x_1, x_2, \ldots, x_n] \qquad (7)$$

Layer 1: Interval type-2 fuzzy weights for the connection between the input and the hidden layer of the neural network (see Fig. 7).

$$\tilde{w}_{ij} = \left[\bar{w}_{ij}, \underline{w}_{ij} \right] \qquad (8)$$

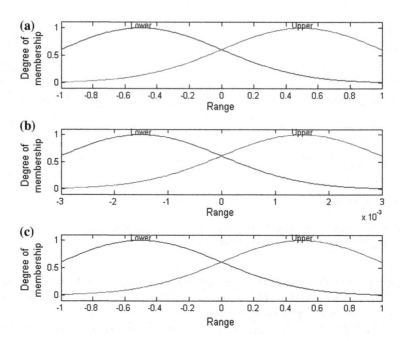

Fig. 7 Inputs (**a**) and (**b**) and output (**c**) of the type-1 fuzzy inference systems that were used to obtain the type-1 fuzzy weights in the hidden layer. **a** Current weight. **b** Change of weight. **c** New weight

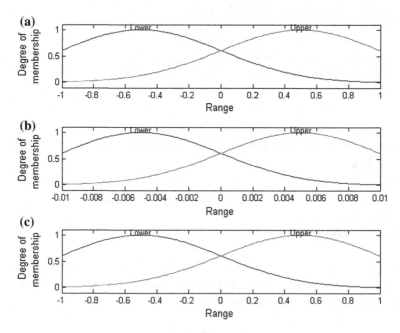

Fig. 8 Inputs (**a**) and (**b**) and output (**c**) of the type-1 fuzzy inference systems that were used to obtain the type-1 fuzzy weights in the output layer. **a** Current weight. **b** Change of weight. **c** New weight

1.	(Current_Weight is lower) and (Change_Weight is lower) then (New_Weight is lower)
2.	(Current_Weight is lower) and (Change_Weight is upper) then (New_Weight is lower)
3.	(Current_Weight is upper) and (Change_Weight is lower) then (New_Weight is upper)
4.	(Current_Weight is upper) and (Change_Weight is upper) then (New_Weight is upper)
5.	(Current_Weight is lower) then (New_Weight is lower)
6.	(Current_Weight is upper) then (New_Weight is upper)

Fig. 9 Rules of the type-1 fuzzy inference system used in the hidden and output layer for the neural network with type-1 fuzzy weights

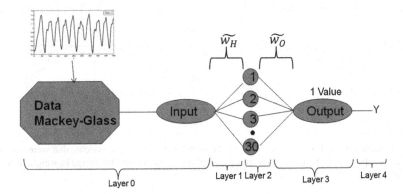

Fig. 10 Proposed neural network architecture with interval type-2 fuzzy weights

where $\tilde{w}_{ij}$ are the weights of the consequents of each rule of the type-2 fuzzy system with inputs (current type-2 fuzzy weight, change of weight) and output (new fuzzy weight) (see Fig. 8).

Layer 2: Equations of the calculations in the hidden neurons using interval type-2 fuzzy weights.

$$Net = \sum_{i=1}^{n} x_i \tilde{w}_{ij} \qquad (9)$$

Layer 3: Equations of the calculations in the output neurons using interval type-2 fuzzy weights.

$$Out = \sum_{i=1}^{n} y_i \tilde{w}_{ij} \qquad (10)$$

Layer 4: Obtain a single output of the neural network.

We applied the same neural network architecture used in the type-1 fuzzy weights for the type-2 fuzzy weights (see Fig. 11).

We used two type-2 fuzzy inference systems to obtain the type-2 fuzzy weights and work in the same way like with the type-1 fuzzy weights.

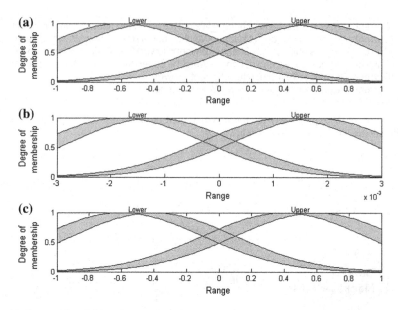

Fig. 11 Inputs (**a**) and (**b**) and output (**c**) of the type-2 fuzzy inference systems that were used to obtain the type-2 fuzzy weights in the hidden layer. **a** Current weight. **b** Change of weight. **c** New weight

The structure and rules (see Fig. 9) of the two type-2 fuzzy inference systems are the same of the type-1 fuzzy inference systems, the difference is in the memberships functions, Gaussian membership functions for type-2 [5, 20, 27, 35].

We used two Gaussian membership functions with their corresponding range for delimiting the inputs and outputs of the two type-2 fuzzy inference systems (see Figs. 11, 12).

We obtain the type-2 fuzzy inference systems incrementing and decrementing 20 % the values of the centers of the Gaussian membership functions and the same standard deviation of the type-1 Gaussians membership functions, we use this method to obtain the footprint of uncertainty (FOU) for the type-2 fuzzy inference systems used in the neural network with type-2 fuzzy weights.

6 Simulation Results

We performed experiments in time-series prediction, specifically for the Mackey–Glass time series ($\tau = 17$) (see Fig. 12).

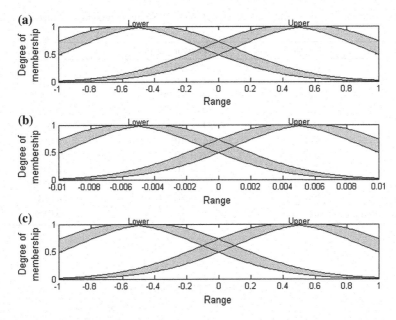

Fig. 12 Inputs (**a**) and (**b**) and output (**c**) of the type-2 fuzzy inference system that were used to obtain the type-2 fuzzy weights in the output layer. **a** Current weight. **b** Change of weight. **c** New weight

We presented the obtained results of the experiments performed with the neural network with type-1 fuzzy weights (NNT1FW) and the neural network with type-1 fuzzy weights (NNT1FW), these results are achieved without optimizing of the neural network and the type-1 fuzzy systems, which means that all parameters of the neural network and the range and values of the membership functions of the type-1 fuzzy systems are established empirically. The average error was obtained of 30 experiments.

In Table 1, we present the prediction error obtained with the results achieved as output of NNT1FW. The best prediction error is of 0.0489 and the average prediction error is of 0.0668.

In Fig. 13 we show prediction data with type-1 fuzzy weights against the test data of the Mackey–Glass time series.

In Table 2, we present the prediction error obtained with the results achieved as output of NNT2FW. The best prediction error is of 0.0413 and the average prediction error is of 0.0579.

In Fig. 14 we show the prediction data with type-2 fuzzy weights against the test data of the Mackey–Glass time series.

Table 1 Prediction error for the neural network with type-1 fuzzy weights for Mackey–Glass time series

No.	Epochs	Network error	Prediction error
E1	100	1×10^{-8}	0.0704
E2	100	1×10^{-8}	0.0554
E3	100	1×10^{-8}	0.0652
E4	*100*	1×10^{-8}	*0.0489*
E5	100	1×10^{-8}	0.0601
E6	100	1×10^{-8}	0.0615
E7	100	1×10^{-8}	0.0684
E8	100	1×10^{-8}	0.0751
E9	100	1×10^{-8}	0.0737
E10	100	1×10^{-8}	0.0711
Average prediction error			0.0668

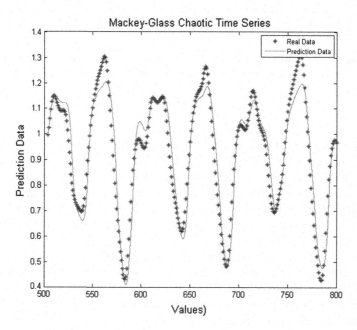

Fig. 13 Plot of the prediction data with NNT1FW against the test data of the Mackey–Glass time series

Table 2 Prediction error for the neural network with interval type-2 fuzzy weights for the Mackey–Glass time series

No.	Epochs	Network error	Prediction error
E1	100	1×10^{-8}	0.0549
E2	100	1×10^{-8}	0.0691
E3	100	1×10^{-8}	0.0557
E4	100	1×10^{-8}	0.0432
E5	100	1×10^{-8}	0.0602
E6	100	1×10^{-8}	0.0673
E7	*100*	1×10^{-8}	*0.0413*
E8	100	1×10^{-8}	0.0618
E9	100	1×10^{-8}	0.0645
E10	100	1×10^{-8}	0.0522
Average prediction error			0.0579

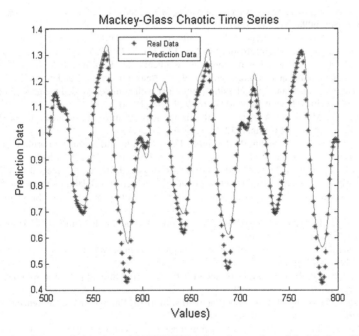

Fig. 14 Plot of the prediction data with NNT2FW against the test data of the Mackey–Glass time series

7 Conclusions

In this chapter, we proposed a new learning method that updates weights (type-1 or type-2 fuzzy weights) in each connection between the neurons of the layers of neural network using a type-1 or type-2 fuzzy inference system with Gaussians membership functions.

Additionally, the neurons work internally with the type-1 or type-2 fuzzy weights and therefore, obtaining results at the output of each neuron of the neural network. The modifications performed in the neural network that allows working with type-1 or type-2 fuzzy weights provide at the neural network greater robustness and less susceptibility at the noise in the data of the time series.

The prediction error of 0.0413 of the neural network with type-2 fuzzy weights for the Mackey–Glass time series is better than the prediction error of 0.0489 of the neural network with type-1 fuzzy weights (as shown in Tables 1 and 2).

This result is good considering that the used parameters for the neural networks at the moment are determined in an empirical way.

References

1. Abiyev, R.: A type-2 fuzzy wavelet neural network for time series prediction. Lect. Notes Comput. Sci. **6098**, 518–527 (2010)
2. Barbounis, T.G., Theocharis, J.B.: Locally recurrent neural networks for wind speed prediction using spatial correlation. Inf. Sci. **177**(24), 5775–5797 (2007)
3. Beale, E.M.L.: A derivation of conjugate gradients. In: Lootsma, F.A. (ed.) Numerical Methods for Nonlinear Optimization, pp. 39–43. Academic Press, London (1972)
4. Casasent, D., Natarajan, S.: A classifier neural net with complex-valued weights and square-law nonlinearities. Neural Networks **8**(6), 989–998 (1995)
5. Castillo, O., Melin, P.: A review on the design and optimization of interval type-2 fuzzy controllers. Appl. Soft Comput. **12**(4), 1267–1278 (2012)
6. Castro, J., Castillo, O., Melin, P., Rodríguez-Díaz, A.: A hybrid learning algorithm for a class of interval type-2 fuzzy neural networks. Inf. Sci. **179**(13), 2175–2193 (2009)
7. Castro J., Castillo O., Melin P., Mendoza O., Rodríguez-Díaz A.: An interval type-2 fuzzy neural network for chaotic time series prediction with cross-validation and akaike test. In: Soft Computing for Intelligent Control and Mobile Robotics, pp. 269–285. Springer, Berlin (2011)
8. Cazorla, M., Escolano, F.: Two bayesian methods for junction detection. IEEE Trans. Image Process. **12**(3), 317–327 (2003)
9. De Wilde, O.: The magnitude of the diagonal elements in neural networks. Neural Networks **10**(3), 499–504 (1997)
10. Draghici, S.: On the capabilities of neural networks using limited precision weights. Neural Networks **15**(3), 395–414 (2002)
11. Feuring T.: Learning in fuzzy neural networks. In: IEEE International Conference on Neural Networks, vol. 2, pp. 1061–1066 (1996)
12. Fletcher, R., Reeves, C.M.: Function minimization by conjugate gradients. Comput. J. **7**, 149–154 (1964)
13. Gaxiola, F., Melin, P., Valdez, F., Castillo, O.: Neural network with type-2 fuzzy weights adjustment for pattern recognition of the human iris biometrics. In: MICAI'12 Proceedings of the 11th Mexican International Conference on Advances in Computational Intelligence, vol. 2, pp. 259–270 (2012)
14. Gaxiola, F., Melin, P., Valdez, F., Castillo, O.: Optimization of type-2 fuzzy weight for neural network using genetic algorithm and particle swarm optimization. In: 5th World Congress on Nature and Biologically Inspired Computing, pp. 22–28 (2013)
15. Gaxiola, F., Melin P., Valdez F.: Backpropagation method with type-2 fuzzy weight adjustment for neural network learning. In: Fuzzy Information Processing Society (NAFIPS), Annual Meeting of the North American, pp. 1–6 (2012)

16. Gaxiola, F., Melin, P., Valdez, F.: Genetic optimization of type-2 fuzzy weight adjustment for backpropagation ensemble neural network. In: Recent Advances on Hybrid Intelligent Systems, pp. 159–171 (2013)
17. Gaxiola, F., Melin, P., Valdez, F.: Neural network with lower and upper type-2 fuzzy weights using the backpropagation learning method. In: IFSA World Congress and NAFIPS Annual Meeting (IFSA/NAFIPS), pp. 637–642 (2013)
18. Gedeon, T.: Additive neural networks and periodic patterns. Neural Networks **12**(4–5), 617–626 (1999)
19. Hagan, M.T., Demuth, H.B., Beale, M.H.: Neural Network Design, p. 736. PWS Publishing, Boston (1996)
20. Hagras, H.: Type-2 fuzzy logic controllers: a way forward for fuzzy systems in real world environments. In: IEEE World Congress on Computational Intelligence, pp. 181–200 (2008)
21. Ishibuchi, H., Morioka, K., Tanaka, H.: A fuzzy neural network with trapezoid fuzzy weights, fuzzy systems. In: IEEE World Congress on Computational Intelligence, vol. 1, pp. 228–233 (1994)
22. Ishibuchi, H., Tanaka, H., Okada, H.: Fuzzy neural networks with fuzzy weights and fuzzy biases. In: IEEE International Conference on Neural Networks, vol. 3, pp. 1650-1655 (1993)
23. Jang, J.S.R., Sun, C.T., Mizutani, E.: Neuro-Fuzzy and Soft Computing: A Computational Approach to Learning and Machine Intelligence, p. 614. Prentice Hall, Englewood Cliffs (1997)
24. Kamarthi, S., Pittner, S.: Accelerating neural network training using weight extrapolations. Neural Networks **12**(9), 1285–1299 (1999)
25. Karnik, N., Mendel, J.: Applications of type-2 fuzzy logic systems to forecasting of time-series. Inf. Sci. **120**(1–4), 89–111 (1999)
26. Martinez, G., Melin, P., Bravo, D., Gonzalez, F., Gonzalez, M.: Modular neural networks and fuzzy sugeno integral for face and fingerprint recognition. Adv. Soft Comput. **34**, 603–618 (2006)
27. Melin, P.: Modular Neural Networks and Type-2 Fuzzy Systems for Pattern Recognition, pp. 1–204. Springer, Berlin (2012)
28. Meltser, M., Shoham, M., Manevitz, L.: Approximating functions by neural networks: a constructive solution in the uniform norm. Neural Networks **9**(6), 965–978 (1996)
29. Moller, M.F.: A scaled conjugate gradient algorithm for fast supervised learning. Neural Networks **6**, 525–533 (1993)
30. Monirul Islam, M.D., Murase, K.: A new algorithm to design compact two-hidden-layer artificial neural networks. Neural Networks **14**(9), 1265–1278 (2001)
31. Neville, R.S., Eldridge, S.: Transformations of sigma–pi nets: obtaining reflected functions by reflecting weight matrices. Neural Networks **15**(3), 375–393 (2002)
32. Phansalkar, V.V., Sastry, P.S.: Analysis of the back-propagation algorithm with momentum. IEEE Trans. Neural Networks **5**(3), 505–506 (1994)
33. Powell, M.J.D.: Restart procedures for the conjugate gradient method. Math. Program. **12**, 241–254 (1977)
34. Salazar, P.A., Melin, P., Castillo, O.: A new biometric recognition technique based on hand geometry and voice using neural networks and fuzzy logic. In: Soft Computing for Hybrid Intelligent Systems, pp. 171–186 (2008)
35. Sepúlveda, R., Castillo, O., Melin, P. Montiel, O.: An efficient computational method to implement type-2 fuzzy logic in control applications. In: Analysis and Design of Intelligent Systems using Soft Computing Techniques, pp. 45–52 (2007)
36. Yam, J., Chow, T.: A weight initialization method for improving training speed in feedforward neural network. Neurocomputing **30**(1–4), 219–232 (2000)
37. Yeung, D., Chan, P., Ng, W.: Radial basis function network learning using localized generalization error bound. Inf. Sci. **179**(19), 3199–3217 (2009)

A Comparative Study of Membership Functions for an Interval Type-2 Fuzzy System used to Dynamic Parameter Adaptation in Particle Swarm Optimization

Frumen Olivas, Fevrier Valdez and Oscar Castillo

Abstract This chapter present an analysis of the effects in quality results that brings the different types of membership functions in an interval type-2 fuzzy system used to adapt some parameters of Particle Swarm Optimization (PSO). Benchmark mathematical functions are used to test the methods and a comparison is performed.

1 Introduction

In this chapter we analyze the effect that brings the different types of membership functions in an interval type-2 fuzzy system to the dynamic parameter adaptation in PSO, which we use in [10], this system is an extension or generalization from a system we use in [9]. So the main contribution of this chapter is the analysis of the effect that the different types of membership functions bring to an interval type-2 fuzzy system.

PSO is a bio-inspired method introduced by Kennedy and Eberhart in 1995 [5, 6], this method maintains a swarm of particles, like a flock of birds or a school of fish, and each particle represent a solution, these particle "fly" in a multidimensional space, using Eq. 1 to update the position of the particle, and Eq. 2 to update the velocity of the particle [1]. The equation of the velocity of the particle is affected by a coefficient component and a social component, where coefficient component is

F. Olivas · F. Valdez (✉) · O. Castillo
Tijuana Institute of Technology, Tijuana, Mexico
e-mail: fevrier@tectijuana.mx

F. Olivas
e-mail: frumen@msn.com

O. Castillo
e-mail: ocastillo@tectijuana.mx

O. Castillo et al. (eds.), *Recent Advances on Hybrid Approaches for Designing Intelligent Systems*, Studies in Computational Intelligence 547,
DOI: 10.1007/978-3-319-05170-3_5, © Springer International Publishing Switzerland 2014

affected by a coefficient factor c_1, and social component is affected by a social factor c_2, these parameters are very important because affect the velocity of the particle, this is when c_1 is larger than c_2 improves that the swarm explore the space of search, and in the other hand, when c_1 is lower than c_2 improves that the swarm exploit the best area found so far of the space of search.

$$x_i(t+1) = x_i(t) + v_i(t+1) \tag{1}$$

$$v_{ij}(t+1) = v_{ij}(t) + c_1 r_1(t)\left[y_{ij}(t) - x_{ij}(t)\right] + c_2 r_{2j}(t)\left[\hat{y}_j(t) - x_{ij}(t)\right] \tag{2}$$

Fuzzy set theory introduced by Zadeh in 1965 [13, 14], help us to model problems with a degree of uncertainty, by using linguistic information about the problem and improves the numerical computation by using linguistic labels represented by membership functions. Type-2 fuzzy logic also introduced by Zadeh [15], help us to model problems with an implicit degree of uncertainty, and this is possible by adding a footprint of uncertainty (FOU) to the membership functions, and this means that even the membership value is fuzzy [7].

Since PSO was introduced in 1995, has had some improvements to the convergence and diversity of the swarm using a fuzzy system, for example: Hongbo and Ajith [3] add a new parameter to the velocity equation, and this new parameter is tuned using a fuzzy logic controller. Shi and Eberhart [11], use a fuzzy system to adjust the inertia weight. Wang et al. [12], use a fuzzy system to determine when to change the velocity of the particles to avoid local minima.

2 Methodology

The change is PSO is the inclusion of a type-2 fuzzy system that dynamically adapts the parameters c1 and c2, the proposal is show in Fig. 1.

We develop an interval type-2 fuzzy system that dynamically adapt some parameters in PSO, this parameters are c_1 and c_2, and we choose these parameters because are important to the change of the velocity of the particle, and we use previous knowledge to make the fuzzy rule set, using the idea that in first iterations PSO will use exploration to search the best area of the space of search, and in final iterations PSO will use exploitation of the best area found so far.

The fuzzy rule set used for parameter adaptation in PSO is shown in Fig. 2. Taking the idea that when c_1 is higher than c_2, this improves the exploration of the space of search, and when c_1 is lower than c_2, this improves the exploitation of the best area from the space of search found so far.

The variables for the interval type-2 fuzzy system for parameter adaptation are:

Inputs:

1. Iteration: this variable is a metric about the iterations of PSO, is a percentage of the iterations elapsed, with a range from [0, 1], and is represented by Eq. 3.

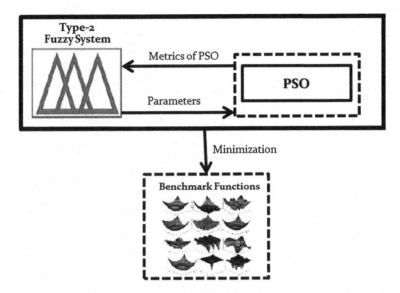

Fig. 1 Proposal of type 2 fuzzy adaptation of PSO parameters

1. If (Iteration is Low) and (Diversity is Low) then	(C₁ is High) and (C₂ is Low)
2. If (Iteration is Low) and (Diversity is Medium) then	(C₁ is MediumHigh) and (C₂ is Medium)
3. If (Iteration is Low) and (Diversity is High) then	(C₁ is MediumHigh) and (C₂ is MediumLow)
4. If (Iteration is Medium) and (Diversity is Low) then	(C₁ is MediumHigh) and (C₂ is MediumLow)
5. If (Iteration is Medium) and (Diversity is Medium) then	(C₁ is Medium) and (C₂ is Medium)
6. If (Iteration is Medium) and (Diversity is High) then	(C₁ is MediumLow) and (C₂ is MediumHigh)
7. If (Iteration is High) and (Diversity is Low) then	(C₁ is Medium) and (C₂ is High)
8. If (Iteration is High) and (Diversity is Medium) then	(C₁ is MediumLow) and (C₂ is MediumHigh)
9. If (Iteration is High) and (Diversity is High) then	(C₁ is Low) and (C₂ is High)

Fig. 2 Fuzzy rule set for parameter adaptation in PSO

2. Diversity: this variable is a metric about the particles of PSO, is the degree of dispersion of the particles in the space of search, diversity is also can be see it as the average of the Euclidian between each particle and the best particle, with a range of [0, 1] (normalized), and is represented by Eq. 4.

Outputs:

1. C1: this variable represent the parameter c_1 on the velocity equation of PSO, is the cognitive factor, and affect the best position of the particle found so far, with a range of [0, 3], but granulated so that only return results in the range of [0.5, 2.5], this range contains the best possible values for c_1 according to [4].
2. C2: this variable represent the parameter c_2 on the velocity equation of PSO, is the social factor, and affect the best position of the best particle in the swarm

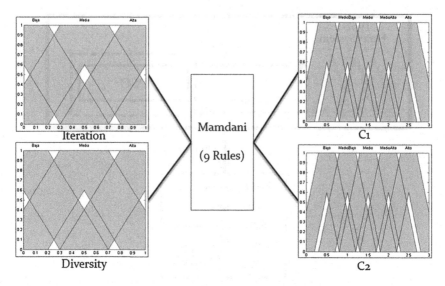

Fig. 3 Interval type-2 fuzzy system used for parameter adaptation in PSO

found so far, with a range of [0, 3], but like the other output, is granulated so that only return results in the range of [0.5, 2.5], this range contains the best possible values for c_2 according to [4].

$$Iteration = \frac{Current\ Iteration}{Maximum\ of\ Iterations} \tag{3}$$

$$Diversity\ (S(t)) = \frac{1}{n_s} \sum_{i=1}^{n_s} \sqrt{\sum_{j=1}^{n_\chi} (\chi_{ij}(t) - \bar{\chi}_j(t))^2} \tag{4}$$

The interval type-2 fuzzy system designed in [10], for the parameter adaptation is shown in Fig. 3, and we develop this system manually, this is, we change the levels of FOU of each point of each membership function, but each point has the same level of FOU, also the input and output variables have only interval type-2 triangular membership functions. However, this system is the best system found by changing the levels of FOU manually.

Thinking that each point of each membership function should have its own level of FOU, that's why we decide to optimize the levels of FOU using an external PSO [10], the result of this optimization is shown in Fig. 4.

The purpose of this chapter is the analysis of using different types of membership functions; we already use interval type-2 triangular membership functions, so the types used in this analysis are: interval type-2 Gaussian membership functions,

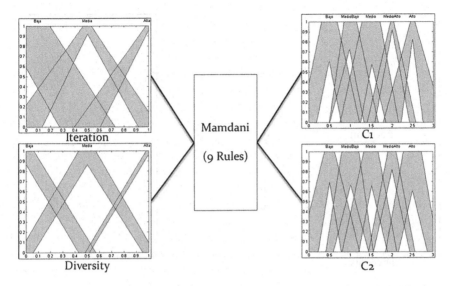

Fig. 4 Optimized interval type-2 fuzzy system used for parameter adaptation in PSO

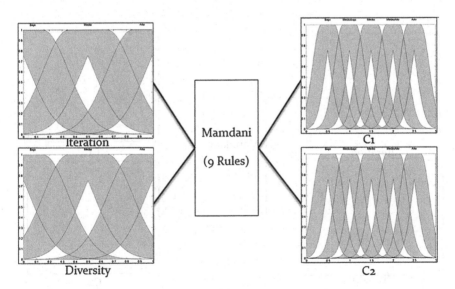

Fig. 5 Interval type-2 fuzzy system with Gaussian membership functions for parameter adaptation in PSO

interval type-2 generalized bell membership functions and interval type-2 trape-zoidal membership functions.

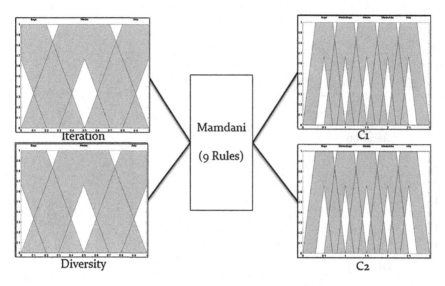

Fig. 6 Interval type-2 fuzzy system with trapezoidal membership functions for parameter adaptation in PSO

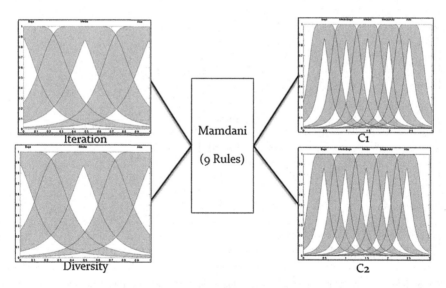

Fig. 7 Interval type-2 fuzzy system with generalized bell membership functions for parameter adaptation in PSO

To switch between types of membership functions, we try to represent the same level of FOU on each type of membership function, taking as reference the interval type-2 triangular membership functions that we already use.

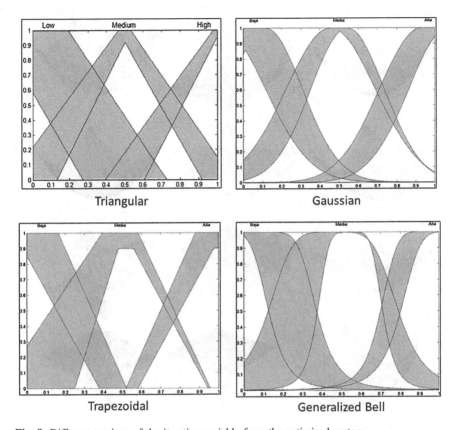

Fig. 8 Different versions of the iteration variable from the optimized system

In Fig. 5 is shown the fuzzy system for parameter adaptation, using interval type-2 Gaussian membership functions. Like the system done manually, now this system has only interval type-2 Gaussian membership functions in input and output variables. We try to maintain the levels of FOU on each membership function as similar to the system that use triangular membership functions, this to see only the effect of a different membership function.

In Fig. 6 is shown the fuzzy system for parameter adaptation, using interval type-2 trapezoidal membership functions. And again we try to maintain the levels of FOU on each membership function as similar to the system that use triangular membership functions.

In Fig. 7 is shown the fuzzy system for parameter adaptation, using interval type-2 generalized bell membership functions. We try to maintain the levels of FOU as in the system that use triangular membership functions, but more likely the system that use trapezoidal membership functions.

We also change the type of membership functions from the optimized interval type-2 fuzzy system, trying to maintain the levels of FOU on each point from each

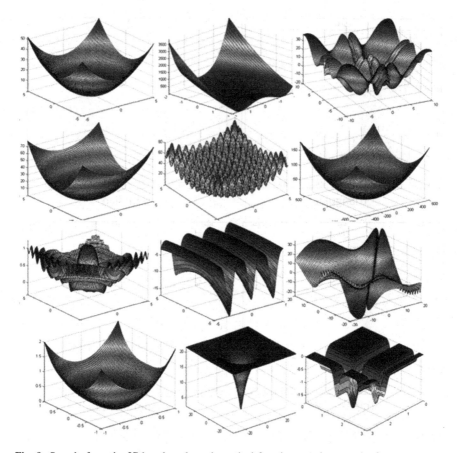

Fig. 9 Sample from the 27 benchmark mathematical functions used to test the fuzzy systems

membership function. In Fig. 8 is shown the iteration variable from the optimized interval type-2 fuzzy system and its variations with the different types of membership functions used. We only show the input iteration from the optimized system because for the other input and outputs is the same idea.

At this point we have 8 systems, these are:

1. FPSO1 the fuzzy system that uses interval type-2 triangular membership functions.
2. FPSO2 the fuzzy system that uses interval type-2 Gaussian membership functions.
3. FPSO3 the fuzzy system that uses interval type-2 trapezoidal membership functions.
4. FPSO4 the fuzzy system that uses interval type-2 generalized bell membership functions.

Table 1 Parameters for PSO

Parameter	Value
Population	30
Iterations	100
c_1 and c_2	dynamic
Inertia and constriction factor	1 (no effect)

Table 2 Results from applying the fuzzy systems for the parameter adaptation on PSO to the minimization of benchmark mathematical functions (Manual)

Fun	FPSO1	FPSO2	FPSO3	FPSO4
f1	1	1	1	1
f2	2.0e−21	1.0e−20	6.2e−21	2.1e−21
f3	1.9e+2	2.0e+2	1.9e+2	2.0e+2
f4	1.4e+4	1.5e+4	1.4e+4	1.5e+4
f5	1.1e+3	1.2e+3	1.0e+3	1.3e+3
f6	−96.64	−98.33	−97.20	−98.90
f7	−18.06	−17.83	−17.47	−18.15
f8	1.4e−13	1.2e−13	5.9e−14	6.2e−14
f9	0.50	0.52	0.48	0.52
f10	−0.52	−0.52	−0.52	−0.52
f11	0.003	0.003	0.002	0.001
f12	−0.335	−0.335	−0.335	−0.335
f13	−16.69	−16.47	−16.59	−16.55
f14	−23.14	−23.08	−23.18	−23.05
f15	4.1e+5	4.1e+5	4.1e+5	4.1e+5
f16	1.4e−12	2.4e−12	1.1e−12	2.0e−12
f17	2.4e−16	3.8e−16	1.3e−16	3.6e−16
f18	3.38	3.41	3.35	3.45
f19	−1.80	−1.80	−1.80	−1.80
f20	11.34	11.29	11.34	11.34
f21	−1	−1	−1	−1
f22	3	3	3	3
f23	−1.0316	−1.0316	−1.0316	−1.0316
f24	2.3694	2.7322	2.5427	2.2306
f25	−0.993	−0.996	−0.996	−0.994
f26	−186.73	−186.73	−186.73	−186.72
f27	8,918	8,970	8,940	9,056
Mean	**16,219.7**	16,276.9	16,252.3	16,266.4

5. FPSO5 the optimized fuzzy system that uses interval type-2 triangular membership functions.
6. FPSO6 the optimized fuzzy system that uses interval type-2 Gaussian membership functions.

Table 3 Results from applying the fuzzy systems for the parameter adaptation on PSO to the minimization of benchmark mathematical functions (optimized)

Fun	FPSO5	FPSO6	FPSO7	FPSO8
f1	1	1	1	1
f2	1.5e−21	7.6e−21	4.1e−21	3.3e−21
f3	2.0e+2	2.0e+2	2.0e+2	2.0e+2
f4	1.4e+4	1.5e+4	1.4e+4	1.5e+4
f5	1.3e+3	1.3e+3	1.2e+3	1.3e+3
f6	−99.09	−98.14	−98.71	−99.09
f7	−18.05	−18.09	−17.58	−17.71
f8	2.8e−13	1.9e−13	1.8e−13	2.3e−13
f9	0.52	0.55	0.52	0.54
f10	−0.52	−0.52	−0.52	−0.52
f11	0.004	0.001	0.002	0.003
f12	−0.335	−0.335	−0.335	−0.335
f13	−16.69	−16.77	−16.63	−16.58
f14	−23.14	−23.23	−23.051	−23.29
f15	4.e+5	4.1e+5	4.1e+5	4.1e+5
f16	1.8e−12	4.2e−12	5.6e−12	6.3e−12
f17	6.3e−16	1.0e−15	1.7e−16	8.8e−16
f18	3.42	3.46	3.43	3.47
f19	−1.80	−1.80	−1.80	−1.80
f20	11.34	11.34	11.34	11.34
f21	−1	−1	−1	−1
f22	3	3	3	3
f23	−1.0316	−1.0316	−1.0316	−1.0316
f24	2.1552	2.3587	2.5278	2.2319
f25	−0.994	−0.995	−0.995	−0.994
f26	−186.72	−186.73	−186.73	−186.73
f27	8,986	8,963	8,991	8,980
Mean	16,258.5	16,268.6	**16,239.5**	16,267.4

7. FPSO7 the optimized fuzzy system that uses interval type-2 trapezoidal membership functions.
8. FPSO8 the optimized fuzzy system that uses interval type-2 generalized bell membership functions.

To test these fuzzy systems we use 27 benchmark mathematical functions obtained from [2, 8], in Fig. 9 is shown a sample of the total of the benchmark functions used to test the systems, we also tested the fuzzy systems using 1,000 dimensions on the benchmark functions.

3 Experimentation with the Fuzzy Systems and the Benchmark Mathematical Functions

Each developed interval type-2 fuzzy system is integrated in a PSO method for dynamically adapt c_1 and c_2 parameters, and each system is tested with the set of benchmark mathematical functions, using the parameters from Table 1. The parameters from Table 1 are used for each fuzzy system integrated in PSO for dynamically adapt the parameters.

The results from Table 2 are from applying the 4 fuzzy systems for the parameter adaptation on PSO to the minimization of the benchmark mathematical functions, with 1,000 dimensions, these results are the average of 100 experiments. These 4 fuzzy systems are the modification of the manual fuzzy system that uses interval type-2 triangular membership functions. And the results from Table 3 are from applying the 4 fuzzy systems which are the modification of the optimized fuzzy system that use interval type-2 triangular membership functions.

The results from Tables 2 and 3 show the performance from each interval type-2 fuzzy system for parameter adaptation, in this case we can see that the results are similar and there isn't a big difference between each other. As we know PSO is a meta-heuristic method and has an explicit randomness, and for that reason each result from Table 2 is the average of 100 experiments, this to avoid randomness in the quality results.

4 Conclusions

In Table 2 is shown that PSO with the interval type-2 fuzzy system with triangular membership functions that we done manually have, on average, better results in terms of quality when compared with the other fuzzy systems using other types of membership functions, even when compared with the optimized fuzzy system from Table 3. In Table 3 is show that the best fuzzy system is with interval type-2 trapezoidal membership functions, but this isn't better than the fuzzy system with triangular membership functions from Table 2.

Also from the results from Tables 2 and 3, we can conclude, that the effect of different types of membership functions, in this kind of problem makes no difference, so we can continue using the interval type-2 triangular membership function, which is a membership function easy to use, because the parameters are more intuitive and easy to set.

This chapter is important because is an study of the effects that brings different types of membership functions, and make this comparison help us to avoid an optimization of the types of membership functions, so we can focus on others optimizations for the interval type-2 fuzzy system used for parameter adaptation in PSO.

References

1. Engelbrecht A.P.: Fundamentals of Computational Swarm Intelligence. University of Pretoria, South Africa (2005)
2. Haupt, R., Haupt, S.: Practical Genetic Algorithms, 2nd edn. A Wiley-Interscience publication, New York (2004)
3. Hongbo, L., Ajith, A.: A fuzzy adaptive turbulent particle swarm optimization. Int. J. Innovative Comput. Appl. **1**(1), 39–47 (2007)
4. Jang, J., Sun, C., Mizutani, E.: Neuro-Fuzzy and Soft Computing: A Computational Approach to Learning and Machine Intelligence. Prentice-Hall, Upper Saddle River (1997)
5. Kennedy, J., Eberhart, R.: Particle swarm optimization. In: Proceedings IEEE International Conference on Neural Networks, IV. Piscataway, NJ: IEEE Service Center, pp. 1942–1948 (1995)
6. Kennedy, J., Eberhart, R.: Swarm Intelligence. Morgan Kaufmann, San Francisco (2001)
7. Liang, Q., Mendel, J.: Interval type-2 fuzzy logic systems: theory and design. IEEE Trans. Fuzzy Syst. **8**(5), 535–550 (2000)
8. Marcin, M., Smutnicki, C.: Test functions for optimization needs (2005)
9. Olivas, F., Castillo, O.: Optimal design of fuzzy classification systems using PSO with dynamic parameter adaptation through fuzzy logic. Elsevier, Expert systems with applications, pp. 3196–3206 (2012)
10. Olivas, F., Valdez, F., Castillo, O.: Particle swarm optimization with dynamic parameter adaptation using interval type-2 fuzzy logic for benchmark mathematical functions. In: 2013 World Congress on Nature and Biologically Inspired Computing (NaBIC), pp. 36–40 (2013)
11. Shi, Y., Eberhart, R.: Fuzzy adaptive particle swarm optimization. In: Evolutionary Computation, pp. 101–106 (2001)
12. Wang, B., Liang, G., ChanLin, W., Yunlong, D.: A new kind of fuzzy particle swarm optimization FUZZY_PSO algorithm. In: 1st International Symposium on Systems and Control in Aerospace and Astronautics. ISSCAA 2006, pp. 309–311 (2006)
13. Zadeh, L.: Fuzzy sets. Inf. Control **8**, 338–353 (1965)
14. Zadeh, L.: Fuzzy logic. IEEE Comput. **8**, 83–92 (1965)
15. Zadeh, L.: The concept of a linguistic variable and its application to approximate reasoning—I. Inform. Sci. **8**, 199–249 (1975)

Genetic Optimization of Type-2 Fuzzy Integrators in Ensembles of ANFIS Models for Time Series Prediction

Jesus Soto and Patricia Melin

Abstract This chapter describes the genetic optimization of interval type-2 fuzzy integrators in Ensembles of ANFIS (adaptive neuro-fuzzy inferences systems) models for the prediction of the Mackey-Glass time series. The considered a chaotic system is he Mackey-Glass time series that is generated from the differential equations, so this benchmarks time series is used for the test of performance of the proposed ensemble architecture. We used the interval type-2 and type-1 fuzzy systems to integrate the output (forecast) of each Ensemble of ANFIS models. Genetic Algorithms (GAs) were used for the optimization of memberships function parameters of each interval type-2 fuzzy integrators. In the experiments we optimized Gaussians, Generalized Bell and Triangular membership functions for each of the fuzzy integrators, thereby increasing the complexity of the training. Simulation results show the effectiveness of the proposed approach.

1 Introduction

The analysis of the time series consists of a (usually mathematical) description of the movements that compose it, then build models using movements to explain the structure and predict the evolution of a variable over time [1, 2]. The fundamental procedure for the analysis of a series of time as described below:

1. Collecting data time series, trying to ensure that these data are reliable.
2. Representing the time series qualitatively noting the presence of long-term trends, cyclical variations and seasonal variations.

J. Soto · P. Melin (✉)
Tijuana Institute of Technology, Tijuana, Mexico
e-mail: pmelin@tectijuana.mx

J. Soto
e-mail: jesvega83@gmail.com

O. Castillo et al. (eds.), *Recent Advances on Hybrid Approaches for Designing Intelligent Systems*, Studies in Computational Intelligence 547,
DOI: 10.1007/978-3-319-05170-3_6, © Springer International Publishing Switzerland 2014

3. Plot a graph or trend line length and obtain the appropriate trend values using either method of least squares.
4. When are present seasonal variations, getting seasonal and adjust the data rate to these seasonal variations (i.e. data seasonally).
5. Adjust the seasonally adjusted trend.
6. Represent cyclical variations obtained in step 5.
7. Combining the results of steps 1–6 and any other useful information to make a prediction (if desired) and if possible discuss the sources of error and their magnitude.

Therefore the above ideas can assist in the important problem of prediction in time series. Along with common sense, experience, skill and judgment of the researcher, such mathematical analysis can, however, be of value for predicting the short, medium and long term [34, 37, 38, 41, 42].

The genetic algorithms are adaptive methods which may be used to solve search and optimization problems. They are based on the genetic process of living organisms. Over generations, the populations evolve in line with the nature of the principles of natural selection and survival of the fittest, postulated by Darwin, in imitation of this process; genetic algorithms are capable of creating solutions to real world problems. The evolution of these solutions to optimal values of the problem depends largely on proper coding them. The basic principles of genetic algorithms were established by Holland [18, 19] and are well described in texts Goldberg [15], Davis [17] and Michalewicz [24]. The evolutionary modeling of fuzzy logic system can be considered as an optimization process where the part or all a fuzzy system parameters constitute a search spaces of model operational (our case), cognitive and structural [8–10].

This chapter reports the results of the simulations, in which the genetic optimization of interval type-2 fuzzy integrators in ensembles of ANFIS models for the Mackey-Glass time series [25, 26], where the results for each ANFIS were evaluated by the method of the root mean square error (RMSE). For the integration of the results of each modular in the ensemble of ANFIS we used the following integration methods: interval type-2 fuzzy systems of kind Mamdani and integrator by interval type-2 FIS optimized [3, 5, 7].

The selection of the time series for the simulations was based on the fact that these time series are widely quoted in the literature by different researchers [1, 2, 20, 26, 28, 39, 40], which allows to compare results with other approaches such as neural networks and linear regression [12, 13, 23, 32].

In the next section we describe the background and basic concepts of ANFIS, Ensemble learning, Interval type-2 fuzzy, Genetic Algorithms and time series. Section 3 presents the proposed architecture of genetic optimization of interval type-2 fuzzy integrators in ensembles of ANFIS models for the time series prediction. Section 4 presents the simulations and the results obtained with different methods of integration that are used in this work. Section 5 presents the conclusions.

2 Background and Basic Concepts

This section presents the basic concepts of ANFIS, Ensemble learning, Interval type-2 fuzzy logic, and Genetic Algorithms.

2.1 ANFIS Models

There are have been proposed systems that have achieved fully the combination of fuzzy systems with neural networks, one of the most intelligent hybrid systems is the ANFIS (Adaptive Neuro Fuzzy Inference System method) as referred to by R. Jang [20, 22] (Fig. 1), which is a method for creating the rule base of a fuzzy system, using the algorithm of backpropagation training from the data collection process. Its architecture is functionally equivalent to a fuzzy inference system of Takagi-Sugeno-Kang.

The basic learning rule of ANFIS is the gradient descent backpropagation [40], which calculates the error rates (defined as the derivative of the squared error for each output node) recursively from the output to the input nodes.

As a result we have a hybrid learning algorithm [14], which combines the gradient descent and least-squares estimation. More specifically in, the forward step of the hybrid learning algorithm, functional signals (output nodes) are processed towards layer 4 and the parameters of consequence are identified by least squares. In the backward step the premise parameters are updated by gradient descent.

2.2 Ensemble Learning

The Ensemble consists of a learning paradigm where multiple component learners are trained for a same task, and the predictions of the component learners are combined for dealing with future instances [35, 36]. Since an Ensemble is often more accurate than its component learners, such a paradigm has become a hot topic in recent years and has already been successfully applied to optical character recognition, face recognition, scientific image analysis, medical diagnosis [45].

2.3 Interval Type-2 Fuzzy

Type-2 fuzzy sets are used to model uncertainty and imprecision; originally they were proposed by Zadeh [43, 44] and they are essentially "fuzzy–fuzzy" sets in which the membership degrees are type-1 fuzzy sets [6, 21, 27] (Fig. 2).

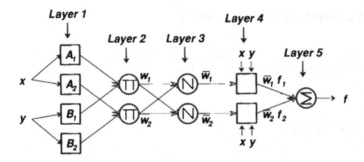

Fig. 1 ANFIS architecture

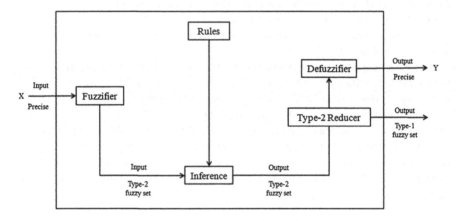

Fig. 2 Basic structure of the interval type-2 fuzzy logic system

The basic structure of a type-2 fuzzy system implements a nonlinear mapping of input to output space. This mapping is achieved through a set of type-2 if–then fuzzy rules, each of which describes the local behavior of the mapping.

The uncertainty is represented by a region called footprint of uncertainty (FOU). When $\mu_{\tilde{A}}(x, u) = 1, \forall u \in l_x \subseteq [0, 1]$; we have an interval type-2 membership function (Fig. 3).

The uniform shading for the FOU represents the entire interval type-2 fuzzy set and it can be described in terms of an upper membership function $\bar{\mu}_{\tilde{A}}(x)$ and a lower membership function $\underline{\mu}_{\tilde{A}}(x)$.

A fuzzy logic systems (FLS) described using at least one type-2 fuzzy set is called a type-2 FLS. Type-1 FLSs are unable to directly handle rule uncertainties, because they use type-1 fuzzy sets that are certain [29–31]. On the other hand, type-2 FLSs are very useful in circumstances where it is difficult to determine an exact certainty value, and there are measurement uncertainties.

Fig. 3 Interval type-2 membership function

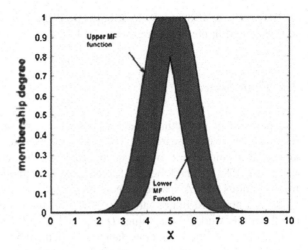

2.4 Genetic Algorithms

Genetic Algorithms (GAs) are adaptive heuristic search algorithms based on the evolutionary ideas of natural selection and the genetic process [16]. The basic principles of GAs were first proposed by John Holland in 1975, inspired by the mechanism of natural selection, where stronger individuals are likely the winners in a competing environment [4]. GAs assumes that the potential solution of any problems an individual and can represented by a set of parameters [11]. These parameters are recharged as the genes of a chromosome and can be structured by string of values in binary form. A positive value, generally known as a fitness value, is used to reflect the degree of "goodness" of the chromosome for the problem which would be highly related with its objective value. The pseudocode of a GAs is as follows:

1. Start with a randomly generated population of n chromosome (candidate a solutions to problem).
2. Calculate the fitness of each chromosome in the problem.
3. Repeat the following steps until n offspring have been created:

   a. Select a pair of parent chromosome from the concurrent population, the probability of selection being an increasing function of fitness. Selection is done with replacement, meaning that the same chromosome can be selected more than once t that the same chromosome can be selected more than once to become a parent.
   b. With the probability (crossover rate), perform crossover to the pair at a randomly chosen point to form two offspring.
   c. Mutate the two offspring at each locus with probability (mutate rate), and place the resulting chromosome in the new population.

4. Replace the current population with the new population.
5. Go to step 2.

The simple procedure just describe above is the basic one for most applications of GAs found in the literature.

2.5 Time Series

The problem of predicting future values of a time series has been a point of reference for many researchers. The aim is to use the values of the time series known at a point $x = t$ to predict the value of the series at some future point $x = t + P$. The standard method for this type of prediction is to create a mapping from D points of a Δ spaced time series, is $(x\,(t - (D - 1)\,\Delta),\, \dots ,\, x\,(t - \Delta),$ $x\,(t))$, to a predicted future value $x\,(t + P)$. To allow a comparison with previous results in this work [6] the values $D = 4$ and $\Delta = P = 6$ were used.

One of the chaotic time series data used is defined by the Mackey-Glass [25, 26] time series, whose differential equation is given by:

$$x(t) = \frac{0.2x(t - \tau)}{1 + x^{10}(t - \tau)} - 0.1x(t - \tau) \tag{1}$$

For obtaining the values of the time series at each point, we applied the Runge-Kutta method for the solution of Eq. (1). The integration step was set at 0.1, with initial condition $x\,(0) = 1.2$, $\tau = 17$, $x\,(t)$ is then obtained for $0 \leq t \leq 1200$, (Fig. 4) (We assume $x\,(t) = 0$ for $t < 0$ in the integration).

From the Mackey-Glass time series we extracted 800 pairs of data points (Fig. 4) [20, 33], similar to [28].

We predict $x\,(t)$ from the three past values of the time series, that is, $x\,(t - 18)$, $x\,(t - 12)$, and $x\,(t - 6)$. Therefore the format of the training data is:

$$[x(t - 18), x(t - 12), x(t - 6)\,; x(t)] \tag{2}$$

where $t = 19$ to 818 and $x(t)$ is the desired prediction of the time series. The first 400 pairs of data are used to train the ANFIS, while the other 400 pairs of data are used to validate the model identification.

3 General Architecture of the Proposed Method

The proposed method combines the ensemble of ANFIS models and the use of interval type-2 fuzzy systems as response integrators (Fig. 5).

This architecture is divided into 5 sections, where the first phase represents the data base to simulate in the Ensemble of ANFIS, which in our case is the data base of the Mackey-Glass [7, 22] time series (used 800 pairs of data points). In the second phase, training (the first 400 pairs of data are used to train the ANFIS) and validation (the second 400 pairs of data are used to value the ANFIS) is performed

Fig. 4 The Mackey-Glass
time series

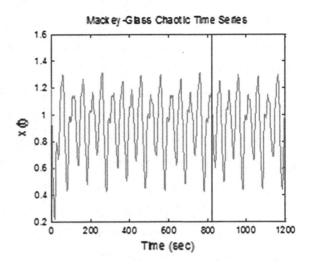

sequentially in each ANFIS model, where the number of ANFIS to be used can be from 1 to n depending on what the user wants to test, but in our case we are dealing with a set of 3 ANFIS in the Ensemble.

In the third phase we have to generate the results of each ANFIS used in the previous section and in the fourth phase we integrate the overall results of each ANFIS, such integration will be done by interval type-2 fuzzy integrators of Mamdani type, but each fuzzy integrators is optimized (GAs) in the membership functions (Gaussians "iguassmtype2", Generalized Bell "iguassmtype2" and Triangular "iguassmtype2") [6, 7]. Finally the outcome or the final prediction of the Ensemble ANFIS learning is obtained.

4 Simulations Results

This section presents the results obtained through experiments on the architecture of genetic optimization of type-2 fuzzy integrators in ensembles of ANFIS models for the time series prediction, which show the performance that was obtained from each experiment to simulate the Mackey-Glass time series.

The design of the fuzzy systems is of Mamdani type and will have 3 inputs (ANFIS1, ANFIS2 and ANFIS3 predictions) and 1 output (forecast), so each input will be assigned two MFs "Small and Large" and the output will be assigned 3 MFs "OutANFIS1, OutANFIS2 and OutANFIS3". The MFs of each interval type-2 fuzzy integrators will be changing to membership functions MFs (Gaussians, Generalized Bell, and Triangular) to observe the behavior of each of them and determine which one provides better forecast of the time series.

The GAs used to optimize the parameters values of the MFs in each interval type-2 fuzzy integrators. The representation of GAs is of Real-Values and the

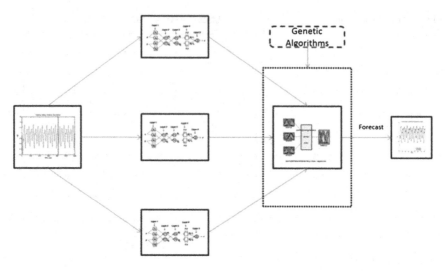

Fig. 5 The architecture of the proposed model

chromosome size will depend of the MFs that are used in each design of the interval type-2 fuzzy integrators.

The objective function is defined to minimize the prediction error as follows:

$$f_{(t)} = \sqrt{\frac{\sum_{t=1}^{n} (a_t - p_t)^2}{n}} \qquad (3)$$

where a, corresponds to the real data of the time series, p corresponds to the output of each fuzzy integrator, t is de sequence time series, and n is the numbers of data points of time series.

In the Table 1 it is illustrated the general representation of the chromosome and represent the utilized fuzzy membership functions. In this Table 1, the first row represented each input and output of the fuzzy inference system, the second row represents the membership function containing each input and output of the fuzzy inference system, the third row represents the MFs parameter "PL = Lower Parameter" where $PL_1 \ldots PL_N$ are the size parameter of the MFs, the fourth row represent the MFs parameter "PL = Upper Parameter" $PU_1 \ldots PU_N$ are the size parameter of the MFs that corresponds to each input and output. The number of parameters varies according to the kind of membership function of the interval type-2 fuzzy (e.g. three parameter are needed to represent a iguassmtype2 membership function of type-2 fuzzy "variance, mean1 and mean2"), therefore the numbers of parameters of each variable of input and output depend of the type of membership functions (Gaussians "iguassmtype2", Generalized Bell "iguassmtype2", and Triangular "iguassmtype2") that will contains each the variables of

Table 1 General representation of the chromosome to the optimization of the interval type-2 fuzzy membership functions

Input 1		Input 2		Input 3		Output		
MF1	MF2	MF1	MF2	MF1	MF2	MF1	MF2	MF3
PL1... PLN	PL1... PLN	PL1... PLN	PL1... PLN	PL1... PLN	PL1... PLN	PL1... PLN	PL1... PLN	PL1... PLN
PU1... PUN	PU1... PUN	PU1... PUN	PU1... PUN	PU1... PUN	PU1... PUN	PU1... PUN	PU1... PUN	PU1... PUN

the fuzzy integrators (our case will use two MFs to each variable of input and using three MFs to the variable output).

The GAs used has the following parameters: 75 individuals or genes, 50 generations and 30 iterations (running the GAs), the selection method are the stochastic universal sampling, the percentage of crossover or recombine is 0.8 and the mutation is 0.1. There are fundamentals parameters for test the performances of the GAs.

4.1 Genetic Optimization of Interval Type-2 Fuzzy Integration Using Gaussian MFs

In the design of the interval type-2 fuzzy integrator will have three input variables and one output variable, so each variable input will have two MFs and the variable output will have three MFs. Therefore the numbers of parameters that one used in the representation of the chromosome is 27, because iguassmtype2 MFs used three parameters (Variance, Mean1 and Mean2) to their representation in the interval type-2 fuzzy systems. The results obtained for the optimization of the iguassmtype2 MFs with GAs are the following: the type-2 fuzzy systems (Fig. 6) generated by the GAs, the structure of interval type-2 fuzzy integrator used three input variable (ANFIS1, ANFIS2 AND ANFIS3) and one output variable (forecast), and have 8 rules. The parameters obtained with the GAs for the type-2 fuzzy MFs iguassmtype2 (Fig. 7). The forecast data (Fig. 8) is generated by optimization of the interval type-2 fuzzy integrators. Therefore the evolution error (Fig. 9) obtained with the GAs for this integration is 0.01843.

4.2 Genetic Optimization of Interval Type-2 Fuzzy Integration Using Generalized Bell MFs

In the design of the interval type-2 fuzzy integrator will have three input variables and one output variable, so each input variable will have two MFs and the output variable will have three MFs. Therefore the number of parameters of the MFs that are used in the representation of the chromosome is 54, because igbelltype2 MFs used six's parameters (a1, b1, c1, a2, b2 and c2) to their representation in the interval type-2 fuzzy systems. The results obtained for the optimization of the igbelltype2 MFs with GAs are the following: the type-2 fuzzy systems (Fig. 10) generated by the GAs, the structure of interval type-2 fuzzy integrator systems used three input variables (ANFIS1, ANFIS2 AND ANFIS3) and one output variable (forecast), and have 9 rules. The parameters obtained with the GAs for the type-2 fuzzy MFs igbelltype2 (Fig. 11). The forecast data (Fig. 12) is generated by optimization of the interval type-2 fuzzy integrators. Therefore the evolution error (Fig. 13) obtained with the GAs for this integration is 0.018154.

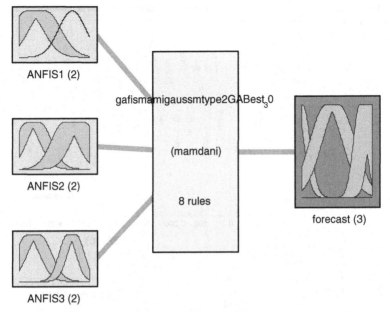

System gafismamigaussmtype2GABest$_3$0: 3 inputs, 1 outputs, 8 rules

Fig. 6 Plot of the fuzzy inference system integrators generated with the GAs

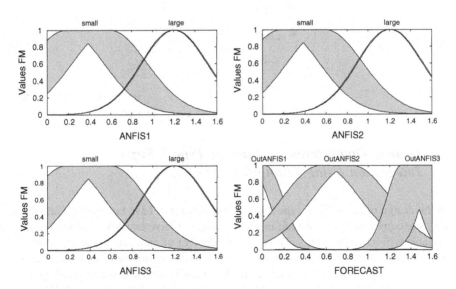

Fig. 7 Plot of the optimization of the Memberships functions (input and output) parameters with the GAs

Fig. 8 Plot of the forecast
generated by the genetic
optimization of type-2 fuzzy
integrators

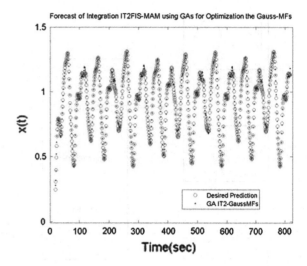

Fig. 9 Plot of the evolution
error generated by GAs

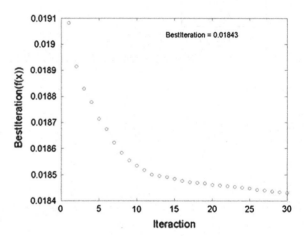

4.3 Genetic Optimization of Interval Type-2 Fuzzy Integration Using Triangular MFs

In the design of the interval type-2 fuzzy integrator will have three input variables and one output variable, so each input variable will have two MFs and the output variable will have three MFs. Therefore the number of parameters that are used in the representation of the chromosome is 54, because itritype2 MFs used six's parameters (a1, b1, c1, a2, b2 and c2) to their representation in the interval type-2 fuzzy systems. The results obtained for the optimization of the itritype2 MFs with GAs are the following: the type-2 fuzzy systems (Fig. 14) generated by the GAs, the structure of interval type-2 fuzzy integrator systems used three input variables (ANFIS1, ANFIS2 AND ANFIS3) and one output variable (forecast), and have 9

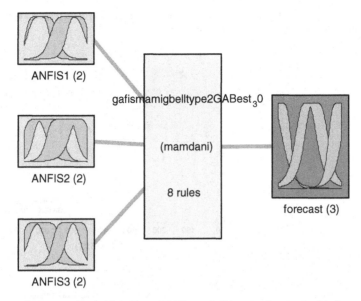

System gafismamigbelltype2GABest$_3$0: 3 inputs, 1 outputs, 8 rules

Fig. 10 Interval type-2 fuzzy systems integrator generated with the GAs

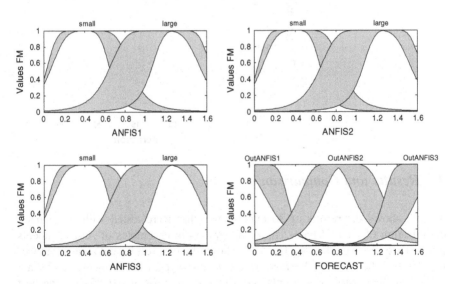

Fig. 11 Optimization of the memberships functions (input and output) parameters with the GAs

rules. The parameters obtained with the GAs for the type-2 fuzzy MFs itritype2 (Fig. 15). The forecast data (Fig. 16) is generated by optimization of the interval type-2 fuzzy integrators. Therefore the evolution error (Fig. 17) obtained with the GAs for this integration is 0.01838.

Fig. 12 Forecast generated
by the genetic optimization of
the type-2 fuzzy integrators

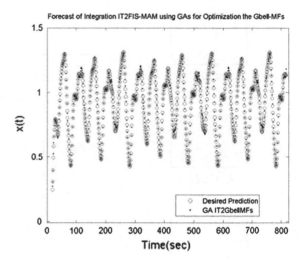

Fig. 13 Plot of the evolution
error generated by GAs

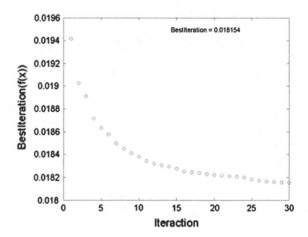

4.4 Results and Comparison

Table 2 shows the results of 30 experiments that were tested with the genetic optimization of interval type-2 fuzzy integrators in ensembles of ANFIS models for the time series prediction. This table shows the best results and mean results of the 30 runs. The Best and Averages results for the Type-1 fuzzy integrator [28] are using Gaussians MFs, which obtained a prediction error of 0.01854 for the best and average prediction error of 0.01954. The Best and Mean results for the interval Type-2 fuzzy integrators using Generalized Bell (igbelltype2) MFs, which obtained a prediction error of 0.01811 for the best and average prediction error of 0.01822.

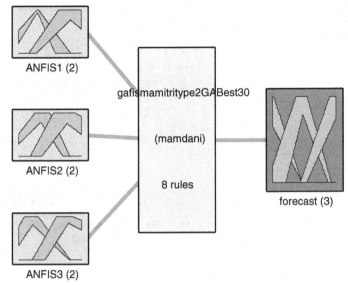

Fig. 14 Interval type-2 fuzzy systems integrator generated with the GAs

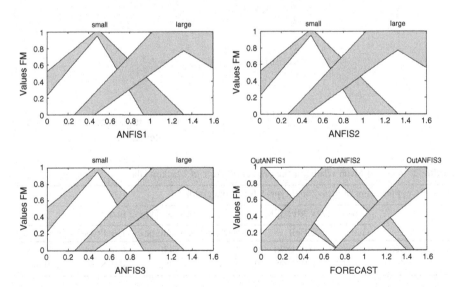

Fig. 15 Optimization of the memberships functions (input and output variables) parameters with the GAs

Fig. 16 Forecast generated
by the genetic optimization of
type-2 fuzzy integrators

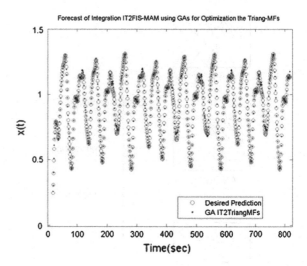

Fig. 17 Plot of the evolution
error generated by GAs

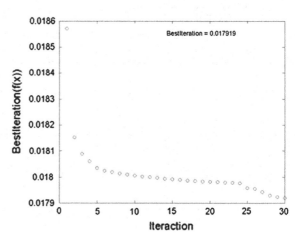

Table 2 Best and Mean results of the prediction error of Mackey-Glass

Genetic optimization of type-2 and type-1 fuzzy integrator

30 Runs	Type-1 fuzzy MFs [28]			Interval Type-2 fuzzy MFs		
	Gauss	GBell	Triangular	igaussmtype2	igbelltype2	itritype2
Best (RMSE)	0.018745	0.021870	0.023313	0.018430	0.018154	0.018388
Mean (RMSE)	0.019013	0.021928	0.023714	0.018552	0.018386	0.018705

5 Conclusion

In conclusion we can say that the results obtained with the proposed architecture of genetic optimization of interval type-2 fuzzy integrators in ensembles of ANFIS models for the time series prediction have been good and positive in predicting time series (like the Mackey-Glass), as it has managed to minimize the prediction error of the time series against the results obtained by other researchers.

We conclude that the results obtained with the architectures of genetic optimization of interval type-2 fuzzy integrators in ensembles of ANFIS models for the time series prediction are good, since we achieved [98.5–99] percent of accuracy in the forecast with the Mackey-Glass time series and on the opposite side it was obtained prediction error of 1.5 %. Therefore, we can conclude that interval type-2 fuzzy integrator is better than the type-1 fuzzy integrator, because in most of the experiments that were performed with the proposed architecture of ensembles of ANFIS, the interval type-2 fuzzy gave better prediction errors than the type-1 fuzzy. Therefore the proposal offers efficient results in the prediction of such time series, which can help us, make decisions and avoid unexpected events in the future.

References

1. Brocklebank, J.C., Dickey, D.A.: SAS for Forecasting Series, pp. 6–140. SAS Institute Inc, Cary (2003)
2. Brockwell, P.D., Richard, A.D.: Introduction to Time Series and Forecasting, pp. 1–219. Springer, New York (2002)
3. Castillo, O., Melin, P.: Optimization of type-2 fuzzy systems based on bio-inspired methods: a concise review. Appl. Soft Comput. 12(4), 1267–1278 (2012)
4. Castillo, O., Melin, P.: Soft Computing for Control of Non-Linear Dynamical Systems. Springer, Heidelberg (2001)
5. Castro, J.R., Castillo, O., Martínez, L.G.: Interval type-2 fuzzy logic toolbox. Eng. Lett. 15(1), 89–98 (2007)
6. Castro, J.R., Castillo, O., Melin, P., Rodriguez, A.: A Hybrid Learning Algorithm for Interval Type-2 Fuzzy Neural Networks: The Case of Time Series Prediction, vol. 15a, pp. 363–386. Springer, Berlin (2008)
7. Castro, J.R., Castillo, O., Melin, P., Rodríguez, A.: Hybrid learning algorithm for interval type-2 fuzzy neural networks. In: GrC, pp. 157–162 (2007)
8. Chua, T.W., Tan, W.W.: Genetically evolved fuzzy rule-based classifiers and application to automotive classification. Lecture Notes in Computer Science, vol. 5361, pp. 101–110 (2008)
9. Cordon, O., Gomide, F., Herrera, F., Hoffmann, F., Magdalena, L.: Ten years of genetic fuzzy systems: current framework and new trends. Fuzzy Sets Syst. 141, 5–31 (2004)
10. Cordon, O., Herrera, F., Hoffmann, F., Magdalena, L.: Genetic Fuzzy Systems: Evolutionary Tuning and Learning of Fuzzy, Knowledge Bases. World Scientific, Singapore (2001)
11. Cordon, O., Herrera, F., Villar, P.: Analysis and guidelines to obtain a good uniform fuzzy partition granularity for fuzzy rule-based systems using simulated annealing. Int. J. Approximate Reasoning 25, 187–215 (2000)

12. Eason, G., Noble, B., Sneddon, I.N.: On certain integrals of Lipschitz-Hankel type involving products of Bessel functions. Phil. Trans. Roy. Soc. London **A247**, 529–551 (1995)
13. Eiben, A.E., Smith, J.E.: Introduction to Evolutionary Computation, pp. 37–69. Springer, Berlin (2003)
14. Filev, D., Yager, R.: On the issue of obtaining OWA operador weights. Fuzzy Sets Syst. **94**(2), 157–169 (1998)
15. Goldberg, D.E., Kalyanmoy, D.: A comparative analysis of selection schemes used in genetic algorithms. In: Rawlins, G.J.E. (ed.) Foundations of Genetic Algorithms, pp. 69–93. Morgan Kaufmann Publishers, San Mateo (1991)
16. Goldberg, D.E.: Genetic Algorithms in Search, Optimization, and Machine Learning. Addison-Wesley, Boston (1989)
17. Goldberg, D.E., Korb, B., Kalyanmoy, D.: Messy genetic algorithms: motivation, analysis, and first results. Complex Syst. **3**, 493–530 (1989)
18. Holland, J.H.: Outline for a logical theory of adaptive systems. J. Assoc. Comput. Mach. **3**, 297–314 (1962)
19. Holland, J.H.: Adaptatioon in Natural and Artificial Systems. University of Michigan Press, Ann Arbor (1975)
20. Jang, J.S.R.: ANFIS: adaptive-network-based fuzzy inference systems. IEEE Trans. Syst. Man Cybernet. **23**, 665–685 (1992)
21. Jang, J.S.R.: Fuzzy modeling using generalized neural networks and Kalman fliter algorithm. In: Proceedings of the Ninth National Conference on Artificial Intelligence. (AAAI-91), pp. 762–767 (1991)
22. Jang, J.S.R.: Rule extraction using generalized neural networks. In: Proceedings of the 4th IFSA Wolrd Congress, pp. 82–86 (1991)
23. Konar, A.: Computational Intelligence: Principles, Techniques and Applications. Springer, Berlin (2005)
24. Lawrence, D.: Handbook of Genetic Algorithms. Van Nostrand Reinhold, New Jersey (1991)
25. Mackey, M.C., Glass, L.: Oscillation and chaos in physiological control systems. Science **197**, 287–289 (1997)
26. Mackey, M.C.: Mackey-Glass. McGill University, Canada. http://www.sholarpedia.org/-article/ Mackey-Glass_equation, 5th Sept 2009
27. Melin, P., Mendoza, O., Castillo, O.: An improved method for edge detection based on interval type-2 fuzzy logic. Expert Syst. Appl. **37**(12), 8527–8535 (2010)
28. Melin, P., Soto, J., Castillo, O., Soria, J.: Optimization of Interval Type-2 and Type-1 fuzzy integrators in ensemble of ANFIS models with genetic algorithms. In: Mexican International Conference on Computer Science, Morelia, Mexico, 30, 31 Oct–1st Nov 2013
29. Mendel, J.M.: Why we need type-2 fuzzy logic systems (Article is provided courtesy of Prentice Hall, By Jerry Mendel) 11 May 2001
30. Mendel, J.M. (ed.): Uncertain Rule-Based Fuzzy Logic Systems: Introduction and New Directions, pp. 25–200. Prentice Hall, USA (2000)
31. Mendel, J.M., Mouzouris, G.C.: Type-2 fuzzy logic systems. IEEE Trans. Fuzzy Syst. **7**, 643–658 (1999)
32. Michalewicz, Z.: Genetic Algorithms + Data Structures=Evolution Programs, vol. AI. Springer-Verlag, New York (1994)
33. Pulido, M., Melin, P., Castillo, O.: Genetic optimization of ensemble neural networks for complex time series prediction. In: Neural Networks International Joint Conference (IJCNN), pp. 202–206 (2011)
34. Rojas, R.: Neural Networks: A Systematic Introduction, pp. 431–450. Springer, Berlin (1996)
35. Rumelhart, D.E., Hinton, G.E., Williams, R.J.: Learning internal representations by error propagation. In: Parallel Distributed Processing: Explorations in the Microstructure of Cognition, vol. 1, pp. 318–362. MIT Press, Cambridge (1986)
36. Takagi, T., Sugeno, M.: Derivation of fuzzy control rules from human operation control actions. In: Proceedings of the IFAC Symposium on Fuzzy Information, Knowledge Representation and Decision Analysis, pp. 55–60 (1983)

37. Takagi, T., Sugeno, M.: Fuzzy identification of systems and its applications to modeling and control. IEEE Trans. Syst. Man Cybernet. **15**, 116–132 (1985)
38. Thomas, G.D.: Machine learning research: four current directions. Artif. Intell. Mag. **18**(4), 97–136 (1997)
39. Wang, C., Zhang, J.P.: Time series prediction based on ensemble ANFIS. In: Proceedings of the Fourth International Conference on Machine Learning and Cybernetics, Guangzhou, 18–21 Aug 2005
40. Werbos, P.: Beyond regression: new tools for prediction and analysis in the behavioral sciences. PhD thesis, Harvard University (1974)
41. Xiaoyu, L., Bing, W., Simon, Y.: Time series prediction based on fuzzy principles. In: Department of Electrical & Computer Engineering FAMU-FSU College of Engineering. Florida State University Tallahassee, FL 32310 (2002)
42. Yager, R., Filev, D.: Essentials of Fuzzy Modeling and Control, p. 388. Wiley, New York (1994)
43. Zadeh, L.A.: Fuzzy Logic. Computer **1**(4), 83–93 (1988)
44. Zadeh, L.A.: Fuzzy logic = computing with words. IEEE Trans. Fuzzy Syst. **4**(2), 103 (1996)
45. Zhou, Z.H., Wu, J., Tang, W.: Ensembling neural networks: many could be better than all. Artif. Intell. **137**(1–2), 239–263 (2002)

Ensemble Neural Network Optimization Using the Particle Swarm Algorithm with Type-1 and Type-2 Fuzzy Integration for Time Series Prediction

Martha Pulido and Patricia Melin

Abstract This chapter describes the design of ensemble neural networks using Particle Swarm Optimization for time series prediction with Type-1 and Type-2 Fuzzy Integration. The time series that is being considered in this work is the Mackey-Glass benchmark time series. Simulation results show that the ensemble approach produces good prediction of the Mackey-Glass time series.

1 Introduction

Time Series is defined as a set of measurements of some phenomenon or experiment recorded sequentially in time. The first step in analyzing a time series is to plot it, this allows: to identify the trends, seasonal components and irregular variations. A classic model for a time series can be expressed as a sum or product of three components: trend, seasonality and random error term.

Time series predictions are very important because based on them we can analyze past events to know the possible behavior of futures events and thus we can take preventive or corrective decisions to help avoid unwanted circumstances.

The contribution of this chapter is the proposed approach for ensemble neural network optimization using particle swarm optimization. The proposed models are also used as a basis for statistical tests [2–5, 11, 12, 14, 17, 18, 23–26].

The rest of the chapter is organized as follows: Sect. 2 describes the concepts of optimization, Sect. 3 describes the concepts of particle swarm optimization, Sect. 4 describes the concepts of Fuzzy Systems as Methods of integration, Sect. 5 describes the problem and the proposed method of solution, Sect. 6 describes the simulation results of the proposed method, and Sect. 7 shows the conclusions.

M. Pulido · P. Melin (✉)
Tijuana Institute of Technology, Tijuana, México
e-mail: epmelin@hafsamx.org; pmelin@tectijuana.mx

O. Castillo et al. (eds.), *Recent Advances on Hybrid Approaches for Designing Intelligent Systems*, Studies in Computational Intelligence 547,
DOI: 10.1007/978-3-319-05170-3_7, © Springer International Publishing Switzerland 2014

2 Optimization

Regarding optimization, we have the following situation in mind: there exists a search space V, and a function:

$$g : V \rightarrow \mathbb{R}$$

and the problem is to find

arg min g.

$$v \in V$$

Here, V *is vector of decision variables,* and g *is the objective function.* In this case we have assumed that the problem is one of minimization, but everything we say can of course be applied *mutatis mutandis* to a maximization problem. Although specified here in an abstract way, this is nonetheless a problem with a huge number of real-world applications.

In many cases the search space is discrete, so that we have the class of *combinatorial optimization problems* (COPs). When the domain of the g function is continuous, a different approach may well be required, although even here we note that in practice, optimization problems are usually solved using a computer, so that in the final analysis the solutions are represented by strings of binary digits (bits).

There are several optimization techniques that can be applied to neural networks, some of these are: evolutionary algorithms [22], ant colony optimization [6] and Particle swarm [8].

3 Particle Swarm Optimization

The Particle Swarm Optimization algorithm maintains a swarm of particles, where each particle represents a potential solution. In analogy with evolutionary computation paradigms, a swarm is a population, while a particle is similar to an individual. In simple terms, the particles are "flown" through a multidimensional search space where the position of each particle is adjusted according to its own experience and that of their neighbors. Let $x_i(t)$ denote the position of particle i in the search space at time step t unless otherwise selected, t denotes discrete time steps. The position of the particle is changed by adding a velocity, $v_i(t)$ to the current position i,e.

$$x_i(t + 1) = x_i(t) + v_i(t + 1) \tag{1}$$

with $x_i(0) \sim U(X_{min}, X_{max})$.

It is the velocity vector the one that drives of the optimization process, and reflects both the experimental knowledge of the particles and the information exchanged in the vicinity of particles. The experimental knowledge of a particle which is generally known as the cognitive component, which is proportional to the distance of the particle from its own best position (hereinafter, the personal best position particles) that are from the first step. Socially exchanged information is known as the social component of the velocity equation.

For the best PSO, the particle velocity is calculated as:

$$v_{ij}(t+1) = v_{ij}(t) + c_1 r_1 [y_{ij}(t) - x_{ij}(t)], + c_2 r_2(t)[\hat{y}_j(t) - x_{ij}(t)] \qquad (2)$$

where $v_{ij}(t)$ is the velocity of the particle i in dimension j at time step t, $c_1 y c_2$ are positive acceleration constants used to scale the contribution of cognitive and social skills, respectively, $yr_{1j}(t), yr_{2j}(t) \sim U(0,1)$ are random values in the range [0,1].

The best personal position in the next time step $t+1$ is calculated as:

$$y_i(t+1) = \begin{cases} y_i(t) & if \ f(x_i(x_i(t+1)) \geq f y_i(t)) \\ x_i(t+1) & if \ f(x_i(x_i(t+1)) > f y_i(t)) \end{cases} \qquad (3)$$

where $f : \mathbb{R}^{nx} \to \mathbb{R}$ is the fitness function, as with EAs, measuring fitness with the function will help find the optimal solution, for example the objective function quantifies the performance, or the quality of a particle (or solution).

The overall best position, $\hat{y}(t)$ at time step t, is defined as:

$$\hat{y}(t) \in \{y_o(t), \ldots, y_{ns}(t)\} f(y(t)) = \min\{f(y_o(t)), \ldots f(y_{ns}(t)),\} \qquad (4)$$

where n_s is the total number of particles in the swarm. Importantly, the above equation defining and establishing $\hat{y}$ the best position is uncovered by either of the particles so far as this is usually calculated from the best position best personal [6, 7, 10].

The overall best position may be selected from the actual swarm particles, in which case:

$$\hat{y}(t) = \min\{f(x_o(t)), \ldots, f(x_{ns}(t)),\} \qquad (5)$$

4 Fuzzy Systems as Methods of Integration

Fuzzy logic was proposed for the first time in the mid-sixties at the University of California Berkeley by the brilliant engineer Lofty A. Zadeh., who proposed what it's called the principle of incompatibility: "As the complexity of system increases, our ability to be precise instructions and build on their behavior decreases to the threshold beyond which the accuracy and meaning are mutually exclusive characteristics." Then introduced the concept of a fuzzy set, under which lies the idea that the elements on which to build human thinking are not numbers but linguistic labels. Fuzzy logic can represent the common knowledge as a form of language that is mostly qualitative and not necessarily a quantity in a mathematical language [29].

Type-1 Fuzzy system theory was first introduced by Castillo and Melin 2] in 2008, and has been applied in many areas such as control, data mining, time series prediction, etc.

The basic structure of a fuzzy inference system consists of three conceptual components: a rule base, which contains a selection of fuzzy rules, a database (or dictionary) which defines the membership functions used in the rules, and reasoning mechanism, which performs the inference procedure (usually fuzzy reasoning) [17].

Type-2 Fuzzy systems were proposed to overcome the limitations of a type-1 FLS, the concept of type-1 fuzzy sets was extended into type-2 fuzzy sets by Zadeh in 1975. These were designed to mathematically represent the vagueness and uncertainty of linguistic problems; thereby obtaining formal tools to work with intrinsic imprecision in different type of problems; it is considered a generalization of the classic set theory. Type-2 fuzzy sets are used for modeling uncertainty and imprecision in a better way [18–20].

5 Problem Statement and Proposed Method

The objective of this work is to develop a model that is based on integrating the responses of an ensemble neural network using type-1 and type-2 fuzzy systems and their optimization. Figure 1 represents the general architecture of the proposed method, where historical data, analyzing data, creation of the ensemble neural network and integrate responses of the ensemble neural network with type-2 fuzzy system integration and finally obtaining the outputs as shown. The information can be historical data, these can be images, time series, etc., in this case we show the application to time series prediction of the Dow Jones where we obtain good results with this series.

Figure 2 shows a type-2 fuzzy system consisting of 5 inputs depending on the number of modules of the neural network ensemble and one output. Each input and output linguistic variable of the fuzzy system uses 2 Gaussian membership functions. The performance of the type-2 fuzzy integrators is analyzed under different levels of uncertainty to find out the best design of the membership functions and consist of 32 rules. For the type-2 fuzzy integrator using 2 membership functions, which are called low prediction and high prediction for each of the inputs and output of the fuzzy system. The membership functions are of Gaussian type, and we consider 3 sizes for the footprint uncertainty 0.3, 0.4 and 0.5 to obtain a better prediction of our time series.

In this Fig. 3 shows the possible rules of a type-2 fuzzy system.

Figure 4 represents the Particle Structure to optimize the ensemble neural network, where the parameters that are optimized are the number of modules, number of layers, number of neurons.

Fig. 1 General architecture
of the proposed method

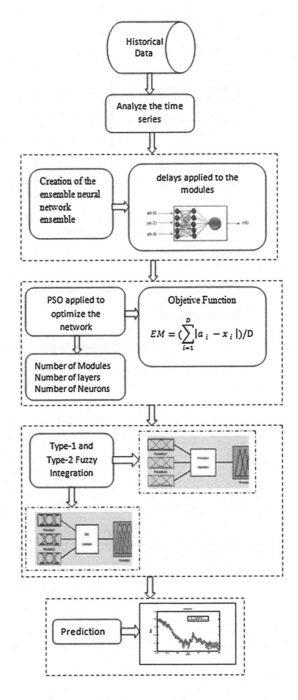

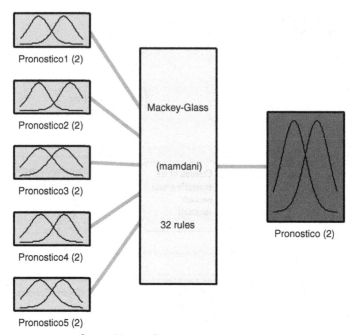

System Mackey-Glass: 5 inputs, 1 outputs, 32 rules

Fig. 2 Type-2 fuzzy system for the Mackey-Glass time series

1. If (Prediction1 is Pred1Low) and (Prediction2 is Pron2Low) and (Prediction3 is Pred3Low) and (Prediction4 is Pred4Low) and (Prediction5 is Pred5Low) then (Prediction is Low)
2. If (Prediction1 is Pred1High) and (Prediction2 is Pron2High) and (Prediction3 is Pred3High) and (Prediction4 is Pred4Low) and (Prediction5 is Pred5High) then (Prediction is High)
3. If (Prediction1 is Pred1Low) and (Prediction2 is Pron2Low) and (Prediction3 is Pred3Low) and (Prediction4 is Pred4Low) and (Prediction5 is Pred5High) then (Prediction is Low)
4. If (Prediction1 is Pred1High) and (Prediction2 is Pron2High) and (Prediction3 is Pred3High) and (Prediction4 is Pred4Low) and (Prediction5 is Pred5Low) then (Prediction is High)
5. If (Prediction1 is Pred1Low) and (Prediction2 is Pron2Low) and (Prediction3 is Pred3Low) and (Prediction4 is Pred4High) and (Prediction5 is Pred5High) then (Prediction is Low)
6. If (Prediction1 is Pred1High) and (Prediction2 is Pron2High) and (Prediction3 is Pred3High) and (Prediction4 is Pred4Low) and (Prediction5 is Pred5Low) then (Prediction is High)
7. If (Prediction1 is Pred1High) and (Prediction2 is Pron2Low) and (Prediction3 is Pred3High) and (Prediction4 is Pred4High) and (Prediction5 is Pred5High) then (Prediction is High)
8. If (Prediction1 is Pred1High) and (Prediction2 is Pron2High) and (Prediction3 is Pred3Low) and (Prediction4 is Pred4Low) and (Prediction5 is Pred5Low) then (Prediction is Low)
9. If (Prediction1 is Pred1Low) and (Prediction2 is Pron2High) and (Prediction3 is Pred3High) and (Prediction4 is Pred4High) and (Prediction5 is Pred5High) then (Prediction is High)
10. If (Prediction1 is Pred1High) and (Prediction2 is Pron2Low) and (Prediction3 is Pred3Low) and (Prediction4 is Pred4Low) and (Prediction5 is Pred5Low) then (Prediction is Low)
11. If (Prediction1 is Pred1Low) and (Prediction2 is Pron2High) and (Prediction3 is Pred3Low) and (Prediction4 is Pred4High) and (Prediction5 is Pred5Low) then (Prediction is Low)
12. If (Prediction1 is Pred1High) and (Prediction2 is Pron2Low) and (Prediction3 is Pred3High) and (Prediction4 is Pred4Low) and (Prediction5 is Pred5High) then (Prediction is High)
13. If (Prediction1 is Pred1Low) and (Prediction2 is Pron2High) and (Prediction3 is Pred3Low) and (Prediction4 is Pred4Low) and (Prediction5 is Pred5Low) then (Prediction is Low)
14. If (Prediction1 is Pred1High) and (Prediction2 is Pron2Low) and (Prediction3 is Pred3High) and (Prediction4 is Pred4Low) and (Prediction5 is Pred5High) then (Prediction is High)
15. If (Prediction1 is Pred1Low) and (Prediction2 is Pron2High) and (Prediction3 is Pred3Low) and (Prediction4 is Pred4Low) and (Prediction5 is Pred5Low) then (Prediction is Low)
16. If (Prediction1 is Pred1High) and (Prediction2 is Pron2Low) and (Prediction3 is Pred3High) and (Prediction4 is Pred4Low) and (Prediction5 is Pred5High) then (Prediction is High)
17. If (Prediction1 is Pred1Low) and (Prediction2 is Pron2Low) and (Prediction3 is Pred3High) and (Prediction4 is Pred4Low) and (Prediction5 is Pred5Low) then (Prediction is Low)
18. If (Prediction1 is Pred1High) and (Prediction2 is Pron2High) and (Prediction3 is Pred3Low) and (Prediction4 is Pred4Low) and (Prediction5 is Pred5High) then (Prediction is High)
19. If (Prediction1 is Pred1Low) and (Prediction2 is Pron2Low) and (Prediction3 is Pred3High) and (Prediction4 is Pred4High) and (Prediction5 is Pred5High) then (Prediction is High)
20. If (Prediction1 is Pred1High) and (Prediction2 is Pron2High) and (Prediction3 is Pred3Low) and (Prediction4 is Pred4Low) and (Prediction5 is Pred5Low) then (Prediction is Low)
21. If (Prediction1 is Pred1Low) and (Prediction2 is Pron2High) and (Prediction3 is Pred3High) and (Prediction4 is Pred4Low) and (Prediction5 is Pred5Low) then (Prediction is Low)
22. If (Prediction1 is Pred1High) and (Prediction2 is Pron2Low) and (Prediction3 is Pred3Low) and (Prediction4 is Pred4High) and (Prediction5 is Pred5High) then (Prediction is High)
23. If (Prediction1 is Pred1Low) and (Prediction2 is Pron2Low) and (Prediction3 is Pred3High) and (Prediction4 is Pred4High) and (Prediction5 is Pred5Low) then (Prediction is Low)
24. If (Prediction1 is Pred1High) and (Prediction2 is Pron2High) and (Prediction3 is Pred3Low) and (Prediction4 is Pred4Low) and (Prediction5 is Pred5High) then (Prediction is High)
25. If (Prediction1 is Pred1Low) and (Prediction2 is Pron2High) and (Prediction3 is Pred3Low) and (Prediction4 is Pred4Low) and (Prediction5 is Pred5Low) then (Prediction is Low)
26. If (Prediction1 is Pred1Low) and (Prediction2 is Pron2High) and (Prediction3 is Pred3Low) and (Prediction4 is Pred4High) and (Prediction5 is Pred5Low) then (Prediction is Low)
27. If (Prediction1 is Pred1Low) and (Prediction2 is Pron2High) and (Prediction3 is Pred3Low) and (Prediction4 is Pred4High) and (Prediction5 is Pred5Low) then (Prediction is Low)
28. If (Prediction1 is Pred1Low) and (Prediction2 is Pron2High) and (Prediction3 is Pred3Low) and (Prediction4 is Pred4High) and (Prediction5 is Pred5Low) then (Prediction is Low)
29. If (Prediction1 is Pred1Low) and (Prediction2 is Pron2Low) and (Prediction3 is Pred3High) and (Prediction4 is Pred4High) and (Prediction5 is Pred5High) then (Prediction is High)
30. If (Prediction1 is Pred1High) and (Prediction2 is Pron2Low) and (Prediction3 is Pred3Low) and (Prediction4 is Pred4Low) and (Prediction5 is Pred5High) then (Prediction is High)
31. If (Prediction1 is Pred1Low) and (Prediction2 is Pron2Low) and (Prediction3 is Pred3High) and (Prediction4 is Pred4Low) and (Prediction5 is Pred5High) then (Prediction is High)
32. If (Prediction1 is Pred1High) and (Prediction2 is Pron2Low) and (Prediction3 is Pred3High) and (Prediction4 is Pred4Low) and (Prediction5 is Pred5Low) then (Prediction is Low)

Fig. 3 Rules of the type-2 fuzzy inference system for the Dow Jones time series

Fig. 4 Particle structure to optimize the ensemble neural network

Number of Modules	Number of Layers 1	Neurons 1		Neurons ... n

Fig. 5 Mackey Glass time series

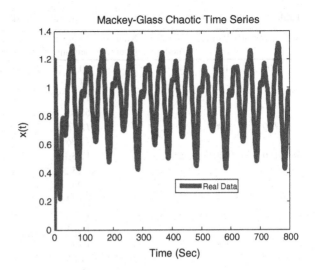

Mackey-Glass Chaotic Time Series

Data of the Mackey-Glass time series was generated using Eq. (6). We are using 800 points of the time series. We use 70 % of the data for the ensemble neural network trainings and 30 % to test the network.

The Mackey-Glass Equation is defined as follows:

$$\dot{x}(t) = \frac{0.2x(t-\tau)}{1+x^{10}(t-\tau)} - 0.1x(t) \tag{6}$$

where it is assumed x(0) = 1.2, τ = 17, τ = 34, and 68 x(t) = 0 for t < 0. Figure 5 shows a plot of the time series for these parameter values.

This time series is chaotic, and there is no clearly defined period. The series does not converge or diverge, and the trajectory is extremely sensitive to the initial conditions. The time series is measured in number of points, and we apply the fourth order Runge–Kutta method to find the numerical solution of the equation [12, 13].

6 Simulation Results

In this section we present the simulation results obtained with the integration of the ensemble neural network with type-2 fuzzy integration and its optimization with the genetic algorithm for the Mackey-Glass time series.

Table 1 shows the particle swarm optimization where the best prediction error is of 0.0063313.

Fuzzy integration is performed initially by implementing a type-1 fuzzy system in which the best result was in the experiment of row number 5 of Table 2 with an error of: 0.1521.

Table 1 Particle swarm results for the ensemble neural network $\tau = 17$

No.	Iterations	Particles	Number modules	Number layers	Number neurons	Duration	Prediction error
1	100	100	4	2	20,14 13,16 17,8 6,26	02:23:13	0.0076048
2	*100*	*100*	*2*	*2*	*12,16 12,26*	*01:45:45*	*0.0063313*
3	100	100	2	3	17,5,18 6,25,24	01:28:42	0.0018838
4	100	100	4	2	7,24 14,22 1,8 15,23	02:40:20	0.0073005
5	100	100	4	2	14,9 11,26 27,16 11,13	02:11:34	0.0081413
6	100	100	4	2	16,16 9,19 6,6 9,12	01:34:05	0.0087983
7	100	100	2	3	11,23,26 15,15,5	02:09:17	0.0076315
8	100	100	2	2	14,10 14,21	01:23:28	0.0061291
9	100	100	3	2	9,5 23,20 22,13	02:17:06	0.0053679
10	100	100	3	3	23,14,16 19,10,13 22,12,11	02:20:04	0.0061983

Table 2 Results of type-1 fuzzy integration for $\tau = 17$

Experiment	Prediction error with fuzzy integration type-1
Experiment 1	0.1879
Experiment 2	0.1789
Experiment 3	0.2221
Experiment 4	0.1888
Experiment 5	*0.1521*
Experiment 6	0.2561
Experiment 7	0.1785
Experiment 8	0.1942
Experiment 9	0.2536
Experiment 10	0.1965

Table 3 Results of type-2 fuzzy integration for $\tau = 17$

Experiment	Prediction error 0.3 uncertainty	Prediction error 0.4 uncertainty	Prediction error 0.5 uncertainty
Experiment 1	0.2385	0.2385	0.3952
Experiment 2	0.2489	0.2231	0.3909
Experiment 3	0.2482	0.2226	0.3642
Experiment 4	0.2214	*0.1658*	0.3856
Experiment 5	0.2658	0.2234	0.3857
Experiment 6	0.2756	0.2592	*0.3134*
Experiment 7	*0.1785*	0.2352	0.3358
Experiment 8	0.1825	0.2546	0.4561
Experiment 9	0.2018	0.2373	0.3394
Experiment 10	0.2076	0.2003	0.3687

Table 4 Particle swarm results for the ensemble neural network for $\tau = 34$

No.	Iterations	Particles	Number modules	Number layers	Number neurons	Duration	Prediction error
1	*100*	*100*	*2*	*3*	*12,23,12 9,19,7*	*02:45:14*	*0.0019726*
2	100	100	4	3	19,11,11 16,11,14 15,24,19 22,13,27	01:28:06	0.0063623
3	100	100	3	2	4,9 9,20 10,11 23,20	02:23:06	0.0046644
4	100	100	4	2	14,18 12,19 20,17 10,6	03:22:13	0.0072153
5	100	100	3	2	7,6 10,15 12,16	01:39:13	0.0075658
6	100	100	3	3	14,20,18 15,21,12 19,17,26	03:08:02	0.0047515
7	100	100	2	3	4,24 9,26	02:00:10	0.003601
8	100	100	2	2	24,17 14,23	02:27:21	0.0065506
9	100	100	3	3	7,11,8 23,21,21 17,8,11	02:03:12	0.0037758
10	100	100	2	3	20,28,15 15,12,24	02:04:18	0.0066375

Table 5 Results of type-1 fuzzy integration for $\tau = 34$

Experiment	Prediction error with fuzzy integration type-1
Experiment 1	0.9587
Experiment 2	*0.4586*
Experiment 3	0.5871
Experiment 4	1.2569
Experiment 5	0.9517
Experiment 6	1.556
Experiment 7	1.0987
Experiment 8	1.9671
Experiment 9	1.698
Experiment 10	1.4626

Table 6 Results of type-2 fuzzy integration for $\tau = 34$

Evolution	Prediction error 0.3 uncertainty	Prediction error 0.4 uncertainty	Prediction error 0.5 uncertainty
Evolution 1	*0.6036*	0.8545	0.4570
Evolution 2	1.5862	1.0021	1.3533
Evolution 3	0.8002	0.6943	*0.3893*
Evolution 4	1.4032	0.9617	0.9665
Evolution 5	0.8658	0.8299	0.6358
Evolution 6	1.3986	0.1052	1.2354
Evolution 7	1.465	1.3566	0.6646
Evolution 8	1.7453	0.8966	0.8241
Evolution 9	0.9866	*0.6524*	0.6661
Evolution 10	1.4552	0.9956	0.7557

Fuzzy integration is performed by implementing a type-1 fuzzy system in which the results were as follows: for the best evolution with a degree of uncertainty of 0.3 a forecast error of 0.1785 was obtained, and with a degree of uncertainty of 0.4 a forecast error of 0.1658 and with a degree of uncertainty of 0.5 a forecast error of 0.3134 was obtained, as shown in Table 3.

Table 4 shows the particle swarm optimization where the best prediction error is of 0.0019726.

Fuzzy integration is performed by implementing a type-1 fuzzy system in which the best result was in the experiment of row number 2 of Table 5 with an error of: 0.4586.

Fuzzy integration is performed by implementing a type-2 fuzzy system in which the results were as follows: for the best evolution with a degree of uncertainty of 0.3 a forecast error of 0.6036 was obtained, and with a degree of uncertainty of 0.4 a forecast error of 0.6524 and with a degree of uncertainty of 0.5 a forecast error of 0.3893 was obtained, as shown in Table 6.

Table 7 Particle swarm results for the ensemble neural network for $\tau = 68$

No.	Iterations	Particles	Number of modules	Number of layers	Number of neurons	Duration	Prediction error
1	*100*	*100*	*2*	*3*	*17,5,18* *6,25,19*	*02:05:53*	*0.0019348*
2	100	100	2	2	7,8 6,20	04:1936	0.0041123
3	100	100	2	3	21,11,16 5,10,10	02:28:02	0.0042367
4	100	100	4	3	15,7,4 11,22,5 24,19,22 4,14,11	02:37:06	0.0050847
5	100	100	3	2	22,23 2,21 10,2	01:5	0.0037132
6	100	100	4	3	10,13,22 24,18,17 13,16,20 7,24,17	02:10:27	0.0057235
7	100	100	2	3	8,20 15,23		0.0033082
8	100	100	3	2	28,6 2,16 18,10	01:40:18	0.0057402
9	100	100	3	2	22,17 10,10 21,12	02:45:31	0.0047309
10	100	100	2	3	22,11,18 27,7,14	01:35:13	0.0044649

Table 7 shows the particle swarm optimization where the prediction error is of 0.0019348.

Fuzzy integration is performed by implementing a type-1 fuzzy system in which the best result was in the experiment of row number 4 of Table 8 with an error of: 0.32546.

Fuzzy integration is also performed by implementing a type-2 fuzzy system in which the results were as follows: for the best evolution with a degree of uncertainty of 0.3 a forecast error of 0.6825 was obtained, and with a degree of uncertainty of 0.4 a forecast error of 0.7652 and with a degree of uncertainty of 0.5 a forecast error of 0.6581 was obtained, as shown in Table 9.

Table 8 Results of type-1 fuzzy integration for $\tau = 68$

Experiment	Prediction error with fuzzy integration type-1
Experiment 1	0.8753
Experiment 2	0.3625
Experiment 3	0.6687
Experiment 4	*0.3254*
Experiment 5	0.5489
Experiment 6	1.3183
Experiment 7	1.8972
Experiment 8	1.6977
Experiment 9	1.5879
Experiment 10	0.9652

Table 9 Results of type-2 fuzzy integration for $\tau = 68$

Evolution	Prediction error 0.3 uncertainty	Prediction error 0.4 uncertainty	Prediction error 0.5 uncertainty
Evolution 1	0.7895	0.9631	0.7365
Evolution 2	0.9875	1.2365	1.564
Evolution 3	0.9874	0.7965	*0.6581*
Evolution 4	1.5325	0.9874	0.9723
Evolution 5	0.7763	0.9723	0.9858
Evolution 6	0.8694	0.9235	1.3697
Evolution 7	*0.6825*	1.4263	0.6646
Evolution 8	1.336	0.8963	0.8288
Evolution 9	0.9852	*0.7652*	0.7234
Evolution 10	1.365	1.4224	1.5984

7 Conclusions

Using the technique of PSO particle we can reach the conclusion that this algorithm is good for reducing the execution time compared to other techniques such as genetic algorithms, and also architectures for ensemble neural network are small and they applied to the time series, as in this case the time series of Mackey-Glass. Also the outputs results obtained integrating the results of the neural network with type-1 and type-2 fuzzy systems and integrated type-2 the best results with type 2 are very good [1, 9, 15, 21].

Acknowledgments We would like to express our gratitude to the CONACYT, Tijuana Institute of Technology for the facilities and resources granted for the development of this research.

References

1. Brockwell, P.D., Richard, A.D.: Introduction to Time Series and Forecasting, pp. 1–219. Springer, New York (2002)
2. Castillo, O., Melin, P.: Type-2 Fuzzy Logic: Theory and Applications. Neural Networks, pp. 30–43. Springer, New York (2008)
3. Castillo, O., Melin, P.: Hybrid intelligent systems for time series prediction using neural networks, fuzzy logic, and fractal theory. IEEE Trans. Neural Netw. **13**(6), 1395–1408 (2002)
4. Castillo, O., Melin, P.: Simulation and forecasting complex economic time series using neural networks and fuzzy logic. In: Proceedings of the International Neural Networks Conference 3, pp. 1805–1810 (2001)
5. Castillo, O., Melin, P.: Simulation and forecasting complex financial time series using neural networks and fuzzy logic. In: Proceedings the IEEE the International Conference on Systems, Man and Cybernetics 4, pp. 2664–2669 (2001)
6. Eberhart, R.C., Kennedy, J.: A new optimizer particle swarm theory. In: Proceedings of the Sixth Symposium on Micromachine and Human Science, pp. 39–43 (1995)
7. Eberhart, R.C.: Fundamentals of Computational Swarm Intelligence, pp. 93–129. Wiley, New York (2005)
8. Jang, J.S.R, Sun, C.T., Mizutani, E.: Neuro-Fuzzy and Soft Computing. Prentice Hall, Englewood Cliffs, 1996
9. Karnik, N., Mendel, M.: Applications of type-2 fuzzy logic systems to forecasting of time-series. Inf. Sci. **120**(1–4), 89–111 (1999)
10. Kennedy, J., Eberhart, R.C.: Particle swarm optimization. In: Proceedings Intelligent Symposium, pp. 80–87, April 2003
11. Krogh, A., Vedelsby, J.: Neural network ensembles, cross validation, and active learning. In: Tesauro, G., Touretzky, D., Leen, T. (eds.) Advances in Neural Information Processing Systems 7, pp. 231–238. Denver, CO, MIT Press, Cambridge (1995)
12. Mackey, M.C.: Adventures in Poland: having fun and doing research with Andrzej Lasota. Mat. Stosow 5–32 (2007)
13. Mackey, M.C., Glass, L.: Oscillation and chaos in physiological control systems. Science **197**, 287–289 (1997)
14. Maguire, L.P., Roche, B., McGinnity, T.M., McDaid, L.J.: Predicting a chaotic time series using a fuzzy neural network **12**(1–4), 125–136 (1998)
15. Melin, P., Castillo, O., Gonzalez, S., Cota, J., Trujillo, W., Osuna, P.: Design of Modular Neural Networks with Fuzzy Integration Applied to Time Series Prediction, vol. 41/2007, pp. 265–273. Springer, Berlin (2007)
16. Multaba, I.M., Hussain, M.A.: Application of Neural Networks and Other Learning Technologies in Process Engineering. Imperial Collage Press, (2001)
17. Plummer, E.A.: Time Series Forecasting with Feed-forward Neural Networks: Guidelines and Limitations, University of Wyoming, July 2000
18. Pulido, M., Mancilla, A., Melin, P.: An ensemble neural network architecture with fuzzy response integration for complex time series prediction. In: Evolutionary Design of Intelligent Systems in Modeling, Simulation and Control, vol. 257/2009, pp. 85–110. Springer, Berlin (2009)
19. Sharkey, A.: One Combining Artificial of Neural Nets, Department of Computer Science University of Sheffield, UK (1996)
20. Sharkey, A.: Combining Artificial Neural Nets: Ensemble and Modular Multi-net Systems. Springer, London (1999)
21. Shimshoni, Y., Intrator, N.: Classification of seismic signal by integrating ensemble of neural networks. IEEE Trans. Signal Process. **461**(5), 1194–1201 (1998)
22. Sollich, P., Krogh, A.: Learning with ensembles: how over-fitting can be useful. In: Touretzky, D.S., Mozer, M.C., Hasselmo, M.E. (eds.) Advances in Neural Information Processing Systems 8, pp. 190–196. Denver CO, MIT Press, Cambridge (1996)

23. Yadav, R.N., Kalra, P.K., John, J.: Time series prediction with single multiplicative neuron model. Soft computing for time series prediction. Appl. Soft Comput. **7**(4), 1157–1163 (2007)
24. Yao, X., Liu, Y.: Making use of population information in evolutionary artificial neural networks. IEEE Trans. Syst. Man Cybern. Part B Cybern. **28**(3), 417–425 (1998)
25. Zhao, L., Yang, Y.: PSO-based single multiplicative neuron model for time series prediction. Expert Syst. Appl. **36**(2), Part 2, 2805–2812 (2009)
26. Zhou, Z.-H., Jiang, Y., Yang, Y.-B., Chen, S.-F.: Lung cancer cell identification based on artificial neural network ensembles. Artif. Intell. Med. **24**(1), 25–36 (2002)

Uncertainty-Based Information Granule Formation

Mauricio A. Sanchez, Oscar Castillo and Juan R. Castro

Abstract A new technique for forming information granules is shown in this chapter. Based on the theory of uncertainty-based information, an approach is proposed which forms Interval Type-2 Fuzzy information granules. This approach captures multiple evaluations of uncertainty from taken samples and uses these models to measure the uncertainty from the difference in these. The proposed approach is tested through multiple benchmark datasets: iris, wine, glass, and a 5th order curve identification.

1 Introduction

Granular computing is concerned with how information is grouped together and how these groups can be used to make decisions [1, 2]. It is inspired by how human cognition manages information. Granular computing is used to improve the final representation of information models by forming information granules which better adapt to the known information. Although granular computing expresses information models, more commonly known as information granules, it can use a variety of representations to express such granules, which could be rough sets [3], quotient space [4], shadowed sets [5], fuzzy sets [6, 33–35], etc.

M. A. Sanchez · J. R. Castro
Autonomous University of Baja California, Tijuana, Mexico
e-mail: mauricio.sanchez@uabc.edu.mx

J. R. Castro
e-mail: jrcastror@uabc.edu.mx

O. Castillo (✉)
Tijuana Institutes of Technology, Tijuana, Mexico
e-mail: ocastillo@tectijuana.mx

O. Castillo et al. (eds.), *Recent Advances on Hybrid Approaches for Designing Intelligent Systems*, Studies in Computational Intelligence 547,
DOI: 10.1007/978-3-319-05170-3_8, © Springer International Publishing Switzerland 2014

Information granules are representations of similar information which can be used for a purpose, typically to model a portion of information. Forming fuzzy information granules is not new, knowing that many representations can be used there have been many approaches which try to solve this: via their relationships [7], optimization of time granularity [8], information granulation [9], with RBF neural networks [10], Interval Type-2 Fuzzy granules [11], non-homogeneous General Type-2 Fuzzy granules [12], etc.

This chapter proposes an approach to information granule formation by capturing, through samples, evaluations of uncertainty where their difference is a direct measure of uncertainty which is used to form Interval Type-2 Fuzzy information granules [25–32].

This chapter is organized as follows: Sect. 2 describes the proposed approach as well as its motivation. Section 3 shows benchmark results alongside the discussion. Finally, Sect. 4 concludes the document.

2 Uncertainty-Based Information Granule Formation Methodology Description

To first understand the main methodology, a review of the motivation is necessary, as it describes the basis for the proposed approach. First, the basis for the proposed approach, which is the theory of uncertainty-based information [13, 14] is described; then, evaluations of uncertainty [15, 16] are described, which defines functions that represent uncertainty measures.

2.1 Uncertainty-Based Information

The concept of uncertainty is closely related to the concept of information. The fundamental characteristic of this relation is that involved uncertainty from any problem-solving situation is a result of information deficiency pertaining to the system within which the situation is conceptualized. This information could be incomplete, imprecise, fragmentary, unreliable, vague, or contradictory.

With the assumption that a certain amount of uncertainty can be measured from a problem-solving situation it is possible that a mathematical theory can be formed.

With another assumption that this amount of uncertainty is reduced by obtaining relevant information as a result of some action (e.g. obtaining experimental results, observing new data, etc.), the amount of obtained information by the action can be measured by that amount of reduced uncertainty. That is, the amount of information related to a given problem-solving situation that is obtained through some action is measured by the difference between a priori uncertainty and a posteriori uncertainty.

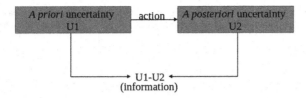

Fig. 1 Diagram of the behavior of the uncertainty-based information where uncertainty is reduced by the difference between two uncertain models of the same information

In Fig. 1, the shown diagram represents the general idea of the behavior of uncertainty-based information; where a reduction of uncertainty can be obtain by the difference of two uncertain models of the same information. That is, the a priori uncertainty model is obtained with a first sample of information, where as the posteriori uncertainty model is obtained with a second sample of information related to the same problem-solving situation.

2.2 Evaluations of Uncertainty

To capture uncertainty, there are two fundamental types of evaluations: Type A and Type B.

Through repeated measurements, an average measured value can infer a standard deviation which forms a Gaussian distribution function, where this functions is a Type A evaluation of uncertainty.

Type B evaluations of uncertainty are represented by a rectangular probability distribution, in other words, a specified interval where the measurements are known to lie in.

2.3 Uncertainty-Based Information Granule Formation

Taking inspiration on uncertainty-based information, this can be interpreted in a manner which forms higher-type information granules where uncertainty can be captured and measured and build Interval Type-2 Fuzzy information granules.

A sample of information can build a model with uncertainty from the complete source of information; this is, since it is impossible to know the complete truth of any given situation, uncertainty will always exist in any sample information which may be taken from it.

Through a first sample of information (D_1), an uncertain model (evaluation of uncertainty) can be created. Through a second sample of information (D_2), another similar uncertain model can be also created. These two models of uncertainty are analogous to the models in the theory of uncertainty-based information, a priori and posteriori uncertainty models.

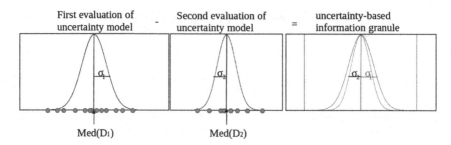

Fig. 2 Explanatory diagram of how the proposed approach measures and defines the uncertainty, and forms an IT2 Fuzzy set with such uncertainty

In a direct comparison with the theory of uncertainty-based information, the proposed approach does not reduce the uncertainty in the model, instead it measures and defines it to be able to use it in an information granule and have an improved representation of the information. The proposed approach is shown in Fig. 2, where a first sample of information obtains an evaluation of uncertainty, in the form of a Gaussian function, or Type-1 Gaussian membership function; and a second sample of information obtained another similar evaluation of uncertainty, of the same form. A difference is found between these two Gaussian membership functions defining the Footprint of Uncertainty (FOU), thus obtaining an IT2 Fuzzy information granule. Here there are three possibilities: (1) the first Gaussian membership function has an σ which is larger than the second; (2) the second Gaussian membership function has an σ which is larger than the first; and (3) the σ from both Gaussian membership functions are the same. For 1 and 2, the FOU which is created defines some uncertainty which has been measured and can now be used by the IT2 Fuzzy System; and for 3, since no uncertainty was measured a T1 Fuzzy Set is created.

To show the viability of the proposed approach in that it captures uncertainty and forms IT2 Fuzzy information granules, an algorithm was created that would allow for results to be obtained. The following steps define the algorithms:

1. Obtain rules and centers. These can be obtained through any clustering algorithm, for the experimental case in this chapter the subcluster algorithm [17] was used.
2. Through a first sample of information (D_1), all σ_1 for all centers are calculated. These were found by calculating the Euclidean n-space distance between each data point and all centers, where the shortest distance defines to which center does that point belong to, afterwards having a set of data points for each cluster, a standard deviation was calculated as to form an evaluation of uncertainty in the form of a Gaussian membership function. For the case of testing, a random sample comprised of 40 % of the dataset was used.
3. Through a second sample of information (D_2), in the same manner as the previous step, all σ_2 for all centers are calculated. A random sample comprised of another 40 % of the dataset was used for this step.

4. Form the IT2 Fuzzy Gaussian information granules as proposed. This only builds the antecedents of a complete IT2 Fuzzy System.
5. The consequents are finally optimized via an evolutionary algorithm, obtaining a complete IT2 Fuzzy System which can be used to acquire results. For this chapter, Interval Takagi-Sugeno-Kang (TSK) [18, 19] consequents were used, they were optimized via a Cuckoo Search algorithm [20].

The next section uses this algorithm to obtain results.

3 Experimental Results and Discussion

For experimental tests, four datasets were used: iris, wince, glass, available from the UCI dataset repository [21], and a 5th order polynomial curve. Where the iris dataset, has 4 input features (petal length, petal width, sepal length, and sepal width), and 3 outputs (iris setosa, iris virginica, and iris versicolor). With 50 samples of each flower type, with a total of 150 elements in the dataset. The wine dataset, with 13 input features of different constituents (Alcohol, malic acid, ash, alcalinity of ash, magnesium, total phenols, flavanoids, nonflavanoid phenols, proanthocyanins, color intensity, hue, OD280/OD315 of diluted wines, and pro-line) identifying 3 distinct Italian locations where the wine came from. With 59, 71, and 48 elements respectively in each class, for a total of 178 elements in the whole dataset. The glass identification dataset, has 9 input variables (refractive index, sodium, magnesium, aluminum, silicon, potassium, calcium, barium, and iron), and 7 classes (building windows float processed, building windows non float processed, vehicle windows float processed, containers, tableware, and head-lamps). With 70, 76, 17, 13, 9, and 29 elements respectively in each class, for a total of 214 elements in the whole dataset.

3.1 Experimental Results

On Table 1, the obtained results are shown, where the after 30 execution runs for each dataset were made to obtain a minimum, maximum, mean, and standard deviation for each dataset.

The following Figs. 3, 4, 5, 6 show one sample of the formed IT2 Fuzzy information granules of each dataset: iris, wine, glass, and 5th order polynomial, respectively.

Table 1 Obtained results for the chosen datasets

	Classification accuracy %			RMSE error
	------	------	------	------
	Iris	Wine	Glass	5th order polynomial
Min	86.66	88.88	44.18	0.36
Mean	93.99	93.05	68.43	0.69
Max	100	97.22	86.04	1.14
Std	5.164	5.982	15.592	0.181

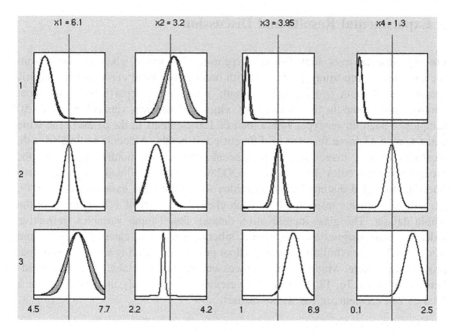

Fig. 3 Sample of the formed IT2 Fuzzy information granules for the Iris dataset

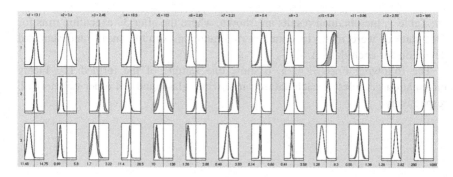

Fig. 4 Sample of the formed IT2 Fuzzy information granules for the Wine dataset

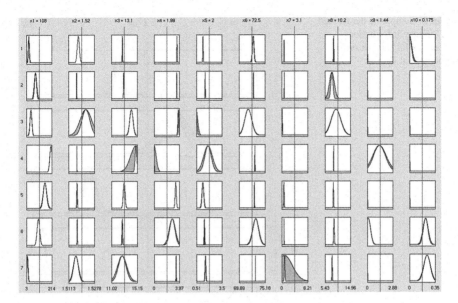

Fig. 5 Sample of the formed IT2 Fuzzy information granules for the Glass dataset

3.2 Results Discussion

The values obtained for the classification accuracy and RMSE error are not the best values obtained in general, yet they are comparable to current algorithms in terms of mean results [22–24]. This is by no manner the best obtainable results this approach can acquire; this is mostly in part to the chosen clustering algorithm as well as the evolutionary algorithm which were used to obtain such results. A better combination as well as tuning should yield better results.

As shown in the formed IT2 Fuzzy information granules, some granules captured more uncertainty than others, in many cases the uncertainty is minimal to the point that there is no measurable uncertainty when forming the evaluation of uncertainty Gaussian function.

Having chosen IT2 Fuzzy Gaussian membership functions as representation for higher type information granules, the characteristics of these is that the center value is the same, and only two values for σ form the FOU. Although results are acceptable, other variations can be used to yield different results as well as different interpretations, for example, where the center is offset and two values for σ are used. Even other types of IT2 Fuzzy membership functions could be used, each one having their own interpretation of the information as well as varying results when the IT2 Fuzzy System is formed and optimized.

Fig. 6 Sample of the formed IT2 Fuzzy information granules for the 5th order polynomial dataset

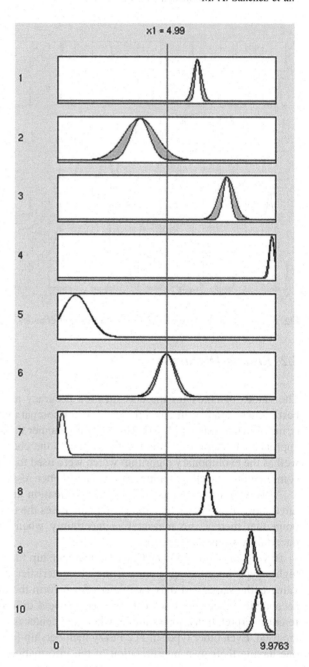

4 Conclusion and Future Work

4.1 Conclusions

Taking inspiration from the uncertainty-based information theory, higher type information granules can be formed which better conceptualize the uncertainty in the information.

The proposed approach reduces the uncertainty in the information model by measuring the uncertainty by means of the difference between two evaluations of uncertainty created by two distinct measurements of information sampling.

By choosing Interval Type-2 Fuzzy sets as the representation of information granules, the proposed approach directly takes the obtained uncertainty measurement and builds higher type information granules.

Any other form of granule representation which can express the uncertainty in the information can be used [36].

4.2 Future Work

Find the optimal amount of samples for each model building step. Although 40 % was used, what is the minimal amount which can be used to obtain acceptable results?

The amount of samples taken could be explored; this chapter only took two samples to form the final information granule. Could taking more samples yield a better result?

Other information granule representations could be used which also support uncertainty. Even though Type A Gaussian evaluations of uncertainty were used, there are other types of functions which could also directly capture uncertainty.

Acknowledgments We thank the MyDCI program of the Division of Graduate Studies and Research, UABC, and Tijuana Institute of Technology the financial support provided by our sponsor CONACYT contract grant number: 314258.

References

1. Bargiela, A., Pedrycz, W.: Toward a theory of granular computing for human-centered information processing. IEEE Trans. Fuzzy Syst. **16**(2), 320–330 (2008)
2. Pedrycz, W.: Granular computing: the emerging paradigm. J. Uncertain Syst. **1**(1), 38–61 (2007)
3. Pawlak, Z.: Rough sets. Int. J. Comput. Inf. Sci. **11**(5), 341–356 (1982)
4. Zhang, L.Z.L., Zhang, B.Z.B.: Quotient space based multi-granular computing, vol. 1 (2005)
5. Pedrycz, W., Vukovich, G.: Granular computing with shadowed sets. Int. J. Intell. Syst. **17**(2), 173–197 (2002)

6. Zhang, Y.Z.Y., Zhu, X.Z.X., Huang, Z.H.Z.: Fuzzy Sets Based Granular Logics for Granular Computing. (2009)
7. Yao, J.T.: Information granulation and granular relationships. In: 2005 IEEE International Conference on Granular Computing, vol. 1 (2005)
8. Yu, F.Y.F., Cai, R.C.R.: Optimized fuzzy information granulation of temporal data. In: Fuzzy Systems and Knowledge Discovery (FSKD), vol. 1 (2010)
9. Yao, J., Yao, Y.Y.: Information granulation for web-based information support systems. In: Proceedings of SPIE, pp. 138–146 (2003)
10. Park, H.-S., Chung, Y.-D., Oh, S.-K., Pedrycz, W., Kim, H.-K.: Design of information granule-oriented RBF neural networks and its application to power supply for high-field magnet. Eng. Appl. Artif. Intell. 24(3), 543–554 (2011)
11. Sanchez, M.A., Castro, J.R., Perez-Ornelas, F., Castillo, O.: A hybrid method for IT2 TSK formation based on the principle of justifiable granularity and PSO for spread optimization. In: 2013 Joint IFSA World Congress and NAFIPS Annual Meeting (IFSA/NAFIPS), pp. 1268–1273 (2013)
12. Sanchez, M.A., Castro, J.R., Castillo, O.: Formation of general type-2 Gaussian membership functions based on the information granule numerical evidence. In: 2013 IEEE Workshop on Hybrid Intelligent Models and Applications (HIMA), pp. 1–6 (2013)
13. Klir, G.J.: Uncertainty and Information: Foundations of Generalized Information Theory, p. 499. Wiley-IEEE Press, New Jersey (2005)
14. Klir, G.J., Wierman, M.J.: Uncertainty-Based Information, vol. 15. Physica-Verlag HD, Heidelberg (1999)
15. Weise, K., Woger, W.: A Bayesian theory of measurement uncertainty. Meas. Sci. Technol. 3, 1–11 (1992)
16. J. C. F. G. I. M. JCGM: Evaluation of measurement data—guide to the expression of uncertainty in measurement. Int. Organ. Stand. Geneva ISBN. 50, 134 (2008)
17. Chiu, S.L.: Fuzzy model identification based on cluster estimation. J. Intell. Fuzzy Syst. 2, 267–278 (1994)
18. Jang, J.-S.R.: Fuzzy modeling using generalized neural networks and Kalman filter algorithm. In: Proceedings of the Ninth National Conference on Artificial Intelligence, pp. 762–767 (1991)
19. Jang, J.S.R.: ANFIS: adaptive-network-based fuzzy inference system. IEEE Trans. Syst. man Cybern. 23(3), 665–685 (1993)
20. Yang, X.-S., Cuckoo search via Lévy flights. In: 2009 World Congress on Nature & Biologically Inspired Computing (NaBIC), pp. 210–214 (2009)
21. Frank A., Asuncion, A.: UCI Machine Learning Repository. University of California Irvine School of Information, School of Information and Computer Sciences, University of California, Irvine, p. 0 (2010) (vol. 2008, no. 14/8)
22. Daneshgar, A., Javadi, R., Razavi, B.S.: Clustering using isoperimetric number of trees. Pattern Recognit. 46(12), 3371–3382 (2012)
23. David, G., Averbuch, A.: SpectralCAT: categorical spectral clustering of numerical and nominal data. Pattern Recognit. 45(1), 416–433 (2012)
24. Thi, L., An, H., Hoai, L., Dinh, P.: New and efficient DCA based algorithms for minimum sum-of-squares clustering. Pattern Recognit. 47, 388–401 (2014)
25. Castillo, O., Huesca, G., Valdez, F.: Evolutionary computing for topology optimization of type-2 fuzzy controllers. Stud. Fuzziness Soft Comput. 208, 163–178 (2008)
26. Castillo, O., Melin, P.: Type-2 Fuzzy Logic: Theory and Applications. Springer-Verlag, Heidelberg (2008)
27. Castillo, O., Aguilar, L.T., Cazarez-Castro, N.R., Cardenas, S.: Systematic design of a stable type-2 fuzzy logic controller. Appl. Soft Comput. J. 8, 1274–1279 (2008)
28. Castillo, O., Martinez-Marroquin, R., Melin, P., Valdez, F., Soria, J.: Comparative study of bio-inspired algorithms applied to the optimization of type-1 and type-2 fuzzy controllers for an autonomous mobile robot. Inf. Sci. 192, 19–38 (2012)

29. Castillo, O., Melin, P., Alanis, A., Montiel, O., Sepulveda, R.: Optimization of interval type-2 fuzzy logic controllers using evolutionary algorithms. J. Soft Comput. **15**(6), 1145–1160 (2011)
30. Castro, J.R., Castillo, O., Melin, P.: An interval type-2 fuzzy logic toolbox for control applications. In: Proceedings of FUZZ-IEEE 2007, London, pp. 1–6 (2007)
31. Castro, J.R., Castillo, O., Martinez, L.G.: Interval type-2 fuzzy logic toolbox. Eng. Lett. **15**(1), 14 (2007)
32. Hidalgo, D., Castillo, O., Melin, P.: Type-1 and type-2 fuzzy inference systems as integration methods in modular neural networks for multimodal biometry and its optimization with genetic algorithms. Inf. Sci. **179**(13), 2123–2145 (2009)
33. Aguilar, L., Melin, P., Castillo, O.: Intelligent control of a stepping motor drive using a hybrid neuro-fuzzy ANFIS approach. J. Appl. Soft Comput. **3**(3), 209–219 (2003)
34. Melin, P., Castillo, O.: Adaptive intelligent control of aircraft systems with a hybrid approach combining neural networks, fuzzy logic and fractal theory. J. Appl. Soft Comput. **3**(4), 353–362 (2003)
35. Montiel, O., Sepulveda, R., Melin, P., Castillo, O., Porta Garcia, M., Meza Sanchez I.: Performance of a simple tuned fuzzy controller and a PID controller on a DC motor. In: FOCI 2007 pp. 531–537
36. Rubio, E., Castillo, O.: Optimization of the interval type-2 fuzzy C-means using particle swarm optimization. In: Proceedings of NABIC 2013, pp. 10–15, Fargo, USA (2013)

A Type 2 Fuzzy Neural Network Ensemble to Estimate Time Increased Probability of Seismic Hazard in North Region of Baja California Peninsula

Victor M. Torres and Oscar Castillo

Abstract A type-2 adaptive fuzzy neural network ensemble approach is presented here to achieve the prediction of seismic events of M_0 magnitude in the north region of the Baja California Peninsula. Three algorithms are used with the ensemble: data analysis, M8 and CN. Seismic data coordinates are used in probabilistic fuzzy sets that are processed in the three fuzzy neural networks that integrate the ensemble to generate an output of a probabilistic set of predictions.

1 Introduction

The forecast methods proposed so far have treated the problem with great uncertainty at medium and long term seismic predictions. While the system is not known directly, it is possible to explore it by indirect methods that allow gaining a general understanding of how earthquakes are generated and propagated.

The study of these movements, their probability of future impacts and consequences is called seismic hazard. Seismic hazard has been gaining more importance every day more as it aims to minimize the social and economic impact of an earthquake.

It has been proposed that is feasible to predict, with a certain probability, the next seismic event [4, 9, 21]. Several methods have been used to reach a better understanding of the genesis of the movements and the emergence of upcoming events. Prediction systems have been proposed based upon statistical methods and neural systems [18], most of them using a single adaptive neural system or an adaptive fuzzy neural system [18, 21].

V. M. Torres · O. Castillo (✉)
Tijuana Institute of Technology, Calzada Tecnologico s/n, Tijuana, Mexico
e-mail: ocastillo@tectijuana.mx

O. Castillo et al. (eds.), *Recent Advances on Hybrid Approaches for Designing Intelligent Systems*, Studies in Computational Intelligence 547,
DOI: 10.1007/978-3-319-05170-3_9, © Springer International Publishing Switzerland 2014

In this chapter an ensemble of fuzzy neural systems is proposed to make predictions of the seismic hazard of the peninsula of Baja California, which is located in the Northwest corner of México.

It is considered that the peninsula was formed due to the movement of the Pacific plate, on which it sits, in relation to the North America plate, which belongs to the territory continent of México. The movement of tectonic plates, which is in the order of 4.5 cm a year, generates large concentrations of effort that occasionally are released in the form of tremors.

In what concerns to the Baja California state, the displacement of tectonic plate has generated several fault systems in which are grouped most of the quakes [1]. Those are: Mexicali Valley, Laguna Salada-Juarez Sierra and San Miguel (see Fig. 1).

In spite of frequent seismic activity in the region, the data was collected consistently only until 2002 [1], year in which the CICESE north-west seismic network went into operation. Before that date, the catalog of tremors only registered seismic movements with magnitude greater than 4 and only a few more of magnitude 3.

2 Neural Networks and Fuzzy Logic

Fuzzy sets were first proposed by Zadeh [20] in 1965 as a way to deal with imprecise or uncertain data. Fuzzy set theory is effective in the linguistic representation of information and incorporates human knowledge dealing with uncertainty and imprecision.

A type-1 fuzzy logic system (FLS) is a system that uses fuzzy set theory to map inputs to outputs. It's contains a fuzzifier block that becomes in fuzzy sets any crisp input data, an inference block that acts in the same way a person do who has do deal with imprecise data, a set of rules or expert knowledge depicted in linguistic values and an defuzzifier block that reduces the output fuzzy set in a crisp output (see Fig. 2).

In 1975, Zadeh [20] introduces type-2 fuzzy inference systems (IT2FLS) as an extension to type-1 fuzzy inference systems. Mendel [5] develops the theory for these sets, which have been shown as more effective than type-1 [14].

A type-2 FLS is very similar to a type 1 FIS but a type-2 FLS incorporates a type-reduction block to reduce an interval type-2 fuzzy set into an interval-valued type-1 fuzzy set [15]. A fuzzy set type-1 reaches the defuzzifier block to produce a crisp output (see Fig. 3) [16].

On the other hand, neural networks are frequently used in predictive systems due to its capacity of learning, parallel processing, and efficiency to approximate continuous functions and time series, and adaptability to changing or unknown conditions [10, 17].

Neural networks are a form of computational multi-threaded involving simple processing elements with a high degree of interconnection and adaptive interaction among them [18] (see Fig. 4).

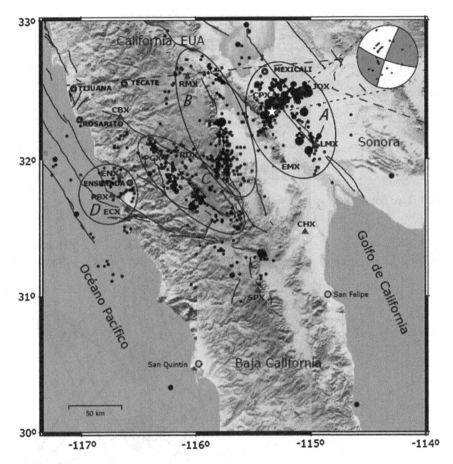

Fig. 1 Spatial distribution of seismic movements. *Source* [1]

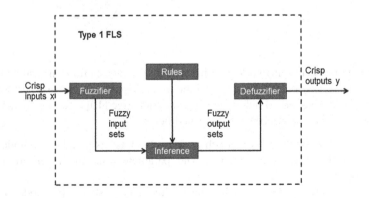

Fig. 2 Type 1 fuzzy logic system diagram

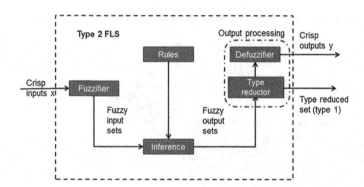

Fig. 3 Type 2 fuzzy logic system diagram

Fig. 4 Neural network
diagram

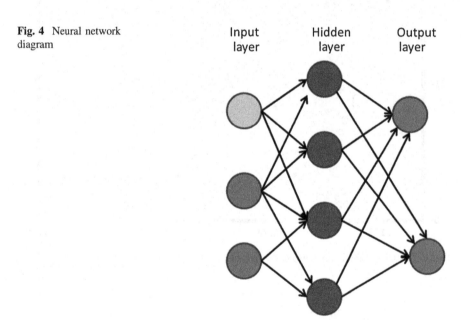

The union of neural networks and fuzzy logic produces a fuzzy neural network. A fuzzy neural network contains a set of nodes arranged on layers that perform the same task independently. Each node is fixed or adaptive [10] (see Fig. 5) and each layer performs different tasks, as described below:

- Layer 1—Fuzzification layer. Each node in this layer is an adaptive node. The nodes in this layer are called *I*. The parameters in this layer are called assumptions.
- Layer 2—Inference layer. Each node in this layer is static. The nodes in this layer are called *P*, whose output is the product of all input signals. The output of each node represents the application of a rule.

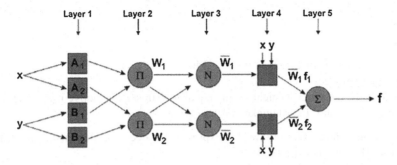

Fig. 5 Fuzzy neural network. *Source* [10]

- Layer 3—Implication layer. Each node in this layer is static and they are called by N. The i-th node calculates the radius of the i-input rules. The output of this layer is called normalization.
- Layer 4—Aggregation layer. Each i-th node in this layer is an Adaptive node. The parameters in this layer are referred to as the consequences.
- Layer 5—The single node in this layer is a static node, called R, which calculates the sum of all signals output.In 2001, Zhou [18] proposes that the use of several systems of neural networks functioning together obtain better predictions than separately. This arrangement of fuzzy neural systems is called an Ensemble. The Ensembles of neural networks are a paradigm of learning in which several neural networks are attached and are trained to solve a problem [3, 21].

3 Prediction Algorithms

In seismology, a tremor is "a sudden and transient shaking of the Earth's crust which releases energy in the form of strong elastic waves" [13]. Each registered tremor has at least the same basic data, which are known as seismic coordinates. These data are:

- Magnitude
- Time of the event
- Latitude and longitude, and
- Depth.After a seismic event, it is common that events of lesser magnitude follow the main movement. These events are known as aftershocks, and decrease with the time.

Gutenberg-Richter law [2] allows to set the scale of the tremors, determining the number of aftershocks that follow a strong land movement in a given region and over a period of time. This relationship is calculated as:

$$N = 10^{(a-b)M} \tag{1}$$

where:

- N is the number of events with $M \geq M0$.
- M is magnitude.
- a and b are constants.Typically, b is equal to 1 in seismic active regions and is equal to the rate of earthquakes in the region.

In the other hand, Omori law [11] expresses the decay of aftershocks in a time interval. This law is expressed as:

$$n(t) = \frac{K}{(c+t)} \tag{2}$$

where:

$n(t)$ is the probability that a seismic event shows up in time t.
c is the days after the main shock.
T is the time of the event.

Utsu [15] in 1961 proposed an amendment to this rule so that the probability of the aftershock decreases exponentially within time. As shown in the next equation.

$$n(t) = \frac{K}{(c+t)^p} \tag{3}$$

Both of these laws are used to determine independent and dependent seismic events.

3.1 Prediction Algorithms

Some predictive algorithms have been developed on the basis of statistical analysis of seismic events. The following are three of them.

3.2 Data Analysis Schema [6]

Strong earthquakes are identified under the $M \geq M_0$ condition. M_0 is the chosen threshold at which the average time between strong-motion is sufficiently long in the considered area. Each interval $(M_0, M_0 + 0.5)$ is analyzed independently.

Forecast tries to find the Time of Increased Probability (TIP) that is a time interval with a significant probability of occurrence.

A seismic region has overlapping areas whose size depends on M_0. In each area the sequence of earthquakes is analyzed. Their average characteristics are determined and are represented by functions defined in time. One or more functions can be used for predictions.

3.3 M8 Algorithm

The M8 algorithm [7, 8] was designed by retrospective analysis of the preceding seismicity of the great movements of the world ($M \geq 8$). This prediction scheme is implemented as follows:

- The territory to be considered is explored by overlapping circles of diameter equal to $D(M_0)$, where D is diameter of the event of M_0 magnitude.
- In each circle, the sequence of earthquakes is considered eliminating their replicas. The sequence is standardized by the lower magnitude $M_{min}(N)$, with N as a standard value.
- Time-dependent functions characterize the sequence that is calculated in a window of time (t, t) and a range of magnitudes $M_0 > M_i \geq M_{min}(N)$. These functions include:

  - $N(t)$, number of main shocks.
  - $L(t)$, standard deviation de $N(t)$ from the long-term trend.

$$L(t) = N(t) - [N_{cum}(t - s)] \cdot \frac{t - t_0}{t - t_0 - s} \qquad (4)$$

  - $N_{cum}(t)$ is the total number of main shocks with $M \geq M_{min}$ from the beginning of the sequence t_0 to t.
  - $Z(t)$ is the concentration of major shocks estimated as the average of the diameter of the source radio l to the average distance r between them.
  - $B(t) = \max\{bi\}$, the maximum number of aftershocks, seen in a time window.
- Each N, L y Z function is calculated for $N = 10$ and $N = 20$. As a result seven functions, including the B-function, describes the earthquake sequence.
- High values in functions are identified by a condition that exceed Q percentiles ($\geq Q$ %).
- An alarm is declared in the following 5 years, when at least six functions, including B, present values that exceed Q. The prediction becomes more stable for two moments, t and t + 05 (in years).

3.4 CN Algorithm

The NC algorithm was designed by retrospective analysis of the seismic activity greater than 6.5 in California and adjacent areas of Nevada. It is described as:

- The selected areas are chosen taking into account the spatial distribution of seismicity.
- Each area is chosen on the basis of at least three (3) significant earthquakes. Aftershocks are removed from this accounting.
- The sequence of tremors is described by nine (9) functions. Three of them are N, Z and B function similar to those described in the M8 algorithm. Others of them represent the proportion of higher magnitudes in the sequence, the different variations of this sequence in time and value average of the source areas, where a slide or rupture happens.
- Once recognized the pattern, the functions become coarsely discrete so that it can differentiate among large, medium and small values.
- TIP warnings are determined by a pair or third functions.

4 The Proposed Approach

We propose the use of an ensemble of type-2 fuzzy neural networks to estimate the time increased probability of seismic movements. Such an ensemble incorporates three FNNs, each dealing with a set of probabilities used by the three prediction algorithm described above: Statistical analysis, M8 algorithm and CN algorithm (see Fig. 6).

In the first phase of the ensemble, a pre-processing of seismic data is required to generate a set of data based upon seismic coordinates. Aftershocks are detected and eliminated from the output set X.

$$X = \{x_1, \ x_2, \ x_3, \ x_4, \ x_5, \ldots, x_n\} \tag{5}$$

where

- X is the output set of seismic coordinates.
- x_i is an seismic coordinate localized in the i region.

In turn, each of the x_i is a vector which contains the following data:

$$x_i = \{m, \ h, \ t, \ l, lg\} \tag{6}$$

where

- x_i is the seismic coordinates vector over i region.
- m is magnitude.
- h is depth.

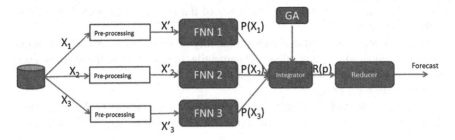

Fig. 6 Seismic hazard predictor ensemble

- t is the time of the event.
- l is latitude.
- lg is longitude.

To eliminate the occurrence of aftershocks in the dataset, the laws of Omori and Gutenberg-Richter are used for pre-processing samples that will yield the required data set by the algorithms. The new data set is called X'. Each occurrence of a seismic event is represented by an x'_i seismic coordinate.

Each of the Fuzzy Neural Networks (FNNs) performs the task of determining the likely responsibility of seismic risk in the same region of study. The output of the FNN is a fuzzy set of probability on the region.—$P(x_i)$-The probability of a seismic event in the region corresponds to the sum of the probabilities of each of the parties that divides the region.

This is a valid approach for the statistical analysis process, a widely used method, but not to the M8 and NC algorithms that overlaps areas of seismicity. For these algorithms, a radial basis function (RBF) can solve the interaction among the nodes.

Once obtained the fuzzy set of probabilities, these are integrated (Integrator block) to determine the fuzzy set of output probabilities—$R(p)$—in the region. A genetic algorithm is used to optimize the output of the integrator block.

The result of this integration is a growing set of probabilities in time, which is reduced (reducing block) through a Q percentile to yield a M_0 prediction.

A statistical method using Markov chain will be used to compare the results of the predictions.

5 Assumptions

We assume that an ensemble of three fuzzy neural networks will provide a better estimate of the probability of emergence of a seismic event. Each fuzzy neural network calculates the probability of events delivering a result set also probabilistic. The integration of the three results will give another set of probability of events which may be reduced on the basis of the magnitude expected in a time frame of 5 years.

The final result set is also a joint probability of events of magnitude M_0 with probability equal to or greater than a Q percentile. In other words, the result is a set of events with magnitude equal or greater than M_0 and a high probability of occurrence in the i-region.

A comparison with the statistical methods that gave origin to the predictive algorithms will allow evaluating the goodness of the ensemble [12, 19, 22].

6 Conclusions

There are no doubts that a seismic event will occur in those regions in which there are seismic and volcanic activities, however, to predict them, in time, location and magnitude, contributes to a better understanding of the phenomenon while allows to citizen and public authorities be better prepared to remedy their impact.

Neural networks and fuzzy systems have been in use to predict seismic events, however, we expect that an ensemble of FNNs will provide with more accurate predictions.

Thus, in order to prove the goodness of the approach proposed, several activities will be undertaken.

- A characterization of seismic activity of the region will be made. Seismic catalog of the Baja California Peninsula is not as accurate as the one from Southern California. Both catalogs are required in order to test the ensemble with real data.
- An implementation of the ensemble will be tested using benchmark data.
- An implementation of an ANFIS will be made to compare end results with the ensemble. BP algorithms are widely used in ANFIS implementation for seismology.
- Real data will feed both systems. We expect that the ensemble behave better than ANFIS.

After an in depth analysis of end results of the ensemble, a prediction will be made for the next 5 years.

References

1. Resnom, G.: Sismicidad de la regiónnorte de Baja California, registradaporresnom en el periodoenero-diciembre de 2002 (CICESE). Unión Geofísica Mexicana, (2002)
2. Gutenberg, B., Richter, C. F.: Seismicity of the earth and associated phenomena, 2nd edn. Princeton University Press, Princeton (1954)

3. Yang, J., Yu, P.S.: Mining asynchronous periodic patterns in time series data. IEEE Trans. Knowl. Data Eng. **15**(3), 613–628 (2003)
4. Kanamori, H.: Earthquake prediction: an overview. Int. Handb. Earthquake Eng. Seismolog. **616**, 1205–1216 (2003)
5. Karnik, N. N., Mendel, J. M.: Introduction to type-2 fuzzy logic systems. In: Proceedings of the 1998 IEEE FUZZ Conference, Anchorage, pp. 915–920 (1998)
6. Keilis-Borok, V.I.: Intermediate-term earthquake prediction (premonitory seismicity patterns/dynamics of seismicity/chaotic systems/instability). In: Proceedings of the National Academy of Sciences USA, vol. 93, pp. 3748–3755. Colloquium paper (1996)
7. Keilis-Borok, V.I., Kossobokov, V.G.: Phys. Earth Planet. Inter. **61**, 73–83 (1990)
8. Keilis-Borok, V.I.: The algorithm M8. Rusian academic of sciences. http://www.mitp.ru/en/m8pred.html(2009)
9. Main, I.: Earthquakes—long odds on prediction. Nature **385**, 19–20 (1997)
10. Monika, A. K.: Comparison of mamdani fuzzy model and neuro fuzzy model for load sensor. Int. J. Eng. Innovative. Technol. (IJEIT) **2**(9), (2013)
11. Omori, F.: On the aftershocks of earthquakes. J Coll. Sci. **7**, 111–200 (1894). Imperial University of Tokyo
12. Pulido, M., Melin, P., Castillo, O.: Optimization of type-2 fuzzy integration in ensemble neural networks for predicting the US Dolar/MX pesos time series. IEEE (978-1-4799-0348-1 2013) (2013)
13. Rumelhart, D.E., Hinton, G.E., Williams, R.J.: Learning representations by back-propagating errors. Nature **323**, 533–536 (1986)
14. Sepulveda, R., Castillo, O., Melin, P., Montiel, O.: An efficient computational method to implement type-2 fuzzy logic in control applications. Adv. Soft Comput. **41**, 45–52 (2007)
15. Sepulveda, R., Castillo, O., Melin, P., Rodriguez-Diaz, A., Montiel, O.: Experimental study of intelligent controllers under uncertainty using type-1 and type-2 fuzzy logic. Inf. Sci. **177**(10), 2023–2048 (2007)
16. Sepulveda, R., Montiel, O., Lizarraga, G., Castillo, O.: Modeling and simulation of the defuzzification stage of a type-2 fuzzy controller using the xilinx system generator and simulink. In: Castillo, O., et al. (eds.) Studies in Computational Intelligence, vol. 257, pp. 309–325. Springer, Heidelberg (2009)
17. Sepulveda, R., Montiel, O., Castillo, O., Melin, P.: Optimizing the MFs in type-2 fuzzy logic controllers, using the human evolutionary model. Int. Rev. Autom. Control **3**(1), 1–10 (2011)
18. Tsunekawa, H.: A fuzzy neural network prediction model of the principal motions of earthquakes based on preliminary tremors. IEEE (0-7803-4503-7) (1998)
19. Utsu, T.: A statistical study of the occurrence of aftershocks. Geophys. Mag. **30**, 521–605 (1961)
20. Zadeh, L.A.: Fuzzy sets. Inf. Control **8**(3), 338–353 (1965)
21. Zamani, A., Sorbi, M.R., Safavi, A.A.: Application of neural network and ANFIS model for earthquake occurrence in Iran. Earth Sci. Inform. **6**, 71–85 (2013). Springer, Heidelberg
22. Zhou, Z., Wu, J., Tang, W.: Ensembling neural networks: many could be better than all. Artif. Intell. **137**(1–2), 239–263 (2002)

Part II
Bio-Inspired Algorithms

Shipwrecked on Fear: Selection of Electives in School Minorities in a University Using Cuckoo Search Algorithm

**Alberto Ochoa-Zezzatti, Oscar Castillo, Patricia Melín,
Nemesio Castillo, Sandra Bustillos and Julio Arreola**

Abstract The purpose of this research is to understand from a Multivariable optimization related with four scholar minorities studies in a University with approximately 87 educational studies on Bachelor level, this sample is composed by: (Safety Sciences, Interior Design, Sports Training and Aeronautics) assuming that any student want to analyze the way in which these minority groups selected electives to complete the set of credits in their respective studies to determine the optimal selection which involve the choice of these materials in educational majority groups to determine the benefit-cost associated with the term professional studies, whose main base the restriction on a small number of subjects in their studies this because such a low minority enrollment, even though this problem has been studied repeatedly by many researchers on the literature have not been established optimal values by supporting bio-inspired algorithms to interact with the different values associated with the achievement of the term loans and the cost-benefit every student to a minority group and comparing their choices of electives with respect the group. There are several factors that can influence the selection of an elective, for our research we propose to use a new bio-inspired algorithm called "Cuckoo search algorithm," which has proven effective for the cohesion of behavior associated with several problems, and when and use restrictions have strategies to keep tempo in the selection of these materials, in our case, a resource such as time gain regarding the subjects studied is represented as the optimal way for the duration of the professional studies with uncertainty not know how long it can last set appropriate conditions for the selection of specialized subjects.

A. Ochoa-Zezzatti (✉) · N. Castillo · S. Bustillos · J. Arreola
Juarez City University, Juarez, México
e-mail: alberto.ochoa@uacj.mx

O. Castillo · P. Melín
Technology Institute of Tijuana, Tijuana, México

O. Castillo et al. (eds.), *Recent Advances on Hybrid Approaches for Designing Intelligent Systems*, Studies in Computational Intelligence 547,
DOI: 10.1007/978-3-319-05170-3_10, © Springer International Publishing Switzerland 2014

1 Introduction

The selection of a specific course is a most important decision on the life student when the people arrive at University; the correct choosing of a course is determinate by: (a) Selecting the best possible course for you is a key decision to make, and often a challenging one.

- There can be intense competition for the most popular courses, and later for graduate jobs.
- Do you want to follow a course offering a clear career path or does your interest lie in one of the more traditional academic subjects?
- There are tens of thousands of courses, and the same title of course will not be taught in the same way, or cover the same material, at every university offering it.

So how do you go about choosing a course? (b) For a few people, choosing a course is simple: they have always wanted to be a brain surgeon or have always had a passion for geography, mountains and glaciers.

- For most, however, there is a bewildering variety of courses, many of which involve subjects that are not taught in schools or colleges.
- Somehow you have to narrow down the thousands of courses to just a few. What should you take into account?

It may seem daunting at first but if you discount courses you do not want to study, Universities you do not want to go to (too far away/close to home, entry requirements out of reach for example) suddenly the field narrows significantly on the University website.

1.1 The Concept of Minority Educational Group

In many universities in Latin America, due to educational learning conditions and technical expertise are generated careers associated with cyclical changes of various paradigms of knowledge, this has generated on average for most universities, at least one of these particular sciences has an enrollment prowling below 200 students, this means offering a syllabus of compulsory subjects and one additional group of electives, which by the small number of students almost always generate excessive costs to deliver within 5–8 class students so that most of these students end up in minority school groups electives other majority races so the graduate profile is far from the original idea, this research aims to describe the implementation of the algorithm using cuckoo search algorithm as you can model the behavior of students from these minority school groups and optimized search form matters require to complete their academic credits, for our research we compare between this proposed bio-inspired algorithm and models of predator–prey game being better bioinspired algorithm, as the proposed in [1].

2 Project Development

To do this research project was developed by dividing into three sections which are modules of application development, implementation of the server and the intelligent module associated with Cuckoo Search Algorithm and Data Mining. Android is the operating system that is growing into Streak 5 from Dell, for this reason we select this mobile dispositive along with other manufacturers are propelling the Latin American landing on Android with inexpensive equipment, and on the other hand, some complain about the fragmentation of the platform due to the different versions. Android is free software, so any developer can download the SDK (development kit) that contains your API [2]. This research tries to improve group travel related with selection of specific courses by Scholar minorities in a large University.

2.1 Cuckoo Search Algorithm

To describe the CS more clearly, the breed behavior of certain cuckoo species is briefly reviewed.

Cuckoo Breeding Behavior

The CS was inspired by the obligate brood parasitism of some cuckoo species by laying their eggs in the nests of host birds. Some cuckoos have evolved in such a way that female parasitic cuckoos can imitate the colors and patterns of the eggs of a few chosen host species. This reduces the probability of the eggs being abandoned and, therefore, increases their re-productivity as in [3]. It is worth mentioning that several host birds engage direct conflict with intruding cuckoos. In this case, if host birds discover the eggs are not their own, they will either throw them away or simply abandon their nests and build new ones, elsewhere. Parasitic cuckoos often choose a nest where the host bird just laid its own eggs. In general, the cuckoo eggs hatch slightly earlier than their host eggs. Once the first cuckoo chick is hatched, his first instinct action is to evict the host eggs by blindly propelling the eggs out of the nest. This action results in increasing the cuckoo chick's share of food provided by its host bird. Moreover, studies show that a cuckoo chick can imitate the call of host chicks to gain access to more feeding opportunity. The CS models such breeding behavior and, thus, can be applied to various optimization problems. In [3], discovered that the performance of the CS can be improved by using Lévy Flights instead of simple random walk.

Lévy Flights

In nature, animals search for food in a random or quasi random manner. Generally, the foraging path of an animal is effectively a random walk because the next move is based on both the current location/state and the transition probability to the next location. The chosen direction implicitly depends on a probability, which can be modeled mathematically. Various studies have shown that the flight behavior of many animals and insects demonstrates the typical characteristics of Lévy flights. A Lévy flight is a random walk in which the step-lengths are distributed according to a heavy-tailed probability distribution. After a large number of steps, the distance from the origin of the random walk tends to a stable distribution.

Cuckoo Search Implementation

Each egg in a nest represents a solution, and a cuckoo egg represents a new solution. The aim is to employ the new and potentially better solutions (cuckoos) to replace not-so-good solutions in the nests. In the simplest form, each nest has one egg. The algorithm can be extended to more complicated cases in which each nest has multiple eggs representing a set of solutions [3]. The Cuckoo Search is based on three idealized rules:

- Each cuckoo lays one egg at a time, and dumps it in a randomly chosen nest;
- The best nests with high quality of eggs (solutions) will carry over to the next generations;
- The number of available host nests is fixed, and a host can discover an alien egg with probability $p_a \in [0,1]$. In this case, the host bird can either throw the egg away or abandon the nest to build a completely new nest in a new location [3].

For simplicity, the last assumption can be approximated by a fraction a p of the n nests being replaced by new nests, having new random solutions. For a maximization problem, the quality or fitness of a solution can simply be proportional to the objective function. Other forms of fitness can be defined in a similar way to the fitness function in genetic algorithms [3]. Based on the above-mentioned rules, the basic steps of the CS can be summarized as the pseudo code, as [3]:

```
begin
    Objective function f(x), x = (x₁,...,xd)ᵀ
    Generate initial population of
        n host nests xi (i = 1,2,...n)
    While (t < MaxGeneration) or (stop criterion)
        Get a Cuckoo randomly by Lévy Flights
        evaluate its quality/fitness Fi
        Choose a nest among n (say.j) randomly
        if (Fi > Fj),
            replace j by the new solution;
        endif
```

A fraction (p_2) of worse nests
    are abandoned and new ones are built;
Keep the best solutions
    (or nests with quality solutions)
    Rank the solutions and find the current best
endwhile
Postprocess results and visualization
end

When generating new solutions $x_i(t + 1)$ for the ith cuckoo, the following Lévy flight is performed

$$X_i(t + 1) = x_i(t) + \alpha \oplus \text{Lévy}(\lambda) \tag{1}$$

where $\alpha > 0$ is the step size, which should be related to the scale of the problem of interest. The product $\oplus$ means entry-wise multiplications [3]. In this research work, we consider a Lévy flight in which the step-lengths are distributed according to the following probability distribution

$$\text{Lévy } u = t^{-\lambda}, 1 < \lambda \leq 3 \tag{2}$$

which has an infinite variance. Here, the consecutive jumps/steps of a cuckoo essentially form a random walk process which obeys a power-law step-length distribution with a heavy tail. It is worth pointing out that, in the real world, if a cuckoo's egg is very similar to a host's eggs, then this cuckoo's egg is less likely to be discovered, thus the fitness should be related to the difference in solutions. Therefore, it is a good idea to do a random walk in a biased way with some random step sizes as in [4].

2.1.1 Improved Cuckoo Search

The parameters pa, λ and α introduced in the CS help the algorithm to find globally and locally improved solutions, respectively. The parameters pa and α are very important parameters in fine-tuning of solution vectors, and can be potentially used in adjusting convergence rate of algorithm. The traditional CS algorithm uses fixed value for both pa and α. These values are set in the initialization step and cannot be changed during new generations. The main drawback of this method appears in the number of iterations to find an optimal solution. If the value of pa is small and the value of α is large, the performance of the algorithm will be poor and leads to considerable increase in number of iterations. If the value of pa is large and the value of α is small, the speed of convergence is high but it may be unable to find the best solutions. The key difference between the ICS and CS is in the way of adjusting pa and α. To improve the performance of the CS algorithm and eliminate the drawbacks lies with fixed values of a and α, the ICS algorithm uses variables pa and α. In the early generations, the values of pa and α must be big enough to

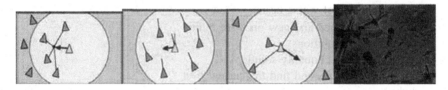

Fig. 1 Representation of cohesion, alignment, separation and escape

enforce the algorithm to increase the diversity of solution vectors. However, these values should be decreased in final generations to result in a better fine-tuning of solution vectors. The values of pa and α are dynamically changed with the number of generation and expressed in Eqs. 3–5, where NI and gn are the number of total iterations and the current iteration, respectively.

$$P_a(gn) = P_{a\,max} - gn/NI(P_{a\,max} - P_{a\,min}) \tag{3}$$

$$\alpha(gn) = \alpha_{max}\exp(c \cdot gn) \tag{4}$$

$$c = (1/NI)\,Ln(\alpha_{min}/\alpha_{max}) \tag{5}$$

3 Implementation of an Intelligent Application

3.1 Group Behavior

The group behavior (Flock Behavior) in our system is based on Cuckoo Search Algorithm, although we have expanded with an additional rule (escape), and most importantly, it has affected the clustering behavior using parameters associated with the values of the bird sensations. The escape rule influences the behavior of each cuckoo to flee the potential danger (mainly predators) in your environment. Therefore, in our model the movements are given by the result of four vector components (see Fig. 1), one for each of the behavior rules:

Cohesion. Attempt to stay near flockmates next.
Alignment. Attempt to match velocity with respect to flockmates next.
Separation. Avoid collisions with flockmates next.
Escape. Run away from potential hazards (e.g. predators).

3.1.1 Implementation

The Implementation of the architecture is of three layers: the agent of the brain, the model of the real world and the environment, the brains of the issues are independent Processes running on a workstation programmed on each agent brain

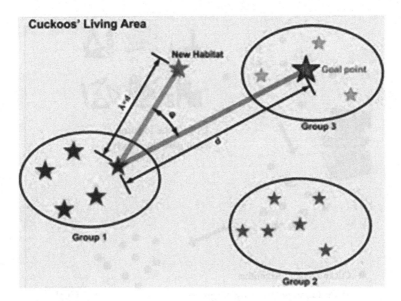

Fig. 2 Conceptual diagram associated with cuckoo search algorithm and its implementation to our research

JBuilder Receives sensory information via the network sockets and sends (via network) the selected action to model the world, container containing the bodies of the agents and the simulation environment. Changes to the model are reflected for each frame of a response scenario trying to achieve the proposed solution. This implementation allows for modularity and extensibility to add and modify the behavior of the bird group. The behavior of the demand for different scenarios became a prototype using Java and support ideas based on Cuckoo Search Algorithm as in Fig. 2.

3.1.2 Experiments and Results

Tests were developed using various designs of environments (search space) in order to test the efficacy of proposed bioinspired algorithm, it was used for different tests to be performed to see the performance of our proposed algorithm. In an effort to give realism to our herd chose a benchamark related with Cuckoo Search Algorithm proposed, for which two experiments were devised in order to assess the conduct of the issues within the environment and contributing to the overall experience. Then we show the results obtained and analyzed in the same way as in [5]. When compare the behavior of group meeting regarding the escape behavior, the latter seems to add a degree of realism perceived through the issues. This is a significant point because the escape behavior is a reaction to something other than other bird species in the environment, and is represented by a marked

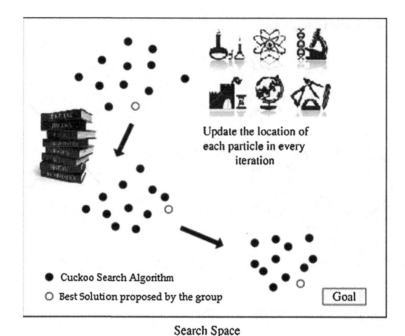

Search Space

Fig. 3 Graphical solution proposed by cuckoo search algorithm related with the selection of specific subjects

change in behavior. Thus, one can say that the escape behavior adds another perceived behavior with our algorithm, rather than refine the existing one if the issues are imbued with both behaviors make the environment more truthful and contribute to realism. It is interesting to observe how behaviors are complementary and are most significant when they occur separately, as the proposed and validated as in [6]. Besides the general effect is greater when combined behaviors escape behavior is refined by the behavior of the excitement, "fear." This result raises a question: why some behaviors refining improves perceived realism, while others do not?

One reason might affect behavior is considered more synthetic meeting. While, adding an emotional component to this escape behavior streamlines and refines the whole grouped behavior, giving them a greater sense of autonomy. For the preparation of graphics, we propose use a class supported with Cuckoo Search Algorithm supported which facilitates to manipulation of data to express visually using different types of graphs, as in the Fig. 3 in where, each iteration proposed is related with a generation. To the 6th Iteration a specific behavior is described to the most adequate optimal subjects where possible take is these by the most of students in the Scholar minority as the proposed in [7] and validated with this research.

To implement the application is installed in operating system devices with Android 2.2 or higher, which tests the system in different scholar minorities of four

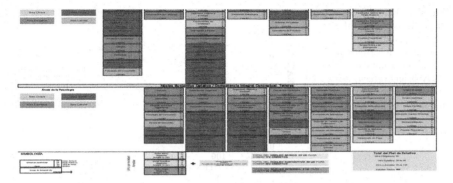

Fig. 4 Hybrid intelligent application based on cuckoo search algorithm and data mining, *orange box* represents elective subjects in a scholar minority

different institutes in a large University in Chihuahua based on the previously research related with Cultural Algorithms on Urban Transport [8], by answering a questionnaire of seven questions to all users related with the use of our intelligent application have elapsed since installing the application, the questions are to raise awareness of the performance, functionality and usability of the system, the demonstrate use of this application is shown in Fig. 4. To understand in an adequate way the functionality of this Intelligent Tool, we proposed evaluate our hybrid approach and compare with only data mining analysis and random select activities to select specific subject in the University, we analyze this information based on the unit named "époques" used in Cuckoo search algorithm, which is a variable time to determine if exist a change in the proposed solution according at different situation of different selection of subjects with better use of restricted resources.

We consider different scenarios to analyze during different time and selection of specific subjects, as is possible see in the Fig. 5, and apply a questionnaire to a sample of users to decide search a specific number of subjects to take class, when the users receive information of another past selection of subjects (Data Mining Analysis)—as in [8] try to improve their space of solution but when we send solutions amalgam our proposal with Cuckoo Search Algorithm and Data Mining was possible determine solutions to improve the solution of scholar group, the use of our proposal solution improve in 77.72 % against a randomly action and 14.56 % against only use Data Mining analysis the possibilities of recommend to select best options in a specific time to take class with specific skills, these selections and their justifications named "narrative script" which permit in the future decrease the possibility of successful selection of subjects in a less time and spend academic resources related with different scholar groups and uncertainty of the adverse academic conditions and the use of limited knowledge resources.

Fig. 5 Solutions proposed to academic problem: (*blue*) hybrid approach using Cuckoo Search Algorithm; (*red*) only using data mining analysis and (*green*) using randomly actions to improve the safety of the users

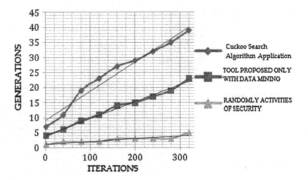

4 Conclusions and Future Work

This work shows that it is feasible to provide jointly based architecture flocking rules more complex architectures of autonomous agents. The results suggest that characterize a search space with an animal group, in our case with vivid features of Cuckoo search algorithm showing enhances the experience, particularly if the animals are concentrated in some autonomous activity. Studies show that the behaviors exhibited group in isolation may have an ambiguous effect on the observer's impression of realism about the environment, but when combined with other behaviors, which react directly to the environment, the impact on the user's pretty improved, the use of pheromones achieves this improvement.

The authors are confident that it is possible to describe population patterns constructed in this proposed bio-inspired algorithm, taking as a basis the interaction of autonomous agents with the environment. Further research is necessary to define these guidelines closely. However, initial studies reported involve interesting relationships between the types of behavior and strength of experience to reach an optimal solution. Future work will be performed to characterize more precisely the changes in the behavior of meeting obtained by this extended architecture. The transitions between the reaction be feeding and predators will be analyzed carefully. The authors plan to validate this work modeling a different kind of problem associated with minorities bird, such as a Po'ouli, respond to predators in a circle and trying to divert the attention of the predator towards defenseless individuals. We believe that our architecture can support extended broadly and more realistic behavior of a flock of this type and, therefore, represents a credible contribution to animals based on bioinspired perspective generated in this algorithm. With the use of this innovative application combine Cuckoo Search Algorithm and Data Mining associated with decisions make for many students in the past is possible determine the adequate subjects where is possible obtain improved knowledge, and share this to others students by an alert sent to a mobile device with GPS, providing statistical information through a Web server that returns the level of successful related with a kind of subjects consulted by the users as in [6], a specific factor decisive in many Universities with scholar minorities in

Mexico. The future research will be to improve the visual representation of this scholar minority group and their interaction inside of a social networking to this we proposed an Intelligent Dyoram—An Intelligent display in 3D used on Social Networking—with real on time information of each one of their integrants and establishment a reactive plan to select subjects including save money related with credits by subject, time in scholar schedules and enjoy with the rest of Scholar minority. The most important contribution is prevent spend time and money in this kind of selection of subjects because is very complicated select the best options related with new subjects to complete fundamental subjects in their Scholar studies, our future research is adequate the information to actualize from the central server of scholar services of the University, to the users, considering that the number of people requires selection of subjects are minorities but are important to the rest of University, because the selection of subjects is very high important activity, this innovative application is possible to use in another Latin American Universities with similar conditions of studies by their respective Scholar minorities, this Intelligent Tool will be used by different kind of people whom requires study together in minority academic groups. In addition this application will be used as Recommender System when travel to another Universities in different societies which is possible take specific courses as the proposed in [9] and explain different scenarios according time, limited resources and location. Another field topic will be benefited with more adequate organization is Logistics of product or service as in [10] which describes the use of Cultural Algorithms to improve a Logistics networking associated with the deliveries of a bottle product. The use of different perspectives of this research will support new implementations as in [11, 12] which improve our daily lives.

Acknowledgments The authors were supported with funds from Social Research Center in Juarez City University and used data from Scholar Services associated with four Academic minorities in UACJ at Chihuahua State whom permits compare the simulation with real selection of subjects realized by these minorities.

References

1. Barbucha, Dariusz: Experimental study of the population parameters settings in cooperative multi-agent system solving instances of the VRP. Trans. Comput. Collective Intell. **9**, 1–28 (2013)
2. Andreu R.A.: Estudio del desarrollo de aplicaciones RA para Android. Trabajo de fin de Carrera. Catalunya, España (2011)
3. Schnitzler, H.-U., Kalko, E.K.V.: Echolocation by insect-eating bats. Bioscience **51**(7), 557–569 (2001)
4. Flake, G.: The Computational Beauty of Nature: Computer Explorations of Fractals, Chaos, Complex Systems, and Adaptation. MIT Press, Cambridge (1999)
5. Kitchell, A., et al.: 10 smell myths: Uncovering the sinkin'truth. Technical report, Department of Biological Sciences, University of South Carolina (1995)
6. Mendoza, R.: Involving cuckoo birds from Kandor in a Karumi representation. In: Karumi Handbook (1995)

 7. Grammer, K.: 5-alpha-androst-16en-3-alpha-one: a male pheromone? Ethol. Sociobiol. **14**(3), 201–207 (1993)
 8. Reyes, L.C., Zezzatti, O.A., et al.: A Cultural Algorithm for the Urban Public Transportation. In: HAIS, pp. 135–142 (2010)
 9. Minsky, M.: The society of mind. Simon & Schuster, New York (1985)
10. Reynolds, C.: Flocks, herds, and schools: a distributed behavioral model. Comput. Graph. **21**(4), 25–34 (1987)
11. Rosenblatt, J.K., et al.: A fine-grained alternative to the subsumption architecture for mobile robot control. In: Proceedings of the IEEE/INNS International Joint Conference on Neural Networks (Washington DC), vol. 2, pp. 317–324, June 1989
12. Velásquez J.: Modelling emotions and other motivations in synthetic agents. In: Proceedings of the Fourteenth National Conference on Artificial Intelligence (AAAI-97) (Providence, RI), MIT/AAAI Press (1997)

An Admission Control and Channel Allocation Algorithm Based on Particle Swarm Optimization for Cognitive Cellular Networks

Anabel Martínez-Vargas, Ángel G. Andrade, Roberto Sepúlveda
and Oscar Montiel-Ross

Abstract During the last few years cellular networks have increased the use of spectrum resources due to the success of mobile broadband services. Mobile devices generate more data than ever before, facing the way cellular networks are deployed today to meet the ever increasing traffic demand. Making new exclusive spectrum available to meet traffic demand is challenging since spectrum resources are finite, therefore costly. Cognitive radio technology is proposed, for conventional cellular networks, as a solution to enlarge the pool of available spectrum resources for mobile users through femtocells (small cells), overlaid on the existing macrocell network, to share a common spectrum. However, by reusing simultaneously spectrum resources from femto networks, potentially destructive interference on macro networks is introduced. In this context, we present a femto-users admission control and channel allocation algorithm based on Particle Swarm Optimization (PSO) to maintain the interference to a required Quality of Service (QoS) level in macro-femto networks while, at the same time, data rate is maximized in the whole system. The proposed approach provides design requirements for deploying future cognitive cellular networks.

A. Martínez-Vargas · R. Sepúlveda · O. Montiel-Ross (✉)
Instituto Politécnico Nacional, Centro de Investigación y Desarrollo de Tecnología Digital
(CITEDI-IPN), Av. Del Parque No. 1310, Mesa de Otay, 22510 Tijuana, Baja California,
México
e-mail: oross@ipn.mx

A. Martínez-Vargas
e-mail: amartinez@citedi.mx

R. Sepúlveda
e-mail: rsepulvedac@ipn.mx

Á. G. Andrade
Autonomous University of Baja California (UABC), Mexicali, Baja California, México
e-mail: aandrade@uabc.edu.mx

O. Castillo et al. (eds.), *Recent Advances on Hybrid Approaches for Designing*   151
Intelligent Systems, Studies in Computational Intelligence 547,
DOI: 10.1007/978-3-319-05170-3_11, © Springer International Publishing Switzerland 2014

1 Introduction

Architecture and topology of cellular networks are undergoing dramatic changes in recent years due to the intense consumer demand for mobile data. In 2010 the amount of global mobile data traffic exceeded the traffic on the entire global internet in 2000 and by 2015 is expected that nearly 1 billion people access the internet exclusively through a mobile wireless device [1]. Making new exclusive spectrum available [for example, for Fourth Generation (4G) cellular network] to meet traffic demand is challenging since spectrum resources are finite, therefore costly. To improve system capacity and meet the increasing growing data demand, more and efficient communications techniques, such as Multiple-Input Multiple-Output (MIMO) and smart antennas, have been applied, working with scheduling schemes designed for resource allocation, such as Orthogonal Frequency-Division Multiple Access (OFDMA) systems. However, these efforts cannot fully solve the bandwidth shortage [2].

Cognitive Radio (CR) technology in conventional cellular networks is proposed as a solution to enlarge the pool of available spectrum resources for mobile users through femtocells (small cells) overlaid on the existing macrocell network, to share a common spectrum. The small cell solution provides localized network capacity and coverage improvement in jam packed areas and big events where people converge, or where wireless signal normally gets congested. However, under such heterogeneous network deployment, reusing resources or spectrum sharing generate potentially destructive interference in macro and femto networks.

Femto Base Stations (femto-BSs) are small, inexpensive, and low-power. They operate in the same frequency bands as the macrocells, and end users can deploy and manage them. Unlike macro Base Stations (macro-BSs), which are installed according to detail network planning, operators only have limited control over femto-BSs, which makes challenging to mitigate the co-channel interference and manage the network resources. The unplanned positions of femto-BSs lead to two kinds of interference between macro and femto cellular networks: cross-tier (the aggressor and the victim of interference belong to different tiers) and intra-tier (the aggressor and the victim of interference belong to the same tier) [3]. Denser, more crowded spectrum will likely make these situations more common. CR technology can address those issues, providing to femto-BS capabilities to sense the environment, interpret the received signaling from the macro-BS and the surrounding femto-BSs, and intelligently allocate spectrum resources while cross-tier and intra-tier interference are controlled. The set of techniques to share and allocate spectrum resources through CR technology is known as Dynamic Spectrum Access. It has a layered network structure with prioritized spectrum access, where the users with low priority are known as secondary users (femtocells) whereas the members of the prioritized user group are named primary users (macrocell) [2]. In order to access to the cellular band, a secondary user performs one of the following dynamic spectrum access techniques: transmit simultaneously with the primary user as long as the resulting interference is constrained (spectrum underlay), or exploit an unused channel of primary user (spectrum overlay) [3].

Interference is a phenomenon that cannot be eliminated in a wireless system, however it can be controlled, and in this sense the regulators have established certain policies for the use of spectrum defining operation rules in devices to access a frequency band. Basically, small cells allow for flexible base station deployment and simple transceiver design due to the limited communications coverage. As the radio environment becomes more and more complicated, using small cells is beneficial for operators to deal with localized coverage and link enhancement while offloading traffic from the macrocells to femtocells. To identify deployment parameters for future cognitive cellular networks helps in the development of regulatory policies that prevent the generation of harmful interference between wireless systems that interact with each other, but mainly to protect primary users. In this chapter, we propose an admission control and channel allocation algorithm based on PSO [4] for cognitive cellular networks using the spectrum underlay access strategy. We aim at maximizing the sum throughput for the number of secondary users that can be admitted in the cognitive cellular network in the presence of primary users with assured QoS for both macro and femto users.

Some works have recently addressed the spectrum sharing problem between macrocells and femtocells. For example, in [3], different spectrum sharing schemes in two-tier cellular network are evaluated to maximize the density of active femto-BSs under QoS constraints. Results conclude that spatial reuse gain is achieved as long as a controlled underlay scheme is performed by the femto-BSs. On the other hand, in [2], a spectrum sharing strategy based on game-theory is proposed. To mitigate interference between macrocell and femto users, a power control approach is employed. However, additional computation time is introduced in the randomized silencing policy in femtocells. In [5], femtocells exploit unoccupied radio resources of the macrocell applying a game-theoretical radio resource management. It achieves the autonomous cross-tier and intra-tier interference mitigation, but, channel sensing introduces an overhead since data transmission and reception cannot be performed within a sensing frame. Work in [6] presents an admission control algorithm to manage interference in two-tier Code-Division Multiple Access (CDMA) cellular network. QoS is provided in the two-tier CDMA network, however, when the network becomes congested, the admission control algorithm converges slowly down.

The rest of the chapter is organized as follows. Section 2 provides the problem formulation. In Sect. 3, the proposed procedure based on PSO is described. Section 4 is devoted to evaluate performance of the proposed algorithm and Sect. 5 concludes the chapter.

2 Problem Formulation

As mentioned before, in CR networks, the coexistence problem is formulated between two independent user groups, primary (macro) and secondary (femto) users. They coexist under a predefined spectrum sharing method that specifies the

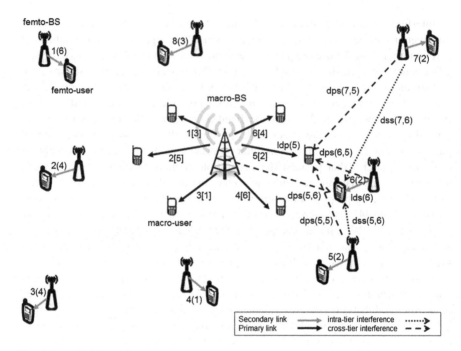

Fig. 1 Downlink scenario

priority in spectrum access and conflict resolution. In terms of femto-macro interference, the downlink is more critical since it is more likely that a mobile subscriber of the macro-BS suffers from downlink interference from nearby femto-BS than a macro-BS suffers from uplink interference from a mobile subscriber of the femto-BS due to lack of equality in cell-size and transmit powers [7]. Under this basis, a downlink scenario as shown in Fig. 1 is considered. The macrocell consists of multiple macro-users and a macro-BS located at the center of a coverage area A. The macrocell is overlaid with femtocells which are randomly distributed on A. A secondary link is represented by the union of a transmitter (femto-BS) and a receiver (femto-user) and it is identified by a number beside the link. Similarly, the union of a transmitter (macro-BS) and a receiver (macro-user) is referred as a primary link. The number of primary links is denoted as *Pl*, in contrast, the number of secondary links is referred as *Sl*. A primary link has a channel to share (the numbers in brackets in Fig. 1) and it can be assigned to several secondary links (the number in braces in Fig. 1), as long as they, together, do not disrupt communication in the primary link. Additionally, the proposed approach assures successful communications for those secondary links that attempt to exploit concurrently a channel with a primary link, that is, a certain QoS is also guaranteed for them. Macro-BS and femto-BSs can transmit at any given channel at full power; therefore, transmission power is maintained constant.

A successful reception of a transmission at a primary link depends on whether the signal-to-interference-plus-noise ratio (SINR) observed by macro-user is larger than an SINR threshold (denoted by β). The SINR at the receiver of primary link v is given by:

$$SINR_v = \frac{P_v/ldp(v)^n}{\sum_{k \in \Phi} P_k/dps(k,v)^n}, 1 \leq v \leq Pl \tag{1}$$

where P_v is the transmit power of primary link v, $ldp(v)$ is the link distance of primary link v, n is the path loss exponent (a value between 2 and 4 can be selected), those parameters characterized the desired signal. On the other hand P_k is the transmit power of secondary link k, $dps(k,v)$ is the distance from transmitter in secondary link k to receiver in primary link v. k is the index of active secondary transmitters. Φ is the set of active secondary transmitters. The aforementioned refers to the aggregated cross-tier interference, that is, the total interference from those secondary links using the same channel that primary link being analyzed. In Fig. 1, $SINR_v$ is computed in primary link 5.

Similarly, the SINR at the receiver of secondary link u is given by:

$$SINR_u = \frac{P_u/lds(u)^n}{\sum_{k \in \Phi} P_k/dss(k,u)^n + P_v/dps(v,u)^n}, 1 \leq u \leq Sl \tag{2}$$

where P_u is the transmit power of secondary link u and $lds(u)$ is the link distance of secondary link u. Meanwhile, P_k is the transmit power of transmitter of secondary link k, $dss(k, u)$ is the distance from transmitter of secondary link k to receiver of secondary link u. The above represents the aggregate intra-tier interference. In contrast, P_v is the transmit power of primary link v, $dps(v, u)$ is the distance from transmitter of primary link v to receiver of secondary link u. That refers the cross-tier interference perceived by a receiver of secondary link u. α represents SINR threshold for secondary links. Figure 1 shows $SINR_u$ computed in secondary link 6.

Data rate contributions of the secondary and primary links are derived from Eqs. (3) and (4), respectively. The data rate depends on channel bandwidth B that secondary and primary links can share and the conditions of the propagation environment (attenuation and interference).

$$c'_u = B \log_2(1 + SINR_u) \tag{3}$$

$$c''_v = B \log_2(1 + SINR_v) \tag{4}$$

The metric considered as a measure of spectral efficiency on the cognitive cellular network is data rate, therefore the objective of resource allocation is to find the maximum data rate of system (5) subject to the SINR requirements of the secondary links (6) and primary links (7), that is:

$$Max \sum_{u=1}^{Sl} c'_u x_u + \sum_{v=1}^{Pl} c''_v \tag{5}$$

$$SINR_u \geq \alpha \tag{6}$$

$$SINR_v \geq \beta \tag{7}$$

$$c'_u > 0, \, u = 1, 2, \ldots, Sl \tag{8}$$

$$c''_v > 0, \, v = 1, 2, \ldots, Pl \tag{9}$$

$$c'_u, \, c''_v \in R^+ \tag{10}$$

$$x_u = \begin{cases} 1, & \text{if } SINR_u \geq \alpha \text{ and } SINR_v \geq \beta \\ 0, & \text{otherwise} \end{cases} \tag{11}$$

where $x_u = 1$ if secondary link u is included in the solution and $x_u = 0$ if it remains out, as indicated in (11).

3 Admission Control and Channel Allocation Algorithm Based on PSO

PSO is a swarm intelligence technique inspired by social sharing of information among birds and fishes [4, 8]. This information exchange mechanism in a population can provide the advantage of finding the best solutions [9]. PSO-based algorithms have shown being efficient, robust and simple to implement thus they have been used in a wide range of applications such as supply chain management problems [10], electromagnetics [11], traffic management [12] and power systems [13].

In PSO, an individual is called *particle* and the number of particles is referred as the *swarm*. Each particle follows two very simple behaviors: the best performing individual *gbest* and the best conditions found by the individual itself *pbest*. This leads to particles converges on one solution [14, 15].

The ith particle of the swarm S can be represented by a D-dimensional vector, $X_i = (x_{i1}, x_{i2}, \ldots, x_{iD})$, $x_{id} \in \{0, 1\}$. The velocity of this particle is represented by another D-dimensional vector $V_i = (v_{i1}, v_{i2}, \ldots, v_{iD})$, $v_{ij} \in [-V_{max}, V_{max}]$ where V_{max} is the maximum velocity. The best previously visited position of the ith particle is denoted as $P_i = (p_{i1}, p_{i2}, \ldots, p_{iD})$, $p_{id} \in \{0, 1\}$. In the admission process given below, g is referred as the index of the best particle in the swarm. For purposes of this work, we add to the basic PSO, a vector P'_i which stores the best channel allocation for secondary links find so far for the ith particle, a vector X'_i which represents a candidate channel allocation for secondary links for the ith particle, and a vector *Spectrum Status* which has the channel allocation for primary links. Figure 2 resumes the above and shows the representation (definition of particles) from the original problem context to PSO.

To find the set of secondary links that can maximize the data rate of cognitive cellular network while the minimum SINRs thresholds of all users can be

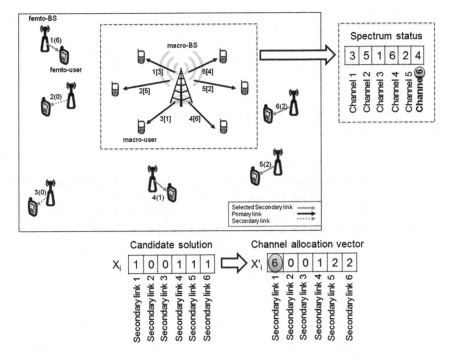

Fig. 2 Representation from the original problem context to PSO

supported, a systematic procedure based on Binary Particle Swarm Optimization (BPSO) [16] is used, in particular, the improved BPSO called Socio-Cognitive Particle Swarm Optimization (SCPSO) [17]. The procedure of the admission control and channel allocation is proposed as follows:

Input: The number of secondary links Sl, the number of primary links Pl, SINR thresholds $\alpha = \beta$, Swarm size S, the number of iterations T_{max}, the number of runs, and maximum velocity V_{max}.

Output: Maximum data rate in the system $f(P_g)$, the set of selected secondary links P_g, channel allocation for primary links *vector Spectrum Status*, the best channel allocation for secondary links P'_g, SINR level at primary links, and SINR level at secondary links.

Initialization stage

    Locate randomly Sl and Pl over the coverage area A

    Initialize randomly candidate solution vector X_i, where $x_{id} \in \{0,1\}$

    Initialize randomly velocity vector V_i, where $v_{id} \in [-V_{max}, V_{max}]$

    Set $P_i = X_i$

    Set $P'_i = X'_i$

    Initialize randomly vector *Spectrum Status* with values from Pl

End Initialization stage

Loop

    For i = 1 to number of particles

        Perform mapping between X'_i and *Spectrum Status* vectors to calculate SINR levels at secondary and primary links.

        Perform mapping between $P'i$ and *Spectrum Status* vectors to calculate SINR levels at secondary and primary links.

        Calculate the fitness, $f(X'_i)$ and $f(P'_i)$.

        If $f(X'_i) > f(P'_i)$ then do

            For d= 1 to D

$$p_{id} = x_{id}$$
$$p'_{id} = x'_{id}$$

        Next d

    End do

    $g = i$

    For j = 1 to number of particles

        Perform mapping between P'_j and *Spectrum Status* vectors to calculate SINR levels at secondary and primary links.

        Calculate the fitness, $f(P'_j)$ and $f(P'_g)$.

            If $f(P'_j) > f(P'_g)$ then $g = j$

    Next j

    For d = 1 to D

$$v_{id} = w \times v_{id} + c_1 r_1 (p_{id} - x_{id}) + c_2 r_2 (p_{gd} - x_{id})$$
$$v_{id} = w' \times v_{id} + c_3 \, (gbest - pbest)$$
$$x_{id} = x_{id} + v_{id}$$
$$x_{id} = x_{id} \; mod \, (2)$$

        if $x_{id} = 1$ then allocate randomly a new channel to x'_{id}.

    Next d

  Next i

  Until stopping criterion met

Return $(P_g, f(P_g))$

When a downlink scenario is analyzed over the procedure of the admission control and channel allocation based on PSO, a candidate solution X_i specifies the possible set of secondary links than may coexist with the primary links. From example illustrated in Fig. 2, particle X_i specifies that secondary links 1, 4, 5, 6 are chosen as a part of the solution; therefore, X'_i contains the channel allocation for those selected secondary links. The mapping of X'_i and *Spectrum Status* provides a possible allocation for secondary and primary links to share a channel. From Fig. 2, secondary link 1 is assigned to exploit channel 6, by verifying *Spectrum Status* it can be observed that channel 6 is occupied by primary link 4, so channel 6 can be exploited simultaneously by primary link 4 and secondary link 1 as long as (6) and (7) are satisfied. In a similar way, the aforementioned is applied to the rest of the selected secondary links in X'_i. If restrictions in (6) and (7) are achieved for secondary and primary links after performing the mapping between X'_i and *Spectrum Status*, the candidate solution X_i is a feasible solution and data rate is calculated for every secondary link involved in the candidate solution. After that, those secondary data rates are added along with the primary data rates, being the total particle's fitness. On the other hand, we penalize an infeasible solution setting total particle's fitness to zero.

4 Simulation Results

This section presents numerical results to demonstrate the performance of the proposed admission control and channel allocation algorithm based on PSO. The simulation parameters used for SCPSO are shown in Table 1.

The network setting is illustrated in Fig. 1, where the primary links have fixed locations and the secondary links are randomly deployed over an area of 5000×5000 m (macrocell range). Secondary links have circles of radius of maximum 30 m. Downlink transmission is considered in all the simulations. Each transmitter either primary or secondary is assumed to employ unit transmission power and the channel strength to be determined by path loss. We evaluate the results of 500 different random deployments (scenarios) at a given SINR threshold; this is defined as an experiment. SINR thresholds are varied to simulate different levels of application requirements (QoS). The stopping criterion for a run is the maximum number of iterations T_{max}. For the ease of reference, the experiment parameters are summarized in Table 2.

From the set of 500 runs that are evaluated for a given experiment at a SINR threshold ($\alpha = \beta$), the run containing the maximum data rate of the system is taken and that information is analyzed and reported in Table 3. Run times show that the proposed approach is not costly computationally since best founds have low run times in the order of some seconds. In general, the higher the SINR threshold, the more the number of algorithm iterations to achieve the goal. As QoS requirements are higher in the cognitive cellular network, it is more challenging to select the set of admitted secondary links that can achieve the SINR thresholds α, β. Throughput (fitness) and number of selected secondary links degrade when SINR thresholds become large. This is due to the fact that, when SINR thresholds become large, the density of active secondary links (femtocells) becomes small to guarantee constraints, leading to a small transmission capacity.

Table 1 Parameters used for particle swarm optimization

Parameters	Values
Swarm size S	40
Maximum number of iterations T_{max}	100
Cognitive, social and socio-cognitive factors c_1, c_2, c_3	2, 2, 12
Inertia weight w	0.721
Maximum velocity V_{max}	[−6,6]

Table 2 Parameters used for experiments

Parameters	Values
Number of secondary links Sl	20
Number of primary links Pl	6
Channels to share	1, 2, 3, 4, 5, 6
Runs	500
SINR thresholds $\alpha = \beta$	4, 6, 8, 10, 12, 14 dB
Channel bandwidth	20 MHz

Table 3 The best found at each experiment

$\alpha = \beta$ (dB)	Fitness (Mbps)	Number of selected secondary links	Run time (s)	Number of algorithm iterations to achieve the goal
4	9,828.3486	16	1.46	32
6	9,376.8802	16	1.31	99
8	9,585.0736	17	1.26	17
10	8,877.6855	16	1.23	32
12	9,295.4222	16	1.28	62
14	8,648.4162	15	1.35	75

In Fig. 3 is shown the evolution of average data rate in the cognitive cellular network under different SINR thresholds (4, 6, 8, 10, 12, 14 dB). A single curve corresponds to one specific experiment in which the average data rate is obtained from the data rate of each of the 500 runs. Figure 3 suggests that the higher the SINR thresholds, the total data rate in the system decreases. This is an expected behavior since as we pointed out in previous observations in Table 3, as the QoS requirements become large; a small number of secondary links is allowed to coexist with primary links which impacts in a lower contribution to the total data rate.

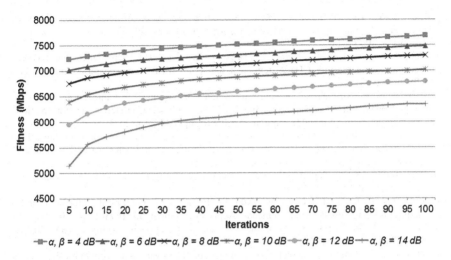

Fig. 3 Data rate evolution

5 Conclusion

In this chapter, we have proposed an admission control and channel allocation algorithm based on PSO to improve the system capacity and manage the interference in two-tier networks. It has been shown that the proposed approach is able to protect the primary links by maintaining their desired SINR requirements, while also allowing secondary links to share channels. A certain QoS is also supported for secondary links in order to provide significant benefits to both primary and secondary users from spectrum sharing. The numerical results demonstrated that the admission control and channel allocation algorithm based on PSO is not costly computationally, and it offers efficient and robust solutions. The presented approach could be extended further by assuming other wireless technologies.

References

1. Andrews, J.G., Claussen, H., Dohler, M., Rangan, S., Reed, M.C.: Femtocells: past, present, and future. IEEE J. Sel. Areas Commun. **30**, 497–508 (2012)
2. Liu, Y., Cai, L.X., Shen, X., Luo, H.: Deploying cognitive cellular networks under dynamic resource management. IEEE Wirel. Commun. **20**, 82–88 (2013)
3. Cheng, S.-M., Ao, W.-C., Tseng, F.-M., Chen, K.-C.: Design and analysis of downlink spectrum sharing in two-tier cognitive femto networks. IEEE Trans. Veh. Technol. **61**, 2194–2207 (2012)
4. Kennedy, J., Eberhart, R.: Particle swarm optimization. In: Proceedings of IEEE International Conference on Neural Networks, 1995, vol. 4, pp. 1942–1948 (1995)

5. Lien, S.-Y., Lin, Y.-Y., Chen, K.-C.: Cognitive and game-theoretical radio resource management for autonomous femtocells with QoS guarantees. IEEE Trans. Wirel. Commun. **10**, 2196–2206 (2011)
6. Ngo, D.T., Le, L.B., Le-Ngoc, T., Hossain, E., Kim, D.I.: Distributed interference management in two-tier CDMA femtocell networks. IEEE Trans. Wirel. Commun. **11**, 979–989 (2012)
7. Palanisamy, P., Nirmala, S.: Downlink interference management in femtocell networks—a comprehensive study and survey. In: 2013 International Conference on Information Communication and Embedded Systems (ICICES), pp. 747–754 (2013)
8. Sombra, A., Valdez, F., Melin, P., Castillo, O.: A new gravitational search algorithm using fuzzy logic to parameter adaptation. In: IEEE Congress on Evolutionary Computation, pp. 1068–1074 (2013)
9. Parsopoulos, K.E.: Particle Swarm Optimization and Intelligence: Advances and Applications. IGI Global (2010)
10. Marinakis, Y., Iordanidou, G.-R., Marinaki, M.: Particle swarm optimization for the vehicle routing problem with stochastic demands. Appl. Soft Comput. **13**, 1693–1704 (2013)
11. Robinson, J., Rahmat-Samii, Y.: Particle swarm optimization in electromagnetics. IEEE Trans. Antennas Propag. **52**, 397–407 (2004)
12. Chan, K.Y., Dillon, T.S., Chang, E.: An intelligent particle swarm optimization for short-term traffic flow forecasting using on-road sensor systems. IEEE Trans. Ind. Electron. **60**, 4714–4725 (2013)
13. Del Valle, Y., Venayagamoorthy, G.K., Mohagheghi, S., Hernandez, J.-C., Harley, R.G.: Particle swarm optimization: basic concepts, variants and applications in power systems. IEEE Trans. Evol. Comput. **12**, 171–195 (2008)
14. Valdez, F., Melin, P., Castillo, O.: Evolutionary method combining particle swarm optimization and genetic algorithms using fuzzy logic for decision making. In: Proceedings of the IEEE International Conference on Fuzzy Systems, pp. 2114–2119 (2009)
15. Valdez, F., Melin, P., Castillo, O.: Parallel particle swarm optimization with parameters adaptation using fuzzy logic. MICAI (2), 374–385 (2012)
16. Kennedy, J., Eberhart, R.C.: A discrete binary version of the particle swarm algorithm. In: IEEE International Conference on Systems, Man, and Cybernetics, 1997. "Computational Cybernetics and Simulation", vol. 5, pp. 4104–4108 (1997)
17. Deep, K., Bansal, J.C.: A socio-cognitive particle swarm optimization for multi-dimensional knapsack problem. In: International Conference on Emerging Trends in Engineering and Technology (ICETET), pp. 355–360 (2008)

Optimization of Fuzzy Controllers Design Using the Bee Colony Algorithm

Camilo Caraveo and Oscar Castillo

Abstract In this chapter we present the application of the optimization method using bee colony (BCO for its acronym in English, Bee Colony Optimization), for optimizing fuzzy controllers, BCO is a heuristic technique inspired by the behavior of honey bees in the nature, to solve optimization problems. This was tested in two BCO optimization problems, one optimized set of mathematical functions for twenty to fifty dimensions, and two fuzzy controllers' optimization. The results are compared with other bio-inspired algorithms state of the art, of which we highlight that there is a lot of competition in terms of quality and consistency in the results, even if the method is one of the latest in the field of collective intelligence. Similarly presents some interesting observations derived from observed performance.

1 Introduction

In an optimization problem, the objective is to find the best alternative among a set of possibilities [1]. The collective intelligence based algorithms constitute a new paradigm of collective intelligence; they are able to find good solutions to optimization problems with reasonable computational cost [2]. It defines collective intelligence as a metaheuristic technique of artificial intelligence based on the study of collective behavior systems present in nature, usually decentralized and self-organizing [3].

The algorithm based on artificial bee colony (BCO for its acronym in English, Bee Colony Optimization), is one of the recently proposed algorithms in the area of collective intelligence [4–7]. There is a chapter focuses on the model proposed

C. Caraveo · O. Castillo (✉)
Tijuana Institute of Technology, Tijuana, México
e-mail: ocastillo@tectijuana.mx

O. Castillo et al. (eds.), *Recent Advances on Hybrid Approaches for Designing Intelligent Systems*, Studies in Computational Intelligence 547,
DOI: 10.1007/978-3-319-05170-3_12, © Springer International Publishing Switzerland 2014

by Lucic and Teodorović in 2001, motivated by observing intelligent behavior of honey bee's swarm [8].

The bee colony optimization (BCO) is a meta-heuristic [9] belongs to the class of nature-inspired algorithms. These algorithms are inspired by various biological and natural processes [10]. Natural systems have become an important source of ideas and models for the development of many artificial systems. The popularity of nature-inspired algorithms is mainly caused by the ability of biological systems to successfully adapt to the continuously variable environment [11].

This method (BCO) uses an analogy based on the way bees do their foraging in nature, and the way in which they apply their search optimization methods find optimal routes to achieve the hive to source food [10]. The basic idea behind the BCO is to build the multi-agent system (artificial bee colony) efficiently able to solve difficult combinatorial optimization problems [8, 9].

2 Fuzzy Logic

A fuzzy logic system (FLS) that is defined entirely in terms of fuzzy sets is known as Fuzzy Logic System (FLS), its elements are defined in Fig. 1 [12].

A fuzzy set in the universe U is characterized by a membership function $u_A(x)$ taking values in the interval $[0, 1]$ and can be represented as a set of ordered pairs $x\ y$ of an element of value belonging to the set [12]:

$$A = \{(x, u_A(x)) \mid x \in U\} \tag{1}$$

A variety of types of membership functions given the usual distribution and is typically known as the bell of Gauss (Gaussian) [13]. The mathematical function is defined in Eq. 2.

$$F(x) = \exp\left(-0.5(x - c)^2\right)/\sigma^2 \tag{2}$$

where c feature parameters representing the mean and σ variance representing

Gaussian distribution is defined by its mean **m** and standard deviation **k > 0** is satisfied that the lower **k**, the narrower the "bell" [3, 12].

Example of a Gaussian type membership function is shown in Fig. 2.

3 Bee Colony Optimization (BCO)

Known as Bee Colony Optimization (BCO) is one of the latest algorithms in the domain of collective intelligence [14]. Created by Lucic and Teodorović in 2001, which was motivated by the intelligent behavior observed in honeybees to bring the foraging process [9, 10].

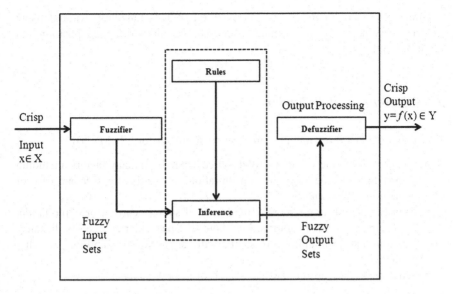

Fig. 1 Architecture of a fuzzy logic system

Fig. 2 Gaussian membership
function

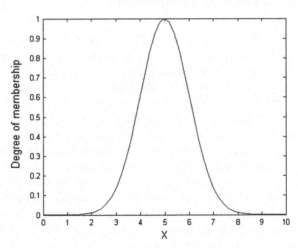

BCO is a combinatorial optimization algorithm based on locations where the solutions of the optimization problem, called food sources are modified by artificial bees, which serve as carriers of variation. The aim of these bees is to discover food sources more nectar [3, 9].

In the BCO algorithm artificial bees move in a multidimensional search space by choosing food sources depending on their past experience and their hive mates or adjusting its position [9].

Some bees (scouts) flying and choose food sources randomly without using experience. When they find a major nectar source, memorize their position and forget the old one.

3.1 Elements and Behavior

The model defines three main components that are set forth below [3]:

- Source of food: the value of a food source depends on many factors, including its proximity to the hive, the concentration of wealth or power and ease of extraction of this energy.
- Bees Employed: are associated with a food source, current or exploitation. They carry with them information about that particular source, its distance, location and profitability to share, with a certain probability, its other partners.
- Scout bees: they are in constant search of a food source.

3.2 Phases of the Algorithm

Population of agents (artificial bees) consisting of B bees collaboratively searches for the optimal solution. Every artificial bee generates one solution to the problem. There are two alternating phases (*forward pass* and *backward pass*) constituting single step in the BCO algorithm. In each forward pass, every artificial bee explores the search space [8]. It applies a predefined number of moves, which construct and/or improve the solution, yielding to a new solution. For example, let bees Bee 1, Bee 2...Bees B participate in the decision-making process on n entities. At each forward pass bees are supposed to select one entity [4, 9]. The possible situation after third forward pass is illustrated on Fig. 3.

Having obtained new partial solutions, the bees go again to the hive and start the second phase, the so-called backward pass. In the backward pass, all artificial bees share information about the quality of their solutions. In nature, bees would return to the hive, perform a dancing ritual, which would inform other bees about the amount of food they have discovered, and the proximity of the patch to the hive. In the search algorithm, the bees announce the quality of the solution, i.e. the value of objective function is computed. Having all solutions evaluated, every bee decides with a certain probability whether it will stay loyal to its solution or not. The bees with better solutions have more chances to keep and advertise their solutions. On the contrary to the bees in nature, artificial bees that are *loyal* to their partial solutions are at the same time *recruiters*, i.e. their solutions would be considered by other bees. Once the solution is abandoned by a bee it becomes *uncommitted* and has to select one of the advertised solutions. This decision is taken with a probability too, so that better advertise solutions have bigger

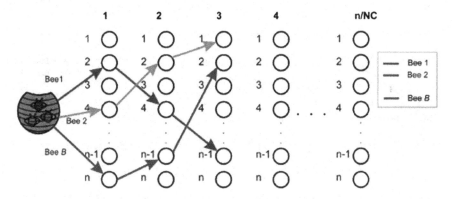

Fig. 3 An illustration of the third forward pass

Fig. 4 Recruiting of uncommitted followers

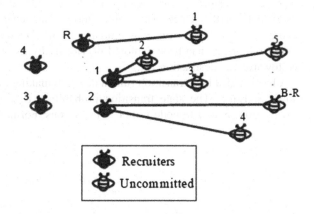

opportunity to be chosen for further exploration. In such a way, within each backward pass all bees are divided into two groups (R recruiters, and remaining B-R uncommitted bees) as it is shown on Fig. 4. Values for R and B-R are changing from one backward pass to another one [8].

After comparing all generated partial solutions, Bee 2, from the previous example decided to abandon already generated solution and to join Bee B (see Fig. 5).

Bee 2 and Bee B "fly together" along the path already generated by the Bee B. In practice, this means that partial solution generated by Bee B is associated (copied) to Bee 2 also [8, 9].

3.3 Path Construction in Artificial Bee

In our proposed model, a bee is allowed to explore and search for a complete tour path. Before leaving the hive, the bee will randomly observe dances performed by

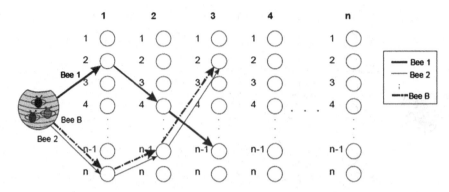

Fig. 5 The possible result of a recruiting process within third backward pass

other bees. The bee is then equipped with an ordered set of moves which are observed from the dance. This set of moves, named as "preferred path" and denoted as θ, will then serve as guidance in its foraging process. θ contains a complete tour that had been explored previously by its mate and it will direct the bee towards the destination [8].

A bee is aided by a transition rule in decision making of the next node to visit is shown in Eq. 3. The state transition probability, $P_{ij,n}$, gives the possibility of moving from node i to node j after N transitions. Formally defined in Eq. 3 [8]

$$p_{ij,n} = \frac{[\rho_{ij,n}]^{\alpha} \cdot \left[\frac{1}{d_{ij}}\right]^{\beta}}{\sum\limits_{j \in A_{i,n}} [\rho_{ij,n}]^{\alpha} \cdot \left[\frac{1}{d_{ij}}\right]^{\beta}} \tag{3}$$

where $\rho_{ij,n}$ is the arc fitness from node i to node j after n transitions and d_{ij} represents the distance between node i and node j. Note that the $P_{ij,n}$ is inversely proportional to the node distance. In other words, the shorter the distance, the higher is the likelihood of that node to be selected. α is a binary variable that turns on or off the arc fitness influence in the model. β is to control the significant level of heuristic distance.

Arc fitness, $\rho_{ij,n}$, is defined as in Eq. (4). Where $|A_{i,n} \cap F_{i,n}|$ is 1 when there is a common instance in both $A_{i,n}$ and $F_{i,n}$, or 0 otherwise. $A_{i,n} - F_{i,n}$ denotes the difference between sets $A_{i,n}$ and $F_{i,n}$. It contains all elements of $A_{i,n}$ that are not present in $F_{i,n}$. When there is only one node left in $A_{i,n}$, $\rho_{ij,n}$ is set to 1 to indicate that the node is the only choice. This happens at the last transition before a bee revisits the start node in order to complete the tour [8]

$$\rho_{ij,n} = \begin{cases} \lambda, & j \in F_{i,n}, |A_{i,n}| > 1 \\ \frac{1 - \lambda |A_{i,n} \cap F_{i,n}|}{|A_{i,n} - F_{i,n}|}, & j \notin F_{i,n}, |A_{i,n}| > 1 \\ 1, & |A_{i,n}| = 1 \end{cases} \left. \begin{array}{l} \forall j \in A_{i,n}, \\ 0 \leq \lambda \leq 1 \end{array} \right\} \tag{4}$$

4 Fuzzy Controller for the Benchmark Water Tank Problem

The problem to be studied is known as the water tank controller which consists of the following: the controller aims at water level in a tank, therefore, have to know the actual water level in the tank with it can be able to set the valve.

Figure 6 shows graphically the way in which it operates the valve opening and hence the filling process in the tank, this will have two variables which is the water level and the speed of opening of the will be output valve tank filling [13, 15, 16].

To evaluate the valve opening in a precise way we rely on fuzzy logic is implemented as a fuzzy controller that performs automated tasks considered as water level and how fast it will be entered to thereby maintain the level of water in the tank in a better way [15].

4.1 Characteristics of the Fuzzy Controller

We present the characteristics of fuzzy controller and its experimental evaluation using base supported by Simulink.

They have two inputs to the fuzzy system: the first call **level** which has three membership functions with high linguistic value, okay and low. The second input variable is called **rate** with three membership functions with linguistic value of negative, good and positive, shown in the Fig. 7 representations of fuzzy variables. The names of the linguistic labels are assigned based on the empirical process of filling behavior of a water tank [13, 15].

The Fuzzy Inference System has an entry called valve which is composed of five all triangular membership functions with the following linguistic values: close_fast, close_slow, no_change, open_slow and open_fast, representation shown in Fig. 8.

The simulation was performed using the following rules:

- If (level is okay) then (valve is no_change).
- If (level is low) then (valve is open_fast).
- If (level is high) then (valve is close_fast).
- If (level is okay) and (rate is positive) then (valve is close_slow).
- If (level is okay) and (rate is negative) then (valve is open_slow).

The combination of rules taken from experimental way according to how the process is performed in a tank filled with water. We start with 5 rules to visualize the behavior of the fuzzy controller. The simulation results are described in the next section.

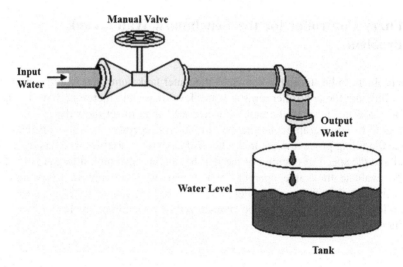

Fig. 6 Graphic representation of the problem to be studied

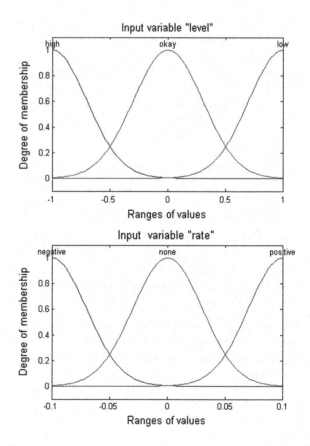

Fig. 7 Fuzzy inference system input variable

Fig. 8 Fuzzy inference
system output variable

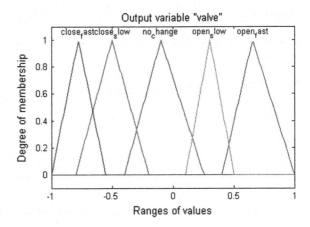

5 Experimental Results

This section presents the experimental results obtained with the method of optimization (BCO) analyzed in this research, experimental tests are done also with
genetic algorithm (GA) [17] with the aim of making a comparison of the performance of both methods.

To validate the proposed method we used a set of 5 benchmark mathematical
functions; all functions were evaluated with different numbers of dimensions, in
this case, the experimental results were obtained with 20 and 50 dimensions [5].

Table 1 shows the definitions of the mathematical functions used in this chapter.

5.1 Performance Analysis of BCO in Mathematics Functions

Tables 2 and 3 shows the experimental results for the benchmark mathematical
functions used in this research using the methods BCO and GA. The Table shows
the experimental results of the evaluations for each function with 20 and 50
dimensions; where it can be seen the best values obtained, and the average of
30 times after executing the method.

As can be seen the method proposed method (BCO) still proving to be a good
method for both small and large optimization problems compared to a genetic
algorithm (GA), as in the experiments shown above for 20 dimensions see
Tables 2 and 3. Continue to show weakness in the Rosenbrock function.

The objective of the proposed method was to apply it to fuzzy control; Table 4
shows the parameter settings of the algorithm (BCO) for experimental tests.

Table 5 shows the experimental results for the problem of the tank and water.
The Table shows the experimental results of the fuzzy controller will be tested in
Simulink model, where you can watch the best values obtained, standard devia-
tion, time, and the average of 30 times after the execution of the method.

Table 1 Mathematical functions

Function	Definition
De Jong's	$f(x) = \sum_{i=1}^{n} x_i^2$
Rosenbrock	$f(x) = \sum_{i=1}^{n-1} \left[100\left(x_{i+1} - x_i^2\right)^2 + (1 - x_i)^2 \right]$
Rastrigin	$f(x) = 10n + \sum_{i=1}^{n} \left[x_i^2 - 10\cos(2\pi x_i) \right]$
Ackley	$f(x) = -a \cdot \exp\left(-b \cdot \sqrt{\frac{1}{n}\sum_{i=1}^{n} x_i^2} \right) - \exp\left(\frac{1}{n}\sum_{i=1}^{n} \cos(cx_i) \right) + a + \exp(1)$

Table 2 Experimental results with 20 dimensions with BCO on GA

Function	Experiments with 20 dimensions					
	BCO			GA		
	Best	Mean	σ	Best	Mean	σ
Rosenbrock	6.45E−05	4.50E−04	5.37E−04	4.43E−11	1.57E−06	1.91E−06
Rastrigin	0.0	0.0	0.0	7.99E−07	2.24E−03	2.60E−03
Ackley	2.98	3.44	0.310	2.98	2.9808	1.86E−04
Sphere	0.0	0.0	0.0	1.09E−06	3.12E−06	1.27E−06

Table 3 Experimental results with 50 dimensions with BCO on GA

Function	Experiments with 50 dimensions					
	BCO			GA		
	Best	Mean	σ	Best	Mean	σ
Rosenbrock	5.77E−04	8.29E−03	9.19E−03	1.32E−08	6.91E−05	1.32E−04
Rastrigin	0.0	0.0	1.38E−10	0.017	0.9862	0.9961
Ackley	3.01	4.56	1.1202	2.98	2.98009	3.00E−04
Sphere	0.0	0.0	0.0	6.65E−05	2.66E−04	7.23E−04

Table 4 Configuration parameters BCO

Configuration the parameters BCO					
Population size	Roles	Iterations	A	β	Selection
120	90/30	200	0.4	4	Rank

Table 5 Experimental results BCO

Experimental results BCO			
Best	Average	σ	Average time
0.0304	0.0368	0.00345	78.65 min

Fig. 9 Variable input level

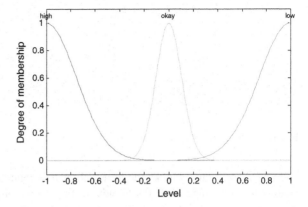

Fig. 10 Variable rate input

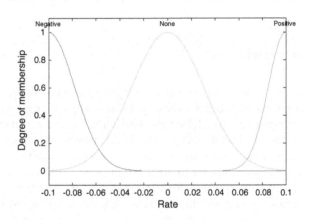

Fig. 11 Variable valve
output

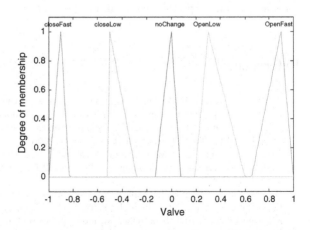

Figures 9, 10, 11, and 12 show the best architecture found by the bee colony
algorithm.

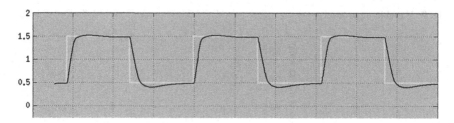

Fig. 12 Performance of fuzzy controller

In previous pictures looks the best architecture found by the proposed method, leading to an average of 0.00345 ends. In Fig. 12 we observe the behavior of the fuzzy logic controller to be tested in the Simulink model.

6 Conclusions

In this chapter we presented the application of a collective intelligence paradigm, the algorithm Bee Colony (BCO) to solve optimization problems applied fuzzy control problems. Based on the results we can say that is a good optimization method to find the parameters of the membership functions and adjust to design high performance fuzzy controllers when modeling them giving us a great advantage compared to other bio-inspired methods. The results obtained showed the BCO be good but not better than most of these bio-inspired algorithms may be improved by performing some adaptations to the method (BCO). Given the above it is concluded that the BCO is able to solve optimization problems and results showing acceptable performance, but not the best.

References

1. Sombra, A., Valdez, F., Melin, P., Castillo, O.: A new gravitational search algorithm using fuzzy logic to parameter adaptation. IEEE Congr. Evol. Comput. 1068–1074 (2013)
2. Aceves, A., Aguilar, J.: A simplified version of Mamdani's fuzzy controller: the natural logic controller. IEEE Trans. Fuzzy Syst. **14**(1), 16–30 (2006)
3. Jang, J.S.R., Sun, C.T., Mizutani, E.: Neuro-fuzzy and soft computing-a computational approach to learning and machine intelligence. IEEE Trans. Autom. Control **42**(10), 1482–1484 (1997)
4. Baykasoglu, A., Özbakýr, L., Tapkan, P.: Artificial bee colony algorithm and its application to generalized assignment problem. In: Felix, T.S.C., Manoj, K.T. (eds.) Swarm Intelligence: focus on Ant and Particle Swarm Optimization, pp. 113–143. Itech Education and Publishing, Vienna (2007)
5. Karaboga, D., Basturk, B.: On the performance of artificial bee colony (ABC) algorithm. Appl. Sof. Comput. **8**(2), 687–697 (2008)

6. Karaboga, D.: An idea based on honey bee swarm for numerical optimization. Technical Report TR06, Computer Engineering Department, Erciyes University, Turkey (2005)
7. Valdez, F., Melin, P., Castillo, O.: Parallel particle swarm optimization with parameters adaptation using fuzzy logic. In: MICAI, vol. 2, pp. 374–385. Mexico (2012)
8. Man, K.F., Tang, K.S., Kwong, S.: Genetic algorithms: concepts and designs. Springer, Berlin (2000)
9. Wong, L.P., Low, M.Y.H., Chong, C.S.: Bee colony optimization with local search for traveling salesman problem. In: Proceedings of the 6th IEEE International Conference on Industrial Informatics, pp. 1019–1025 (2008)
10. Elvia, R., Ramírez, A.: Optimización de Funciones de Membresía en controladores Difusos estables por medio de Algoritmos Genéticos, Tesis Maestría. Instituto Tecnológico de Tijuana (2012)
11. Valdez, F., Melin, P., Castillo, O.: Evolutionary method combining particle swarm optimization and genetic algorithms using fuzzy logic for decision making. In: Proceedings of the IEEE International Conference on Fuzzy Systems, pp. 2114–2119 (2009)
12. Lučić, P., Teodorović, D.: Transportation modeling: an artificial life approach. In: Proceedings of the 14th IEEE International Conference on Tools with Artificial Intelligence, pp. 216–223, Washington (2002)
13. Neyoy, H., Castillo, O., Soria, J.: Dynamic fuzzy logic parameter tuning for ACO and its application in TSP problems. Recent Advances on Hybrid Intelligent Systems, vol. 451, pp. 259–271. Springer, Heidelberg (2013)
14. Flores Mendoza, J.I.: Propuesta de optimización mediante Cúmulos de Partículas para Espacios Restringidos, Tesis de Maestría en Ciencias de la Computación, LANIA Xalapa, Ver., Octubre (2007)
15. Amador-Angulo, L., Castillo, O., Pulido, M.: Comparison of fuzzy controllers for the water tank with type-1 and type-2 fuzzy logic. In: IFSA World Congress and NAFIPS Annual Meeting (IFSA/NAFIPS), 2013 Joint, pp. 1062–1067. IEEE, June 2013
16. Fierro, R., Castillo, O. (eds.): Design of fuzzy different PSO variants, recent advances on hybrid intelligent systems, pp. 81–88. Springer, Heidelberg (2013)
17. Karaboga, D., Akay, B., Ozturk, C.: Artificial bee colony (ABC) optimization algorithm for training feed-forward neural networks, pp. 318–329. Springer, Heidelberg (2007)

Optimization of Benchmark Mathematical Functions Using the Firefly Algorithm

Cinthya Solano-Aragón and Oscar Castillo

Abstract Nature-inspired algorithms are more relevant today, such as PSO and ACO, which have been used in various types of problems such as the optimization of neural networks, fuzzy systems, control, and others showing good results. There are other methods that have been proposed more recently, the firefly algorithm is one of them, this paper will explain the algorithm and describe how it behaves. In this chapter the firefly algorithm was applied in optimizing benchmark functions and comparing the results of the same functions with genetic algorithms.

1 Introduction

The main problem of optimization is to find the values of the variables of a function to be optimized. These types of problems exist in many disciplines. Despite the fact that there are many methods of solution, there are many problems that need special attention and are difficult to solve using deterministic solution methods. In contrast to the deterministic algorithms, meta-heuristic methods are not affected by the behavior of the optimization problem. This makes the algorithms more widely usable.

The meaning of optimization is to find the parameter values in a function that makes a better solution. All appropriate values are possible solutions and the best value is the optimal solution [1]. The classification of optimization algorithms can be performed in many ways. A simple way is to study the nature of the algorithms, and this algorithm divides into two categories: deterministic algorithms and stochastic algorithms.

C. Solano-Aragón · O. Castillo (✉)
Tijuana Institute of Technology, Tijuana, Mexico
e-mail: ocastillo@tectijuana.mx

O. Castillo et al. (eds.), *Recent Advances on Hybrid Approaches for Designing Intelligent Systems*, Studies in Computational Intelligence 547,
DOI: 10.1007/978-3-319-05170-3_13, © Springer International Publishing Switzerland 2014

The deterministic algorithms follow a rigorous procedure, and its path and values for both the design variables and functions are repeatable. For stochastic algorithms, we generally have two types: heuristics and meta-heuristics.

Meta heuristic algorithms inspired by nature are increasingly powerful in solving modern problems of global optimization. All meta-heuristic algorithms use a certain balance between randomization and local search [2–4]. Stochastic algorithms often have a deterministic component and a random component [5–7].

The stochastic component can take many forms, such as simple random assignment by random sampling of the search space or random walks. Randomization provides a good way to move out of local search for the global search. More algorithms can be considered as stochastic meta-heuristics and a good example is the Genetic Algorithm (GA) [8]. Many modern meta-heuristic algorithms were developed based on swarm intelligence in nature, such as PSO [9–11].

The rest of the chapter is organized as follows. Section 2 we describe the Firefly Algorithm developed by Xin-She Yang. In Sect. 3, talks about Benchmark Functions that are evaluated. In Sect. 4, shows the experimental results with Genetic Algorithm of Benchmark Functions. In Sect. 5, shows the experimental results with Firefly Algorithm of Benchmark Functions. Finally Sect. 6, shows the conclusions.

2 Firefly Algorithm

The Firefly Algorithm (FA) is a meta-heuristic, nature-inspired, optimization algorithm which is based on the social (flashing) behavior of fireflies. The flashing light of fireflies is an amazing sight in the summer sky in the tropical and temperate regions. The primary purpose for a firefly's flash is to act as a signal system to attract other fireflies. In addition, flashing may also serve as a protective warning mechanism.

For simplicity, the flashing characteristics of fireflies are idealized in the following three rules [1, 12, 13]:

- All fireflies are unisex, so that one firefly is attracted to other fireflies regardless of their sex.
- Attractiveness is proportional to their brightness, thus for any two flashing fireflies, the less bright one will move towards the brighter one. The attractiveness is proportional to the brightness and they both decrease as their distance increases. If no one is brighter than a particular firefly, it moves randomly.
- The brightness of a firefly is affected or determined by the landscape of the objective function to be optimized.

Based on these three rules, the basic steps of the firefly algorithm can be summarized as the pseudo code shown in Fig. 1.

Fig. 1 Pseudo code of the firefly algorithm

Firefly Algorithm

Objective function $f(x)$, $x = (x_1, ..., x_d)^T$
Generate initial population of fireflies x_i ($i=1,2,...,n$)
Light intensity I_i at x_i is determined by $f(x_i)$
Define light absorption coefficient γ
while (*$t <$ MaxGeneration*)
for *$i = 1 : n$ all n fireflies*
  for *$j = 1 : n$ all n fireflies (inner loop)*
  if *($I_i < I_j$), Move firefly i towards j;* **end if**
    Vary attractiveness with distance r via exp[-γr]
    Evaluate new solutions and update light intensity
  end for *j*
end for *i*
Rank the fireflies and find the current global best **g.**
end while
Postprocess results and visualization

In the firefly algorithm, there are two important issues: the variation of light intensity and the formulation of the attractiveness. For simplicity, it is assumed that the attractiveness of a firefly is determined by its brightness which in turn is associated with the encoded objective function.

2.1 Attractiveness

The form of the attractiveness function of a firefly is the following monotonically decreasing function [12]:

$$\beta(r) = \beta_0 e^{-\gamma r^m} (m \geq 1) \tag{1}$$

where r is the distance between any two fireflies, β_0 is the attractiveness at $r = 0$ and γ is a fixed light absorption coefficient.

2.2 Distance

The distance between any two fireflies i and j at X_i and X_j, respectively, is the Cartesian distance as follows:

$$r_{ij} = \| X_i - X_j \| = \sqrt{\sum_{k=1}^{d} \left(x_{i,k} - x_{j,k} \right)^2} \tag{2}$$

where $x_{i,k}$ is the kth component of the spatial coordinate X_i of ith firefly and d is the number of dimensions.

2.3 Movement

The movement of a firefly i is attracted to another more attractive (brighter) firefly j is determined by following equation:

$$X_i = X_i + \beta_0 e^{-\gamma r_{ij}^2}(X_j - X_i) + \alpha(\text{rand} - 0.5) \tag{3}$$

where the second term is due to the attraction while the third term is randomization with α being the randomization parameter, rand is a random number generator uniformly distributed in [0, 1]. For most cases in the implementation, $\beta_0 = 1$ *and* $\alpha \in [0, 1]$.

3 Benchmark Functions

In the field of evolutionary computation, it is common to compare different algorithms using a large test set, especially when the test set involves function optimization. However, the effectiveness of an algorithm against another algorithm cannot be measured by the number of problems that it solves better. If we compare two searching algorithms with all possible functions, the performance of any two algorithms will be, on average, the same. As a result, attempting to design a perfect test set where all the functions are present in order to determine whether an algorithm is better than other for every function. This is the reason why, when an algorithm is evaluated, we must look for the kind of problems where its performance is good, in order to characterize the type of problems for which the algorithm is suitable. In this way, we have made a previous study of the functions to be optimized for constructing a test set with six benchmark functions and a better selection. This allows us to obtain conclusions of the performance of the algorithm depending on the type of function. The mathematical functions analyzed in this chapter are in the table. All the functions were evaluated considering 20 variables [14] (Table 1).

4 Experimental Results

The results obtained after applying the genetic algorithm to the mathematical functions are shown on Tables 2, 3, 4, 5, 6, 7.

Parameters of Tables 2, 3, 4, 5, 6, 7:

GEN = Generations number.
POP = Population size.
CROSS = Percent of crossover.
MUT = Percent of mutation.

Table 1 Mathematical functions

Function	Expression
Rastrigin	$f(x) = 10n + \sum_{i=1}^{n} \left(x^2 - 10\cos(2\pi x_i)\right)$
Rosenbrock	$f(x) = \sum_{i=1}^{n-1} \left[100\left(x_i^2 - x_i^2 + 1\right)^2 + (x_i - 1)^2\right]$
Ackley	$f(x) = 20 + e - 20e^{-1/5}\sqrt{1/n\sum_{i=1}^{n} x_i^2} - e^{1/n}\sum_{i=1}^{n}\cos(2\pi x_1)$
Shubert	$f(x) = \left(\sum_{i=1}^{5} i\cos((i+1)x_1) + i\right)\left(\sum_{i=1}^{5} i\cos((i+1)x_2 + i\right)$
Sphere	$f(x) = \sum_{i=1}^{n} x_i^2$
Griewank	$f(x) = \sum_{i=1}^{n}\frac{x_i^2}{4,000} - \prod_{i=1}^{n}\cos\left(x_i/\sqrt{i}\right) + 1$

Table 2 Results obtained after applying the GA to the Rastrigin's function

GEN	POP	CROSS	MUT	SEL	BEST	MEAN
100	100	0.8	0.4	Rou	6.68E−03	7.28E−03
150	100	0.6	0.4	Rou	6.57E−03	7.50E−03
100	80	0.7	0.5	Rou	5.68E−03	1.19E−02
100	120	0.5	0.5	Rou	*5.58E−03*	*6.45E−03*

Table 3 Results obtained after applying the GA to the Rosenbrock's function

GEN	POP	CROSS	MUT	SEL	BEST	MEAN
100	80	0.9	0.6	Rou	6.06E−02	*3.93E−01*
150	100	0.6	0.4	Rou	*3.95E−02*	4.69E−01
150	80	0.9	0.6	Rou	1.98E−01	5.67E−01
100	120	0.5	0.8	Rou	1.17E−01	4.31E−01

Table 4 Results obtained after applying the GA to the Ackley's function

GEN	POP	CROSS	MUT	SEL	BEST	MEAN
100	100	0.8	0.4	Rou	1.04E−01	4.01E−01
150	100	0.6	0.4	Rou	5.69E−02	4.63E−01
100	80	0.7	0.5	Rou	*6.83E−03*	*3.99E−01*
100	120	0.5	0.5	Rou	2.74E−02	4.83E−01

Table 5 Results obtained after applying the GA to the Shubert's function

GEN	POP	CROSS	MUT	SEL	BEST	MEAN
100	100	0.8	0.4	Rou	186.71	*186.41*
150	100	0.6	0.4	Rou	186.72	185.62
100	80	0.7	0.5	Rou	*186.73*	185.30
100	120	0.5	0.5	Rou	186.72	183.01

Table 6 Results obtained after applying the GA to the Sphere's function

GEN	POP	CROSS	MUT	SEL	BEST	MEAN
100	100	0.8	0.4	Rou	5.08E−02	4.36E−01
150	100	0.6	0.4	Rou	1.15E−01	*3.88E−01*
100	80	0.7	0.5	Rou	9.61E−02	5.48E−01
100	120	0.5	0.5	Rou	*3.75E−02*	5.29E−01

Table 7 Results obtained after applying the GA to the Griewank's function

GEN	POP	CROSS	MUT	SEL	BEST	MEAN
100	100	0.8	0.4	Rou	2.48E−03	1.22E−01
150	100	0.6	0.4	Rou	1.89E−02	*6.07E−02*
100	80	0.7	0.5	Rou	1.23E−02	1.01E−01
100	120	0.5	0.5	Rou	*1.73E−03*	1.87E−01

SEL = Selection method.
BEST = Best fitness value.
MEAN = Mean of 10 test.

4.1 Results Obtained After Applying the Genetic Algorithm to the Rastrigin's Function

Table 2 shows four different cases of the GA by modifying some of their parameters and running 10 times in each case, showing the best objective value of the four cases having a value of 5.58E−03, highlighted in Table 2 (experiment 4).

The best average is 6.45E−03 and is obtained in experiment 4.

4.2 Results Obtained After Applying the Genetic Algorithm to the Rosenbrock's Function

Table 3 shows four different cases of the GA by modifying some of their parameters and running 10 times in each case, showing the best objective value of the four cases having a value of 3.95E−02, highlighted in Table 3 (experiment 2).

The best average is 3.93E−01 and is obtained in experiment 1.

4.3 Results Obtained After Applying the Genetic Algorithm to the Ackley's Function

Table 4 shows four different cases of the GA by modifying some of their parameters and running 10 times in each case, showing the best objective value of the four cases having a value of 6.83−E03, highlighted in Table 4 (experiment 3). The best average is 3.99E−01 and is obtained in experiment 3.

4.4 Results Obtained After Applying the Genetic Algorithm to the Shubert's Function

Table 5 shows four different cases of the GA by modifying some of their parameters and running 10 times in each case, showing the best objective value of the four cases having a value of 186.73, highlighted in Table 5 (experiment 3). The best average is 186.41 and is obtained in experiment 1.

4.5 Results Obtained After Applying the Genetic Algorithm to the Sphere's Function

Table 6 shows four different cases of the GA by modifying some of their parameters and running 10 times in each case, showing the best objective value of the four cases having a value of 3.75E−02, highlighted in Table 6 (experiment 4). The best average is 3.88E−01 and is obtained in experiment 2.

4.6 Results Obtained After Applying the Genetic Algorithm to the Griewank's Function

Table 7 shows four different cases of the GA by modifying some of their parameters and running 10 times in each case, showing the best objective value of the four cases having a value of 1.73E−03, highlighted in Table 7 (experiment 4). The best average is 6.07E−02 and is obtained in experiment 2.

5 Experimental Results With the Firefly Algorithm

The results obtained after applying the Firefly Algorithm to the mathematical functions are shown on Tables 8, 9, 10, 11, 12, 13.

Parameters of Tables 8, 9, 10, 11, 12, 13:

GEN = Number of iterations.
FF = Number of Fireflies.
BEST = Best fitness value.
MEAN = Mean of 10 test.

5.1 Results Obtained After Applying the Firefly Algorithm to the Rastrigin's Function

Table 8 shows four different cases of the FA by modifying some of their parameters and running 10 times in each case, showing the best objective value of the four cases having a value of 3.18E−04, highlighted in Table 8 (experiment 2).

The best average is 3.47E−02 and is obtained in experiment 1.

5.2 Results Obtained After Applying the Firefly Algorithm to the Rosenbrock's Function

Table 9 shows four different cases of the FA by modifying some of their parameters and running 10 times in each case, showing the best objective value of the four cases having a value of 2.21E−04, highlighted in Table 9 (experiment 1).

The best average is 7.39E−03 and is obtained in experiment 1.

5.3 Results Obtained After Applying the Firefly Algorithm to the Ackley's Function

Table 10 shows four different cases of the FA by modifying some of their parameters and running 10 times in each case, showing the best objective value of the four cases having a value of 3.18E−04, highlighted in Table 10 (experiment 1).

The best average is 1.89E−02 and is obtained in experiment 1.

Table 8 Results obtained after applying the FA to the Rastrigin's function

GEN	FF	α	γ	β	BEST	MEAN
50	12	0.20	1.0	0.20	4.73E−03	*3.47E−02*
150	100	0.20	1.0	0.20	*3.18E−04*	4.70E−02
100	80	0.20	1.0	0.20	5.68E−02	6.45E−02
100	120	0.20	1.0	0.20	5.58E−02	6.45E−02

Table 9 Results obtained after applying the FA to the Rosenbrock's function

GEN	FF	α	γ	β	BEST	MEAN
80	25	0.20	1.0	0.20	*2.21E−04*	*7.39E−03*
150	100	0.20	1.0	0.20	3.95E−02	4.69E−01
150	80	0.20	1.0	0.20	1.98E−01	5.67E−01
100	120	0.20	1.0	0.20	1.17E−01	4.31E−01

Table 10 Results obtained after applying the FA to the Ackley's function

GEN	FF	α	γ	β	BEST	MEAN
60	25	0.20	1.0	0.20	*3.18E−04*	*1.89E−02*
150	100	0.20	1.0	0.20	5.69E−02	4.63E−01
100	80	0.20	1.0	0.20	6.83E−03	3.99E−01
100	120	0.20	1.0	0.20	2.74E−02	4.83E−01

5.4 Results Obtained After Applying the Firefly Algorithm to the Shubert's Function

Table 11 shows four different cases of the FA by modifying some of their parameters and running 10 times in each case, showing the best objective value of the four cases having a value of −186.730701, highlighted in Table 11 (experiment 1).

The best average is −186.587857 and is obtained in experiment 1.

5.5 Results Obtained After Applying the Firefly Algorithm to the Sphere's Function

Table 12 shows four different cases of the FA by modifying some of their parameters and running 10 times in each case, showing the best objective value of the four cases having a value of 1.21E–06, highlighted in Table 12 (experiment 1).

The best average is 7.55E−04 and is obtained in experiment 1.

Table 11 Results obtained after applying the FA to the Shubert's function

GEN	FF	α	γ	β	BEST	MEAN
100	30	0.20	1.0	0.20	−186.730701	−186.587857
150	100	0.20	1.0	0.20	−186.720403	−185.616052
100	80	0.20	1.0	0.20	−186.717087	−185.301134
100	120	0.20	1.0	0.20	−186.724626	−183.009441

Table 12 Results obtained after applying the FA to the Sphere's function

GEN	FF	α	γ	β	BEST	MEAN
40	12	0.20	1.0	0.20	1.21E−06	7.55E−04
150	100	0.20	1.0	0.20	1.15E−01	3.88E−01
100	80	0.20	1.0	0.20	9.61E−02	5.48E−01
100	120	0.20	1.0	0.20	3.75E−02	5.29E−01

Table 13 Results obtained after applying the FA to the Griewank's function

GEN	FF	α	γ	β	BEST	MEAN
100	50	0.20	1.0	0.20	1.01E−05	1.67E−02
150	100	0.20	1.0	0.20	1.89E−02	6.07E−02
100	80	0.20	1.0	0.20	1.23E−02	1.01E−01
100	120	0.20	1.0	0.20	1.73E−03	1.87E−01

5.6 Results Obtained After Applying the Firefly Algorithm to the Griewank's Function

Table 13 shows four different cases of the FA by modifying some of their parameters and running 10 times in each case, showing the best objective value of the four cases having a value of 1.01E−05, highlighted in Table 13 (experiment 1).

The best average is 1.67E−02 and is obtained in experiment 1.

6 Conclusion

The analysis of the simulation results of the two optimization methods considered in this chapter, in this case the Genetic Algorithm (GA) and the Firefly Algorithm (FA), lead us to the conclusion that for the problems of optimization of these 6 mathematical functions, in all cases one can say that the two proposed methods work correctly and they can be applied for this type problems.

After studying the two methods of evolutionary computing, we reach the conclusion that for the optimization of these 6 mathematical functions, GA and FA

Table 14 Comparison between GA and FA

Mathematical functions	GA	FA	Numbers of variables	Objective value
Rastrigin	6.45E−03	3.47E −02	20	0
Rosenbrock	3.93E−01	7.39E−03	20	0
Ackley	3.99E−01	1.89E−02	20	0
Shubert	−186.41	−186.58	20	−186.73
Sphere	3.88E−01	7.55E−04	20	0
Griewank	6.07E−02	1.67E−02	20	0

Table 15 Statistical test between FA and GA

Parameters	Value
level of significance	95 %
Alpha	0.05 %
H_a	$\mu 1 < \mu 2$
H_0	$\mu 1 \geq \mu 2$
Critical value	−4.04
P	0.0002
Evidence	Significant

evolved in a similar form and near the objectives, so they are recommended for the solution of these problems.

The advantage to use FA is that there are few parameters used for the implementation. The genetic algorithm uses more parameters for its implementation [15–30].

From Table 14 it can be appreciated that after executing the GA and FA, the comparison of the results obtained between these two evolutionary algorithms for the optimization of the 6 mathematical functions used of this chapter. The table shows the results of the simulations, it can be appreciated that only in one function, the GA was better than the FA, for the Rastrigin's function. In other cases, the GA was better than the FA, specifically for the Rosenbrock's function, Ackley's function, Shubert's function, Sphere's function and Griewank's function.

Statistical t-student comparison is made between Genetic Algorithm and Firefly Algorithm resulting with a critical value of −4.04 showing that firefly algorithm is better than genetic algorithm in this application; in Table 8 the test parameters are shown. As future work we will consider the application of FA in optimizing fuzzy controllers [31–33] (Table 15).

References

1. Liu, Y., Passino, K.M.: Swarm intelligence: a survey. In: International Conference of Swarm Intelligence (2005)
2. Li, L.X., Shao, Z.J., Qian, J.X.: An optimizing method based on autonomous animals: fish swarm algorithm. Syst. Eng. Theory Pract (2002)

3. Yang, X.S.: Nature-Inspired Metaheuristic Algorithms. Luniver Press (2008)
4. Yang, X.S.: Firefly algorithms for multimodal optimization. In: Stochastic Algorithms Foundations and Applications, Stochastic Algorithms: foundations and Applications (SAGA'09). Lecture Notes in Computing Sciences, vol. 5792, pp. 169–178. Springer, Berlin (2009)
5. Sombra, A., Valdez, F., Melin, P., Castillo, O.: A new gravitational search algorithm using fuzzy logic to parameter adaptation. IEEE Congr. Evol. Comput. 1068–1074 (2013)
6. Valdez, F., Melin, P., Castillo, O.: Evolutionary method combining particle swarm optimization and genetic algorithms using fuzzy logic for decision making. In: Proceedings of the IEEE International Conference on Fuzzy Systems, pp. 2114–2119. (2009)
7. Valdez, F., Melin, P., Castillo, O.: Parallel particle swarm optimization with parameters adaptation using fuzzy logic. In: MICAI, vol. 2, pp. 374–385 (2012)
8. Holland, H.: Adaptation in Natural and Artificial Systems. University of Michigan Press, Ann Arbor (1975)
9. Kennedy, J., Eberhart, R.C.: Particle swarm optimization. In: IEEE International Conference on Neural Networks, pp. 942–1948. Piscataway, NJ (1995)
10. Koza, J.R.: Genetic Programming: on the Programming of Computers by Means of Natural Selection. MIT Press, Cambridge (1992)
11. Dorigo, M., Gambardella, L.M.: Ant colony system: a cooperative learning approach to the traveling salesman problem. IEEE Trans. Evol. Comput. 1(1997), 53–66 (1997)
12. Yang, X.S.: Firefly algorithm stochastic test functions and design optimization. Int. J. Bio-Inspired Comput. 2(2), 78–84 (2010)
13. Yang, X.-S.: Firefly algorithm, Lévy flights and global optimization. In: Bramer, M., Ellis, R., Petridis, M. (eds.) Research and Development in Intelligent Systems, vol. XXVI, pp. 209–218. Springer, London (2010)
14. Valdez, F., Melin, P.: Comparative study of particle swarm optimization and genetic algorithms for mathematical complex functions. J. Autom. Mob. Rob. Intell. Syst. JAMRIS (2008)
15. Melendez, A., Castillo, O.: Evolutionary optimization of the fuzzy integrator in a navigation system for a mobile robot. Recent Adv. Hybrid Intell. Syst. 21–31 (2013)
16. Goldberg, D.E.: Genetic algorithms in search, optimization and machine learning, reading, mass. Addison Wesley, Boston (1989)
17. Simon, D.: Biogeography-based optimization. IEEE Trans. Evol. Comput. 12(6), 702–713 (2008)
18. Valdez, F., Melin, P., Castillo, O.: An improved evolutionary method with fuzzy logic for combining particle swarm optimization and genetic algorithms. Appl. Soft Comput. 11(2), 2625–2632 (2011)
19. Valdez, F., Melin, P., Castillo, O.: Bio-inspired optimization methods on graphic processing unit for minimization of complex mathematical functions. Recent Adv. Hybrid Intell. Syst. 313–322 (2013)
20. Kennedy, J., Eberhart, J.R., Shi, Y.: Swarm intelligence. Academic Press, Massachusetts (2001)
21. Rodriguez, K.V.: Multiobjective evolutionary algorithms in non-linear system identification, in automatic control and systems engineering, p. 185. The University of Sheffield, Sheffield (1999)
22. Zadeh, L.A.: Foreword. In: Cordon, O., Herrera, F., Hoffman, F., Magdalena, y L. (eds.) Genetic Fuzzy Systems: evolutionary Tuning And Learning Of Fuzzy Knowledge Bases. (2001)
23. Astudillo, L., Melin, P., Castillo, O.: Optimization of a fuzzy tracking controller for an autonomous mobile robot under perturbed torques by means of a chemical optimization paradigm. Recent Adv Hybrid Intell. Syst. 3–20 (2013)
24. Cervantes, L., Castillo, O.: Genetic optimization of membership functions in modular fuzzy controllers for complex problems. Recent Adv. Hybrid Intell. Syst. 51–62 (2013)

25. Dorigo, M.: Optimization, learning and natural algorithms. Ph.D. Thesis, Politecnico di Milano, Milano (1992)
26. Erol, O.K., Eksin, I.: A new optimization method: big bang-big crunch. Adv. Eng. Softw. **37**, 106–111 (2006)
27. Melin, P., Olivas, F., Castillo, O., Valdez, F., Soria, J., García, J.M.: Valdez: optimal design of fuzzy classification systems using PSO with dynamic parameter adaptation through fuzzy logic. Expert Syst. Appl. **40**(8), 3196–3206 (2013)
28. Baeck, T., Fogel, D.B., Michalewicz, Z.: Handbook of Evolutionary Computation. Taylor & Francis, UK (1997)
29. Yang, X.S.: Engineering Optimization: an Introduction with Metaheuristic Applications. Wiley, New Jersey (2010)
30. Maldonado, Y., Castillo, O., Melin, P.: Optimization of membership functions for an incremental fuzzy PD control based on genetic algorithms. Soft Comput. Intell. Control Mob. Rob. 195–211 (2011)
31. Montiel, O., Sepulveda, R., Melin, P., Castillo, O., Porta, M., Meza, I.: Performance of a simple tuned fuzzy controller and a PID controller on a DC motor. FOCI. 531–537 (2007)
32. Castillo, O., Martinez, A.I., Martinez, A.C.: Evolutionary computing for topology optimization of type-2 fuzzy systems. Adv. Soft Comput. **41**, 63–75 (2007)
33. Castillo, O., Huesca, G., Valdez, F.: Evolutionary computing for topology optimization of type-2 fuzzy controllers. Stud. Fuzziness Soft Comput. **208**, 163–178 (2008)

Optimization of Fuzzy Control Systems for Mobile Robots Based on PSO

David de la O, Oscar Castillo and Abraham Meléndez

Abstract This paper describes the optimization of a navigation controller system for a mobile autonomous robot using the PSO algorithm to adjust the parameters of each fuzzy controller, the navigation system is composed of 2 main controllers, a tracking controller and a reactive controller, plus an integrator block control that combines both fuzzy inference systems (FIS). The integrator block is called Weighted Fuzzy Inference System (WFIS) and assigns weights to the responses in each block of behavior in order to combine them into a single response. A comparison with the results obtained with genetic algorithms is also performed.

1 Introduction

The mobile robots must be able to operate in a real environment, and navigate in an autonomous manner. In this case an intelligent control strategy that can handle the uncertainty can be implemented by the working environment, while comparing with the performance in real time of a relatively low computational load.

One of the applications of fuzzy logic is the design of fuzzy control systems. The success of this control lies in the correct selection of the parameters of fuzzy controller; it is here where the Particle Swarm Optimization (PSO) metaheuristic will be applied, which is one of the most used for optimization with real

D. de la O · O. Castillo (✉) · A. Meléndez
Tijuana Institute of Technology, Tijuana, Mexico
e-mail: ocastillo@tectijuana.mx

D. de la O
e-mail: ddsh@live.com

A. Meléndez
e-mail: abraham.ms@gmail.com

O. Castillo et al. (eds.), *Recent Advances on Hybrid Approaches for Designing Intelligent Systems*, Studies in Computational Intelligence 547,
DOI: 10.1007/978-3-319-05170-3_14, © Springer International Publishing Switzerland 2014

parameters. PSO, because of their ease of implementation, converges faster than the Evolutionary Algorithms (EA) has better performance [9–30].

The present research work deals with the problem of mobile robot autonomy, for which a fuzzy control system is developed using fuzzy logic and the PSO algorithm [31–52].

This paper is organized into four sections as follows: Sect. 2: the points are developed and progress of the research work is shown. Sections 3 and 4: In part disclose the theory underlying the present work, in which issues such as fuzzy logic, PSO algorithm and a bit on the operation of autonomous mobile robot are discussed. Section 5 shows the results of the simulations are presented.

2 Mobile Robots

The particular mobile robot considered in this work is based on the description of the Simulation toolbox for mobile robots [1], which assumes a wheeled mobile robot consisting of one conventional, steered, unactuated and not-sensed wheel, and two conventional, actuated, and sensed wheels (conventional wheel chair model). This type of chassis provides two DOF (degrees of freedom) locomotion by two actuated conventional non-steered wheels and one unactuated steered wheel. The Robot has two degrees of freedom (DOFs): y-translation and either x-translation or z-rotation [1]. Figure 1 shows the robot's configuration, it has 2 independent motors located on each side of the robot and one castor wheel for support located at the front of the robot.

The kinematic equations of the mobile robot are as follows:

Equation 1 shows the sensed forward velocity solution [2]

$$\begin{pmatrix} V_{Bx} \\ V_{By} \\ \omega_{Bz} \end{pmatrix} = \frac{R}{2l_a} \begin{bmatrix} -l_b & l_b \\ -l_a & -l_a \\ -1 & -1 \end{bmatrix} \begin{pmatrix} \omega_{W1} \\ \omega_{W2} \end{pmatrix} \tag{1}$$

Equation 2 shows the Actuated Inverse Velocity Solution [3]

$$\begin{pmatrix} \omega_{W1} \\ \omega_{W2} \end{pmatrix} = \frac{1}{R(l_b^2 + 1)} \begin{bmatrix} -l_a l_b & -l_b^2 - 1 & -l_a \\ l_a l_b & -l_b^2 - 1 & l_a \end{bmatrix} \begin{pmatrix} V_{Bx} \\ V_{By} \\ \omega_{Bz} \end{pmatrix} \tag{2}$$

Under the Metric system are define as:

V_{Bx}, V_{By} Translational velocities $\left[\frac{m}{s}\right]$,

ω_{Bz} Robot z-rotational velocity $\left[\frac{rad}{s}\right]$,

ω_{W1}, ω_{W1} Wheel rotational velocities $\left[\frac{rad}{s}\right]$,

R Actuated wheel radius [m],

l_a, l_b Distances of wheels from robot's axes [m].

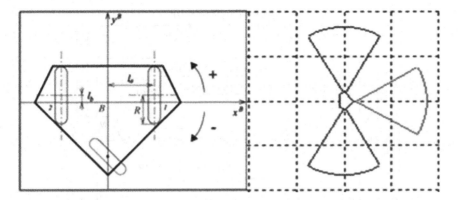

Fig. 1 Kinematic coordinate system assignments [2]

3 Navigation Control System

The proposed control system consists of three main fuzzy blocks: two controllers are based on the behavior of the robot (tracking and reactive) and one is in charge of combining the responses of the other two drivers, called block integration.

Each controller has a specific behavior, the problem is that they seem to be in conflict with each other when an unexpected obstacle appears, because if the route is planned when obstacles are present, the route can be designed to avoid them but when there are obstacles we do not realize that the two behaviors are in contradiction, one is designed to prevent the collision object and the other to keep the robot on the path.

The most common solution is to simply change among drivers as needed. However, this method is not very effective because of the lack of knowledge of the two controllers on to one another, the reagent remains the robot free from a collision, but may redirect the robot further from your destination to a point at which the tracking controller can no longer find their way back to the reference, or the tracking controller can directly guide the robot to an obstacle if the control reagent provides no actuation time. The proposed reference for navigation control always has both active controls and responses combined to create the movement of the robot. The integration is performed with another block called diffuse WFIS [2] (Weight-Fuzzy Inference System) so that this controller assigning weights is made responsive to each of the response values of the drivers.

The inputs are gathered from the information we can collect from the robot (sensors) or the environment by other means (cameras) and from this we need to create the knowledge rule base to give higher activation values to the response. If we want to take the lead on the robot movement one example of the rule is the following (if Front_Sensor_Distance is Close Then TranckingWeight is Medium and ReactiveWeight is Medium), both of our controls provide the right and left

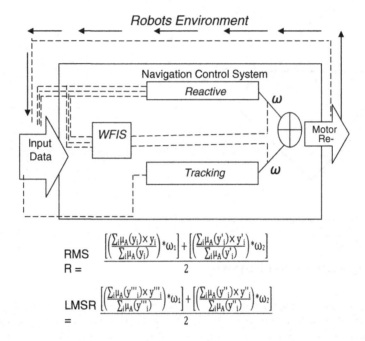

$$\text{RMS} \atop R =\quad \frac{\left[\left(\frac{\sum_i\mu_A(y_i)\times y_i}{\sum_i\mu_A(y_i)}\right)*\omega_1\right]+\left[\left(\frac{\sum_i\mu_A(y'_i)\times y'_i}{\sum_i\mu_A(y'_i)}\right)*\omega_2\right]}{2}$$

$$\text{LMSR} \atop =\quad \frac{\left[\left(\frac{\sum_i\mu_A(y'''_i)\times y'''_i}{\sum_i\mu_A(y'''_i)}\right)*\omega_1\right]+\left[\left(\frac{\sum_i\mu_A(y''_i)\times y''_i}{\sum_i\mu_A(y''_i)}\right)*\omega_2\right]}{2}$$

Fig. 2 Navigation control system [2]

motor speed and we combine each one with the weight given by the WFIS block. Figure 2 shows the proposed navigation control [2].

4 PSO Algorithms

The Particle Swarm Optimization Algorithm (PSO) was applied to each of the design problems in order to find the best fuzzy controller reactive and tracking. The purpose of using a behavior-based method is to find the best controllers of each type and this can be achieved by PSO as it searches along the search space of the solution, which combines the knowledge of the best controllers (particles), and we can handle the exploration and exploitation throughout the iterations. The main task of the algorithm is to convert the particle into a FIS and then evaluate each particle to determine its performance. Figure 3 shows the flowchart of the PSO.

4.1 Particle Encoding

The particle consists of 60 real-valued vectors, representing the parameters for the triangular membership function; we use five membership functions for each variable. This encoding is shown Fig. 4.

Fig. 3 Flowchart of the PSO

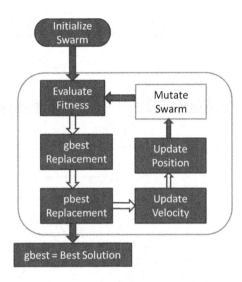

4.2 Reactive Controller

The role of reactive control is to apply the same capacity when we are driving, i.e. to react to unexpected situations, traffic jams, traffic lights, etc., but in a more basic concept. The objective is to guide the robot through the maze avoiding any collision. It is our goal to optimize the robot to find the exit of the maze, we used a maze to optimize the reactive control because of the feature that conditions the simulation, i.e. it is a closed space in which the robot cannot move easily and each wall is considered an obstacle to the robot to avoid while moving. The FIS is Mamdani type; each consisting of 3 entries, namely the distances obtained by the sensors of the robots described in Sect. 2, and 2 outputs that control the speed of the servo motors on the robot, all this information is encoded in each particle.

4.3 Tracking Controller

The tracking controller has the responsibility to keep the robot on the right track, given a reference; the robot will move on the reference and keep it on the road, allowing moving from point A to B without obstacles present in the environment.

The controller will work keeping the error (Δep, $\Delta\theta$) in the minimum values, Fig. 5, these minimum values are the relative position error and the relative error of the orientation of the front of the robot, the Mamdani fuzzy system and its 2 inputs are (Δep, $\Delta\theta$) and two outputs which control the speed of each servomotor of the robot (Fig. 6).

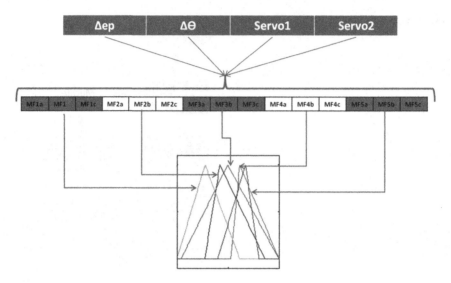

Fig. 4 Particle encoding

Controller performance is measured using the equation of the mean square error between the reference and the robot trajectory. We perform each test three times and take the average, above, below and random reference.

In Fig. 7 we can see the overall flowchart of the PSO. At the point of evaluation of the particle, we measure the effectiveness of the tracking controller FIS (Inference System Fuzzy) in our toolbox of evidence, which will be in a closed circuit with a given reference by a straight line [2, 4–7] environment.

4.4 WFIS Controller

The function of the WFIS control is to correctly combine the two behaviors of tracking and reaction and obtain a new global behavior that resembles the same ability that we apply when we are driving, that is to react to unexpected objects, but in a more basic concept and ability, to the problem that is the navigation of the robot. A forward moving behavior response out of the global control is desired. The objective is to guide the robot through the reference avoiding any collision with any obstacle present. It's not our objective to optimize the robot to find the maze exit. We use a closed space where the robot cannot easily wonder off and each wall is considered an obstacle to the robot that it must avoid while it moves around. The FISs are Mamdani fuzzy systems [8], each consisting of three inputs, which are the distances obtained by the robots sensors described on Sect. 2, and two outputs that are the weights that will be used to integrate the responses of the other two controllers. All this information is encoded into each particle.

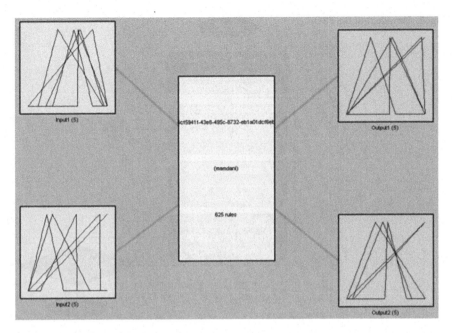

Fig. 5 Fuzzy controller inputs ep, eθ

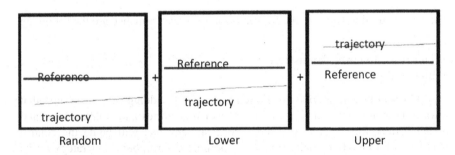

Fig. 6 Calculation of controller performance

4.5 Objective Function

The PSO will be generating particles that will need to be evaluated and assigned a crisp value that will represent the controller performance on each of the criteria that we want to improve. For this, we need to provide the PSO with a good evaluation scheme that will penalize unwanted behaviors and reward with higher fitness values those individuals that provide the performance we are looking for in our controller; if we fail to provide a proper evaluation method we can guide the population to suboptimal solutions or no solution at all [2, 4–7].

Fig. 7 Flowchart of the PSO

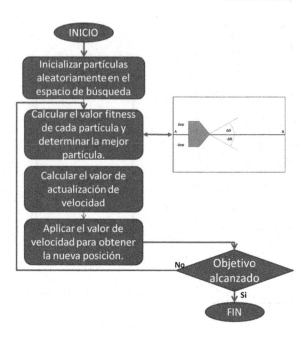

4.5.1 Reactive Controller Objective Function

The criteria used to measure the Reactive controller performance are the following

- Covered Distance
- Time used to cover the distance
- Battery life.

A Fitness FIS will provide the desired fitness value, adding very basic rules that reward the controller that provided the longer trajectories and smaller times and higher battery life. This seems like a good strategy that will guide the control population into evolving and provide the optimal control, but this strategy on its own is not capable of doing just that: it needs to have a supervisor on the robots trajectory to make sure it is a forward moving trajectory and that they does not contain any looping parts. For this, a Neural Network (NN), is used to detect cycle trajectories that do not have the desired forward moving behavior by giving low activation value and higher activation values for the ones that are cycle free. The NN has two inputs and one output, and 2 hidden layers, see Fig. 8.

The evaluation method for the reactive controller has integrated both parts of the FIS and the NN where the fitness value for each individual is calculated with Eq. 3. Based on the response of the NN the peak activation value is set to 0.35, this meaning that any activation lower than 0.35 will penalize the fitness given by the FIS [2, 4–7].

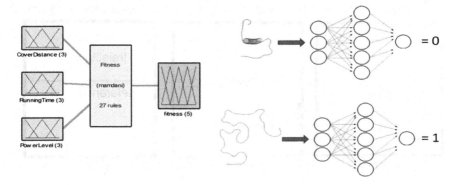

Fig. 8 Fitness function for the reactive controller

Equation 3 expresses how to calculate the fitness value of each individual

$$f(i) = \begin{cases} fv * nnv & nnv < 0.35 \\ fv & nnv \geq 0.35 \end{cases} \tag{3}$$

where:
fi Fitness value of the i-th individual,
fv Crisp value out of the fitness FIS,
nnv Looping trajectory activation value.

4.5.2 Tracking Controller Objective Function

The Tracking controller performance is measured with the RMSE between the reference and the robots trajectory; we apply the test three times and take the average on each of the three tests. The robot and the reference vertical position is random, but it's ensured that on one test the robots' vertical position is above the reference and on another test is below it. We do this to ensure the controller works properly for any case the robot may need it when its above or below (Fig. 9) [2, 4–7].

4.5.3 WFIS Controller Objective Function

The WFIS controller performance is measured by the RMSE between the reference and the robot's trajectory. We apply the test three times and take the average. On each of the three tests the robot's and the reference vertical position are random, but we make sure that on one test the robot's vertical position is above the reference and on another test is below it. We do this to ensure the controller works properly for any case the robot may need may need to deal with (Fig. 10) [2, 4–7].

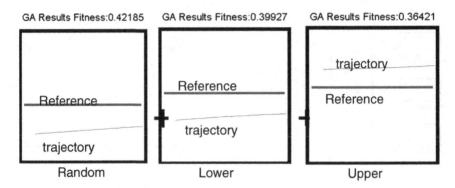

Fig. 9 Fitness function for the tracking controller

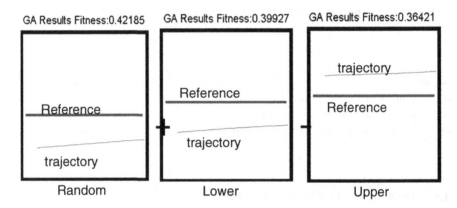

Fig. 10 Fitness functions for the WFIS controller

5 Simulation Results

This section presents the results of experiments carried out for the robot control system, with individual optimization of each behavior (tracking and reactive), necessary to obtain shown WFIS controller results.

The results are divided in 3 main principals:

- Tracking Controller
- Reactive Controller
- WFIS Controller

The tools that were used to conduct the experiments are Matlab and simulation tool Simrobot.

Table 1 Summary of tracking results

Particle	Iteration	C1	C2	Inertia weight	Constraint factor
20	500	1.4962	1.4962	LD	D
	Fitness		**Fitness**		**Fitness**
1	0.3206	11	0.3266	21	0.3293
2	0.3256	12	0.3148	22	0.3229
3	0.3014	13	0.3266	23	0.3117
4	0.3219	14	0.3255	24	0.3102
5	0.3266	15	0.2944	25	0.3117
6	0.3229	16	**0.3313**	26	0.3233
7	0.3164	17	0.2557	27	0.3136
8	0.3271	18	0.2806	28	0.3233
9	0.2919	19	0.2926	29	0.3313
10	0.2796	20	0.3154		
Average					0.312924138
Best					0.2557
Poor					0.3313
Std dev					0.000334708

5.1 Tracking Controller

Table 1 shows the configuration of the PSO and the results, in which we have the fitness value obtained in each experiment. It also shows the mean, variance, as well as the best and worst value obtained. In optimizing the weight of the inertia controller a constraint factor was used, not at the same time, but when the weight iteration inertia torque used as odd iteration a constraint factor is used.

Figure 11 shows the best path simulation during cycle PSO obtained for the tracking controller. The reference is defined by the red line and the smallest dot line is the trajectory of the robot, we can also see the graph of the FIS where input1 and input2 indicate the error on the position and orientation respectively, output1 and output2 speed to be applied to each actuator.

5.2 Reactive Controller

In this section, we show test of reactive controller, which includes the creation of a PSO algorithm to optimize the controller. The fitness of each controller is determined by their performance in the simulation tool. The robot should react in a closed environment (maze), avoiding obstacles present (walls) and must perform movements and avoid repeated.

Table 2 shows the configuration of PSO and displays the results, where we have the fitness value obtained in each experiment. It also shows the mean, variance, the best and worst value obtained. In this controller optimization constraint factor was used.

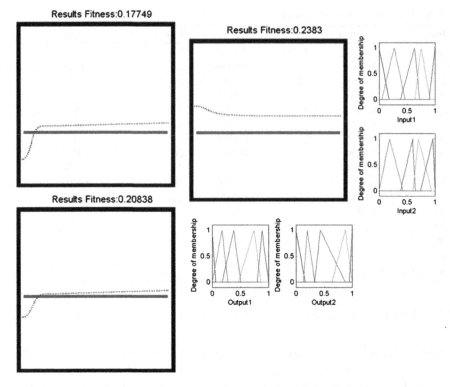

Fig. 11 Tracking controller results

Table 2 Summary of reactive controls results

Particle	Iteration	C1	C2	Inertia weight	Constraint factor
20	500	1.4962	1.4962	–	D
		Rank	Fitness		
		1	0.36249		
		2	0.36079		
		3	0.36219		
		4	0.35533		
		5	**0.35518**		
		6	0.39059		
		7	0.39018		
		8	0.37154		
		9	**0.41107**		
		10	0.39917		
Average					0.375853
Best					0.41107
Poor					0.35518
Std dev					0.0201816

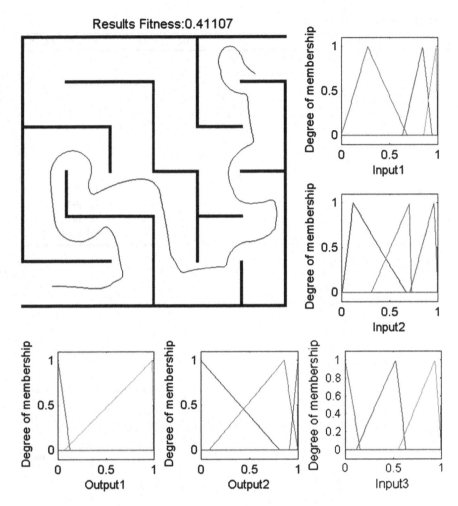

Fig. 12 Reactive controller results

Figure 12 shows the best robot path during the simulation of the PSO cycle obtained for the reactive controller. We can also see the graph of the FIS, where input1, input2 and input3 refer to the reading of the sensors robot orientation (straight, left and front), respectively, output1 and output2 speed to be applied to each servomotor.

5.3 WFIS Controller

In this section, we discuss the WFIS test driver. The best of each type of reagent and tracking controller are used as blocks in the behavioral integration system, which include the creation of a PSO algorithm to optimize the controller.

Table 3 Summary of WFIS results

Particle	Iteration	C1	C2	Inertia weight	Constraint factor
20	500	1.4962	1.4962	–	D
		Rank	Fitness		
		1	0.3497		
		2	0.3497		
		3	0.3497		
		4	0.3497		
		5	0.3497		
		6	0.3497		
		7	0.3497		
Average					0.3497
Best					0.3497
Poor					0.3497
Std dev					5.99589E−17

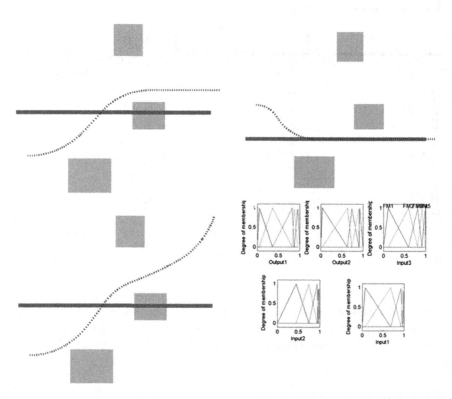

Fig. 13 WFIS controller results

The fitness of each controller is determined by its performance in the simulation tool, in which the robot must start from a random starting point and move forward on the reference line to avoid any collision. This test is performed 3 times forcing each robot controller to start at least once above and below the reference line.

Table 3 shows the configuration of PSO and results, in which we have the fitness value obtained in each experiment. It also shows the mean, variance, the best and worst value obtained. In this controller optimization a constraint factor was used.

Figure 13 shows the best robot path during the simulation of the PSO cycle obtained for the WFIS controller. We can also see the graph of the FIS, where input1, input2 and input3 refer to the reading of the sensors robot orientation (straight, left and front), respectively. Also output1 and output2 fuzzy weights will be applied to the controllers.

6 Conclusions

In this paper, the improved PSO algorithm used to tune the parameters of the fuzzy controller for the Reactive and Tracking Controllers, and we are currently working on the WFIS Optimization.

The fuzzy controllers both provide good results as they are able to guide the robot through the maze without hitting any wall and keep the robot on track. In comparison with the GA, only one of the two controllers in the tests that performed with the PSO proved to be better than the GA. Also, in the PSO less iterations have been performed, therefore consuming less computational resources.

Acknowledgments We would like to express our gratitude to CONACYT, and Tijuana Institute of Technology for the facilities and resources granted for the development of this research.

References

1. Measurement and Instrumentation, Faculty of Electrical Engineering and Computer Science, Brno University of Technology, Czech Republic Department of Control: Autonomous Mobile Robotics Toolboxfor Matlab 5 (2001). Online. http://www.uamt.feec.vutbr.cz/robotics/simulations/amrt/simrobot%20en.html
2. Melendez, A., Castillo, O.: Hierarchical genetic optimization of the fuzzy integrator for navigation of a mobile robot. In: Melin, P., Castillo, O. (eds.) Soft Computing Applications in Optimization, Control, and Recognition. Volume 294 of Studies in Fuzziness and Soft Computing, pp. 77–96. Springer, Berlin (2013)
3. Astudillo, L., Melin, P., Castillo, O.: Nature optimization applied to design a type-2 fuzzy controller for an autonomous mobile robot. In: 2012 Fourth World Congress on Nature and Biologically Inspired Computing (NaBIC), pp. 212, 217, 5–9 Nov 2012. doi: 10.1109/NaBIC. 2012.6402264

4. Melendez, A., Castillo, O.: Optimization of type-2 fuzzy reactive controllers for an autonomous mobile robot. In: 2012 Fourth World Congress on Nature and Biologically Inspired Computing (NaBIC), pp. 207–211 (2012)
5. Melendez, A., Castillo, O.: Evolutionary optimization of the fuzzy integrator in a navigation system for a mobile robot. In: Castillo, O., Melin, P., Kacprzyk, J. (eds.) Recent Advances on Hybrid Intelligent Systems, volume 451 of Studies in Computational Intelligence, pp. 21–31. Springer, Berlin (2013)
6. Melendez, A., Castillo, O., Soria, J.: Reactive control of a mobile robot in a distributed environment using fuzzy logic. In: Annual Meeting of the North American on Fuzzy Information Processing Society, 2008. NAFIPS 2008, pp. 1–5 (2008)
7. Melendez, A., Castillo, O., Garza, A., Soria, J.: Reactive and tracking control of a mobile robot in a distributed environment using fuzzy logic. In: FUZZ-IEEE, pp. 1–5 (2010)
8. Astudillo, L., Castillo, O., Aguilar, L.: Intelligent control of an autonomous mobile robot using type-2 fuzzy logic. In: IC-AI 2006, pp. 565–570
9. Adika, C.O., Wang, L.: Short term energy consumption prediction using bio-inspired fuzzy systems. In: North American Power Symposium (NAPS), 2012, pp. 1, 6, 9–11 Sept 2012
10. Amin, S., Adriansyah, A.: Particle swarm fuzzy controller for behavior-based mobile robot. In: 9th International Conference on Control, Automation, Robotics and Vision, 2006. ICARCV '06, pp. 1, 6, 5–8 Dec 2006. doi: 10.1109/ICARCV.2006.345293
11. Engelbrecht, A.P.: Fundamentals of Computational Swarm Intelligence. Wiley, New York, 2006
12. Astudillo, L., Castillo, O., Aguilar, L., Martínez, R.: Hybrid Control for an Autonomous Wheeled Mobile Robot under Perturbed Torques. IFSA (1), 594–603 (2007)
13. Cardenas, S., Garibaldi, J., Aguilar, L., Castillo, O.: Intelligent Planning and Control of Robots using Genetic Algorithms and Fuzzy Logic. In: IC-AI 2005, pp. 412–418
14. Castillo, O., Martinez, R., Melin, P., Valdez, F., Soria, J.: Comparative study of bio-inspired algorithms applied to the optimization of type-1 and type-2 fuzzy controllers for an autonomous mobile robot. Inf. Sci. 192, 19–38 (2012)
15. Cervantes, L., Castillo, O.: Design of a fuzzy system for the longitudinal control of an F-14 airplane. In: Soft Computing for Intelligent Control and Mobile Robotics, pp. 213–224 (2011)
16. Tsai, C.C., Tsai, K.I., Su, C.T.: Cascaded fuzzy-PID control using PSO-EP algorithm for air source heat pumps. In: 2012 International Conference on Fuzzy Theory and it's Applications (iFUZZY), pp. 163, 168, 16–18 Nov 2012
17. De Santis, E., Rizzi, A., Sadeghiany, A., Mascioli, F.M.F.: Genetic optimization of a fuzzy control system for energy flow management in micro-grids. In: IFSA World Congress and NAFIPS Annual Meeting (IFSA/NAFIPS), 2013 Joint, pp. 418, 423, 24–28 June 2013. doi: 10.1109/IFSA-NAFIPS.2013.6608437
18. Wang, D., Wang, G., Hu, R.: Parameters optimization of fuzzy controller based on PSO. In: 3rd International Conference on Intelligent System and Knowledge Engineering, 2008. ISKE 2008, vol. 1, pp. 599, 603, 17–19 Nov 2008
19. Eberhart, R.C., Shi, Y.: Comparing inertia weights and constriction factors in particle swarm optimization. In: Proceedings of the 2000 Congress on Evolutionary Computation, 2000, vol. 1, pp. 84, 88 (2000). doi: 10.1109/CEC.2000.870279
20. Esmin, A.A.A., Aoki, A.R., Lambert-Torres, G.: Particle swarm optimization for fuzzy membership functions optimization. In: 2002 IEEE International Conference on Systems, Man and Cybernetics, vol. 3, 6 pp., 6–9 Oct 2002
21. Fierro, R., Castillo, O.: Design of fuzzy control systems with different PSO variants. In: Recent Advances on Hybrid Intelligent Systems, pp. 81–88 (2013)
22. Fang, G., Kwok, N.M., Ha, Q.: Automatic fuzzy membership function tuning using the particle swarm optimization. In: Pacific-Asia Workshop on Computational Intelligence and Industrial Application, 2008. PACIIA '08, vol. 2, pp. 324, 328, 19–20 Dec 2008
23. Hassen, T., Ahmed, M., Mohamed, E.: Pso-belbic scheme for two-coupled distillation column process. J. Adv. Res. 2(1), 73–83 (2011)

24. Chen, J., Xu, L.: Road-junction traffic signal timing optimization by an adaptive particle swarm algorithm. In: 9th International Conference on Control, Automation, Robotics and Vision, 2006. ICARCV '06, pp. 1, 7, 5–8 Dec 2006

25. Kamejima, T., Phimmasone, V., Kondo, Y., Miyatake, M.: The optimization of control parameters of PSO based MPPT for photovoltaics. In: 2011 IEEE Ninth International Conference on Power Electronics and Drive Systems (PEDS), pp. 881, 883, 5–8 Dec 2011

26. Astudillo, L., Melin, P., Castillo, O.: Optimization of a fuzzy tracking controller for an autonomous mobile robot under perturbed torques by means of a chemical optimization paradigm. In: Recent Advances on Hybrid Intelligent Systems, pp. 3–20 (2013)

27. Wang, L., Kang, Q., Qiao, F., Wu, Q.: Fuzzy logic based multi-optimum programming in particle swarm optimization. In: Proceedings. 2005 IEEE Networking, Sensing and Control, 2005, pp. 473, 477, 19–22 March 2005

28. Mahendiran, T.V., Thanushkodi, K., Thangam, P., Gunapriya, B.: Speed control of three phase switched reluctance motor using particle swarm optimization. In: 2012 International Conference on Advances in Engineering, Science and Management (ICAESM), pp. 315, 319, 30–31 March 2012

29. Martínez, R., Castillo, O., Soria, J.: Particle swarm optimization applied to the design of type-1 and type-2 fuzzy controllers for an autonomous mobile robot. In: Bio-inspired Hybrid Intelligent Systems for Image Analysis and Pattern Recognition, pp. 247–262 (2009)

30. Martínez, R., Castillo, O., Aguilar, L.: Optimization of interval type-2 fuzzy logic controllers for a perturbed autonomous wheeled mobile robot using genetic algorithms. Inf. Sci. **179**(13), 2158–2174 (2009)

31. Martinez, R., Castillo, O., Aguilar, L., Baruch, I.: Bio-inspired optimization of fuzzy logic controllers for autonomous mobile robots. In: 2012 Annual Meeting of the North American on Fuzzy Information Processing Society (NAFIPS), pp. 1–6 (2012)

32. Martínez, R., Castillo, O., Aguilar, L., Melin, P.: Fuzzy logic controllers optimization using genetic algorithms and particle swarm optimization. MICAI **2**, 475–486 (2010)

33. Melin, P., Astudillo, L., Castillo, O., Valdez, F., Garcia, M.: Optimal design of type-2 and type-1 fuzzy tracking controllers for autonomous mobile robots under perturbed torques using a new chemical optimization paradigm. Expert Syst. Appl. **40**(8), 3185–3195 (2013)

34. García, M.A.P., Montiel, O., Castillo, O., Sepúlveda, R.: Optimal path planning for autonomous mobile robot navigation using ant colony optimization and a fuzzy cost function evaluation. In: Analysis and Design of Intelligent Systems using Soft Computing Techniques, pp. 790–798 (2007)

35. Milla, F., Sáez, D., Cortés, C.E., Cipriano, A.: Bus-stop control strategies based on fuzzy rules for the operation of a public transport system. In: IEEE Transactions on Intelligent Transportation Systems, vol. 13, no. 3, pp. 1394, 1403, Sept 2012

36. Yang, M., Wang, X.: Fuzzy PID controller using adaptive weighted PSO for permanent magnet synchronous motor drives. In: Second International Conference on Intelligent Computation Technology and Automation, 2009. ICICTA '09, vol. 2, pp. 736, 739, 10–11 Oct 2009

37. Montiel, O., Camacho, J., Sepúlveda, R., Castillo, O.: Fuzzy system to control the movement of a wheeled mobile robot. In: Soft Computing for Intelligent Control and Mobile Robotics, pp. 445–463 (2011)

38. Porta, M., Montiel, O., Castillo, O., Sepúlveda, R., Melin, P.: Path planning for autonomous mobile robot navigation with ant colony optimization and fuzzy cost function evaluation. Appl. Soft Comput. **9**(3), 1102–1110 (2009)

39. Martínez, R., Castillo, O., Aguilar, L.: Optimization of interval type-2 fuzzy logic controllers for a perturbed autonomous wheeled mobile robot using genetic algorithms. In: Soft Computing for Hybrid Intelligent Systems, pp. 3–18 (2008)

40. Rajeswari, K., Lakshmi, P.: PSO optimized fuzzy logic controller for active suspension system. In: 2010 International Conference on Advances in Recent Technologies in Communication and Computing (ARTCom), pp. 278, 283, 16–17 Oct 2010

41. Vaneshani, S., Jazayeri-Rad, H.: Optimized fuzzy control by particle swarm optimization technique for control of CSTR **5**(11), 464, 470 (2011)
42. Aguas-Marmolejo, S.J., Castillo, O.: Optimization of membership functions for type-1 and type 2 fuzzy controllers of an autonomous mobile robot using PSO. In: Recent Advances on Hybrid Intelligent Systems, pp. 97–104 (2013)
43. Selvakumaran, S., Parthasarathy, S., Karthigaivel, R., Rajasekaran, V.: Optimal decentralized load frequency control in a parallel ac-dc interconnected power system through fHVDCg link using fPSOg algorithm. Energy Procedia **14**(0), 1849, 1854 (2012). In: 2011 2nd International Conference on Advances in Energy Engineering (ICAEE)
44. Singh, R., Hanumandlu, M., Khatoon, S., Ibraheem, I.: An adaptive particle swarm optimization based fuzzy logic controller for line of sight stabilization tracking and pointing application. In: 2011 World Congress on Information and Communication Technologies (WICT), pp. 1259, 1264, 11–14 Dec 2011
45. Talbi, N.; Belarbi, K.: Fuzzy rule base optimization of fuzzy controller using hybrid tabu search and particle swarm optimization learning algorithm. In: 2011 World Congress on Information and Communication Technologies (WICT), pp. 1139, 1143, 11–14 Dec 2011
46. Valdez, F., Melin, P., Castillo, O.: Fuzzy control of parameters to dynamically adapt the PSO and GA Algorithms. In: FUZZ-IEEE 2010, pp. 1–8
47. Vázquez, J., Valdez, F., Melin, P.: Comparative study of particle swarm optimization variants in complex mathematics functions. In: Recent Advances on Hybrid Intelligent Systems, pp. 223–235 (2013)
48. Venayagamoorthy, G., Doctor, S.: Navigation of mobile sensors using PSO and embedded PSO in a fuzzy logic controller. In: Industry Applications Conference, 2004. 39th IAS Annual Meeting. Conference Record of the 2004 IEEE, vol. 2, pp. 1200, 1206, 3–7 Oct 2004
49. Wong, S., Hamouda, A.: Optimization of fuzzy rules design using genetic algorithm. Adv. Eng. Softw. **31**(4), 251–262 (2000). ISSN 0965-9978, http://dx.doi.org/10.1016/S0965-9978(99)00054-X
50. Yen, J, Langari R.: Fuzzy Logic: Intelligence, Control, and Information. Prentice Hall, Englewood Cliffs (1999)
51. Liu, Y., Zhu, X., Zhang, J., Wang, S.: Application of particle swarm optimization algorithm for weighted fuzzy rule-based system. In: 30th Annual Conference of IEEE Industrial Electronics Society, 2004. IECON 2004, vol. 3, pp. 2188, 2191, 2–6 Nov 2004
52. Zafer, B., Oğuzhan, K.: A fuzzy logic controller tuned with PSO for 2 DOF robot trajectory control. Expert Syst. Appl. **38**(1), 1017–1031 (2011). ISSN 0957-4174, http://dx.doi.org/10.1016/j.eswa.2010.07.131

Design of a Fuzzy System for Flight Control of an F-16 Airplane

Leticia Cervantes and Oscar Castillo

Abstract In this paper the main idea is to control the flight of an F-16 airplane using fuzzy system and PID controller to achieve the control. In general, to control the total flight is necessary to control the angle of the elevator, angle of the aileron and the angle of the rudder. For this reason, 3 fuzzy systems are used to control the respective angles. In this paper the fuzzy systems are presented with results using the simulation plant of the airplane.

1 Introduction

This paper focuses on the field of fuzzy logic and fuzzy control. In this paper the case of study is flight control of an airplane F-16, to maintain the airplane in stable position is necessary to control the elevators, ailerons and rudder [1–7]. To move the angle of elevator is necessary to move the wheel (pull it and push it), to move the angle of ailerons is necessary to move the wheel (left or right) and to control de angle of the rudder the pedals need to be push (one at a time left or right) [8–15]. In the first part of the paper the fuzzy systems [16–18] are explained and presented, then simulation plant is presented with the results also the behavior of the airplane with 40,000 and 20,000 ft in altitude and 650 and 900 ft/s in velocity is presented. This paper is organized as follows, Sect. 2 presents some basic concepts, in Sect. 3 problem description is described, Sect. 4 presents simulation results and finally Sect. 5 the conclusions.

L. Cervantes · O. Castillo (✉)
Tijuana Institute of Technology, Tijuana, Mexico
e-mail: ocastillo@tectijuana.mx

L. Cervantes
e-mail: Lettyy2685@hotmail.com

O. Castillo et al. (eds.), *Recent Advances on Hybrid Approaches for Designing*   209
Intelligent Systems, Studies in Computational Intelligence 547,
DOI: 10.1007/978-3-319-05170-3_15, © Springer International Publishing Switzerland 2014

Fig. 1 Flight maneuvers

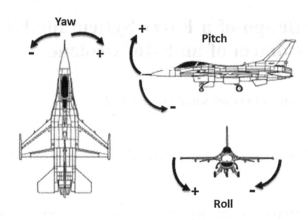

2 Background and Basic Concepts

Some basic concepts are presented in this section to understand in a better way the content of this paper. Any aircraft is capable of performing 3 possible rotations around 3 perpendicular axes whose point of intersection is above the center of gravity of the airplane [19–22]. These rotations are Pitch, Roll and Yaw, Pitch is the type of rotation that is done on the Y axis of the plane and it can be represented as an ascent or descent of its nose. Roll is a movement around the longitudinal axis of the plane (X axis) and can be seen as a lateral inclination thereof. Yaw is a rotation about the vertical axis (Z) of the plane, and produces a change in the horizontal direction of flight [23–27]. To control each rotation is necessary to use three types of control (side control, longitudinal control and direction control). The side control is responsible for maintaining control of the ailerons, longitudinal control is responsible for maintaining control of the elevators and direction control is responsible of the rudder [28–32].

3 Problem Description

In this case the main problem is to control the flight of an airplane F-16 in 3 axes (pitch, roll and yaw) see Fig. 1. To control the pitch is necessary to push or pull the wheel and these movement causes the elevators to go up and down the plane, to control the yaw is necessary to push the pedals (right or left), if the driver presses the left pedal, thus turning the rudder surface to the left and if right pedal is presses thus turning right. To control the roll wheel we need to move it (right or left) depending on the direction required.

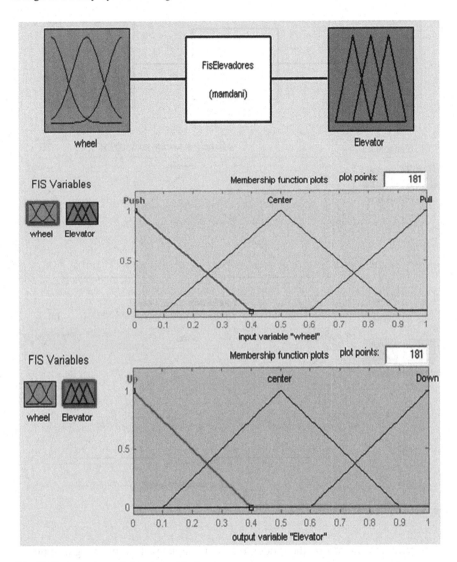

Fig. 2 Fuzzy system of elevator control

3.1 Fuzzy System

Three fuzzy systems were used to enforce the flight control. The first fuzzy system is used to control the elevator of the airplane, in this case the input of the fuzzy system is the wheel (push or pull) and the output is the angle of the elevators. In the second fuzzy system the ailerons need to be controlled, in this part the input is the wheel (move left or right) and the output is the angle of the ailerons. The last

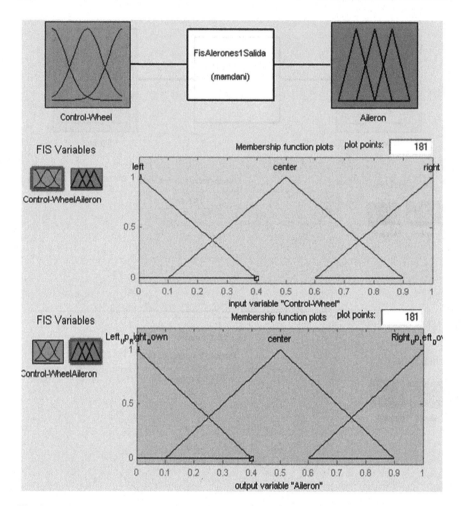

Fig. 3 Fuzzy system of aileron control

fuzzy system is controlling the rudder, the input are the pedals (press right or press left) and the output is the angle of the rudder when it is moving right or left depending on the desire direction. The fuzzy systems are shown in Figs. 2, 3 and 4.

Each fuzzy system has its own rules, and the rules of each fuzzy system are shown in Figs. 5, 6 and 7.

Having the fuzzy systems and the rules, the simulation was performed, and the simulation plant is shown in Fig. 8.

The original simulation plant in the block of the pilot has a PID controller that provides the control of the elevator, ailerons and the rudder to maintain the flight control. In this work the block of the pilot was changed for the fuzzy systems and a

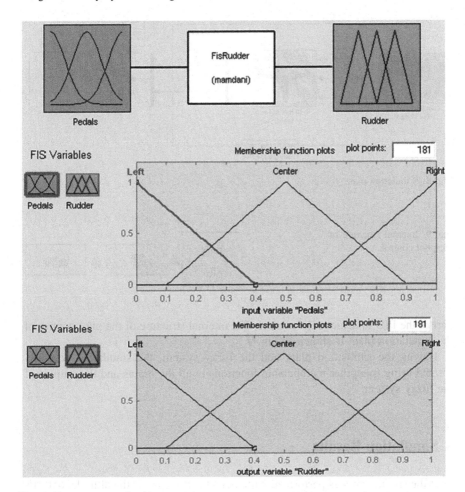

Fig. 4 Fuzzy system of rudder control

Fig. 5 Rules of the elevator
fuzzy system

1. If (wheel is Push) then (Elevator is Down) (1)
2. If (wheel is Center) then (Elevator is center) (1)
3. If (wheel is Pull) then (Elevator is Up) (1)

Fig. 6 Rules of the ailerons
fuzzy system

1. If (Control-Wheel is left) then (Aileron is Left_Up_Right_Down) (1)
2. If (Control-Wheel is center) then (Aileron is center) (1)
3. If (Control-Wheel is right) then (Aileron is Right_Up_Left_Down) (1)

Fig. 7 Rules of the rudder
fuzzy system

1. If (Pedals is Left) then (Rudder is Left) (1)
2. If (Pedals is Center) then (Rudder is Center) (1)
3. If (Pedals is Right) then (Rudder is Right) (1)

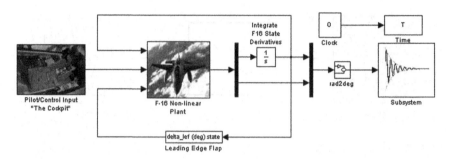

Fig. 8 Simulation plant

Fig. 9 Internal structure of
the pilot control

part of the PID to achieve the control. The internal structure of the pilot control of
the simulation plant is shown in Fig. 9.

Having the simulation plant and the fuzzy system, the simulation was performed using triangular membership functions in all the inputs and the outputs of
the fuzzy system.

4 Simulation Results

First the simulation was performed with a 20,000 ft value for the altitude, 650 ft/s
in velocity, 10° for the elevator disturbance deflection, in aileron disturbance
deflection of 10° and in rudder disturbance deflection of 10°. And the condition is
steady wings-level flight. The behavior of the elevator, aileron and the rudder are
shown Figs. 10, 11 and 12, the yellow line is the reference and the purple line is
the behavior.

The errors of the control are shown in Table 1.

The behavior in the flight control when the fuzzy system is applied is shown in
Figs. 13, 14, 15 and 16.

The previous figures show the changes during the 30 s, changes in the pitch,
roll, yaw, the behavior in the north and east position. After having results with the
simulation plant with 20,000 ft in altitude and 650 ft/s in velocity, then the altitude
was increased from 20,000 to 40,000 ft and the velocity was increased from 650 to
900 ft/s, this is to observe the behavior of the airplane when the velocity and the
altitude are increased and to observe if the fuzzy system can maintain the stability
of the airplane while it is on flight.

Fig. 10 Behavior of the elevators

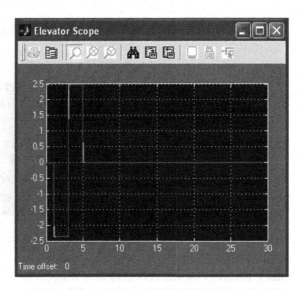

Fig. 11 Behavior of the ailerons

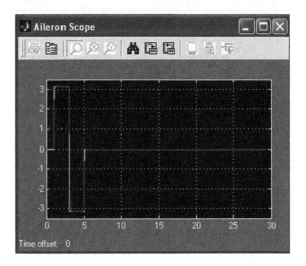

In this case $-15°$ in elevator disturbance deflection was used to observe the behavior. The behavior of the elevator, aileron and the rudder are shown in Figs. 17, 18 and 19. The yellow lines are the references and the purple lines are the behavior.

When the altitude and the velocity are modified the behavior of the airplane is different because it needs to modify its angles of the elevator, aileron and the rudder to have a stable flight. The results for this case are shown in Table 2.

Fig. 12 Behavior of the
rudder

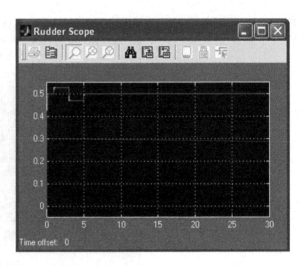

Table 1 Results for the simulation plant using triangular membership functions

	Elevator	Aileron	Ruder
Error	1.9721E−35	1.9611E−035	4.44E−20

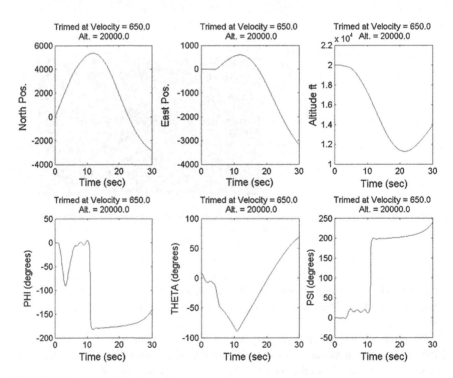

Fig. 13 Behavior in flight control

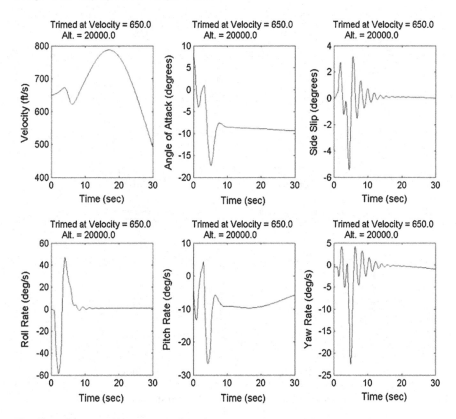

Fig. 14 Behavior in flight control of the 3 axes

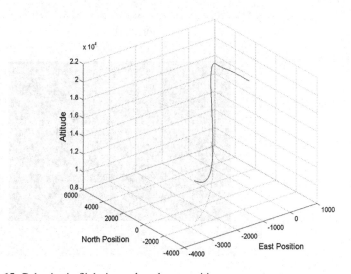

Fig. 15 Behavior in flight in north and east position

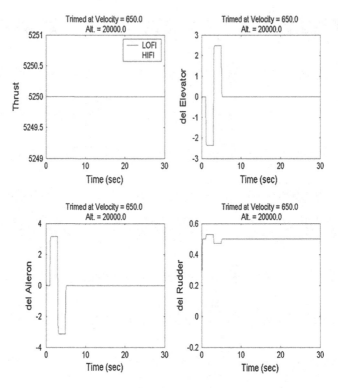

Fig. 16 Behavior of the elevators, ailerons, rudder and in thrust

Fig. 17 Behavior of the elevators

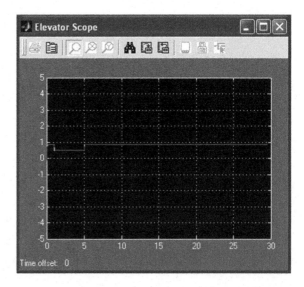

Fig. 18 Behavior of the
ailerons

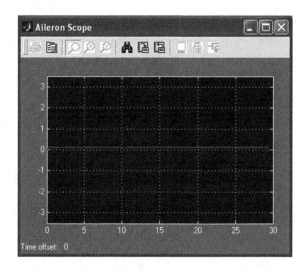

Fig. 19 Behavior of the
rudder

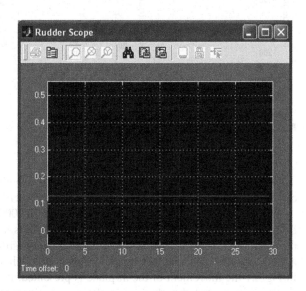

Table 2 Results for the simulation plant using triangular membership functions

	Elevator	Aileron	Ruder
Error	8.88E−20	2.22E−20	2.11E−20

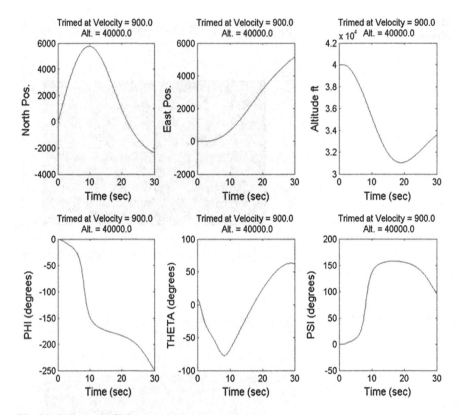

Fig. 20 Behavior in flight control

During the flight the angles and the behavior of the elevator, aileron and rudder can change and these changes and movement are shown in Figs. 20, 21, 22 and 23.

This last figure presented the simulation with 40,000 ft in altitude and changes in the velocity and in the perturbations in the angle is noteworthy the behavior in the positions in the 30 s and the difference when simulation used 2,000 ft in altitude is noticeable, and also the behavior of the pitch, roll and yaw are different. In Fig. 21 the movement of the airplane is presented, it shows the behavior in the north position, and the changes in the east position depending on the time of simulation and the altitude in this case the altitude started in 40,000 ft, and in the Fig. 21 the behavior is shown during the 30 s that drive the simulation.

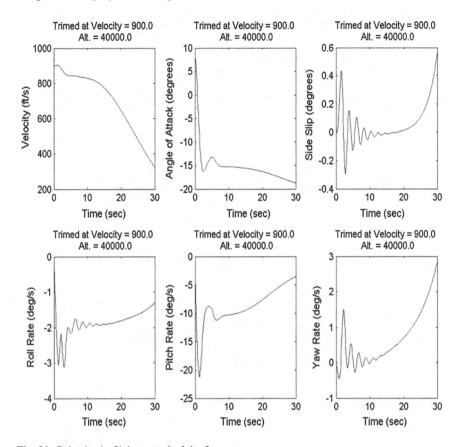

Fig. 21 Behavior in flight control of the 3 axes

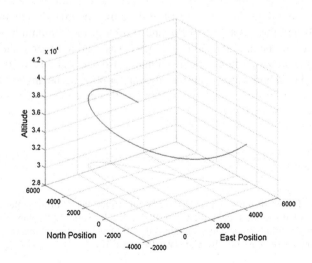

Fig. 22 Behavior in flight in north and east position

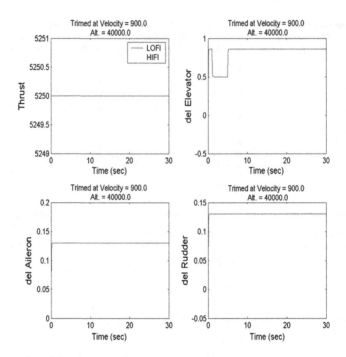

Fig. 23 Behavior of the elevators, ailerons, rudder and in thrust

5 Conclusions

Having obtained the previous results it is important to mention that when fuzzy systems are applied to achieve the flight control, the error in the simulation is very small, and in this case the fuzzy systems helped to maintain the stability of the airplane during the flight. In the first part of the simulation de altitude in the flight was 20,000 ft and the velocity 650 ft/s and results in the simulation were good, then an increase in the altitude and the velocity was used to observe the behavior, in this case the fuzzy systems could maintain also the stability of the airplane during the flight. In both cases disturbances were applied in the angles of elevator, aileron and rudder; however the fuzzy systems in this case of study maintained the stability of the plane in flight.

References

1. Castillo, O.: Design of Interval Type-2 Fuzzy Logic Controllers. Springer, Berlin (2012)
2. Castillo, O., Martinez-Marroquin, R., Melin, P., Valdez, F., Soria, J.: "Comparative study of bio-inspired algorithms applied to the optimization of type-1 and type-2 fuzzy controllers for an autonomous mobile robot". Inf. Sci. **192**, 19–38 (2012)

3. Castillo, O., Melin, P.: A review on the design and optimization of interval type-2 fuzzy controllers. Appl. Soft. Comput. **12**(4), 1267–1278 (2012)
4. Castillo, O., Melin, P.: New fuzzy-fractal-genetic method for automated mathematical modelling and simulation of robotic dynamic systems. IEEE International Conference on Fuzzy Systems 2. pp. 1182–1187, 1998
5. Castillo, O., Melin, P., Montiel, O., Sepúlveda, R.: Optimization of interval type-2 fuzzy logic controllers using evolutionary algorithms. Soft. Comput. **15**(6), 1145–1160 (2011)
6. Castillo, O., Kacprzyk, J., Pedrycz, W.: Soft Computing for Intelligent Control and Mobile Robotics. Springer, Berlin (2011)
7. Cázarez, N., Aguilar, L., Castillo, O.: Fuzzy logic control with genetic membership function parameters optimization for the output regulation of a servomechanism with nonlinear backlash. Expert Syst. Appl. **37**(6), 4368–4378 (2010)
8. Cervantes, L., Castillo, O.: Design of a fuzzy system for the longitudinal control of an F-14 airplane. In: Soft Computing for Intelligent Control and Mobile Robotics, vol 318/2011, pp. 213–224. Springer, Berlin (2011)
9. Cervantes, L., Castillo, O.: Intelligent control of nonlinear dynamic plants using a hierarchical modular approach and type-2 fuzzy logic. In: Lecture Notes in Computer Science, vol. 7095/2011, pp. 1–12. , Springer, Berlin (2011)
10. Cervantes, L., Castillo, O.: Hierarchical genetic algorithms for optimal type-2 fuzzy system design. Annual Meeting of the North American Fuzzy Information Processing Society, pp 324–329 (2011)
11. Cervantes, L., Castillo O.: Automatic design of fuzzy systems for control of aircraft dynamic systems with genetic optimization. World Congress and AFSS International Conference, pp. OS-413-1–OS-413-7 (2011)
12. Cervantes, L., Castillo, O.: Comparative study of type-1 and type-2 fuzzy systems for the three tank water control problem. LNAI 7630, pp. 362–373. Springer, Berlin (2013)
13. Coley, A.: An Introduction to Genetic Algorithms for Scientists and Engineers. World Scientific, Singapore (1999)
14. Dubois, D., Prade, H.: Fuzzy sets and systems: theory and applications. Academic Press, New York (1980)
15. Gibbens, P., Boyle, D.: Introductory flight mechanics and performance. University of Sydney, Australia (1999)
16. Haupt, R., Haupt, S.: Practical Genetic Algorithm. Wiley InterScience, Hoboken (2004)
17. Hidalgo, D., Melin, P., Castillo, O.: An optimization method for designing type-2 fuzzy inference systems based on the footprint of uncertainty using genetic algorithms. Expert Syst. Appl. **39**(4), 4590–4598 (2012)
18. Martinez-Soto, R., Castillo, O., Aguilar, L.: "Optimization of interval type-2 fuzzy logic controllers for a perturbed autonomous wheeled mobile robot using genetic algorithms". Inf. Sci. **179**(13), 2158–2174 (2009)
19. Melin, P., Castillo, O.: Adaptive intelligent control of aircraft systems with a hybrid approach combining neural networks, fuzzy logic and fractal theory. J. Appl. Soft Comput. **3**(4), 353–362 (2003)
20. Mitchell, M.: An Introduction to Genetic Algorithms. Massachusetts Institute of Technology Book, Cambridge (1999)
21. Rachman, E., Jaam, J., Hasnah, A.: Non-linear simulation of controller for longitudinal control augmentation system of F-16 using numerical approach. Inf. Sci. J. **164**(1–4) 47–60 (2004)
22. Reiner, J., Balas, G., Garrard, W.: Flight control design using robust dynamic inversion and time- scale separation. Autom. J. **32**(11), 1493–1504 (1996)
23. Sanchez, E., Becerra, H., Velez, C.: Combining fuzzy, PID and regulation control for an autonomous mini-helicopter. J. Inform. Sci. **177**(10), 1999–2022 (2007)
24. Sefer, K., Omer, C., Okyay, K.: Adaptive neuro-fuzzy inference system based autonomous flight control of unmanned air vehicles. Exp. Syst. Appl. J. **37**(2), 1229–1234 (2010)

25. Song, Y., Wang, H.: Design of flight control system for a small unmanned tilt rotor aircraft. Chin. J. Aeronaut. **22**(3), 250–256 (2009)
26. Walker, D.J.: Multivariable control of the longitudinal and lateral dynamics of a fly by wire helicopter. Control Eng. Pract. **11**(7), 781–795 (2003)
27. Zadeh, L.: Fuzzy Sets and Fuzzy Information Granulation Theory. Beijing Normal University Press, Beijing (2000)
28. Zadeh, L.: Fuzzy sets. Inf. Control. **8**(3), 338–353 (1965)
29. Zadeh, L.: Shadows of fuzzy sets. Probl. Peredachi Inf. **2**(1), 37–44 (1966)
30. Zadeh, L.: Fuzzy logic, neural networks and soft computing. ACM Commun. **37**(3), 77–84 (1994)
31. Zadeh, L.A.: Some reflections on soft computing, granular computing and their roles in the conception, design and utilization of information/intelligent systems. Soft. Comput. **2**, 23–25 (1998)
32. Zadeh, L.A.: Outline of a new approach to the analysis of complex systems and decision processes. IEEE Trans. Syst. Man Cybern. **SMC-3**, 28–44 (1973)

Bat Algorithm Comparison with Genetic Algorithm Using Benchmark Functions

Jonathan Pérez, Fevrier Valdez and Oscar Castillo

Abstract We describe in this chapter a Bat Algorithm and Genetic Algorithm (GA) conducting a performance comparison of the two algorithms Benchmark testing them in mathematical functions, parameters adjustment is done manually for both algorithms in 6 math functions, including some references on work done with the bat and area algorithm optimization with mathematical functions.

1 Introduction

We describe the algorithm of the bat in its original form by making a comparison with the genetic algorithm. In both algorithms parameters adjustment is performed manually based on the existing literature. These algorithms were applied to benchmark mathematical functions and comparative tables showing the optimization of results between Genetic Algorithm and bat algorithm are presented.

The comparative study of the two algorithms is performed in order to show the effectiveness of bat algorithm versus optimization problems, in the same manner prove effective against genetic algorithm.

The chapter is organized as follows: in Sect. 2 a description about Bat Algorithm, in Sect. 3 a description about Genetic Algorithm, in Sect. 4 description the mathematical functions, in Sect. 5 the simulations results are described and we can appreciate a comparison between bat algorithm and genetic algorithm, in Sect. 6 the conclusions obtained after the study of the two algorithms versus mathematical functions.

In the current literature there are papers where the algorithm has worked with the bat, like in the paper A New Metaheuristic Bat-Inspired Algorithm [17]. In this

J. Pérez · F. Valdez (✉) · O. Castillo
Tijuana Institute of Technology, Tijuana, BC, México
e-mail: fevrier@tectijuana.mx

O. Castillo et al. (eds.), *Recent Advances on Hybrid Approaches for Designing Intelligent Systems*, Studies in Computational Intelligence 547,
DOI: 10.1007/978-3-319-05170-3_16, © Springer International Publishing Switzerland 2014

chapter, we propose a new metaheuristic method, the Bat Algorithm, making comparison with the PSO and GA algorithms by applying them to Benchmark functions, in Article Performance of Firefly and Bat Algorithm for Unconstrained Optimization Problems [4]. In this chapter we compare the algorithm against firefly algorithm using bat Benchmark functions, the article proposes the Bat Algorithm: Literature Review and Applications [18], perform detailed explanation of the bat algorithm and the applications thereof and variants existing at present. In the paper A Comparison of BA, GA, PSO, BP and LM for Training Feed forward Neural Networks in e-Learning Context [8], a comparison of algorithms for training feed forward neural networks was done. Tests were done on two gradient descent algorithms: Backpropagation and Levenberg–Marquardt, and three population based heuristic: Bat Algorithm, Genetic Algorithm, and Particle Swarm Optimization. Experimental results show that bat algorithm (BA) outperforms all other algorithms in training feed forward neural networks [8], A Binary Bat Algorithm for Feature Selection [12], performed applying the bat algorithm for feature selection using a binary version of the algorithm, in Article Local Memory Search Bat Algorithm for Grey Economic Dynamic System [19], in this chapter, the LMSBA is introduced in economic control field, test and simulation results are ideal, and programming of method is concise. This algorithm is suitable for numerical solution in practical dynamic economic control, providing numerical theoretical foundation for steady, healthy and optimal economic growth [12]. In the article Solving Multi-Stage MultiMachine Multi-Product Scheduling Problem Using Bat Algorithm, the algorithm takes into account the just in time production philosophy by aiming to minimise the combination of earliness and tardiness penalty costs [11]. In the article Use of Fuzzy Systems and Bat Algorithm for Energy Modeling in a Gas Turbine Generator, The purpose of this chapter has been to demonstrate the use of fuzzy methods to capture variation of energy destruction in a GTG [1].

In other items listed below on the use of bat algorithm we have, in the article Chaotic bat algorithm [3], performed in which the aggregation of chaos in the standard version of bat algorithm. In the article A bat-inspired algorithm for structural optimization [6], performing the comparison of several algorithms showing affection bat algorithm in resolving problems braced dome truss bar among others. In the article Bat inspired algorithm for discrete size optimization of steel frames [5], The objective in this study is to investigate efficiency of the bat algorithm in discrete sizing optimization problems of steel frames [5]. In the article A New Meta-heuristic Bat Inspired Classification Approach for Microarray Data [10], the bat algorithm successfully formulated and is used to update the weight of the FLANN classifier. In the article A wrapper approach for feature selection based on Bat Algorithm and Optimum-Path Forest [13], In this chapter, the authors presented a wrapper feature selection approach based on the Bat Algorithm and Optimum-Path Forest classifier, which combines an exploration of the search space and an intense local analysis by exploiting the neighborhood of a good solution to reduce the feature space dimensionality [13], in the article Bat algorithm for the fuel arrangement optimization of reactor core [7], for the first time, the bat optimization algorithm is applied for the LPO problem. Prior to perform the LPO, the developed BA was validated against a test function obtaining the exact minimum value during various iterations.

2 Bat Algorithm

If we idealize some of the echolocation characteristics of microbats, we can develop various bat-inspired algorithms or bat algorithms. For simplicity, we now use the following approximate or idealized rules [17]:

1. All bats use echolocation to sense distance, ant they also 'know' the difference between food/prey and background barriers in some magical way.
2. Bats fly randomly witch velocity v_i at position x_i witch a fixed frequency f_{min}, varying wavelength λ and loudness A_0 to search for prey. They can automatically adjust the wavelength (or frequency) of their emitted pulses and adjust the rate of pulse emission $r \in [0, 1]$, depending on the proximity of their target.
3. Although loudness can vary in many ways, we assume that the loudness varies from a large (positive) A_0 to a minimum constant value A_{min}.

For simplicity, the frequency $f \in [0, f_{max}]$, the new solutions x_i^t and velocity v_i^t at a specific time step t are represented by a random vector drawn from a uniform distribution [4].

2.1 Pseudo Code for Bat Algorithm

Objective function f(x), $x = (x_1, ..., xd)^T$
Initialize the bat population x_i (i = 1, 2, ..., n) and v_i
Define pulse frequency f_i at x_i
Initialize pulse rates r_i and the loudness A_i
***While** (t < Max numbers of iterations)*
Generate new solutions by adjusting frequency and updating velocities and locations/solutions [equations (1) to (3)]
  ***if**(rand > r_i)*
  Select a solution among the best solutions
  Generate a local solution around the selected best solution
  end if
  Generate a new solutions by flying randomly
  ***if** (rand < A_i & f(x_i) < f($x*$))*
  Accept the new solutions and increase r_i and reduce A_i
  end if

Rank the bats and find the current best x_*
end while
Postprocess results and visualization

2.2 Movements in Bat Algorithm

Each bat is associated with a velocity v_i^t and location x_i^t, alt iteration t, in a dimensional search or solution space. Among all the bats, there exist a current best solution x_*. Therefore, the above three rules can be translated into the updating equations for x_i^t and velocities v_i^t:

$$f_i = f_{min} + (f_{max} - f_{min})\beta, \tag{1}$$

$$v_i^t = v_i^{t-1} + (x_i^{t-1} - x_*)f_i, \tag{2}$$

$$x_i^t = x_i^{t-1} + v_i^t, \tag{3}$$

where $\beta \in [0, 1]$ is a random vector drawn from a uniform distribution [13].

As mentioned earlier, we can either use wavelengths or frequencies for implementation, we will use $f_{min} = 0$ and $f_{max} = 1$, depending on the domain size of the problem of interest. Initially, each bat is randomly assigned a frequency which is drawn uniformly from $[f_{min} - f_{max}]$. The loudness and pulse emission rates essentially provide a mechanism for automatic control and auto zooming into the region with promising solutions [18].

2.3 Loudness and Pulse Rates

In order to provide an affective mechanism to control the exploration and exploitation and switch to exploitation stage when necessary, we have to vary the loudness A_i and the rate r_i of pulse emission during the iterations. Since the loudness usually decreases once a bat has found its prey, while the rate of pulse emission increases, the loudness can be chosen as any value of convenience, between A_{min} and A_{max}, assuming $A_{min} = 0$ means that a bat has just found the prey and temporarily stop emitting any sound. With these assumptions, we have

$$A_i^{t+1} = \alpha A_i^t, \quad r_i^{t+1} = r_i^0[1 - \exp(-\gamma^t)], \tag{4}$$

where α and γ are constants. In essence, here α is similar to the cooling factor of a cooling schedule in simulated annealing. For any $0 < \alpha < 1$ and $\gamma > 0$, we have

$$A_i^t \to 0, \ r_i^t \to r_i^0, \quad \text{as } t \to \infty. \tag{5}$$

In the simplest case, we can use $\alpha = \gamma$, and we have used $\alpha = \gamma = 0.9$–0.98 in our simulations [4].

3 Genetic Algorithms

Genetic algorithms (GA) emulate genetic evolution. The characteristics of individuals are therefore expressed using genotypes. The original form of the GA, as illustrated by John Holland in 1975, had distinct features: (1) a bit string representation, (2) proportional selection, and (3) cross-over as the primary method to produce new individuals. Since then, several variations to the original Holland GA have been developed, using different representation schemes, selection, cross-over, mutation and elitism operators [2].

3.1 Representation

The classical representation scheme for GAs is a binary vectors of fixed length. In the case of an n_x-dimensional search space, each individual consists on n_x variables with each variable encoded as a bit string. If variables have binary values, the length of each chromosome is n_x bits. In the case of nominal-valued variables, each nominal value can be encoded as an n_d-dimensional bit vectors where 2nd is the total numbers of discrete nominal values for that variable. Each n_d-bit string represents a different nominal value. In the case of continuous-valued variables, each variable should be mapped to an n_d-dimensional bit vector,

$$\phi : R \rightarrow (0,1)^{n_d} \qquad (6)$$

The range of continuous space needs to be restricted to a finite range, $[\mathbf{x_{min}}, \mathbf{x_{max}}]$. Using standard binary decoding, each continuous variable x_{ij} of chromosome x_i is encoded using a fixed length bit string.

Gas have also been developed that use integer or real-valued representations and order-based representations where the order of variables in a chromosome plays an important role. Also, it is not necessary that chromosomes be of fixed length [2].

3.2 Crossover Operations

Several crossover operators have been developed for GA's depending on the format in which individuals are represented. For binary representations, uniform crossover, one-point crossover and two-point crossover are the most popular:

- **Uniform Crossover**, where corresponding bit positions are randomly exchanged between the two parents to produce two offspring.
- **One-Point Crossover**, where a random bit position is selected, and the bit substrings after the selected bit are swapped between the two parents to produce two offspring.

- **Two-Point Crossover**, where two bit positions are randomly selected and the bit substrings between the selected bit positions are swapped between the two parents.

For continuous valued genes, arithmetic crossover can be used:

$$x_{ij} = r_j x_{1j} + (1.0 - r_j) x_{2j} \qquad (7)$$

where $r_j \sim U(0, 1)$ and x_i is the offspring produced from parents x_1 and x_2 [2].

3.3 Mutation

The mutation scheme used in a GA depends on the representation scheme. In the case of bit string representations,

- **Random Mutation**, randomly negates bits, while
- **In-Order Mutation**, performs random mutation between two randomly selected bit positions.

For discrete genes with more than two possible values that a gene can assume, random mutation selects a random value from the finite domain of the gene. In the case of continuous valued genes, a random value sampled from a Gaussian distribution with zero mean and small deviation is usually added to the current gene value. As an alternative, random noise can be sampled from a Cauchy distribution [2].

4 Benchmark Mathematical Functions

This section list a number of the classical benchmark functions used to evaluate the performance of optimization algorithms.

In the area of optimization using mathematical functions have been carried out work as mentioned below: A new gravitational search algorithm using fuzzy logic to parameter adaptation [14], Optimal design of fuzzy classification systems using PSO with dynamic parameter adaptation through fuzzy logic [9], Parallel Particle Swarm Optimization with Parameters Adaptation Using Fuzzy Logic [15, 16].

The mathematical functions are shown below:

- **Spherical**

$$f(x) = \sum_{j=1}^{n_x} x_j^2 \qquad (8)$$

Witch $x_j \in [-100, 100]$ and $f^*(x) = 0.0$

- **Rosenbrock**

$$f(x) = \sum_{j=1}^{n_z/2} [100(x_{2j} - x_{2j-1}^2)^2 + (1 - x_{2j-1})^2] \tag{9}$$

Witch $x_j \in [-2.048, 2.048]$ and $f^*(x) = 0.0$

- **Rastrigin**

$$f(x) = \sum_{j=1}^{n_x} (x_j^2 - 10\cos(2\pi x_j) + 10) \tag{10}$$

With $x_j \in [-5.12, 5.12]$ and $f^*(x) = 0.0$

- **Ackley**

$$f(x) = -20e^{-0.2\sqrt{\frac{1}{n_x}\sum_{j=1}^{n_x} x_j^2 - \frac{1}{n_x}\sum_{j=1}^{n_x}\cos(2\pi x_j)}} + 20 + e \tag{11}$$

With $x_j \in [-30, 30]$ and $f^*(x) = 0.0$

- **Zakharov**

$$f(x) = \sum_{i=1}^{n} x_i^2 + (\sum_{i=1}^{n} 0.5ix_i)^2 + (\sum_{i=1}^{n} 0.5ix_i)^4 \tag{12}$$

Witch $x_i \in [-5, 10]$ and $f^*(x) = 0.0$

- **Sum Square**

$$f(x) = \sum_{i=1}^{n} ix_i^2 \tag{13}$$

Witch $x_i \in [-2, 2]$ and $f^*(x) = 0.0$

5 Simulation Results

In this section the comparison Bat algorithm is made against the genetic algorithm, in each of the algorithms integrating 6 math functions Benchmark was performed separately likewise a dimension of 10 variables were managed with 30 tests for each function varying the parameters of the algorithms.

The parameters in the handled Bat algorithm were:

- Population size: 2–40 Bats
- Volume: 0.5–1
- Pulse frequency: 0.5–1
- Frequency min.: 0–2
- Frequency max.: 0–2.

The parameters for the genetic algorithm are shown below:

- Number of Individuals: 4–40
- Selection: Stochastic, Remainder, Uniform, Roulette
- Crossover: Scattered, Single Point, Two Point, Heuristic, Arithmetic
- Mutation: Gaussian, Uniform.

5.1 Simulation Results with Bat Algorithm

In this section we show the results obtained by the bat algorithm in separate tables of the math functions.

From Table 1 it can be appreciated that after executing the Bat Algorithm 30 times, with different parameters, we can see the best, average and worst results for the Spherical function.

From Table 2 it can be appreciated that after executing the Bat Algorithm 30 times, with different parameters, we can see the best, average and worst results for the Rosenbrock function.

From Table 3 it can be appreciated that after executing the Bat Algorithm 30 times, with different parameters, we can see the best, average and worst results for the Rastrigin function.

From Table 4 it can be appreciated that after executing the Bat Algorithm 30 times, with different parameters, we can see the best, average and worst results for the Ackley function.

From Table 5 it can be appreciated that after executing the Bat Algorithm 30 times, with different parameters, we can see the best, average and worst results for the Zakharov function.

From Table 6 it can be appreciated that after executing the Bat Algorithm 30 times, with different parameters, we can see the best, average and worst results for the Sum Square function.

Table 1 Simulation results for the Spherical function

Number of bats	Best	Worst	Mean
2	0.008573311	0.628709060	−0.398181821
5	0.003604271	0.515240072	0.029093107
10	0.000028268	0.000734987	0.000073050
20	0.000134559	0.000475252	−0.000017147
30	0.000012139	0.000396452	0.000108109
40	0.000008976	0.000551191	0.000004764

Table 2 Simulation results for the Rosenbrock function

Number of bats	Best	Worst	Mean
2	0.005489045	1.485626894	0.200876688
5	0.272515093	1.634311040	0.097148588
10	0.303810519	1.795474228	−0.048411515
20	0.137302657	1.678337846	0.023740171
30	0.389096958	1.783597538	0.628591616
40	0.251285348	1.335020023	0.178971864

Table 3 Simulation results for the Rastrigin

Number of bats	Best	Worst	Mean
2	0.073011031	1.860011728	0.610017174
5	0.063675308	1.041547381	−0.035659748
10	0.049162432	1.972491336	0.530195263
20	0.008711488	1.186999703	0.153645344
30	0.003517482	1.008732703	−0.054677168
40	0.043943569	0.910166115	−0.281693828

Table 4 Simulation results for the Ackley function

Number of bats	Best	Worst	Mean
2	0.00060716	1.314174678	0.006825389
5	0.00040240	0.987404904	0.098504521
10	0.00001427	0.991924651	−0.396498938
20	0.00018141	1.985907809	0.198576524
30	0.00017769	1.983460152	−0.09910227
40	0.00003844	0.982687927	0.098321367

Table 5 Simulation results for the Zakharov function

Number of bats	Best	Worst	Mean
2	0.02818104	1.646383953	0.404631305
5	0.02152469	0.727428301	−0.063022496
10	0.00010811	0.000507957	−5.31073E−05
20	0.00005543	0.000550302	−8.59865E−05
30	0.00000521	0.000503802	1.72884E−05
40	0.00000858	0.000701617	0.000115979

Table 6 Simulation results for the Sum Square function

Number of bats	Best	Worst	Mean
2	0.01027493	1.518982322	0.455767428
5	0.00129802	0.993277001	0.063612226
10	0.00010129	0.000507828	0.000143938
20	0.00002421	0.000632709	8.70312E−05
30	0.00000613	0.000523433	0.000199962
40	0.00003690	0.000490066	−0.000101807

Table 7 Simulation results for the Sphere function

Population	Best	Worst	Mean
4	0.000655746	0.867154805	0.501445118
5	0.016158419	0.735175081	0.297672568
10	0.029477858	0.985900891	0.616489628
20	0.125851757	1.018067545	0.250640697
30	0.050431819	0.928690136	0.521845011
40	0.004944109	1.847289399	0.558953430

5.2 Simulation Results with Genetic Algorithm

In this section we show the results obtained by the genetic algorithm in separate tables of the math functions.

From Table 7 it can be appreciated that after executing the Genetic Algorithm 30 times, with different parameters, we can see the best, average and worst results for the Sphere function.

From Table 8 it can be appreciated that after executing the Genetic Algorithm 30 times, with different parameters, we can see the best, average and worst results for the Rosenbrock function.

From Table 9 it can be appreciated that after executing the Genetic Algorithm 30 times, with different parameters, we can see the best, average and worst results for the Rastrigin function.

From Table 10 it can be appreciated that after executing the Genetic Algorithm 30 times, with different parameters, we can see the best, average and worst results for the Ackley function.

From Table 11 it can be appreciated that after executing the Genetic Algorithm 30 times, with different parameters, we can note the best, average and worst results for the Zakharov function.

From Table 12 it can be appreciated that after executing the Genetic Algorithm 30 times, with different parameters, we can see the best, average and worst results for the Sum Square function.

Table 8 Simulation results for the Rosenbrock function

Population	Best	Worst	Mean
4	0.089847334	0.802024183	0.561886727
5	0.045156568	0.878087097	0.476680695
10	0.026476082	0.788665597	0.212878969
20	0.008755795	0.654965394	0.151633433
30	0.001220403	0.292128413	0.050677876
40	0.000245092	0.843183891	0.242982579

Table 9 Simulation results for the Rastrigin function

Population	Best	Worst	Mean
4	0.014939893	0.997649164	0.532383503
5	0.025389969	1.023785562	0.484650675
10	0.000983143	1.991943912	0.591455815
20	0.005860446	0.973098541	0.416048173
30	0.001461684	1.030270667	0.402938563
40	0.0017108	1.011176844	0.304818043

Table 10 Simulation results for the Ackley function

Population	Best	Worst	Mean
4	0.003764969	0.974296237	0.346336685
5	0.034816548	1.008082845	0.555913584
10	0.00068874	1.006617458	0.309683405
20	0.00000943	0.999864848	0.217317909
30	0.0024992	0.045123886	0.01309672
40	0.00110442	0.963070253	0.229423519

Table 11 Simulation results for the Zakharov function

Population	Best	Worst	Mean
4	0.00291835	0.974030463	0.442892629
5	0.003929128	2.494950163	0.368494675
10	0.000621031	0.999709009	0.475767117
20	0.00754580	0.787983798	0.444793802
30	0.00074412	4.568706165	1.518534814
40	0.01262987	2.859991576	1.30142878

Table 12 Simulation results for the Sum Square function

Population	Best	Worst	Mean
4	0.029476905	0.834760272	0.379557704
5	0.009588642	0.441979109	0.172141341
10	0.000848578	0.045665911	−0.002597231
20	0.01018758	0.668265573	0.194550999
30	0.00259567	0.226922795	0.103921866
40	0.00638302	0.406954403	0.149111035

6 Conclusions

In the simulation analysis for the comparative study of genetic algorithms and the effectiveness of the bat algorithm, bat algorithm during the modification of parameters by trial and error is shown, the rate of convergence of the genetic algorithm is much faster, the comparison was made with the original versions of the two algorithms with the recommended literature parameters, this conclusion is based on 6 Benchmark math functions results may vary according to mathematical or depending on the values set in the parameters of the algorithm.

The application of various problems in bat algorithm has a very wide field where the revised items its effectiveness is demonstrated in various applications, their use can be mentioned in the processing digital pictures, search for optimal values, neural networks, and many applications.

Acknowledgments We would like to express our gratitude to the CONACYT and Tijuana Institute of Technology for the facilities and resources granted for the development of this research.

References

1. Alemu, T., Mohd, F.: Use of Fuzzy Systems and Bat Algorithm for Exergy Modeling in a Gas Turbine Generator, Tamiru Alemu Lemma. Department of Mechanical Engineering, Malaysia (2011)
2. Engelbrecht, A.: Fundamentals of Computational Swarm Intelligence, pp. 25–26 and 66–70. Wiley, Chichester (2005)
3. Gandomi, A., Yang, X.: Chaotic bat algorithm. Department of Civil Engineering, The University of Akron, USA (2013)
4. Goel, N., Gupta, D., Goel, S.: Performance of Firefly and Bat Algorithm for Unconstrained Optimization Problems. Department of Computer Science, Maharaja Surajmal Institute of Technology GGSIP University C-4, Janakpuri (2013)
5. Hasançebi, O., Carbas, S.: Bat Inspired Algorithm for Discrete Size Optimization of Steel Frames. Department of Civil Engineering, Middle East Technical University, Ankara (2013)
6. Hasançebi, O., Teke, T., Pekcan, O.: A Bat-Inspired Algorithm for Structural Optimization. Department of Civil Engineering, Middle East Technical University, Ankara (2013)
7. Kashi, S., Minuchehr, A., Poursalehi, N., Zolfaghari, A.: Bat algorithm for the fuel arrangement optimization of reactor core. Nuclear Engineering Department, Shahid Beheshti University, Tehran (2013)
8. Khan, K., Sahai, A.: A Comparison of BA, GA, PSO, BP and LM for Training Feed forward Neural Networks in e-Learning Context. Department of Computing and Information Technology, University of the West Indies, St. Augustine (2012)
9. Melin, P., Olivas, F., Castillo, O., Valdez, F., Soria, J., Garcia, J.: Optimal design of fuzzy classification systems using PSO with dynamic parameter adaptation through fuzzy logic. Expert Syst. Appl. **40**(8), 3196–3206 (2013)
10. Mishra, S., Shaw, K., Mishra, D.: A New Meta-heuristic Bat Inspired Classification Approach for Microarray Data. Institute of Technical Education and Research, Siksha O Anusandhan Deemed to be University, Bhubaneswar (2011)

11. Musikapun, P., Pongcharoen, P.: Solving Multi-Stage MultiMachine Multi-Product Scheduling Problem Using Bat Algorithm. Department of Industrial Engineering, Faculty of Engineering, Naresuan University, Thailand (2012)
12. Nakamura, R., Pereira, L., Costa, K., Rodrigues, D., Papa J., BBA: A Binary Bat Algorithm for Feature Selection. Department of Computing Sao Paulo State University Bauru, Brazil (2012)
13. Rodrigues, D., Pereira, L., Nakamura, R., Costa, K., Yang, X., Souza, A., Papa, J.P.: A wrapper approach for feature selection based on Bat Algorithm and Optimum-Path Forest. Department of Computing, Universidade Estadual Paulista, Bauru (2013)
14. Sombra, A., Valdez, F., Melin, P., Castillo, O.: A new gravitational search algorithm using fuzzy logic to parameter adaptation. IEEE Congress on Evolutionary Computation, pp. 1068–1074. (2013)
15. Valdez, F., Melin, P., Castillo, O.: Evolutionary method combining particle swarm optimization and genetic algorithms using fuzzy logic for decision making. In: Proceedings of the IEEE International Conference on Fuzzy Systems, pp. 2114–2119 (2009)
16. Valdez, F., Melin, P., Castillo, O.: Parallel Particle Swarm Optimization with Parameters Adaptation Using Fuzzy Logic. MICAI, vol. 2, pp. 374–385. (2012)
17. Yang, X.: A New Metaheuristic Bat-Inspired Algorithm. Department of Engineering, University of Cambridge, Cambridge (2010)
18. Yang, X.: Bat Algorithm: Literature Review and Applications. School of Science and Technology, Middlesex University, London (2013)
19. Yuanbin, M., Xinquan, Z., Sujian, X.: Local Memory Search Bat Algorithm for Grey Economic Dynamic System. Statistics and Mathematics Institute (2013)

Comparative Study of Social Network Structures in PSO

Juan Carlos Vazquez, Fevrier Valdez and Patricia Melin

Abstract In this chapter a comparative study of social network structures in Particle Swarm Optimization is performed. The social networks employed by the *gbest* PSO and *lbest* PSO algorithms are star, ring, Von Neumann and random topologies. Each topology is implemented on four benchmark functions. The objective is knows the performance between each topology with different dimensions. Benchmark functions were used such as Rastrigin, Griewank, Rosenbrock and Sphere.

1 Introduction

The particle swarm algorithm [1–3] is based on a social psychological model of social influence and social learning. A population of candidate problem solutions, randomly initialized in a high-dimensional search space, discovers optimal regions of the space through a process of individuals' emulation of the successes of their neighbors. Each particle represents a candidate solution [4–6]. The performance of each particle is measured using a fitness function that varies depending on the optimization problem. The PSO algorithm has been applied successfully to a number of problems, including standard function optimization problems [7–12].

Different social networks (topologies) which differ in the size of their neighborhood have been developed [13, 14] from two common PSO algorithms [1, 3, 15]; those two algorithms are the global best (*gbest*) and local best(*lbest*). In this

J. C. Vazquez · F. Valdez (✉) · P. Melin
Tijuana Institute of Technology, Tijuana, Mexico
e-mail: fevrier@tectijuana.mx

J. C. Vazquez
e-mail: juancarlos_jcs27@hotmail.com

P. Melin
e-mail: pmelin@tectijuana.edu.mx

O. Castillo et al. (eds.), *Recent Advances on Hybrid Approaches for Designing Intelligent Systems*, Studies in Computational Intelligence 547,
DOI: 10.1007/978-3-319-05170-3_17, © Springer International Publishing Switzerland 2014

chapter, a comparative study of social network structures are implemented on four benchmark functions.

The rest of chapter is organized as follows: Sect. 2 describes the standard PSO algorithm, Sect. 3 shows an overview of neighborhoods topologies, Sect. 4 provides the parameters of the algorithm, Sect. 5 presents the results of the experiments and finally, conclusions are summarized in Sect. 6.

2 Standard PSO Algorithm

Particle Swarm Optimization is one of the bio-inspired computation techniques developed by Eberhart and Kennedy in 1995 [1–3] based on the social behaviors of birds flocking or fish schooling, biologically inspired computational search and optimization method [16]. The process of the PSO algorithm in finding optimal values follows the work of this animal society. This social algorithm maintains a swarm of particles, where each particle represents a potential solution. In analogy with evolutionary computation paradigms, a *swarm* is similar to a population, while a *particle* is similar to an individual. In simple terms, the particles are flown through a multidimensional search space, where the position of each particle is adjusted according to its own experience and that of its neighbors. Let $x_i(t)$ denote the position of particle i in the search space at time step t. The position of the particle is changed by adding a velocity, $v_i(t)$ to the current position, with following equation

$$x_i(t+1) = x_{ij}(t) + v_{ij}(t+1) \tag{1}$$

It is the velocity vector that drives the optimization process, and reflects both the experiential knowledge of the particle and socially exchanged information from the particle's neighborhood.

2.1 Global Best PSO

The neighborhood for each particle is the entire swarm. The social network employed by the *gbest* PSO reflects the star topology (refer to Sect. 3.1). The social information is the best position found by the swarm, referred to as PSO consist of a swarm of particles moving in an n dimensional. Every particle has a position vector ($x_{ij}(t)$) encoding a candidate solution to the problem and velocity vector ($v_{ij}(t+1)$). Moreover, each particle contain as $\hat{y}(t)$. For the *gbest* PSO, the velocity of particle i is calculated with following equation

$$v_{ij}(t+1) = vi_{ij}(t) + c_1 r_{1j}(t)[y_{ij}(t) - x_{ij}(t)] + c_2 r_{2j}(t)[\hat{y}_j(t) - x_{ij}(t)] \tag{2}$$

where $v_{ij}(t)$ is the velocity of particle i in dimension $j = 1, \ldots, n_x$ at time step t, $x_{ij}(t)$ is the position of particle i in dimension j at time step t, $y_{ij}(t)$ is as stated before, the best position seen so far by particle, $\hat{y}_j(t)$ is the global best position obtained so far by any particle,c_1 and c_2 are positive acceleration constants used to scale the contribution of the cognitive and social components respectively, and $r_{1j}(t), r_{1j}(t) \sim U(0, 1)$ are random values in the range [0, 1], sampled from a uniform distribution. These random values introduce a stochastic element to the algorithm. The personal best position, v_i, associated with particle i is the best position the particle has visited since the first time step.

2.2 Local Best PSO

The local best PSO uses a ring social network topology (refer to Sect. 3.2) where smaller neighborhoods are defined for each particle. The social component reflects information exchanged within the neighborhood of the particle, reflecting local knowledge of the environment. With reference to the velocity equation, the social contribution to particle velocity is proportional to the distance between a particle and the best position found by the neighborhood of particles. The velocity is calculated with following equation

$$v_{ij}(t+1) = vi_{ij}(t) + c_1 r_{1j}(t)[y_{ij}(t) - x_{ij}(t)] + c_2 r_{2j}(t)[\hat{y}_{ij}(t) - x_{ij}(t)] \qquad (3)$$

where $\hat{y}_{ij}$ is the best position, found by the neighborhood of particle i in dimension j. The local best particle position, $\hat{y}_i$.

3 Neighborhood Topologies

Various types of neighborhood topologies are investigated and presented in literature [13, 14, 17]. The neighborhood topologies, which are considered in this work, are:

- Star topology.
- Ring topology.
- Von Neumann topology.
- Random topology.

3.1 Star Topology

Known as *gbest*, where all particles are interconnected. Each particle can therefore communicate with every other particle. In this case each particle is attracted towards the best solution found be the entire swarm. Each particle therefore imitates the overall best solution. Using the *gbest* model the propagation is very

Fig. 1 Representation
diagram for star topology

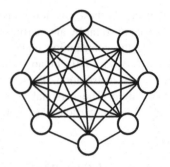

fast. This fast propagation may result in the premature convergence problem. This
occurs when some poor individuals attract the population due to a local optimum
or bad initialization preventing further exploration of the search space. Fig. 1
illustrates the star neighborhood topology.

3.2 Ring Topology

Known as *lbest*, connects each particle to its K immediate neighbors, e.g. K = 2
(left and right particles). The flow of information in ring topology is heavily
reduced compared to the star topology. Using the ring topology will slow down the
convergence rate because the best solution found has to propagate through several
neighborhoods before affecting all particles in the swarm. This slow propagation
will enable the particles to explore more areas in the search space and thus
decreases the chance of premature convergence. In *lbest* model, particles update
their velocity using Eq. (3). Figure 2 illustrates the ring neighborhood topologies.

3.3 Von Neumann Topology

Von Neumann is also a type of *lbest* model. However, in the Von Neumann
topology, particles are connected using a grid network (2-dimensional lattice)
where each particle is connected to its four neighbor particles (above, below, right,
and left particles) [14]. In the Von Neumann topology, particles update their
velocity using Eq. (3). However, not like Ring topology, *lbest* here represent the
best value obtained so far by any particle of the neighbors (above, below, right,
and left particles). Like Ring topology, using Von Neumann topology will slow
down the convergence rate. Slow propagation will enable the particles to explore
more areas in the search space and thus decreases the chance of premature convergence. Figure 3 illustrates the Von Neumann topology.

Fig. 2 Representation
diagram for ring topology

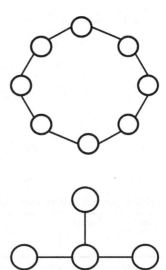

Fig. 3 Representation
diagram for Von Neumann
topology

3.4 Random Topology

Random topology is also a type of *lbest* model. However, in the Random topology, randomly connect every particle with other particles of size neighborhood. Neighborhood connections are not symmetric, meaning if particle A is a neighbor of particle B, particle B does not necessarily have to be a neighbor of particle A. Figure 4 illustrates the Random topology.

4 Parameters of the Algorithm

Within this section the particle swarm optimization algorithm (Eqs. (1)–(3)) is implemented to the optimization of four mathematical functions commonly known in the literature. All elements of algorithm simulation parameters and results are described.

The functions, the number of dimensions (D) and the optimal functions are summarized in Table 1. The goal is to reach the optimum of the functions. The maximum iteration number is fixed to 100,000. Swarm sizes of 300 particles are implemented. Learning factor cognitive (c_1) and Learning factor social (c_2) are static with $c_1 = c_2 = 2$. Inertia weight is static with $W = 0.2$. Each optimization experiment was run 30 times with random initial values of x in the range [x_{min}, x_{max}].

Fig. 4 Representation
diagram for random topology

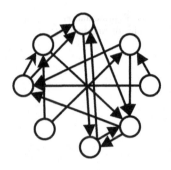

Table 1 General definition for benchmark functions

Name	Formula	D	Range $[x_{min}, x_{max}]$	Optimal f	Sketch
Rastringin	$f_1(x) = 10N + \sum_{i=1}^{N}\left[x_i^2 - 10\cos(2\pi x_i)\right]$	N	$[-100, 100]$	0	
Sphere	$f_2(x) = \sum_{i=1}^{N} i * x_i^2$	N	$[-100, 100]$	0	
Griewank	$f_3(x) = \sum_{i=1}^{N} \frac{x_i^2}{4{,}000} - \prod_{i=1}^{N}\cos\left(\frac{x_i}{\sqrt{i}}\right) + 1$	N	$[-100, 100]$	0	
Rosenbrock	$f_4(x) = \sum_{i=1}^{N-1}\left\{100\left[x_{i+1} - x_i^2\right]^2 + [1 - x_i]^2\right\}$	N	$[-100, 100]$	0	

The PSO algorithm is described in the form of flowchart in Fig. 5.

5 Simulation Results

This section shows the results of every function of Table 1 using the PSO algorithm of Fig. 5.

5.1 Rastrigin Function

Table 2 presents optimization results of different topologies (Star, Ring, Von Neumann and Random) for the Rastrigin function with two dimensions. These topologies show its great capacity to find to optimal solution with full success rate.

Table 3 presents optimization results of different topologies (Star, Ring, Von Neumann and Random) for the Rastrigin function with four dimensions. These topologies show its great capacity to find to optimal solution with full success rate.

Fig. 5 Flowchart of PSO
algorithm for each topology

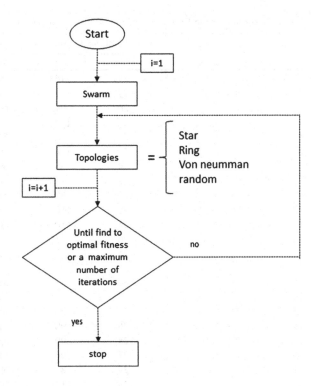

Figure 6 illustrates results from Tables 2 (Fig. 6a) and 3 (Fig. 6b), where both show a good behavior between different topologies.

Figure 7 illustrates convergence graphs of one experiment from Tables 2 (Fig. 7a) and 3 (Fig. 7b), where both show a good behavior between different topologies.

5.2 Griewank Function

Table 4 presents the optimization results of different topologies (Star, Ring, Von Neumann and Random) for the Griewank function with two dimensions. These topologies show its great capacity to find to optimal solution with full success rate.

Table 5 presents optimization results of different topologies for the Griewank function with four dimensions. Ring, Von Neumann and Random topologies show its great capacity to find to optimal solution with full success rate; while the Star topology gets trapped in a local minimum.

Figure 8 illustrates convergence graphs of one experiment from Tables 4 (Fig. 8a) and 5 (Fig. 8b). The graph (a) has shown a good behavior; while the graph (b) has shown that the Star topology gets trapped in a local minimum repeatedly.

Figure 9 illustrates convergence graph of one experiment from Table 4; where graph has shown a good behavior between different topologies.

Table 2 Results for the Rastrigin function with two dimensions

Experiment	Iterations to complete			
	Star	Ring	Von Neumann	Random
1	24	115	63	110
2	27	91	64	113
3	26	94	64	120
4	23	104	67	107
5	29	72	54	102
6	31	128	50	101
7	24	85	69	100
8	26	76	58	113
9	28	80	70	108
10	27	110	48	115
11	25	110	56	97
12	26	91	53	109
13	25	97	69	96
14	34	120	63	116
15	31	83	65	116
16	30	77	54	95
17	28	86	61	106
18	26	90	71	104
19	28	99	68	109
20	27	96	57	102
21	26	84	60	119
22	28	113	72	106
23	31	110	62	102
24	26	84	47	95
25	25	90	66	106
26	24	94	69	97
27	33	109	61	109
28	27	80	60	99
29	27	93	71	109
30	28	94	62	108
Iteration to completes (average)	27.33	95.16	61.83	106.36
Success rate	30/30	30/30	30/30	30/30

Table 3 Results for the Rastrigin function with four dimensions

Topology	Iterations to complete (average)	Success rate
Star	84.26	30/30
Ring	417.96	30/30
Von Neumman	189.53	30/30
Random	535.56	30/30

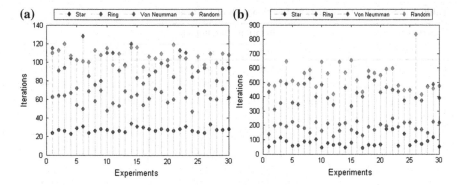

Fig. 6 Comparative evaluation of different topologies of Rastrigin's function with two (**a**), and four (**b**) dimensions

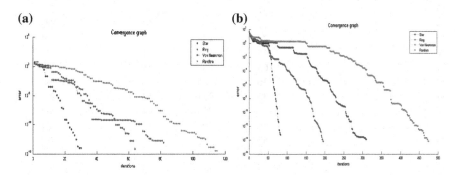

Fig. 7 Convergence graph of different topologies for the Rastrigin function (**a**) and (**b**)

Figure 10 illustrates convergence graphs of two experiments from Table 5. The graph (a) has shown a good behavior; while the graph (b) has shown that Star topology gets trapped in a local minimum repeatedly.

5.3 Sphere Function

Table 6 presents the optimization results of different topologies (Star, Ring, Von Neumann and Random) for the Sphere function with two dimensions. These topologies show its great capacity to find to optimal solution with full success rate.

Table 7 presents optimization results of different topologies for the Sphere function with four dimensions. These topologies show its great capacity to find to optimal solution with full success rate.

Table 4 Results for the Griewank function with two dimensions

Experiment	Iterations to complete			
	Star	Ring	Von Neumman	Random
1	115	84	65	169
2	49	92	80	248
3	96	124	54	309
4	23	124	63	251
5	28	160	72	250
6	756	101	58	312
7	33	112	50	255
8	29	145	66	176
9	23	145	55	237
10	29	217	55	284
11	63	87	66	251
12	37	83	84	223
13	737	77	74	237
14	28	101	64	210
15	24	149	76	278
16	45	98	81	238
17	163	87	78	203
18	36	82	46	248
19	33	171	53	221
20	46	253	64	188
21	36	92	62	200
22	52	132	77	265
23	35	205	57	256
24	24	180	75	286
25	26	119	58	213
26	48	99	56	198
27	32	156	70	193
28	44	213	98	173
29	28	110	73	232
30	147	96	49	289
Iteration to completes (average)	95.54	129.81	65.96	236.43
Success rate	30/30	30/30	30/30	30/30

Table 5 Results for the Griewank function with four dimensions

Topology	Iterations to complete (average)	Success rate
Star	1,673.42	**15**/30
Ring	3,995.12	30/30
Von Neumman	1,127.26	30/30
Random	2,409.96	30/30

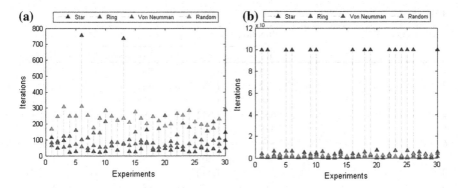

Fig. 8 Comparative evaluation of different topologies of Griewank's function with two (**a**), and four (**b**) dimensions

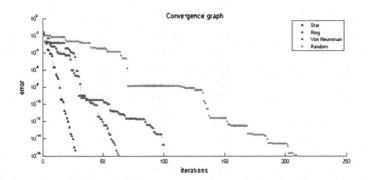

Fig. 9 Convergence graph of different topologies for the Griewank function

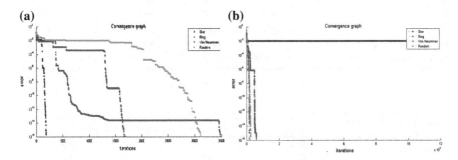

Fig. 10 Convergence graph of different experiments for the Rastrigin function

Figure 11 illustrates convergence graphs of one experiment from Tables 6 (Fig. 11a) and 7 (Fig. 11b), where both show a good behavior between different topologies.

Table 6 Results for the Sphere function with two dimensions

Experiment	Iterations to complete			
	Star	Ring	Von Neumman	Random
1	554	1,362	888	1,088
2	553	1,359	883	1,080
3	556	1,367	875	1,088
4	557	1,342	881	1,089
5	556	1,345	890	1,089
6	556	1,333	886	1,088
7	554	1,369	881	1,090
8	557	1,340	874	1,082
9	556	1,365	882	1,098
10	557	1,342	886	1,090
11	552	1,333	872	1,081
12	558	1,311	890	1,081
13	558	1,366	878	1,080
14	551	1,359	896	1,086
15	557	1,330	888	1,082
16	555	1,355	875	1,080
17	550	1,338	880	1,077
18	558	1,349	876	1,091
19	558	1,348	886	1,084
20	560	1,352	872	1,073
21	560	1,356	880	1,069
22	558	1,364	882	1,091
23	555	1,358	861	1,088
24	555	1,354	879	1,083
25	560	1,361	877	1,076
26	555	1,337	874	1,076
27	558	1,325	872	1,081
28	563	1,363	883	1,076
29	556	1,348	894	1,080
30	553	1,356	884	1,073
Iteration to completes (average)	556.2	1,349.56	880.83	1,083.12
Success rate	30/30	30/30	30/30	30/30

Table 7 Results for the Spherefunction with four dimensions

Topology	Iterations to complete (average)	Success rate
Star	635.32	30/30
Ring	3,287.73	30/30
Von Neumman	1,687.57	30/30
Random	2,768.16	30/30

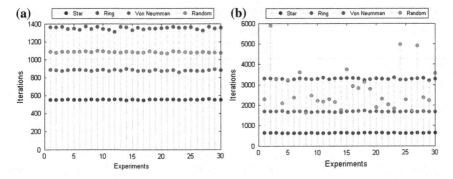

Fig. 11 Comparative evaluation of different topologies of Sphere's function with two (**a**), and four (**b**) dimensions

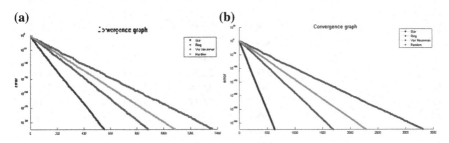

Fig. 12 Convergence graph of different experiments for the Sphere function (**a**) and (**b**)

Figure 12 illustrates convergence graphs of one experiment from Tables 6 (Fig. 12a) and 7 (Fig. 12b), where both show a good behavior between different topologies.

5.4 Rosenbrock Function

Table 8 presents the optimization results of different topologies (Star, Ring, Von Neumann and Random) for the Rosenbrock function with two dimensions. These topologies show its great capacity to find to optimal solution with full success rate.

Figure 13 illustrates convergence graphs of one experiment from Table 8; where it has shown a good behavior between different topologies.

Figure 14 illustrates convergence graph of one experiment from Table 8; where graph has shown a good behavior between different topologies.

Table 8 Results for the Rosenbrock function with two dimensions

Experiment	Iterations to complete			
	Star	Ring	Von Neumman	Random
1	308	2,629	1,450	1,460
2	334	2,710	1,251	1,486
3	332	2,689	1,348	1,475
4	374	2,762	1,369	1,479
5	318	2,669	1,339	1,469
6	367	2,765	1,347	1,400
7	302	2,928	1,275	1,421
8	339	2,257	1,274	1,412
9	304	2,506	1,289	1,484
10	267	2,474	1,455	1,368
11	345	2,480	1,294	1,401
12	316	2,456	1,273	1,419
13	249	2,612	1,315	1,474
14	255	2,447	1,320	1,445
15	343	2,650	1,343	1,518
16	335	2,695	1,224	1,477
17	337	2,230	1,259	1,466
18	348	2,586	1,230	1,348
19	289	2,637	1,333	1,445
20	298	2,785	1,304	1,467
21	298	2,535	1,391	1,323
22	348	2,073	1,372	1,546
23	326	2,702	1,240	1,421
24	338	2,966	1,309	1,517
25	301	2,619	1,255	1,443
26	381	2,745	1,314	1,513
27	325	2,566	1,393	1,462
28	350	2,949	1,416	1,394
29	298	2,657	1,256	1,407
30	371	2,083	1,250	1,384
Iteration to completes (average)	323.24	2,595.43	1,316.26	1,444.13
Success rate	30/30	30/30	30/30	30/30

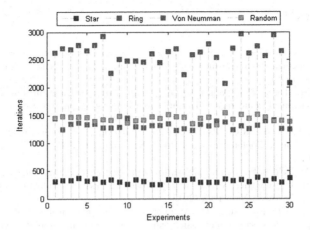

Fig. 13 Comparative evaluation of different topologies for the Rosenbrock function

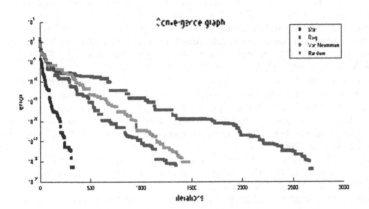

Fig. 14 Convergence graph of different topologies for the Griewankfunction

6 Conclusions

In this work multi types of topologies are implemented for a comparative study. The topology type is defined depending on how the particles are interconnected. Star topology uses global best algorithm (*gbest*) and Ring, Von Neumann and Random topologies use local best algorithm (*lbest*). The *gbest* algorithm converges faster to optimal fitness but gets trapped increases the chance of premature convergence. The *lbest* algorithm converges slower to optimal fitness without get trapped decreases the chance of premature convergence. Future works such as tuning of PSO parameters for high dimensions benchmark functions and apply other topologies.

References

1. Eberhart, R.C., Kennedy, J.: A new optimizer using particle swarm theory. In: Procedings of the Sixth International Symposium on Micro Machine and Human Science, pp. 39–43. IEEE, New York (1995)
2. Kennedy, J.: The particle swarm: social adaptation of knowledge. In: Proceedings of the IEEE International Conference on Evolutionary Computation, pp. 303–308. IEEE, New York (1997)
3. Kennedy, J., EberhartR.C.: Particle swarm optimization. In: Proceedings of the IEEE International Joint Conference on Neuronal Networks, pp. 1942–1948. IEEE Press, New York (1995)
4. Sombra, A., Valdez, F., Melin, P., Castillo, O.: A new gravitational search algorithm using fuzzy logic to parameter adaptation. In: IEEE Congress on Evolutionary Computation, pp. 1068–1074. IEEE, New York (2013)
5. Valdez, F., Melin, P., Castillo, O.: An improved evolutionary method with fuzzy logic for combining Particle Swarm Optimization and Genetic Algorithms. Appl. Sof. Comput. 11(2), 2625–2632 (2011)
6. Valdez, F., Melin, P., Castillo, O.: Parallel particle swarm optimization with parameters adaptation using fuzzy logic. In: MICAI, vol. 2, pp. 374–385. Springer, Mexico (2012)
7. Clerc, M., Kennedy, J.: The particle swarm: explosion, stability, and convergence in a multi-dimensional complex space. IEEE Trans. Evol. Comput. 6(1), 58–73 (2002)
8. Kennedy, J., Spears, W.: Matching algorithms to problems: an experimental test of the particle swarm and some genetic algorithms on the multimodal problem generator. In: Proceedings of the IEEE Congress on Evolutionary Computation, pp. 78–83. IEEE Press, New York, May 1998
9. Shi, Y., Eberhart, R.: Empirical study of particle swarm optimization. In: Proceedings of the 1999 Congress on Evolutionary Computation 1999, CEC 99. IEEE, Washington (1999)
10. Valdez, F., Melin, P., Castillo, O.: Evolutionary method combining particle swarm optimization and genetic algorithms using fuzzy logic for decision making. In: Proceedings of the IEEE International Conference on Fuzzy Systems, pp. 2114–2119. IEEE, New York (2009)
11. Vazquez, J.C., Valdez, F., Melin, P.: Fuzzy logic for dynamic adaptation in PSO with multiple topologies. In: IFSA World Congress and NAFIPS Annual Meeting (IFSA/NAFIPS), Joint. Albeta, Canada (2013)
12. Vazquez, J.C., Valdez, F., Melin, P.: Comparative study of particle swarm optimization variants in complex mathematics functions. In: Recent Advances on Hybrid Intelligent Systems. Springer, Mexico (2013)
13. Kennedy, J., Mendes, R.: Population structure and particle swarm performance. In: Proceedings of the IEEE Congress on Evolutionary Computation, pp. 1671–1676. IEEE, Hawaii 2002
14. Kennedy, J.: Small worlds and megaminds: effects of neighborhood topology on particle swarm performance. In: Proceedings of IEEE Congress on Evolutionary Computation, pp. 1931–1938. IEEE, Washington D.C. 1999
15. Shi, Y., Eberhart, R.: Parameter selection in particle swarm optimization. In: Proceedings of Evolutionary Programming 98, pp. 591–600. Springer 1998
16. Shi, Y., Eberhart, R.C.: Fuzzy adaptive particle swarm optimization. In: Proceedings of the IEEE Congress on Evolutionary Computation, vol. 1, pp. 101–106. IEEE Press, New York, May 2001
17. Kennedy, R., Mendes, R., Neighborhood topologies in fully-informed and best-of-neighborhood particle swarms. IEEE Syst. Man Cybern. Soc. 36(4), 515–519 (2006)

Comparison of the Optimal Design of Fuzzy Controllers for the Water Tank Using Ant Colony Optimization

Leticia Amador-Angulo and Oscar Castillo

Abstract A study of the behavior and evaluation of the Ant Colony Optimization algorithm (ACO) in Type-1 and Type-2 Fuzzy Controller design is presented in this chapter. The main objective of the work is based on the main reasons in tuning membership functions for the optimization Fuzzy Controllers of the benchmark problem known as the Water Tank with the algorithm of Ant Colony Optimization. For the design of Type-1 and Type-2 Fuzzy Controllers for particular applications, the use of bio-inspired optimization methods have helped in the complex task of finding the appropriate values of the parameters and the structure of fuzzy systems. In this research we consider the application of ACO as the paradigm that aids in the optimal design of Type-1 and Type-2 Fuzzy Controllers. We also analyzed that in evaluating the uncertainty, the results in the simulation are better with Type-2 Fuzzy Controllers. Finally, provide a comparison of the different methods for the case of designing Type-1 and Type-2 Fuzzy Controllers with Ant Colony Optimization.

1 Introduction

Nowadays, fuzzy logic is one of the most used methods of computational intelligence and with the best future; this is possible thanks to the efficiency and simplicity of Fuzzy Systems since they use linguistic terms similar to those that human beings use [20, 21, 30, 31].

Fuzzy control systems combine information of human experts (natural language) with measurements and mathematical models. Fuzzy systems transform the knowledge base in a mathematical formulation that has proven to be very efficient [1, 27–29, 32, 33]. This is why it has been shown that bio-inspired algorithms in

L. Amador-Angulo · O. Castillo (✉)
Tijuana Institute of Technology, Tijuana, Mexico
e-mail: ocastillo@hafsamx.org; ocastillo@tectijuana.mx

O. Castillo et al. (eds.), *Recent Advances on Hybrid Approaches for Designing Intelligent Systems*, Studies in Computational Intelligence 547,
DOI: 10.1007/978-3-319-05170-3_18, © Springer International Publishing Switzerland 2014

nature, for this research S-ACO (Simple Ant Colony Optimization) is developed for allow finding the values of the membership functions of the fuzzy controller of an intelligent and collaborative simulation by the way ants behave in their natural environment. With this technique we can design and find a better fuzzy controller for the problem [24, 25].

At present, algorithms inspired by nature have become motivated when used as a tool to find the best design of fuzzy controller; the results have shown the ACO algorithm to implement fuzzy controller design form the methodology for how they help the algorithm in finding the optimal fuzzy controller to find good results [22, 23].

The complexity for developing fuzzy systems can be found at the time of deciding which are the best parameters of the membership functions, the number of rules or even the best granularity that could give us the best solution for the problem that we want to solve [21].

A solution for the above mentioned problem is the application of evolutionary algorithms for the optimization of fuzzy systems. Evolutionary algorithms can be a useful tool since its capabilities of solving nonlinear problems, well-constrained or even NP-hard problems. Among the most used methods of bio-inspired algorithms we can find: Genetic Algorithms, Ant Colony Optimization, Particle Swarm Optimization, Bee Colony Optimization, etc. [4–6, 18–20].

This chapter describes the application of bio-inspired algorithms, such as the Ant Colony Optimization as a method of optimization of the parameters of the membership functions of the FLC in order to find the best intelligent controller for the problem benchmark of Water Tank implementing Type-1 and Type-2 Fuzzy Logic [1–3, 7, 13–17].

This chapter is organized as follows: Sect. 2 shows the concept of Ant Colony Optimization and a description of S-ACO which is the technique that was applied for optimization. Section 3 presents the problem statement. Section 4 shows the fuzzy logic controller proposed and in Sect. 5 it's described the development of the evolutionary method. In Sect. 6 the simulation results are shown. Finally, Sect. 7 shows the conclusions.

2 S-ACO Algorithm

Ant Colony Optimization (ACO) is a probabilistic technique that can be used for solving problems that can be reduced to finding good path along graphs [11, 12, 26]. This method is inspired on the behavior presented by ants in finding paths from the nest or colony to the food source [8–10].

The S-ACO is an algorithmic implementation that adapts the behavior of real ants to solutions of minimum cost path problems on graphs [1]. A number of artificial ants build solutions for a certain optimization problem and exchange information about the quality of these solutions making allusion to the communication systems of the real ants [20].

Let us define the graph $G = (V, E)$, where V is the set of nodes and E is the matrix of the links between nodes. G has $n_G = |V|$ nodes. Let us define L^K as the number of hops in the path built by the ant k from the origin node to the destiny node. Therefore, it is necessary to find:

$$Q = \{q_a, \ldots, q_f | q_1 \in C\} \tag{1}$$

where Q is the set of nodes representing a continuous path with no obstacles; $q_a, \ldots, q_f$ are former nodes of the path and C is the set of possible configurations of the free space. If $x^k(t)$ denotes a Q solution in time t, $f(x^k(t))$ expresses the quality of the solution. The general steps of S-ACO are the followings: [20, 21]

- Each link (i, j) is associated with a pheromone concentration denoted as τ_{ij}.
- A number $k = 1, 2, \ldots, n_k$ are placed in the nest.
- On each iteration all ants build a path to the food source (destiny node). For selecting the next node a probabilistic equation is used:

$$p_{ij}^k(t) \begin{cases} \dfrac{\tau_{ij}^k}{\sum_{j \in N_i^k} \tau_{ij}^\alpha(t)} & if \quad j \in N_i^k \\ 0 & if \quad j \notin N_i^k \end{cases} \tag{2}$$

where, N_i^k is the set of feasible nodes (in a neighborhood) connected to node i with respect to ant k, τ_{ij} is the total pheromone concentration of link ij, and α is a positive constant used as again for the pheromone influence.

- Remove cycles and compute each route weight $f(x^k(t))$. A cycle could be generated when there are no feasible candidates nodes, that is, for any i and any k, $N_i^k = \emptyset$; then the predecessor of that node is included as a former node of the path.
- Pheromone evaporation is calculated with Eq. (3):

$$\tau_{ij}(t) \leftarrow (1 - \rho)\tau_{ij}(t) \tag{3}$$

where $\rho \in [0, 1]$ is the evaporation rate value of the pheromone trail. The evaporation is added to the algorithm in order to force the exploration of the ants, and avoid premature convergence to sub-optimal solutions [20]. For $\rho = 1$ the search becomes completely random [20].

- The update of the pheromone concentration is realized using Eq. (4):

$$\tau_{ij}(t + 1) = \tau_{ij}(t) + \sum_{k=1}^{n_k} \Delta \tau_{ij}^k(t) \tag{4}$$

where $\Delta \tau_{ij}^k$ is the amount of pheromone that an ant k deposits in a link ij in a time t.

- Finally, the algorithm can be ended in three different ways:

  - When a maximum number of epochs has been reached.
  - When it has been found an acceptable solution, with $f(x_k(t)) < \varepsilon$.
  - When all ants follow the same path.

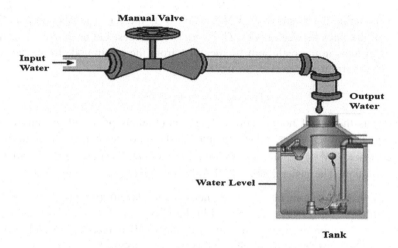

Fig. 1 Graphic representation of the problem to be studied

3 Problem Statement

The problem to be studied is known as the water tank controller, which aims at achieving the water level in a tank, therefore, we have to know the actual water level in the tank and with it has to be able to set the valve.

Figure 1 shows graphically the way in which this system operates the valve opening and hence the filling process in the tank, and this will have two variables, which is the water level and the speed of opening of the will be output valve tank filling.

To evaluate the valve opening in a precise way we rely on fuzzy logic, which is implemented as a fuzzy controller that performs automated tasks considered as water level and how fast is be entered to thereby maintain the level of water in the tank in a better way.

3.1 Model Equations of the Water Tank

The process of filling the water tank is presented as a differential equation for the height of water in the tank, H, is given by:

$$\frac{d}{dt}Vol = A\frac{dH}{dt} = bV - a\sqrt{H} \tag{5}$$

where Vol is the volume of water in the tank, A is the cross-sectional area of the tank, b is a constant related to the flow rate into the tank, and a is a constant related to the flow rate out of the tank. The equation describes the height of water, H, as a

Fig. 2 Graphical representation of the mathematical equation of the water tank filler

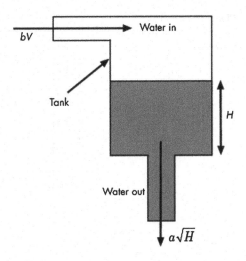

function of time, due to the difference between flow rates into and out of the tank. In Fig. 2 shows graphically the mathematical model.

4 Fuzzy Logic Control Design

We present the characteristics of the Type-1 and Type-2 fuzzy controller, besides the results of model evaluation.

4.1 Characteristics of Type-1 Fuzzy Controller

Membership functions are defined for the two inputs to the fuzzy system: the first is called **level**, which has three membership functions with linguistic value of *high*, *okay* and *low*. The second input variable is called **rate** with three membership functions with linguistic value of *negative*, *none* and *positive*, shown in Fig. 3 representations of fuzzy variables. The names of the linguistic labels are assigned based on the empirical process of filling behavior of a water tank.

The Type-1 Fuzzy Inference System has an input called **valve**, which is composed of five triangular membership functions with the following linguistic values: *close_fast*, *close_slow*, *no_change*, *open_slow* and *open_fast*, representation shown in Fig. 4.

The rules are listed below and are based on the behavior that the water tank to be filled. The simulation shows that five rules are sufficient, which are detailed below:

Fig. 3 Type-1 fuzzy
inference system inputs
variables

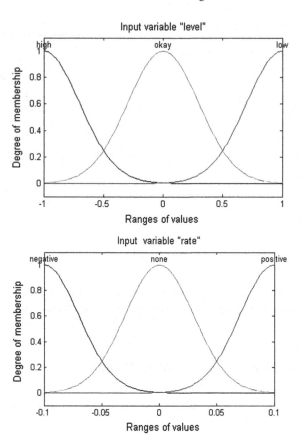

- *If* (*level is okay*) *then* (*valve is not_change*).
- *If* (*level is low*) *then* (*valve is open_fast*).
- *If* (*level is high*) *then* (*valve is close_fast*).
- *If* (*level is okay*) *and* (*rate is positive*) *then* (*valve is close_slow*).
- *If* (*level is okay*) *and* (*rate is negative*) *then* (*valve is open_slow*).

The combination of rules was taken from experimental knowledge according to how the process is performed in a tank filled with water. We start with 5 rules to visualize the behavior of the Type-1 fuzzy controller.

4.2 Mean Square Error (MSE)

The metric on which the evaluation is being made for the fuzzy controller is by using the mean square error for measuring the behavior of the controller by reference to the error tends to be 0 (zero), the errors for the PID controller and fuzzy

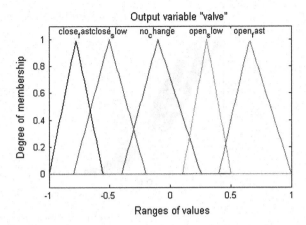

Fig. 4 Type-1 fuzzy inference system output variable

controller are used. The MSE is the sum of the variance and the squared deviation of the estimator (reference). The mathematical definition is presented in the following Equation.

$$MSE = \frac{1}{n} \sum_{i=1}^{n} (\bar{Y}_i - Y_i)^2 \qquad (6)$$

4.3 Characteristics of Type-2 Fuzzy Controller

The Type-2 Fuzzy Controller used in the initial experiments was designed based on the best Type-1 Fuzzy Controller that S-ACO found, the variables linguistic and rules are the same. The shapes of the membership functions of the inputs are shown in Fig. 5.

The distribution of the membership functions of the output **valve** were done taking into account the distribution of the membership functions of the Type-1 fuzzy controller, as shown in Fig. 6.

After that S-ACO finding the design the Type-2 Fuzzy Controller we begin with the simulations in the block diagram. The representation of S-ACO architecture in the problem is shown in Sect. 5.

5 ACO Architecture

A S-ACO algorithm was applied for the optimization of the membership functions for the fuzzy logic controller. For developing the architecture of the algorithm it was necessary to follow the following steps:

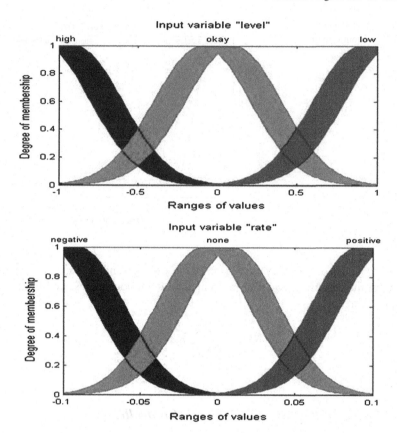

Fig. 5 Type-2 fuzzy inference system input variables

Fig. 6 Type-2 fuzzy inference system output variable

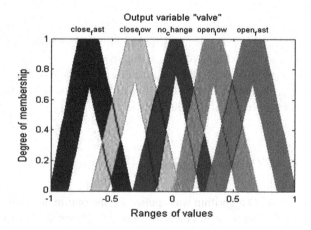

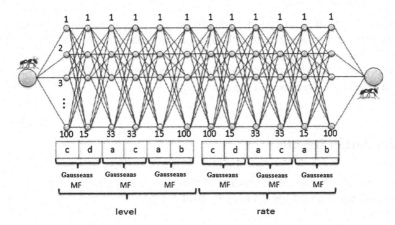

Fig. 7 S-ACO architecture

1. Representing the architecture of the FLC as a graph that artificial ants could traverse.
2. Achieving an adequate handling of the pheromone but permitting the algorithm to evolve by itself.

5.1 Limiting the Problem and Graph Representation

One of problems found on the development of the S-ACO algorithm was to make a good representation of FLC. First we reduced the number of elements that the method needed to find by deleting the elements whose could be not necessary.

Figure 7 shows the search graph for the proposed S-ACO algorithm, the graph can be viewed as a tree where the root is the nest and the last node is the food source.

5.2 Updating Pheromone trail

An important issue is that the update of pheromone trail be applied in the best way possible. In this sense we need to handle the evaporation (Eq. 3), and increase or deposit of pheromone (Eq. 4), where the key parameter in evaporation is denoted by ρ that represents the rate of evaporation and in deposit of pheromone is denoted by $\Delta\tau$ that represents the amount of pheromone that an ant k deposits in a link ij in a time t. For ρ we assign a random value and Eq. (7) shows the way how the increase of pheromone is calculated [18].

$$\Delta\tau = \frac{(e_{\max} - e_k)}{e_{\max}} \tag{7}$$

where $e_{max} = 10$ the maximum error of control is permitted and e_k is error of control generated by a complete path of an ant k. We decided to allocate $e_{max} = 10$ in order to stand $\Delta\tau \in [0, 1]$.

Once analyzed the above characteristics of the behavior of S-ACO. The simulation results are described in the next section.

6 Simulation Results

In this section we present the results of the proposed controller to find the best design of Type-1 and Type-2 Fuzzy Controller.

6.1 Evaluation of Control Diagram

Figure 8 shows that there is a closed-loop control, the aim is to make the plant output to follow the input r, the adder is applied to the system, it is used for a controller in the first the output is connected directly to one of two inputs of the adder, in the second situation, the output and the model is perturbed with noise in order to introduce uncertainty in the data feedback. The noise is a disturbance that helps the model with the objective that the ACO algorithm further explores its search space and show better results. For a Fuzzy Controller of an aircraft, the noise can be interpreted as the turbulence that may arise.

Finally, at the output of the adder we have the error signal, which is applied to the fuzzy controller together with a signal derived from this, which is a change in the error signal over time.

Figure 9 shows the simulation of the behavior of the Type-1 fuzzy controller denoted in pink with the PID controller output denoted in black. These simulations are made without noise.

The simulation model is shown in the Fig. 8 where the black line denotes the reference model and the pink line the output of the model using a Type-1 fuzzy system.

6.2 Implementation of Noise in the Model

Different noise levels were applied as a disturbance in the signal processing to evaluate the target Type-1 fuzzy controller and to visualize the results in the model. Figure 10 shows the representation of the simulation with a noise level of 0.15, is displayed in black for the PID controller behavior and in pink for the Type-1 fuzzy controller with the above disturbance.

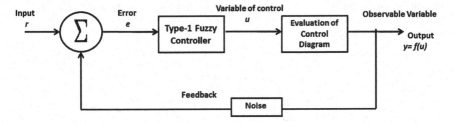

Fig. 8 Evaluation of control diagram with Type-1 fuzzy inference system

Fig. 9 Simulation model
using Type-1 fuzzy controller

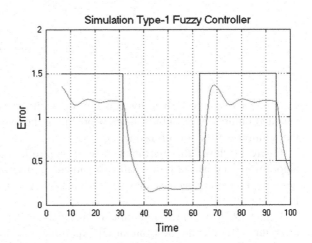

There were some variations of the Type-1 fuzzy system, changing the types of membership functions in inputs, to observe the behavior of the controller; the results will be discussed in the section of results and comparisons.

In order to better examine the fuzzy controller; it was decided to change the type of the membership functions of the input variable fuzzy controller, and thus be able to better examine the effect of this change.

6.3 S-ACO Algorithm Results for the Optimization of the Type-1 FLC

Simulation for 30 experiments have been performed with different characteristics of the algorithm, by changing the kinds of membership functions at the input and the numbers of rules in the fuzzy controller, the best experiments are presented in Table 1.

Table 1 shows the results in which the computational time increases when there are more iterations and the number of ants is increased in the search space,

Fig. 10 Simulation model
using Type-1 fuzzy controller
with a noise level of 0.15

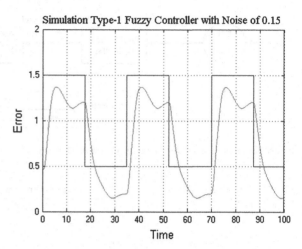

experiments were performed by changing the value of the pheromone, number of ants and number of iterations, thereby obtaining the experiment 3 is the one with the minimum error for all testing.

Figure 11 shows the first input of fuzzy controller called **level** than the ACO algorithm found as the best of all experiments performed, the membership functions are Triangular because they were the ones that showed the error of simulation lowest of all.

Figure 12 shows the second input of fuzzy controller called **rate** than the ACO algorithm found as the best one of all experiments performed.

Simulating the fuzzy controller shown in Fig. 13, representing the yellow color the reference value and control the behavior pink optimized fuzzy controller, the simulation was performed in 100 iterations.

S-ACO convergence of the algorithm is presented in Fig. 14, showing the behavior of the control reference to the behavior of the algorithm; was performed for 100 generations in the algorithm.

Finally in Table 2 shows a comparison with the experiments performed with the fuzzy controller that solve the problem of optimized water tank and without optimized.

Table 2 shows that having triangular membership functions give a better performance in the fuzzy controller, and the results that were found when the fuzzy controller is optimized is less error due to the intelligent ants together to find a better solution. All experiments were done for 100 iterations of the fuzzy controller.

It is shown that by increasing the number of fuzzy rules in the controller tends to explore much more, which is why experiments show that without optimize the controller gets the minimum error in the triangular membership functions with 9 rules, whereas the optimized controller has better results with only 5 rules, due to exports made by the ant colony algorithm in the search space. That is why, based

Table 1 S-ACO results for Type-1 FLC optimization

No.	Ants	Gen	Pheromone	Alfa α	Beta β	Average Error	S-ACO Time (min.)
1	10	50	0.7	2	5	0.085	18:15
2	30	100	0.7	2	5	0.096	25:20
3	**10**	**100**	**0.7**	**2**	**5**	**0.049**	**20:45**
4	10	100	0.5	2	5	0.084	23:03
5	50	50	0.5	2	5	0.091	49:58

Fig. 11 Distribution of membership functions of the fuzzy controller input call level

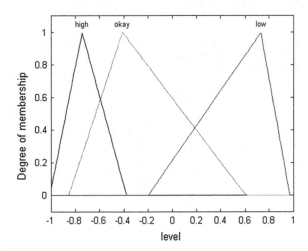

Fig. 12 Distribution of membership functions of the fuzzy controller input called rate

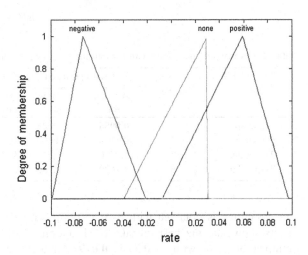

on experimentation shows that the simulation error is less when using bio-inspired S-ACO algorithm in fuzzy controller design.

The results shown in Table 2, it is envisioned that the S-ACO algorithm computing time increases if the controller is not being optimized, yet it is much

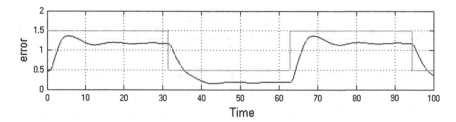

Fig. 13 Simulation using Type-1 fuzzy controller optimization with ACO

Fig. 14 Optimization
behavior for the S-ACO on
type-1 FLC

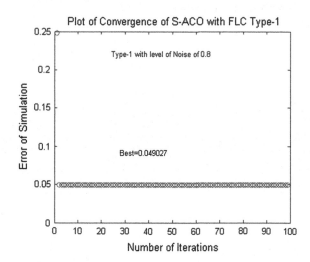

Table 2 Comparison the results with the optimized fuzzy controllers and without optimization

No.	Type of FM's (input)	No. rules	Without optimized		Optimized	
			Average error	Time (s)	Average error	S-ACO time (min)
1	Gaussian	5	**0.0986**	05:45	0.085	43:15
2	Gaussian	9	0.1178	08:15	0.096	25:45
3	Triangular	5	1.2770	09:13	**0.049**	22:45
4	Triangular	9	1.2629	12:36	0.084	23:03

less error than is obtained with S-ACO optimization in fuzzy controller. These results show that the best mistake NOT optimized is **0.0986**, whereas the best optimization error with S-ACO is **0.049**. Another variant is that changing the number of rules in the fuzzy controller does not present a major impact on the outcome. However, to further explore the different ways to design a fuzzy controller leads to better results in their evaluation.

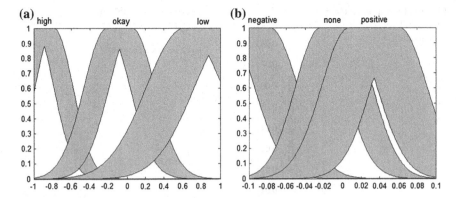

Fig. 15 **a** Level, and **b** rate optimized by S-ACO algorithm

6.4 S-ACO Algorithm Results for the Optimization of the Type-2 FLC

The design of Type-2 Fuzzy Controller obtained by the S-ACO algorithm shows the error minimization in the FLC as the uncertainty is analyzed more specific way. Figure 15 shows the membership functions of the Type-2 FLC obtained by S-ACO algorithm.

The behavior the S-ACO is observed in Fig. 16, finding the minimum error of the Fuzzy Controller with a value of **0.0035**, thereby observing the evaluation of Type-2 Fuzzy Controller gives better results than Type-1.

Experiments were performed by varying several parameters of the Fuzzy Controller, such as the type of membership function, the number of rules, well as the values the Beta and Alpha parameters of S-ACO algorithm to analyze the behavior of the algorithm, the best results are shown in Table 3 with the Type-1 FLC and Type-2 FLC.

Figure 17 shows the block diagram used for the FLC that obtained the best results of the water tank benchmark problem.

6.5 Statistical Test

Statistical analysis to evaluate the results mathematically with the Type-1 and Type-2 Fuzzy Controller is performed; the values of the samples, standard deviation are shown in Table 4.

The results show that there is sufficient statistical evidence of significance with a value of 95 % that H_A is true therefore Type-2 FLC produces the best results in comparison with Type-1 FLC.

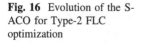

Fig. 16 Evolution of the S-ACO for Type-2 FLC optimization

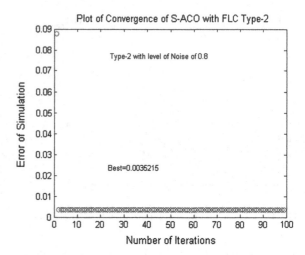

Table 3 S-ACO results of simulations for the Type-1 and Type-2 FLC optimization

Comparison of Type-1 and Type-2 FLC with S-ACO

No.	Type of FM's (Inputs)	No. of rules	Alpha α	Beta β	Type-1 FLC S-ACO		Type-2 FLC S-ACO		Noise (0.8)
					Minimum error	Time (s)	Minimum error	Time (min)	
1	Gaussian	5	2	5	**0.0617**	25:45	0.06438	55:15	No
2	Gaussian	9	2	5	0.2481	28:15	0.1888	45:45	No
3	Triangular	5	2	5	0.0617	30:13	**0.049**	36:45	No
4	Triangular	9	2	5	0.0893	42:36	0.247	53:03	No
5	Trapezoidal	5	2	5	0.1383	10:45	0.7370	1:02:15	No
6	Trapezoidal	9	2	5	0.1999	47:32	0.09652	45:45	No
7	Gaussian	5	3	7	0.0851	18:15	0.042145	1:10:15	No
8	Gaussian	9	3	7	0.0963	36:20	0.02602	59:45	No
9	Triangular	5	3	7	0.0367	20:45	0.36521	58:45	No
10	Triangular	9	3	7	0.0844	23:03	0.04854	48:03	Yes
11	Trapezoidal	5	3	7	0.0916	49:58	0.065842	43:42	Yes
12	Trapezoidal	9	3	7	0.2999	38:44	0.07197	1:05:47	Yes
13	Gaussian	5	5	9	0.0898	41:20	**0.00352**	1:10:15	Yes
14	Gaussian	9	5	9	0.2095	32:18	0.02797	59:45	Yes
15	Triangular	5	5	9	**0.0308**	20:45	0.05016	46:45	Yes
16	Triangular	9	5	9	0.22194	28:15	0.07370	53:03	Yes
17	Trapezoidal	5	5	9	0.1749	29:16	0.06854	1:12:15	Yes
18	Trapezoidal	9	5	9	0.1745	30:45	0.01265	57:45	Yes

Finally Fig. 18 shows the grouping of data errors found in the simulations with the S-ACO in the statistical test.

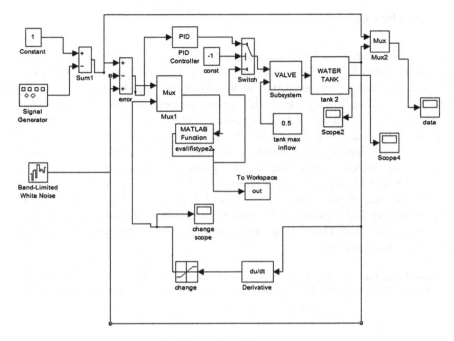

Fig. 17 Block diagram for the simulation of the FLC

Table 4 Statistical test of Type-1 and Type-2 FLC optimization

Type-1 FLC S-ACO versus Type-2 FLC S-ACO			
$\alpha = 0.05$	μ	σ	Samples
Type-1 FLC S-ACO	0.0371196	0.00319528	$\bar{x}_1 = 0.00319528$	$n_1 = 60$
Type-2 FLC S-ACO	0.00840587	0.0546	$\bar{x}_2 = 0.0345379$	$n_2 = 60$
$P = 0.075$		$Zc = 3.71$	
μ_1: Type-1 FLC S-ACO		μ_2: Type-2 FLC S-ACO	
H_O: $\mu_1 \geq \mu_2$ H_A: $\mu_2 < \mu_1$ (statement)			

Fig. 18 Blocks of data of Type-1 and Type-2 FLC

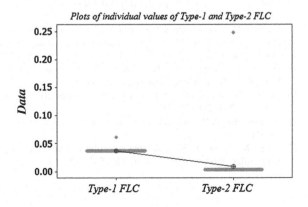

Plots of individual values of Type-1 and Type-2 FLC

7 Conclusions

This chapter proposes a bio-inspired algorithm in nature that solves combinatorial problems called S-ACO for the design of optimal Fuzzy Controllers. This algorithm allows finding the parameters of the appropriate membership functions for the behavior of the fuzzy controller has a better performance. Several experiments were performed with the S-ACO algorithm, thereby finding that the results of optimized fuzzy controller have lower errors, but the computational time is higher. To fix this problem is to implement the robustness of Type-2 fuzzy logic as will analyze the problem of uncertainty that usually occur in the fuzzy controllers. The following will be considered as a future work for this research, further explore everything that involves the use of Type-2 fuzzy logic, because the S-ACO algorithm may find a longer search space in the footprint of uncertainty presented in the problems with Type-2 fuzzy logic, and find better design of fuzzy controller for the problem studied.

Acknowledgments We would like to express our gratitude to the CONACYT and Tijuana Institute of Technology for the facilities and resources granted for the development of this research.

References

1. Amador-Angulo, L., Castillo, O.: Comparison of fuzzy controllers for the water tank with Type-1 and Type-2 fuzzy logic. In: NAFIPS 2013, Edmonton, Canada, pp. 1–6
2. Cázarez-Castro, N., Aguilar, L.T., Castillo, O.: Designing Type-1 and Type-2 fuzzy logic controllers via fuzzy lyapunov synthesis for nonsmooth mechanical systems. Eng Appl AI (EAAI) **25**, 971–979 (2012)
3. Cervantes, L., Castillo, O., Melin, P.: Intelligent control of nonlinear dynamic plants using a hierarchical modular approach and Type-1 fuzzy logic. MICAI, 2011, pp. 1–12 (2011)
4. Castillo, O., Melin, P.: A review on the design and optimization of interval Type-2 fuzzy controllers. Appl. Soft Comput. (ASC) **12**(4), 1267–1278 (2012)
5. Castillo, O., Melin, P.: Optimization of type-2 fuzzy systems based on bio-inspired methods: a concise review. Inf. Sci. (ISCI) **205**, 1–19 (2012)
6. Castillo, O.: Interval Type-2 mamdani fuzzy systems for intelligent control. Combining Experimentation Theor. pp. 163–177 (2012)
7. Castillo, O.: Type-2 fuzzy logic in intelligent control applications. Springer, Berlin (2012)
8. Dorigo, M., Birattari, M., Stützle, T.: Ant colony optimization. IEEE Comput. Intell. Mag. **1**, 28–39 (2006)
9. Dorigo, M., Stützle, T.: Ant Colony Optimization. Bradford, Cambridge Massachusetts Institute of Technology, Cambridge (2004)
10. Dorigo, M., Blum, M.: Ant colony optimization theory: a survey. Theoret. Comput. Sci. **344**(2–3), 243–278 (2005)
11. Engelbrecht, A.P.: Fundamentals of Computational Swarm Intelligence. Wiley, England (2005)
12. Galea, M., Shen, Q.: Simultaneous ant colony optimization algorithms for learning linguistic fuzzy rules. Stud. Comput. Intell. **34**, 75–99 (2006)

13. Juang, C.F., Hsu, C.H., Chuang, C.F.: Reinforcement self-organizing interval type-2 fuzzy system with ant colony optimization. In: Proceedings of IEEE International Conference on Systems, Man and Cybernetics, San Antonio, pp. 771–776 (2009)
14. Karnik, N.N., Mendel, J.M.: Operations on Type-2 fuzzy sets. Int. J. Fuzzy Sets Syst. **122**, 327–348 (2001)
15. Klir, G.J., Yuan, B.: Fuzzy Sets and Fuzzy Logic: Theory and Applications. Practice Hall, New Jersey (1995)
16. Melendez, A., Castillo, O.: Optimization of type-2 fuzzy reactive controllers for an autonomous mobile robot. In: NaBIC 2012, pp. 207–211 (2012)
17. Mendel, J.M.: Uncertain Rule-Based Fuzzy Logic System: Introduction and New Directions. Practice Hall, New Jersey (2001)
18. Mendel, J.M.: Computing derivatives in interval Type-2 fuzzy logic systems. IEEE Trans. Fuzzy Syst. **12**, 84–98 (2004)
19. Mendel, J.M., Mouzouris, G.C.: Type-2 fuzzy logic system. IEEE Trans. Fuzzy Syst. **7**, 642–658 (1999)
20. Martinez, R., Castillo, O., Soria, J.: Optimization of membership functions of a fuzzy logic controller for an autonomous wheeled mobile robot using ant colony optimization, pp. 4–6 (2006)
21. Martínez, R., Castillo, O., Aguilar, L.: Intelligent control for a perturbed autonomous wheeled mobile robot using type-2 fuzzy logic and genetic algorithms. JAMRIS **2**(1), 1–11 (2008)
22. Mizumoto, M., Tanaka, K.: Some properties of fuzzy sets of Type-2. Inf. Control **31**, 312–340 (1976)
23. Naredo, E., Castillo, O.: ACO-tuning of a fuzzy controller for the ball and beam problem. In: MICAI 2011, pp. 58–69 (2011)
24. Neyoy, J., Castillo, O., Soria, J.: Dynamic Fuzzy Logic Parameter Tuning for ACO and Its Application in TSP Problems, Recent Advances on Hybrid Intelligent Systems, pp. 259–271. Springer, Berlin (2013)
25. Porta-García, M., Montiel, O., Sepulveda, R.: An ACO path planner using a FIS for a path selection adjusted with a simple tuning algorithm. JAMRIS **2**(1), 1–11 (2008)
26. Sepulveda, R., Castillo, O., Melin, P., Montiel, O.: An efficient computational method to implement type-2 fuzzy logic in control applications. Adv. Soft Comput. **41**, 45–52 (2007)
27. Sepulveda, R., Castillo, O., Melin, P., Rodriguez-Diaz, A., Montiel, O.: Experimental study of intelligent controllers under uncertainty using type-1 and type-2 fuzzy logic. Inf. Sci. **177**(10), 2023–2048 (2007)
28. Sepulveda, R., Montiel, O., Lizarraga, G., Castillo, O.: Modeling and simulation of the defuzzification stage of a type-2 fuzzy controller using the Xilinx system generator and Simulink. Stud. Comput. Intell. **257**, 309–325 (2009)
29. Sepulveda, R., Montiel, O., Castillo, O., Melin, P.: Optimizing the MFs in type-2 fuzzy logic controllers, using the human evolutionary model. Int. Rev. Autom. Control **3**(1), 1–10 (2011)
30. Zadeh, L.A.: The concept of a lingüistic variable and its application to approximate reasoning, Part I. Inf. Sci. **8**, 199–249 (1975)
31. Zadeh, L.A.: The concept of a lingüistic variable and its application to approximate reasoning, Part II. Inf. Sci. **8**, 301–357 (1975)
32. Zadeh, L.A.: Fuzzy sets. Inf. Control **8**, 338–353 (1965)
33. Zadeh, L.A.: Toward a theory of fuzzy information granulation and its centrality in reasoning and fuzzy logic. Fuzzy Sets Syst. Elsevier **90**, 117 (1997)

Differential Evolution with Dynamic Adaptation of Parameters for the Optimization of Fuzzy Controllers

Patricia Ochoa, Oscar Castillo and José Soria

Abstract The proposal described in this chapter uses the Differential Evolution (DE) algorithm as an optimization method in which we want to dynamically adapt its parameters using fuzzy logic control systems, with the goal that the fuzzy system gives the optimal parameter of the DE algorithm to find better results, depending on the type of problems the DE is applied.

1 Introduction

The use of fuzzy logic in evolutionary computing is becoming a common approach to improve the performance of the algorithms [15–17]. Currently the parameters involved in the algorithms are determined by trial and error. In this aspect we propose the application of fuzzy logic which is responsible in performing the dynamic adjustment of mutation and crossover parameters in the Differential Evolution (DE) algorithm. This has the goal of providing better performance to Differential Evolution.

Fuzzy logic or multi-valued logic is based on fuzzy set theory proposed by Zadeh in 1965 which helps us in modeling knowledge, through the use of if-then fuzzy rules. The fuzzy set theory provides a systematic calculus to deal with linguistic information, and that improves the numerical computation by using linguistic labels stipulated by membership functions [12]. Differential Evolution (DE) is one of the latest evolutionary algorithms that have been proposed. It was created in 1994 by Price and Storn in, attempts to resolve the problem of Chebychev polynomial. The following year these two authors proposed the DE for optimization of nonlinear and non-differentiable functions on continuous spaces.

P. Ochoa · O. Castillo (✉) · J. Soria
Tijuana Institute of Technology, Tijuana, Mexico
e-mail: ocastillo@tectijuana.mx

O. Castillo et al. (eds.), *Recent Advances on Hybrid Approaches for Designing Intelligent Systems*, Studies in Computational Intelligence 547,
DOI: 10.1007/978-3-319-05170-3_19, © Springer International Publishing Switzerland 2014

The DE algorithm is a stochastic method of direct search, which has proven effective, efficient and robust in a wide variety of applications such as learning of a neural network, a filter design of IIR, aerodynamically optimized. The DE has a number of important features which make it attractive for solving global optimization problems, among them are the following: it has the ability to handle non-differentiable, nonlinear and multimodal objective functions, usually converges to the optimal uses with few control parameters, etc.

The DE belongs to the class of evolutionary algorithms that is based on populations. It uses two evolutionary mechanisms for the generation of descendants: mutation and crossover; finally a replacement mechanism, which is applied between the vector father and son vector determining who survive into the next generation. There exist works where they currently use fuzzy logic to optimize the performance of the algorithms, to name a few articles such as:

Optimization of Membership Functions for Type-1 and Type 2 Fuzzy Controllers of an Autonomous Mobile Robot Using PSO [1], Optimization of a Fuzzy Tracking Controller for an Autonomous Mobile Robot under Perturbed Torques by Means of a Chemical Optimization Paradigm [2], Design of Fuzzy Control Systems with Different PSO Variants [4], A Method to Solve the Traveling Salesman Problem Using Ant Colony Optimization Variants with Ant Set Partitioning [6], Evolutionary Optimization of the Fuzzy Integrator in a Navigation System for a Mobile Robot [7], Optimal design of fuzzy classification systems using PSO with dynamic parameter adaptation through fuzzy logic [8]: Dynamic Fuzzy Logic Parameter Tuning for ACO and Its Application in TSP Problems [10], Bio-inspired Optimization Methods on Graphic Processing Unit for Minimization of Complex Mathematical Functions [18].

Similarly as there are papers on Differential Evolution (DE) applications that uses this algorithm to solve real problems. To mention a few:

A fuzzy logic control using a differential evolution algorithm aimed at modelling the financial market dynamics [5], Design of optimized cascade fuzzy controller based on differential evolution: Simulation studies and practical insights [11], Eliciting transparent fuzzy model using differential evolution [3], Assessment of human operator functional state using a novel differential evolution optimization based adaptive fuzzy model [14].

This chapter is organized as follows: Sect. 2 shows the concept of the differential evolution algorithm as applied to the technique for parameter optimization. Section 3 describes the proposed methods. Section 4 shows the simulation results. Section 5 offers the conclusions.

2 Differential Evolution

Differential Evolution (DE) is an optimization method belonging to the category of evolutionary computation applied in solving complex optimization problems.

DE is composed of four steps:

- Initialization.
- Mutation.
- Crossover.
- Selection.

This is a non-deterministic technique based on the evolution of a vector population (individuals) of real values representing the solutions in the search space. The generation of new individuals is carried out by differential crossover and mutation operators [13].

The operation of the algorithm is explained below:

2.1 Population Structure

The differential evolution algorithm maintains a pair of vector populations, both of which contain Np D-dimensional vectors of real-valued parameters [8].

$$P_{x,g} = (x_{i,g}), \quad i = 0, 1, \ldots, Np, \quad g = 0, 1, \ldots, g_{max} \tag{1}$$

$$x_{i,g} = (x_{j,i,g}), \quad j = 0, 1, \ldots, D - 1 \tag{2}$$

where:

P_x = current population
g_{max} = maximum number of iterations
i = index population
j = parameters within the vector

Once the vectors are initialized, three individuals are selected randomly to produce an intermediate population, $P_{v,g}$, of Np mutant vectors, $v_{i,g}$.

$$P_{v,g} = (v_{i,g}), \quad i = 0, 1, \ldots, Np - 1, \quad g = 0, 1, \ldots, g_{max} \tag{3}$$

$$v_{i,g} = (v_{j,I,g}), \quad j = 0, 1, \ldots, D - 1 \tag{4}$$

Each vector in the current population are recombined with a mutant vector to produce a trial population, P_u, the NP, mutant vector $u_{i,g}$:

$$P_{v,g} = (u_{i,g}), \quad i = 0, 1, \ldots, Np - 1, \quad g = 0, 1, \ldots, g_{max} \tag{5}$$

$$u_{i,g} = (u_{j,I,g}), \quad j = 0, 1, \ldots, D - 1 \tag{6}$$

2.2 Initialization

Before initializing the population, the upper and lower limits for each parameter must be specified. These 2D values can be collected by two initialized vectors, D-dimensional, b_L and b_U, to which subscripts L and U indicate the lower and upper limits respectively. Once the initialization limits have been specified number generator randomly assigns each parameter in every vector a value within the set range. For example, the initial value (g = 0) of the jth vector parameter is i-th:

$$x_{j,i,0} = \ \text{rand}_j(0,1) \ \cdot \ (b_{j,U} - b_{j,L}) + b_{j,L} \tag{7}$$

2.3 Mutation

In particular, the differential mutation adds a random sample equation showing how to combine three different vectors chosen randomly to create a mutant vector.

$$\mathbf{v}_{i,g} = \mathbf{x}_{r0,g} + \ F \ \cdot \ (\mathbf{x}_{r1,g} - \mathbf{x}_{r2,g}) \tag{8}$$

The scale factor, $F \in (0,1)$ is a positive real number that controls the rate at which the population evolves. While there is no upper limit on F, the values are rarely greater than 1.0.

2.4 Crossover

To complement the differential mutation search strategy, DE also uses uniform crossover. Sometimes known as discrete recombination (dual). In particular, DE crosses each vector with a mutant vector:

$$U_{i,g} = (u_{j,i,g}) = \begin{cases} v_{j,i,g} & if\,(rand_j(0,1) \le Cr \ or \ j = j_{rand}) \\ x_{j,i,g} & otherwise \end{cases} \tag{9}$$

2.5 Selection

If the test vector, $U_{i,g}$ has a value of the objective function equal to or less than its target vector, $X_{i,g}$. It replaces the target vector in the next generation; otherwise, the target retains its place in population for at least another generation [2]

$$X_{i,g+1} = \left\{ \begin{array}{ll} U_{i,g} & if\ f\left(U_{i,g}\right) \le f\left(X_{i,g}\right) \\ X_{i,g} & otherwise \end{array} \right\} \tag{10}$$

The process of mutation, recombination and selection are repeated until the optimum is found, or terminating pre criteria specified is satisfied. DE is a simple, but powerful search engine that simulates natural evolution combined with a mechanism to generate multiple search directions based on the distribution of solutions in the current population. Each vector i in the population at generation G, xi,G, called at this moment of reproduction as the target vector will be able to generate one offspring, called trial vector (ui,G). This trial vector is generated as follows: First of all, a search direction is defined by calculating the difference between a pair of vectors $r1$ and $r2$, called *"differential vectors"*, both of them chosen at random from the population. This difference vector is also scaled by using a user defined parameter called *"F ≥ 0"*. This scaled difference vector is then added to a third vector $r3$, called *"base vector"*. As a result, a new vector is obtained, known as the mutation vector. After that, this mutation vector is recombined with the target vector (also called parent vector) by using discrete recombination (usually binomial crossover) controlled by a crossover parameter $0 \le CR \le 1$ whose value determines how similar the trial vector will be with respect to the target vector. There are several DE variants. However, the most known and used is DE/rand/1/bin, where the base vector is chosen at random, there is only a pair of differential vectors and a binomial crossover is used. The detailed pseudocode of this variant is presented in Fig. 1 [9].

2.6 Illustrative Example of the Classic DE Algorithm

A simple numerical example adopted is presented to illustrate the classic DE algorithm. Let us consider the following objective function for optimization:

$$\text{Minimize } f(x) = x_1 + x_2 + x_3$$

The initial population is chosen randomly between the bounds of decision variables, in this case $x1$, $x2$ and $x \in [0, 1]$. The population along with its respective objective function values is shown in Table 1. The first member of the population, "Individual 1", is set as the target vector. In order to generate the mutated vector, three individuals ("Individual 2", "Individual 4" and "Individual 6") from the population size are selected randomly (ignoring "Individual 1", since it is set as the target vector). The weighted difference between "Individual 2" and "Individual 4" is added to the third randomly chosen vector "Individual 6" to generate the mutated vector. The weighting factor F is chosen as 0.80 and the weighted difference vector is obtained in Table 2 and the mutated vector in Table 3 [19].

Fig. 1 "DE/rand/1/bin" pseudocode rand [0, 1) is a function that returns a real number between 0 and 1. Randint (min, max) is a function that returns an integer number between min and max. *NP, MAX GEN, CR* and *F* are user-defined parameters *n* is the dimensionality of the problem [9]

```
Begin
  G=0
  Create a random initial population x_{i,G} ∀i, i = 1,...,NP
  Evaluate f(x_{i,G}) ∀i, i = 1,...,NP
  For G=1 to MAX_GEN Do
    For i=1 to NP Do
      Select randomly r₁ ≠ r₂ ≠ r₃ :
      j_{rand} = randint(1,D)
      For j=1 to n Do
        If (rand_j[0,1) < CR or j = j_{rand}) Then
          u_{i,j,G+1} = x_{r₃,j,G} + F(x_{r₁,j,G} - x_{r₂,j,G})
        Else
          u_{i,j,G+1} = x_{i,j,G}
        End If
      End For
      If (f(u_{i,G+1}) ≤ f(x_{i,G})) Then
        x_{i,G+1} = u_{i,G+1}
      Else
        x_{i,G+1} = x_{i,G}
      End If
    End For
    G = G+1
  End For
End
```

Table 1 An illustrative example [19]

Population size NP = 6 (user defined), D = 3

	1	2	3	4	5	6
x_1	0.68	0.92	0.22	0.12	0.40	0.94
x_2	0.89	0.92	0.14	0.09	0.81	0.63
x_3	0.04	0.33	0.40	0.05	0.83	0.13
$f(x)$	1.61	2.17	0.76	0.26	2.04	1.70

Table 2 Calculation of the weighted difference vector for the illustrative example [19]

	Individual 2	Individual 4	Difference vector		Weighted difference vector
x_1	0.92		0.12	= 0.80		= 0.64
x_2	0.92	−	0.09	= 0.83	xF	= 0.66
					(F = 0.80)	
x_3	0.33		0.05	= 0.28		= 0.22

Table 3 Calculation of the mutated vector for the illustrative example [19]

	Weighted difference vector		Individual 6	Mutated vector
x_1	0.64		0.94	= 1.58
x_2	0.66	+	0.63	= 1.29
x_3	0.22		0.13	= 0.35

Table 4 Generation of the trial vector for the illustrative example [19]

	Target vector		Mutated vector	Trial vector
x_1	0.68		**1.58**	= 1.58
x_2	**0.89**	Crossover	1.29	= 0.89
x_3	**0.04**	(CR = 0.50)	0.35	= 0.04
$f(x)$	1.61		3.22	2.51

Table 5 New populations for the next generation in the illustrative example [19]

	New population for the next generation					
	Individual 1	Individual 2	Individual 3	Individual 4	Individual 5	Individual 6
x_1	0.68					
x_2	0.89					
x_3	0.04					
$f(x)$	1.61					

The mutated vector does a crossover with the target vector to generate the trial vector, as shown in Table 4. This is carried out by (1) generating random numbers equal to the dimension of the problem (2) for each of the dimensions: if random number $>CR$; copy the value from the target vector, else copy the value from the mutated vector into the trial vector. In this example, the crossover constant CR is chosen as 0.50. The bold values in Table 4 indicate selected numbers.

The objective function of the trial vector is compared with that of the target vector and the vector with the lowest value of the two (minimization problem) becomes "Individual 1" for the next generation. To evolve "Individual 2" for the next generation, the second member of the population is set as target vector (see Table 5) and the above process is repeated. This process is repeated NP times until the new population set array is filled, which completes one generation. Once the termination criterion is met, the algorithm ends.

3 Proposed Method

The Differential Evolution (DE) Algorithm is a powerful search technique used for solving optimization problems. In this chapter a new algorithm called Fuzzy Differential Evolution (FDE) with dynamic adjustment of parameters for the optimization of controllers is proposed. The main objective is that the fuzzy system will provides us with the optimal parameters for the best performance of the DE algorithm. In addition the parameters that the fuzzy system optimizes are the crossover and mutation, as shown in Fig. 2.

Algoritmo Evolución Diferencial

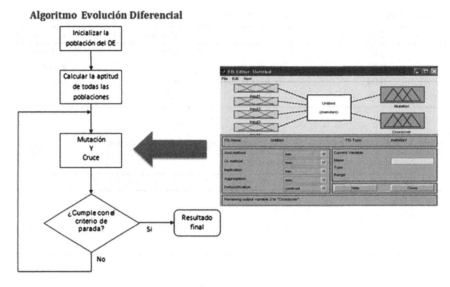

Fig. 2 The proposed is an algorithm of differential evolution (DE) by integrating a fuzzy system to dynamically adapt parameters

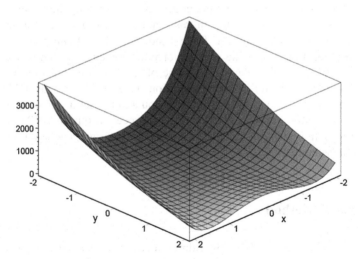

Fig. 3 Rosenbrock's in 2D, $f(x,y) = 100(y - x^2)^2 + (1 - x)^2$

4 Simulation Results

This chapter presents the current results obtained from the experiments using the scale factor F from the Differential Evolution (DE) algorithm. This helps us visualize how the algorithm performs modifications to this variant which helps

Table 6 Demonstrates the results with F = 0.01

Experimento	Promedio	Mejor	Peor
1	0.01099565	5.3599E−07	0.03150259
2	0.01131575	4.47E−12	0.1149224
3	0.00889237	1.77E−08	0.06249856
4	0.01571152	1.86E−09	0.08654202
5	0.01677769	9.11E−08	0.1480454
6	0.02614949	5.00E−10	0.37007527
7	0.01057668	7.23E−07	0.09474443
8	0.01099565	5.36E−07	0.16184713
9	0.01131575	4.47E−12	0.1149224
10	0.00889237	1.77E−08	0.06249856
11	0.01571152	1.86E−09	0.08654202
12	0.01677769	5.14E−07	0.1480454
13	0.02614949	3.48E−08	0.09871141
14	0.01099565	5.79E−07	0.16184713
15	0.01131575	4.47E−12	0.1149224
16	0.00889237	4.47E−06	0.06249856
17	0.01571152	1.86E−09	0.08654202
18	0.01677769	9.11E−08	0.06250744
19	0.02614949	5.00E−10	0.37007527
20	0.01099565	5.79E−07	0.16184713
21	0.01131575	4.47E−12	0.1149224
22	0.00889237	1.77E−08	0.06249856
23	0.01571152	1.43E−11	0.08654202
24	0.01677769	9.11E−08	0.1480454
25	0.02614949	5.00E−10	0.11850916
26	0.01057668	7.23E−07	0.09474443
27	0.01931193	4.17E−08	0.10814125
28	0.01402754	3.80E−09	0.03090101
29	0.01099565	5.79E−07	0.16184713
30	0.01131575	4.47E−12	0.1149224

create mutations. This helps us visualize how the algorithm performs modifications to this variant which helps create mutations. The Rosenbrock mathematical function was used to carry out these experiments. Rosenbrock is a classic optimization problem, also known as banana function or the second function of De Jong. The global optimum lies inside a long, narrow, parabolic shaped flat valley. To find the valley is trivial, however convergence to the global optimum is difficult and hence this problem has been frequently used to test the performance of optimization algorithms. Function has the following definition.

$$f(x) = \sum_{i=1}^{n-1} [100(x_{i+1} - x_i^2)^2 + (1 - x_i)^2]. \tag{11}$$

Table 7 Demonstrates the results with F = 0.02

Experimento	Promedio	Mejor	Peor
1	0.01099565	5.3599E−07	0.03150259
2	0.01131575	4.47E−12	0.1149224
3	0.00889237	1.77E−08	0.06249856
4	0.01571152	1.86E−09	0.08654202
5	0.01677769	9.11E−08	0.1480454
6	0.02614949	5.00E−10	0.37007527
7	0.01057668	7.23E−07	0.09474443
8	0.01099565	5.36E−07	0.16184713
9	0.01131575	4.47E−12	0.1149224
10	0.00889237	1.77E−08	0.06249856
11	0.01571152	1.86E−09	0.08654202
12	0.01677769	5.14E−07	0.1480454
13	0.02614949	3.48E−08	0.09871141
14	0.01099565	5.79E−07	0.16184713
15	0.01131575	4.47E−12	0.1149224
16	0.00889237	4.47E−06	0.06249856
17	0.01571152	1.86E−09	0.08654202
18	0.01677769	9.11E−08	0.06250744
19	0.02614949	5.00E−10	0.37007527
20	0.01099565	5.79E−07	0.16184713
21	0.01131575	4.47E−12	0.1149224
22	0.00889237	1.77E−08	0.06249856
23	0.01571152	1.43E−11	0.08654202
24	0.01677769	9.11E−08	0.1480454
25	0.02614949	5.00E−10	0.11850916
26	0.01057668	7.23E−07	0.09474443
27	0.01931193	4.17E−08	0.10814125
28	0.01402754	3.80E−09	0.03090101
29	0.01099565	5.79E−07	0.16184713
30	0.01131575	4.47E−12	0.1149224

Test area is usually restricted to hypercube $-2.048 \leq x_i \leq 2.048$, $i = 1, \ldots, n$. Its global minimum equal $f(x) = 0$ is obtainable for $x_i = 0$, $i = 1, \ldots, n$ (Fig. 3).

Table 8 Demonstrates the results with F = 0.03

Experimento	Promedio	Mejor	Peor
1	6.66E−04	1.73E−17	0.01785224
2	1.39E−02	5.39E−18	0.01226117
3	1.06E−02	3.04E−14	0.00209507
4	2.62E−03	5.02E−13	0.05327532
5	6.66E−04	1.73E−17	0.01785224
6	1.15E−03	1.59E−17	0.01705839
7	1.34E−03	5.49E−17	0.01487419
8	1.39E−02	1.61E−20	0.38046604
9	9.61E−03	6.21E−15	0.17429945
10	6.36E−03	8.17E−17	0.06436387
11	1.06E−02	3.06E−15	0.31396925
12	9.78E−04	9.51E−15	0.02470255
13	6.66E−04	1.73E−17	0.01785224
14	2.62E−03	4.85E−16	0.05327532
15	1.39E−02	6.56E−17	0.38046604
16	1.06E−02	7.07E−13	0.31396925
17	2.62E−03	4.85E−16	0.05327532
18	6.66E−04	1.73E−17	0.01785224
19	1.15E−03	1.59E−17	0.01705839
20	1.39E−02	5.39E−18	0.01226117
21	1.06E−02	3.06E−15	0.31396925
22	2.62E−03	4.85E−16	0.05327532
23	1.15E−03	1.59E−17	0.01705839
24	1.34E−03	5.49E−17	0.01487419
25	1.39E−02	6.56E−17	0.38046604
26	9.61E−03	6.21E−15	0.17429945
27	6.36E−03	8.17E−17	0.06436387
28	1.06E−02	3.06E−15	0.31396925
29	9.78E−04	9.37E−15	0.02470255
30	3.73E−04	4.85E−17	0.00940052

Table 9 Demonstrates the results with F = 0.04	Experimento	Promedio	Mejor	Peor
	1	0.01179166	4.34E−13	0.12649579
	2	0.00689591	3.79E−11	0.07022387
	3	0.00884897	8.91E−16	0.02545798
	4	0.00786076	3.11E−12	0.08015746
	5	0.01261989	4.97E−11	0.27651624
	6	0.01652249	2.22E−10	0.25001424
	7	0.01179166	4.34E−13	0.12649579
	8	0.00884897	8.91E−16	0.17181307
	9	0.00786076	3.11E−12	0.08015746
	10	0.01261989	4.97E−11	0.27651624
	11	0.01652249	3.49E−11	0.25001424
	12	0.02675393	5.16E−13	0.17059178
	13	0.00347209	1.08E−13	0.02806518
	14	0.04041566	1.80E−15	0.74575319
	15	0.00468022	9.61E−13	0.05179923
	16	0.01471826	1.73E−12	0.21308918
	17	0.01179166	4.34E−13	0.06084437
	18	0.00884897	8.91E−16	0.02545798
	19	0.00786076	3.11E−12	0.08015746
	20	0.01652249	3.49E−11	0.25001424
	21	0.02675393	5.16E−13	0.3559589
	22	0.00347209	1.08E−13	0.02806518
	23	0.04041566	1.80E−15	0.36500605
	24	0.00468022	9.61E−13	0.05179923
	25	0.01471826	3.93E−09	0.21308918
	26	0.00247429	7.88E−13	0.01730106
	27	0.02497659	2.84E−10	0.0582611
	28	0.00639566	3.73E−13	0.04982843
	29	0.0372501	6.99E−11	0.61885365
	30	0.00866303	7.95E−12	0.11939131

30 experiments were done each one 30 times, using the following parameters:

Parameters
D = 50
NP = 250
F = 0.1
CR = 0.1
GEN = 6,000
L = −500
H = 500

Where:

D = Vector dimension
NP = Size of population
F = scale factor

CR = crossover
GEN = Maximum number of generations
L = Lower limits
H = Upper limits

Table 6 shows the simulation results when F = 0.01. Table 7 shows the results when F = 0.02. The experiments with F = 0.03 and 0.04 are shown on Tables 8 an 9, respectively.

5 Conclusions

To conclude this chapter, the preliminary results will help us better understand with more clarity the Differential Evolution (DE) algorithm. In the same way we can discover the way of exploring and then exploitating the algorithm. The results will be observed to see if the majority are good although in very few cases the errors are very high. We can tentatively conclude when Fuzzy logic is applied we can get better results from Differential Evolution (DE) avoiding high errors, like those obtained when trial and error is used.

References

1. Aguas-Marmolejo, S. J., Castillo, O.: Optimization of membership functions for type-1 and type 2 fuzzy controllers of an autonomous mobile robot using PSO. In: Recent Advances on Hybrid Intelligent Systems, pp. 97–104. Springer, Heidelberg (2013)
2. Astudillo, L., Melin, P., Castillo O.: Optimization of a fuzzy tracking controller for an autonomous mobile robot under perturbed torques by means of a chemical optimization paradigm. In: Recent Advances on Hybrid Intelligent Systems, pp. 3–20. Springer, Berlin (2013)
3. Eftekhari, M., Katebi, S.D., Karimi, M., Jahanmir, A.H.: Eliciting transparent fuzzy model using differential evolution. Appl. Soft Comput. 8(1), 466–476 (2008)
4. Fierro, R., Castillo, O.: Design of fuzzy control systems with different PSO variants. In: Recent Advances on Hybrid Intelligent Systems, pp. 81–88. Springer, Berlin (2013)
5. Hachicha, N., Jarboui, B., Siarry, P.: A fuzzy logic control using a differential evolution algorithm aimed at modelling the financial market dynamics. Inf. Sci. 181, 79–91 (2011)
6. Lizárraga, E., Castillo, O., Soria, J.: A method to solve the traveling salesman problem using ant colony optimization variants with ant set partitioning. In: Recent Advances on Hybrid Intelligent Systems, pp. 237–246. Springer, Berlin (2013)
7. Melendez, A., Castillo, O.: Evolutionary optimization of the fuzzy integrator in a navigation system for a mobile robot. In: Recent Advances on Hybrid Intelligent Systems, pp. 21–31. Springer, Berlin (2013)
8. Melin, P., Olivas, F., Castillo, O., Valdez, F., Soria, J., García, J.: Optimal design of fuzzy classification systems using PSO with dynamic parameter adaptation through fuzzy logic. Expert Syst. Appl. 40(8), 3196–3206 (2013)
9. Mezura-Montes, E., Palomeque-Ortiz, A.: Self-adaptive and Deterministic Parameter Control in Differential Evolution for Constrained Optimization. Laboratorio Nacional de Informatica Avanzada (LANIA A.C.), Rebsamen 80, Centro, Xalapa, Veracruz, 91000, MEXICO (2009)

10. Neyoy, H., Castillo, O., Soria, J.: Dynamic fuzzy logic parameter tuning for ACO and its application in TSP problems. Recent Advances on Hybrid Intelligent Systems, pp. 259–271. Springer, Berlin (2013)
11. Oh, S.-K., Kim, W.-D., Pedrycz, W.: Design of optimized cascade fuzzy controller based on differential evolution: simulation studies and practical insights. Eng. Appl. Artif. Intell. **25**, 520–532 (2012)
12. Olivas, F., Castillo, O.: Particle swarm optimization with dynamic parameter adaptation using fuzzy logic for benchmark mathematical functions. In: Recent Advances on Hybrid Intelligent Systems, pp. 247–258, Springer, Berlin (2013)
13. Price, K.V., Storn R., Lampinen. J.A.: Differential Evolution. Springer, Berlin (2005)
14. Raofen, W., Zhang, J., Zhang, Y., Wang, X.: Assessment of human operator functional state using a novel differential evolution optimization based adaptive fuzzy model. Biomed. Signal Process. Control **7**, 490–498 (2012)
15. Sombra, A., Valdez, F., Melin, P., Castillo, O.: A new gravitational search algorithm using fuzzy logic to parameter adaptation. In: IEEE Congress on Evolutionary Computation, pp. 1068–1074. IEEE, New York (2013)
16. Valdez, F., Melin, P., Castillo, O.: Evolutionary method combining particle swarm optimization and genetic algorithms using fuzzy logic for decision making. In: Proceedings of the IEEE International Conference on Fuzzy Systems, pp. 2114–2119. IEEE, New York (2009)
17. Valdez, F., Melin, P., Castillo, O.: Parallel particle swarm optimization with parameters adaptation using fuzzy logic. In: MICAI, part 2, pp. 374–385. Springer, Berlin (2012)
18. Valdez, F., Melin, P., Castillo, O.: Bio-inspired optimization methods on graphic processing unit for minimization of complex mathematical functions. In: Recent Advances on Hybrid Intelligent Systems, pp. 313–322. Springer, Berlin (2013)
19. Vucetic, D.: Fuzzy Differential Evolution Algorithm. The University of Western Ontario, London (2012)

A Fuzzy Control Design
for an Autonomous Mobile
Robot Using Ant Colony Optimization

Evelia Lizarraga, Oscar Castillo, José Soria and Fevrier Valdez

Abstract In this chapter we describe the methodology to design an optimized fuzzy logic controller for an autonomous mobile robot, using Ant Colony Optimization (ACO). This is achieved by applying a systematic and hierarchical optimization modifying the conventional ACO algorithm using ants partition. The simulations results proved that the proposed algorithm performs even better that the classic ACO algorithm when optimizing membership functions of FLC, parameters and fuzzy rules.

1 Introduction

There are many evolutionary and swarm intelligence algorithms that have been used to optimize Fuzzy Logic Controllers (FLC) this is particularly because this meta-heuristics have a good approach to a combinatorial problems, such as find the right parameters of a FLC. This adjustment is usually performed by trial and error taking a lot of time for a designer; this is why several bio-inspired and evolutionary techniques have been used, such as: Genetic Algorithms (GA), Particle Swarm Optimization (PSO) [7, 17], Differential Evolution (DE) [16] and Ant Colony Optimization (ACO) [4], the latter being the one explored in this chapter [10, 11, 13]. ACO works based in the analogy of working ants that search for food. The analogy presents a cooperative group of ants that follow a path depending on the level of pheromone; the path with more pheromone is the shortest path or the path with less cost. The implementation of this algorithm requires deconstructing the problem

E. Lizarraga · O. Castillo · J. Soria · F. Valdez (✉)
Tijuana Institute of Technology, Tijuana, Mexico
e-mail: fevrier@tectijuana.mx

O. Castillo
e-mail: ocastillo@tectijuana.mx

O. Castillo et al. (eds.), *Recent Advances on Hybrid Approaches for Designing Intelligent Systems*, Studies in Computational Intelligence 547,
DOI: 10.1007/978-3-319-05170-3_20, © Springer International Publishing Switzerland 2014

into a graph representation given that ACO works with this kind of scheme. ACO has been applied successfully in several combinatorial problems like the Traveling Salesman Problem (TSP) [1], data mining [2], network routing [3] among others. It also has been applied in FLC design; however in this chapter we proposed a new approach for ACO. The proposed ACO algorithm, that we are called ACO Variants Subset Evaluation (AVSE), conducts a general modification of the classic ACO algorithm. To explain this, the proposed algorithm is seen as a series of steps that are performed hierarchically and sequentially, allowing a faster optimization and better results than the classical ACO algorithm.

AVSE works dividing the total number of ants, equivalently, between the five main ACO variants: Ant System (AS), Elitist Ant System (EAS), Rank Based Ant System (ASRank), Man-Min Ant System (MMAS), Ant Colony System (ACS). To improve the performance a stagnation mechanism was added; this mechanism will stop the variant that is not giving good performance (minimal error or threshold minimal error), and as result the algorithm will be faster. This approach has been applied previously for the Traveling Salesman Problem (TSP) [12, 18] and in the Ball and Beam FLC design with satisfactory results.

We made several experiments applying the proposed method to an autonomous mobile unicycle robot [14]. The proposed AVSE algorithm optimized type of membership functions, parameters of membership functions and fuzzy rules [15, 20].

This chapter is organized as follows. Section 2 reviews the basic concepts of a fuzzy system. Section 3 describes the basic concepts of Ant Colony Optimization. Section 4 generally describes the problem of a mobile robot. Section 5 details the proposed approach AVSE and its graphic representation. Section 6, simulations results will be given for results analysis. In Sect. 7, a conclusion will be drawn.

2 Basic Concepts of a Fuzzy System

Our proposal covers three parts of the fuzzy system, and we must define each, to understand the next steps.

Type of membership function

A fuzzy set expresses the degree to which an element belongs to a set and is characterized by its membership function. A convenient way to express it is as a mathematical formula [9]. If X is a collection of objects, denoted by x, then a fuzzy set A in X is defined as a set or ordered pairs:

$$A = \{(X, \mu_A(X))|x \in X\} \tag{1}$$

where $\mu_A(x)$ is called the membership function (MF) for the fuzzy set. The MF maps each element of X to a membership grade between 0 and 1.

A membership function can take different shapes that we refer as *type of membership function*. There are four main shapes: triangular, trapezoidal, Gaussian and generalized bell as Fig. 1 illustrates.

Fig. 1 Types of membership functions

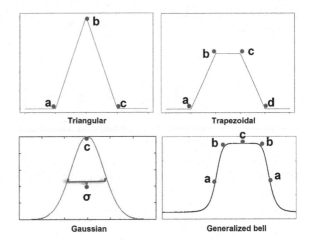

Table 1 Relation between parameters and type of MF

Type of membership function	Parameters definition Parameters
Triangular	{a,b,c} a < b < c
Trapezoidal	{a,b,c,d} a < b≤c ≤ d
Gaussian	{c,σ}
Generalized bell	{a,b,c}

Parameters of membership function: as we stated before, a MF can take many shapes which are defined by *the membership function parameters*. These parameters determinate the coordinates of each underlying edge and the location of the membership function. The set of parameters that define each shape are shown in Table 1. Figure 2 shows the relation between the shape of membership function and its parameters.

We use (2) to calculate the Gaussian curve, where c represents the center, the variance σ determines the curve and x is a point in the curve. The width of the curve for the generalized bell is calculated with parameters a and b using (3) where c is the parameter that locates the center of the curve. The parameter b should be positive,

$$f(x) = \exp\left(\frac{-0.5(x-c)^2}{\sigma^2}\right), \tag{2}$$

$$f(x) = \frac{1}{1 + \left|\frac{x-c}{a}\right|^{2b}}. \tag{3}$$

Moving the parameters can affect the shape of the MF, making it wider or narrower. A wider shape covers more possible fuzzy values, unlike a more narrow

Fig. 2 Defined parameters
for type of MF

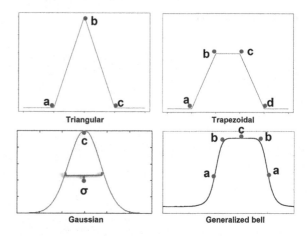

shape, which covers few fuzzy values. Optimizing type and shape (parameters), in a membership function, can help to find more suitable fuzzy values for a particular problem.

Fuzzy rules: Also known as fuzzy if-then rules assumes the form:

$$\text{if } x \text{ is } A \text{ then } y \text{ is } B \tag{4}$$

where "x is A" is called antecedent and "y is B" is called consequence. The interpretation of a fuzzy rule, is a three-parts process: (a) Assign the antecedents a membership grade from 0 to 1; (b) if there are multiples antecedents, apply fuzzy logical operators to reduce the antecedent to a single number between 0 and 1. The result will be the support for the entire rule to shape the fuzzy set output. (c) Apply the implication method; this is, if the rule has more than one antecedent, a fuzzy operator is applied to obtain a number representing the result of applying said rule [25, 26, 27].

3 Basic Concepts of Ant Colony Optimization

Proposed as an ant colony analogy by Dorigo et al. in 1990s [5], ACO is defined biologically speaking by ants who aim to find food visiting potential food places [6]. To communicate, ants use stigmergy; a biological mechanism that can transmit information through the environment using the pheromones that ants issued. Artificially speaking, each ant is a possible solution to the problem, where the set of possible solutions is represented as a graph [8].

In the ACO algorithm an ant k visits each node. To select the next node j, is applied a stochastic probabilistic rule (5), which is determined by using information of the amount of pheromone τ_{ij} in node i, within a feasible neighborhood N_i^k,

$$p_{ij}^k(t) = \begin{cases} \dfrac{[\tau_{ij}(t)]^\alpha [\eta_{ij}]^\beta}{\sum\limits_{k \in N_i^k} [\tau_{ik}(t)]^\alpha [\eta_{ik}]^\beta} & \text{si } j \in N_i^k \\ 0 & \text{otherwise} \end{cases} \tag{5}$$

The heuristic information η_{ij} is provided by the mean squared error (MSE) given by:

$$MSE = \frac{1}{N} \sum_{q=1}^{N} (u_q - \tilde{u}_q)^2 \tag{6}$$

where $\hat{u}$ is the control signal, u the reference and N the number of observed points. Parameters α and β are relatives weights of pheromone and heuristic information, respectively [21, 22, 23, 24].

Once the path is constructed it will be evaluated to determine the cost of the path. Depending on the cost of the path is the amount of pheromone $\Delta \tau_{ij}$ that an ant will deposit on the node (7). The better the path, the more amount of pheromone will be deposited.

$$\tau_{ij} = \rho \tau_{ij} + \Delta \tau_{ij} \tag{7}$$

where ρ is a parameter that represents the evaporation coefficient, $0 < 1 - \rho < 1$.

The algorithm terminates when the path created by each ant have been evaluated.

4 Fuzzy Logic Controller of an Autonomous Robot

We considered the following model (8) of a unicycle mobile robot to test our optimized fuzzy system [19]. Where the adequate torque is τ, the position is $q(t)$ in a time t and the reference trajectory is $q_a(t)$ in a time t;

$$\lim_{\tau \to \infty} \|q_d(t) - q(t)\| = 0 \tag{8}$$

The optimization goal is to design a FLC that gives an appropriate torque τ to be applied.

We designed a Mamdani FLC where the linguistic variables are in the input and output. To determine the movement of the torques (outputs) right (Torque_R) and left (Torque_L) is required as inputs: the linear (v) and angular (w) velocities. Therefore our Fuzzy Logic Controller of the autonomous is as Fig. 4 depict, using initially triangular membership functions before further optimization is made. The FLC uses 9 basic set of rules, number that will be modified in the optimization process. These rules were constructed taking in consideration the error and change of error of the inputs v and w, as well of the output (Torque_R) and (Torque_L)

Fig. 3 Autonomous mobile robot FLC

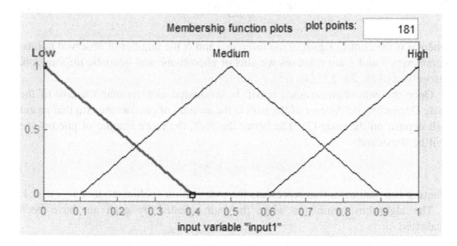

Fig. 4 Triangular membership functions

with linguistic values of Negative Error (NE), No change Error(NCE), Positive Error (PE), illustrated in Tables 2 and 3 (Fig. 3).

As for the type of membership functions for our basic FLC we selected triangular, considering that in the optimization process these can be changed selecting the four available types of membership function (triangular, trapezoidal, Gaussian, Generalized Bell)

5 Proposed Approach

We divide the proposed method in four different phases.

Phase 1: Automatic parameter extraction

One of the main contributions of this method is to generalize the optimization of fuzzy systems. To this end, we have developed two graphical interfaces in

Table 2 Fuzzy rule set

v/w	NE	NCE	PE
NE	NE/NE	NCE/NE	PE/NE
NCE	NE/NCE	NCE/NCE	NCE/PE
PE	NE/PE	NCE/PE	PE/PE

Table 3 Fuzzy rules defined

Rule number	Rule
1	If (v is NE) and (w is NE) then ($\tau1$ is NE)($\tau2$ is NE)
2	If (v is NE) and (w is NCE) then ($\tau1$ is N)($\tau2$ is NCE)
3	If (v is NE) and (w is PE) then ($\tau1$ is NE)($\tau2$ is PE)
4	If (v is NCE) and (w is NE) then ($\tau1$ is NCE)($\tau2$ is NE)
5	If (v is NCE) and (w is NCE) then ($\tau1$ is NCE)($\tau2$ is NCE)
6	If (v is NCE) and (w is PE) then ($\tau1$ is NCE)($\tau2$ is PE)
7	If (v is PE) and (w is NE) then ($\tau1$ is P)($\tau2$ is N)
8	If (v is PE) and (w is NCE) then ($\tau1$ is PE)($\tau2$ is NCE)
9	If (v is PE) and (w is PE) then ($\tau1$ is PE)($\tau2$ is PE)

MATLAB. The first one is called Fuzzy Parameter Extraction (FPE) (Fig. 5) which allows extraction of fuzzy parameters needed to construct the pheromone matrix such as:

- Total number of inputs
- Total number of outputs
- Total number of MF per input and output
- Type of membership function per input and output
- Range per input and output.

In the said interface, the user loads the fuzzy system to be optimized and the correspondent plant, and also can select what type of optimization will be made, choosing at least one. This will make the optimization automatic, regardless of the type of fuzzy system, the number of inputs or outputs, nor its membership function types, allowing freedom and easiness to make experiments.

The second interface is called Pheromone Matrix Constructor (Fig. 6) and is used to enter the necessary ACO parameters to construct the pheromone matrix and de solution nodes. The user enters the following data:

- Total number of ants (that will be divided in n subset of ants). Where n is the number of variations selected
- Maximum number of iterations
- Alpha
- Beta

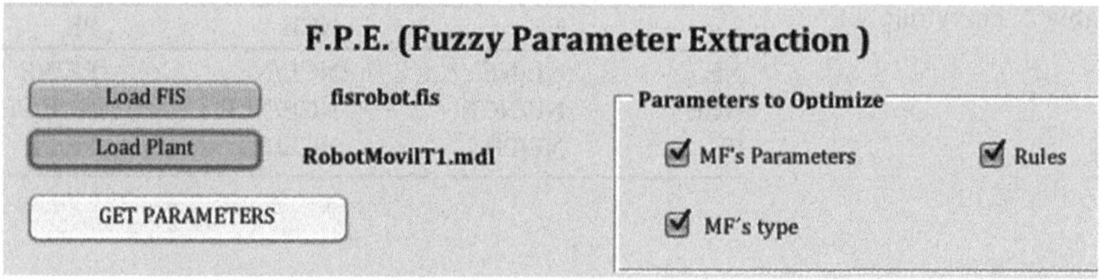

Fig. 5 Fuzzy parameter extractor interface

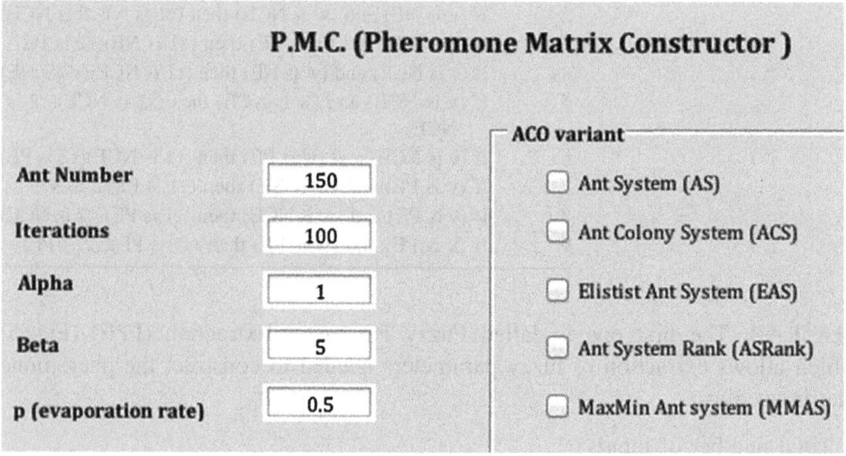

Fig. 6 Pheromone matrix constructor interface

- Rho (evaporation rate)
- Variations to use in optimization (AS, EAS, ASRank, MMAS, ACS) with a minimum of two.
- Depending of the chosen variation, the user can introduce other parameters; the e constant for the Elitist Ant System and the w constant for the Rank-Based Ant System.

Phase 2: ACO Variants Subset Evaluation (AVSE)

Having a set of ants m the method equivalently divides the total number of ants in five different subsets and each one is evaluated separately by the corresponding variation of ACO(AS,EAS, ASRanks, ACS and MMAS) as Fig. 7 shows. The evaluation of the each variant (ant subset) is made sequentially in the same iteration i. Subsequently, the best ant of each partition is compared with each other; obtaining the best global ant i (global best i) as illustrated in Fig. 8.

Therefore, in each iteration the best global ant i is compared with the best global ant $i-1$ following the conventional ACO algorithm. This allows us to compare in one iteration different variations, saving time doing tests and less

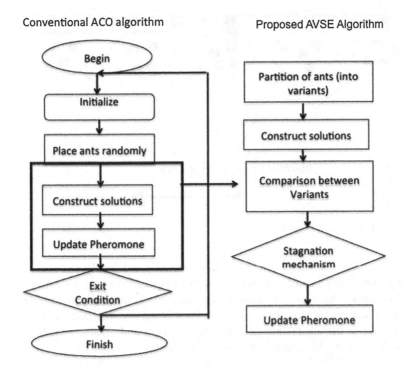

Fig. 7 Proposed AVSE algorithm

overhead in comparison with the conventional method, which uses just one variation of ACO in all iterations. This means n (iterations) $\times$ 5 (variations) number of tests comparison with the proposed method which only needs n tests. Because the proposed method selects one ant per iteration that represents also a variant of ACO; the end result provides the most used variant hence the one with best performance.

To improve the performance of the proposed method AVSE, we added a stagnation mechanism, which allows to stop a variant that has not been given good results after 5 consecutive iterations. It is understood as a good result if a variant reach a local best, per iteration. It has been noted in several preliminary experiments that if after an average of 5 iterations a variant does not reach a local best, it usually stagnate. The use of this mechanism allows a faster and proficient performance, using only the variants that get to obtain better results.

Phase 3: Hierarchical Optimization

Because the proposed optimization is made in three parts of the fuzzy system (type, parameters and rules) we must decide which comes first. Thus our optimization is sequential and hierarchical. Sequential, because is made successively, and not parallel; using the same number of iterations and ants for each kind of optimization. It is hierarchical, because we optimize firstly, types of membership

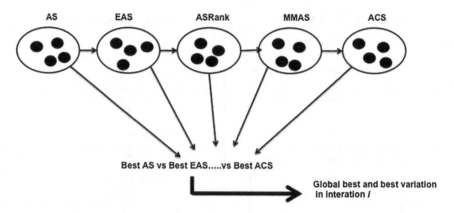

Fig. 8 Ants evaluation in subsets

functions, then parameters of membership functions and finally, the optimization of fuzzy rules. The justification for this, is that in order to optimize parameters, we need to know what types of membership function has the fuzzy system. Optimization of fuzzy rules was done at the end, as it is the one that takes less time and don't affect in the other kinds of optimizations (type and parameters). The Fig. 9 shows how the optimization sequence is structured.

First, the AVSE algorithm optimize the type of membership function; when a best global is reached, the algorithm saves the best global as type-global-best and also saves the best fuzzy system obtained as type-best-fuzzy. Once completed the maximum number of iterations, the algorithm continues with the optimization of parameters of membership functions setting the same maximum of iterations and ants, using the best resulting fuzzy system obtained in the previous optimization (type-best-fuzzy), it also sets the global best as type-global-best. Similarly, when performing parameter optimization the algorithm will save the best fuzzy system found as param-best-fuzzy and the global best as param-global-best. As the previous optimization, the methodology for optimizing fuzzy rules is similar; saving the best fuzzy system as rules-best-fuzzy and if reached rules-global-best. Remaining in the latter, the best fuzzy system found result of the three kinds of optimization.

Phase 4: Graphic Representation

Because of the nature of the ACO algorithm, the optimization problem has to be represented as a graph. This requires to represent all the possible solutions for each type of optimization (parameters, type and rules) and each type of optimization will have a different graph representation. Each resulting graph will generate a three-dimensional matrix, which will vary in dimension.

(a) Graph representation of type of membership function optimization:

Rows: as we defined before, we will use the 4 main types of membership functions, this is 4 rows for the pheromone matrix.

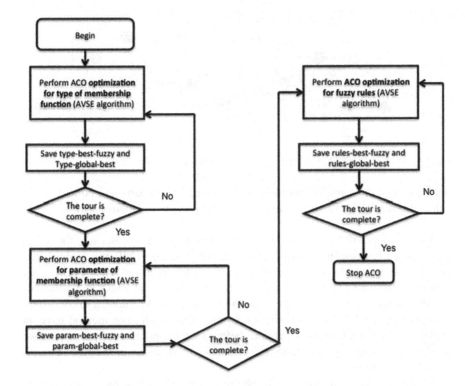

Fig. 9 Hierarchical sequence for AVSE optimization

Columns: this depends on the number of membership functions per input and output, which is an information provided by the FPE interface. For the Ball and Beam problem the total number of MF per input are 8. Because the Ball and Beam fuzzy system has the architecture of TSK there is no need of optimize the type of membership function, given that the outputs are mathematical equations.

Third dimension: this will be given by the number of ants in each subset, not the total of ants. For example if the total number of ants are 150 and we use three variations to optimize, the third dimension of the matrix will be 150/3 = 50 ants.

The resulting graph for the Ball and Beam problem is illustrated in Fig. 10.

(b) Graph representation of parameters of membership function optimization

As the optimization of membership function types, the universe of discourse can be represented as a graph. Where the rows of the pheromone matrix are the number of possible values for each parameter. Because the range value in the

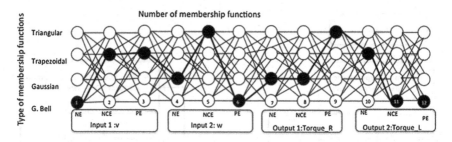

Fig. 10 Graph representation of type of membership function optimization

inputs are not the same we decided to discretized in 100 parts and normalize the values from 0.01 to 1. So this means that the pheromone matrix will have 100 rows.

The columns will be the total of parameters to be optimized. Because the points to be moved (parameters) depend on type and location of a MF, we established a general way to decide which points are going to be optimized using Fig. 11 as a pattern. Where the solid circle is the fixed point (point that won't be optimized) and the empty circle represent the point to be optimized. As a note, a fuzzy system requires that the MF intersect with each other, this is why our method have an internal validation of points. Of course, before we can optimize the parameters we need to have what type of membership functions will result in the *type of membership function optimization.*

Finally, the third dimension of the pheromone matrix will be the number of ants as calculated on type of membership function optimization.

Graph representation of fuzzy rules optimization.

To perform the optimization in the fuzzy rules, we have to consider how many antecedents will be (number of membership functions). As in the previous optimizations, is necessary to represents the possible combinations in a graph, making the artificial ant build a tour, each tour will be a set of fuzzy rules.

The basic actions of this part of the algorithm are as follows:

1. Combine antecedents
2. Assign a consequent to each combination of antecedents etc. rules
3. Activate the necessary rules.

The first step is to create the graph for all the possible combinations of X antecedents and consequent Y as follows:

If $X_{1,1}$ and $X_{1,2}$ and $X_{1,3}$ ... X_{nj} then $Y_{1,1}$
If $X_{2,1}$ and $X_{2,2}$ and $X_{2,3}$ ... X_{nj} then, $Y_{1,2}$
If $X_{3,1}$ and $X_{3,2}$ and $X_{3,3}$ ... X_{nj} then $Y_{1,3}$
If $X_{1,J}$ and $X_{2,j}$ and $X_{3,j}$ then $X_{n,j}$ then $Y_{o,h}$

where i is the number of inputs, j are the linguistic labels (membership functions) per input, o is the number of outputs and h are the linguistic labels per output.

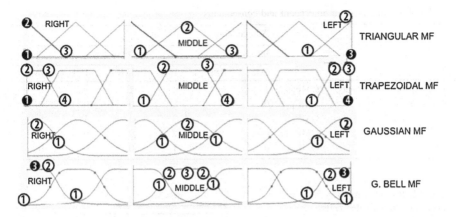

Fig. 11 Relation between type and location of a membership function

Table 4 shows a representation of how we can combine antecedents and consequents, where the highlighted bold cells with the number 1 are the activated antecedents.

The construction of the rules is taking row by row from top to bottom as Fig. 12 shows:

Rule 1: If input$_1$ is $\mathbf{MF_{1,1}}$ and input$_2$ is $\mathbf{MF_{2,1}}$, then output$_1$ is $\mathbf{OMF_{1,1}}$.
Rule 2: If input$_1$ is $\mathbf{MF_{1,1}}$ and input$_2$ is $\mathbf{MF_{2,2}}$, then output$_1$ is $\mathbf{OMF_{1,1}}$.

6 Simulation Results

In this section we present the results of the optimized all and beam FLC. We performed 30 experiments with the proposed method AVSE and 30 experiments using each variant independently as Table 5 shows. AGB stands for Average Global Best, given by (10), where GB is the global best per iteration and *it* is the maximum number of iterations and ATPE stands for Average Time Per Experiment.

$$AGB = \frac{GB}{\sum_{i=1}^{it} GB_i} \tag{10}$$

EGB stand for Experiment Global Best, calculated using (11), where n is the total number of experiments (30),

$$EGB = \frac{AGB}{\sum_{x=1}^{n} AGB_x}. \tag{11}$$

Table 4 Combination of antecedent and consequents

			Antecedents					
			Input$_1$:			Input$_2$:		
			IMF$_{1,1}$	IMF$_{1,2}$	IMF$_{1,j}$	IMF$_{2,1}$	IMF$_{2,2}$	IMF$_{i,n}$
Consequents	Output$_o$:	OMF$_{1,1}$	1	0	0	1	1	0
		OMF$_{1,2}$	0	1	0	0	0	0
		OMF$_{1,3}$	0	0	0	0	1	0
		OMF$_{1,4}$	0	0	0	0	0	0
		OMF$_{o,h}$	0	0	1	0	0	0

Fig. 12 Rule construction according to activated cells

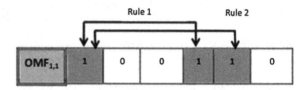

The values of the parameters are: alpha = 1, beta = 3, rho = 0.7, w = 6, e = 6 for optimization of type and e = 1.1 for the optimization of parameters. We used these parameters according to literature [1, 4]. We tested several values within the recommended range. The number of iterations were 100 and number of ants 50, and were chosen based on previous experiments.

We also compared our results with two other related works [2, 14] as Table 6 represents.

Table 5 shows that the proposed algorithm has a smaller error than the reached for each variant individually. This is because the "competition" between variants provide for the better variant that "won" in each iteration.

7 Conclusions

Our proposal was the creation of a modified ACO Algorithm AVSE that divides ants equivalently into 5 different ACO variations. Simulations where performed optimizing the Autonomous Mobile Robot FLC using the proposed algorithm AVSE and each ACO variant individually. Table 5 shows that the global best average for the total of 30 experiments is smaller using AVSE than using the variants individually. We can notice that experiments takes less time using AVSE, than those experiments made independently, which means that we obtained a smaller error in less time using the proposed algorithm. We also compared our results with two others approaches, both works optimizing only parameters of

Table 5 Experiments using AVSE algorithm compared with each variant individually

Optimization type	Variant	Experiments	AGB	EGB	Time (h)	ATPE (min)
Type + parameters + rules	AVSE: (AS, ACS, EAS, MMAS, ASRank)	30	0.0156	0.0577	10:11	0:20
Type + parameters + rules	AS	30	0.322	0.471	3:11	0:06
Type + parameters + rules	ACS	30	0.0243	0.0876	5:42	0:11
Type + parameters + rules	EAS	30	0.0488	0.1913	4:39	0:09
Type + parameters + rules	ASRank	30	0.0853	0.0725	2:16	0:04
Type + parameters + rules	MMAS	30	0.0211	0.0634	2:58	0:06

Table 6 Comparisons with related works versus AVSE

Optimization type	Variant	Experiments	AGB	GB	Time (h)	ATPE (min)
Type of MF + parameters + rules	AVSE: (AS, ACS, EAS, MMAS, ASRank)	30	0.0071	0.00009	9:14	18:28
Parameters * [2]	S-ACO	28	0.118	0.0982	6:25	16:45
Parameters * [15]	ASRank + CONVCONT	30	0.0131	0.00029	NA	NA

membership functions; Table 6 illustrates said comparison, in which AVSE shows a smaller error than [2, 14]. As a note the Ref. [17] used several ACO variations, the results presented in Table 6 show only the best variant combined with their modified algorithm (CONVCONT).

References

1. Brownlee, J.: Clever Algorithms, Nature-Inspired Programming Recipes. Creative Commons Publisher, Melbourne, Australia (2011)
2. Castillo, O., Neyoy, H., Soria, J., García, M., Valdez, F.: Dynamic fuzzy logic parameter tuning for ACO and its application in the fuzzy logic control of an autonomous mobile robot. Int. J. Adv. Rob. Syst., **10**, (2013)
3. Chang, Y.-H., Chang, C.-W., Lin, H.-W., Tao, C.W.: Fuzzy Controller Design for Ball and Beam System with an Improved Ant Colony Optimization, pp. 616–621. World Academy of Science, Engineering and Technology (2009)
4. Chaparro, I., Valdez, F.: Variants of Ant Colony Optimization: A Metaheuristic for Solving the Traveling Salesman Problem. Recent Adv. on Hybrid Intell. Syst., 323–331 (2013)
5. Dorigo, M., Birattari, M., Stützle, T.: Ant colony optimization. IEEE Computational Intelligence Magazine, pp. 28–39 (2006)
6. Dorigo, M., Stützle, T.: Ant Colony Optimization. Massachusetts Institute of Technology, Bradford, Cambridge (2004)
7. Eberhart, R.C., Kennedy, J.: A new optimizer using particle swarm theory. In: Proceedings of the Sixth International Symposium on Micro Machine and Human Science, pp. 39–43. IEEE Service Center, Piscataway, 1995

304 E. Lizarraga et al.

8. Engelbrecht, A.: Fundamentals of Computational Swarm Intelligence, pp. 85–131. Wiley, London (2005)
9. Galea, M., Shen, Q.: Simultaneous ant colony optimisation algorithms for learning linguistic fuzzy rules. In: Abraham, A., Grosan, C., Ramos, V. (eds.) Swarm Intelligence in Data Mining, pp. 75–99. Springer, Berlin (2006)
10. Jang, J.-S.R., Mizutani, E., Sun, C.-T.: Neuro-fuzzy and soft computing: a computational approach to learning and machine intelligence. Prentice Hall, Upper Saddle River (1997)
11. Lam, H.K., Leung, F.H.F., Tam, P.K.S.: Design of a fuzzy controller for stabilizing a ball-and-beam system, industrial electronics society. In: IECON'99 Proceedings. The 25th Annual Conference of the IEEE, vol. 2, pp. 520–524 (1999)
12. Langari R., Yen, J., Fuzzy Logic: Intelligence, Control and Information, Center of Fuzzy Logic, Robotics, and Intelligent Systems. Texas A&M University, Prentice Hall, USA (1998)
13. Lizárraga, E., Castillo, O., Soria, J.: A method to solve the traveling salesman problem using ant colony optimization variants with ant set partitioning. In: Castillo, O., Melin, P., Kacprzyk, J., Pedrycz, W. (eds.) Recent Advances on Hybrid Intelligent Systems (Studies in Computational Intelligence), pp. 237–247. Springer, Berlin (2013)
14. Melendez, A., Castillo, O.: Evolutionary optimization of the fuzzy integrator in a navigation system for a mobile robot. In: Recent Advances on Hybrid Intelligent Systems, pp. 21–31. Springer, Berlin (2013)
15. Melin, P., Olivas, F., Castillo, O., Valdez, F., Soria, J., García, J.: Optimal design of fuzzy classification systems using PSO with dynamic parameter adaptation through fuzzy logic. Expert Syst. Appl. **40**(8), 3196–3206 (2013)
16. Mezura-Montes, E., Palomeque-Ortiz, A.: Self-adaptive and Deterministic Parameter Control in Differential Evolution for Constrained Optimization. Efren Mezura-Montes, Laboratorio Nacional de Informática Avanzada (LANIA A.C.), Rébsamen 80, Centro, Xalapa, Veracruz, 91000, Mexico (2009)
17. Naredo, E., Castillo, O.: ACO-tuning of a fuzzy controller for the ball and beam problem. MICAI **2**, 58–69 (2011)
18. Mendel, J.: Uncertain Rule-Based Fuzzy Logic Systems, Introduction and new directions, PH PTR (2001)
19. Neyoy, H., Castillo, O., Soria, J.: Dynamic fuzzy logic parameter tuning for ACO and its application in TSP problems. In: Kacprzyk, J. Recent Advances on Hybrid Intelligent Systems (Studies in Computational Intelligence), pp. 259–273. Springer, Berlin (2013)
20. Olivas, F., Castillo, O.: Particle swarm optimization with dynamic parameter adaptation using fuzzy logic for benchmark mathematical functions, pp. 247–258. Recent Advances on Hybrid Intelligent Systems. Springer, Berlin (2013)
21. Stützle, T., Hoos, H.H.: MAX-MIN ant system and local search for combinatorial optimization problems. In: Voss, S., Martello, S., Osman, I., Roucairol, C. (eds.) Meta-Heuristics: Advances and Trends in Local Search Paradigms for Optimization, pp. 137–154. Kluwer Academic Publishers, Dordrecht (1999)
22. Stützle, T., Hoos, H.H.: MAX-MIN ant system. Future Generation Comput. Syst. **16**(8), 889–914 (2000)
23. Stützle, T., Linke, S.: Experiments with variants of ant algorithms. Mathw. Soft. Comput. **9**(2–3), 193–207 (2002)
24. Stützle, T., Hoos, H.H.: Improving the ant system: a detailed report on the MAXMIN ant system. Technical report AIDA-96-12, FG Intellektik, FB Informatik. TU Darmstadt, Germany (1996)
25. Tzen, S-T.: GA approach for designing fuzzy control with nonlinear ball-and-beam. The International Conference on Electrical Engineering, pp. 05–09. Japan (2008)
26. Yen, J., Langari, R.: Fuzzy Logic: Intelligence, Control and Information. Prentice Hall, Upper Saddle River (2003)
27. Zadeh, L.A.: Fuzzy Logic. IEEE Computer, pp. 338–353. (1965)

Part III
Neural Networks

Optimization of Modular Neural Networks with the LVQ Algorithm for Classification of Arrhythmias Using Particle Swarm Optimization

Jonathan Amezcua and Patricia Melin

Abstract In this chapter we describe the application of a full model of PSO as an optimization method for modular neural networks with the LVQ algorithm in order to find the optimal parameters of a modular architecture for the classification of arrhythmias. Simulation results show that this modular model optimized with PSO achieves acceptable classification rates for the MIT-BIH arrhythmia database with 15 classes.

1 Introduction

LVQ networks [1, 8, 9, 22, 26] are an adaptive learning method for the classification of data. Although an LVQ network uses supervised training, the LVQ algorithm employs techniques of unsupervised data clustering to preprocess the dataset and obtain the centers of the clusters.

The LVQ algorithm consists of two stages: a method of unsupervised learning for data clustering used for positioning the centers of the clusters without using class information, and, in stage two, the class information is used to refine the cluster centers [7].

After the learning process, a network with the LVQ algorithm classifies an input vector by assigning it to the class that has the closest cluster center at the same input vector. Below a sequential description of the LVQ method is presented:

1. Initialize the cluster centers using a clustering method.
2. Label each cluster by the voting method.
3. Randomly select an input vector $\mathbf{x}$ for training, and find $\mathbf{k}$ such that $\|\mathbf{x} - \mathbf{w_k}\|$ is minimized.

J. Amezcua · P. Melin (✉)
Tijuana Institute of Technology, Tijuana, México
e-mail: epmelin@hafsamx.org; pmelin@tectijuana.mx

O. Castillo et al. (eds.), *Recent Advances on Hybrid Approaches for Designing Intelligent Systems*, Studies in Computational Intelligence 547,
DOI: 10.1007/978-3-319-05170-3_21, © Springer International Publishing Switzerland 2014

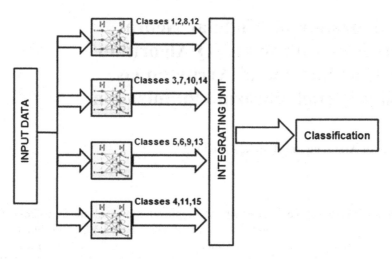

Fig. 1 LVQ network architecture

4. If x and w_k belongs to the same class, update w_k by: $\Delta w_k = -\eta(x - w_k)$, otherwise, update w_k by: $\Delta w_k = -\eta(x - w_k)$ where η is the learning rate, a positive constant to be decremented in each iteration.
5. If the maximum number of iterations is reached, stop. Otherwise return to step 3.

For this work a network architecture of 4 modules was optimized for the classification of arrhythmias [5, 6, 18] using the MIT-BIH database [16], and *the winner takes all* as the integrating unit [1]. Figure 1 shows the neural network architecture diagram.

The classes used in this work are: *Normal (1), LBBB (2), RBBB (3), PVC (4), Fusion Paced and Normal (5), Paced (6), Nodal (7), Fusion Ventricular and Normal (8), Atrial (9), Ventricular flutter wave (10), Paced maker fusion (11), Aberrated APC (12), Blocked APC (13), Atrial Escape (14), Fusion PVC (15).*

Particle Swarm Optimization (PSO) was used as an optimization method for the parameters of a modular neural network with the LVQ algorithm in order to find the best parameters to classify arrhythmias. We considered the optimization based on the epochs, clusters and learning rate of the LVQ neural network.

This chapter is organized as follows: Sect. 2 briefly describes the concept of Particle Swarm Optimization which is the algorithm that was applied for optimization. Section 3 presents the problem statement for the optimization of parameters for the modular LVQ network architecture. Section 4 describes the development of the PSO algorithm. In Sect. 5 results for the non-optimized and optimized architecture are shown. Section 6 shows the Conclusions.

2 PSO Algorithm

Particle swarm optimization (PSO) is a stochastic optimization technique based on the social behaviors observed in animals or insects. PSO has gained increasing popularity among researchers as a robust and efficient technique for solving optimization problems [3, 10, 24, 25]. In PSO, individual particles of a swarm represent potential solutions, which move through the problem search space seeking an optimal solution [4, 17, 23].

The position of each particle is adjusted according to its velocity and the difference between its current position respectively the best position found by its neighbors, and the best position it has found so far. The velocity of each particle is modified iteratively by its personal best position, and the best position found by particles in its neighborhood [2].

The velocity of particle i is calculated as:

$$v_{ij}(t+1) = v_{ij}(t) + c_1 r_{1j}(t)\left[y_{ij}(t) - x_{ij}(t)\right] + c_2 r_{2j}(t)\left[\hat{y}_j(t) - x_{ij}(t)\right] \quad (1)$$

where $v_{ij}(t)$ is the velocity of particle i in dimension $j = 1,..., n_x$ at time step t, $x_{ij}(t)$ is the position of particle i in dimension j at time step t; c_1 and c_2 are positive acceleration constants used to scale the contribution of the cognitive and social components respectively and $r_{1j}(t)$, $r_{2j}(t) \sim U(0,1)$ are random values in the range [0, 1]. These random values introduce a stochastic element to the algorithm.

The personal best position, y_i, is the best position the particle has visited since the first time step. Considering minimization problems, the personal best position at the next time step, $t + 1$, is calculated as

$$y_i(t+1) = \begin{cases} y_i(t), & f(x_i(t+1) \geq f(y_i(t)) \\ x_i(t+1), & f(x_i(t+1) < f(y_i(t)) \end{cases} \quad (2)$$

where f: $R^{n_x} \rightarrow$ R is the fitness function. The fitness function measures how close the corresponding solution is to the optimum, i.e. the fitness function quantifies the performance of a particle (or solution).

The global best position, $\hat{y}(t)$, at time step t, is defined as

$$\hat{y}(t) \in \{y_0(t), ..., y_{n_s}(t)\}|f(\hat{y}(t)) = \min \{f(y_0(t), ...,f(y_{n_s}(t))\} \quad (3)$$

where n_s is the total number of particles in the swarm. It is important to note that the definition in Eq. (3) states that $\hat{y}$ is the best position discovered by any of the particles so far. The global best position can also be selected from the particles of the current swarm.

The algorithm for this PSO model is as follows. In the notation S. x_i is used to denote the position of particle i in swarm S.

Create and initialize an n_x-dimensional swarm, S;

repeat
    for each particle $i = 1,...,S.n_s$ **do**
    //set the personal best position
        if $f(S.x_i) < f(S.y_i)$ **then**
           $S.y_i = S.x_i;$
        end
    //set the global best position
        if $f(S.y_i) < f(S.\hat{y})$ **then**
           $S.\hat{y} = S.y_i;$
        end
    end
    for each particle $i = 1,...,S.n_s$ **do**
    update the velocity;
    update the position;
    end
until *stopping condition is true*;

3 Problem Statement

For this work, a modular neural network with the LVQ algorithm was used. In previous work, an architecture of 4 modules was developed for classification of 15 classes of arrhythmias, having more than 2 classes in each module. The architecture and the combination of classes in each module are shown in Fig. 1 in Sect. 1. The combination of the classes in each module is random.

Other modular architectures were developed, some with 3 modules and 5 classes per module, another even with 2 modules with 8 and 7 classes in each module, however, with the 4 modules architecture presented in this paper is the one that threw the more high classification rate (99.16 %), so it was decided to optimize this architecture to try to further minimize the error.

In Fig. 2 the distribution of the data is shown by each of the modules in our architecture; one remarkable aspect is that there is a high degree of similarity between the data of the various classes. The higher the number of classes being handled by module, the classification rate tends to decrease.

4 PSO Architecture

In this work, only the number of clusters, number of epochs and learning rate of the LVQ network were optimized. Therefore, to develop the architecture of the PSO algorithm the following aspects were considered:

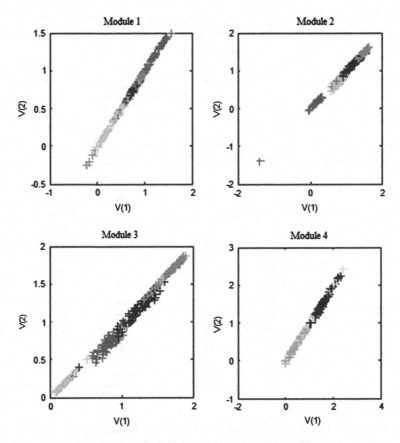

Fig. 2 Data distribution for the arrhythmia dataset in a 4 module architecture

Table 1 Parameters of the LVQ network included in the PSO search	Parameter	Minimal value	Maximal value
	Clusters	5	30
	Epochs	50	100
	Learning rate	0.001	0.1

- First, it was necessary to have a minimum and a maximum limit for each of the parameters to be optimized.
- The representation of these parameters for each individual of the population of PSO. In this work, a representation by real numbers was chosen.

In Table 1, the optimized parameters as well as the minimum and maximum values that were used are shown. These values were obtained from manual configurations that were made to the LVQ neural network algorithm, it was with these range of values with which the best classification results were obtained.

Table 2 Example of the population structure for PSO

	Clusters	Epochs	Learning rate
1	26	57	0.0709
2	28	71	0.0042
3	8	96	0.0284
4	25	58	0.0195

Table 3 Classification results for the LVQ network without PSO

Exp.	Clusters	Epochs	LR	Error
1	45	1,500	0.1	0.84
2	45	552	0.1	0.84
3	45	1,299	0.1	0.84
4	45	1,500	0.1	0.84
5	45	17	0.1	0.84
6	45	370	0.1	0.84
7	45	1,180	0.1	0.84
8	45	24	0.1	0.84
9	45	1,500	0.1	0.84
10	45	946	0.1	0.84
11	45	109	0.1	0.84
12	45	38	0.1	0.84
13	45	1,500	0.1	0.84
14	45	61	0.1	0.84
15	45	1,500	0.1	0.84

Table 2 shows an example of the structure of the population generated for the optimization of the above parameters, for a modular neural network with the LVQ algorithm.

It is noteworthy that the optimization of the parameters is made generally on all modules, this is, the parameters obtained from each particle in the population, were applied to the four modules of the architecture of the neural network simultaneously. In this way we obtained good results. In this study, 15 experiments with 20 iterations each were performed using a total of 15 particles for the population. The following section describes in more detail the results.

5 Simulation Results

In this section we present the results of the optimization of parameters for a modular neural network with the LVQ algorithm. First, Table 3 shows the classification results of the modular neural network without optimizing. It has a total of 15 experiments, in all experiments the same classification rate was obtained but in different number of epochs.

Table 4 Classification results for LVQ network with PSO algorithm

Exp.	Clusters	Epochs	LR	Error
1	20	54	0.0939	1.83×10^{-13}
2	19	60	0.0703	1.67×10^{-10}
3	14	72	0.0827	1.38×10^{-14}
4	11	71	0.0994	1.25×10^{-15}
5	13	97	0.0429	1.83×10^{-13}
6	11	94	0.0158	1.41×10^{-12}
7	26	84	0.0797	1.83×10^{-13}
8	14	52	0.0426	1.69×10^{-10}
9	28	64	0.0427	1.25×10^{-15}
10	11	92	0.0455	1.48×10^{-12}
11	30	87	0.0388	1.48×10^{-12}
12	12	80	0.0208	1.35×10^{-14}
13	30	87	0.0388	1.51×10^{-11}
14	12	80	0.0208	1.67×10^{-10}
15	11	87	0.0135	1.54×10^{-11}

Table 4 shows the set of experiments for the neural network with the LVQ algorithm optimized by PSO. Comparing to the results without optimization, we were able to notice a decrease to the classification error.

6 Conclusions

This chapter presents the application of an optimization method based on particle swarm optimization (PSO) for a modular neural network with the LVQ algorithm for the classification of arrhythmias. Although the neural network yielded a good percentage of classification without optimization, the PSO was able to reduce the error even further. As shown in the result table before optimization (see Table 3), it was difficult for the neural network to minimize the classification error when parameters are stationary, however, by applying the optimization, it is more likely to generate better classification results due to the variety of parameters that are generated from a random population. Future work could include combining type 2 fuzzy logic [19–21] with the neural network model, as in other similar hybrid approaches [11–15].

References

1. Biehl, M., Ghosh, A., Hammer, B.: Learning vector quantization: the dynamics of winner-takes-all algorithms. Neurocomputing **69**(7–9), 660–670 (2006)
2. Blum, C., Merkle, D.: Swarm Intelligence. Introduction and Applications, Part I, pp. 3–101. Springer, Berlin (2008)
3. Egelbrecht, A.P.: Fundamentals of Computational Swarm Intelligence, pp. 94–105. Wiley, New York (2005)

4. Fikret, M.: Particle swarm optimization and other metaheuristic methods in hybrid flow shop scheduling problem. Part Swarm Opt, 155–168 (2009)
5. Hu, Y.H., Palreddy, S., Tompkins, W.: A patient adaptable ECG beat classifier using a mixture of experts approach. IEEE Trans. Biomed. Eng, 891–900 (1997)
6. Hu, Y.H., Tompkins, W., Urrusti J L., Afonso, V.X.: Applications of ann for ecg signal detection and classification. J. Electrocardiology. 28, 66–73
7. Jang, J., Sun, C., Mizutani, E.: Neuro-Fuzzy and Soft Computing. Prentice Hall, New Jersey (1997)
8. Kohonen, T.: Improved versions of learning vector quantization. In: International Joint Conference on Neural Networks, vol. 1, pp. 545–550. San Diego (1990)
9. Kohonen, T.: Self-organization and associate memory, 3rd edn. Springer, London (1989)
10. Ciarelli, P M., Krohling, R.A., Oliveira, E.: Particle swarm optimization applied to parameters learning of probabilistic neural networks for classification of economic activities. Part. Swarm Opt. 313–328 (2009)
11. Melin, P., Castillo, O.: Hybrid Intelligent Systems for Pattern Recognition. Springer, Heidelberg (2005)
12. Melin, P., Castillo, O.: An intelligent hybrid approach for industrial quality control combining neural networks, fuzzy logic and fractal theory. Inf. Sci. 177, 1543–1557 (2007)
13. Mendoza, O., Melin, P., Castillo, O., Licea, G.: Type-2 fuzzy logic for improving training data and response integration in modular neural networks for image recognition. Lect. Notes Artif. Intell. 4529, 604–612 (2007)
14. Mendoza, O., Melin, P., Castillo, O.: Interval type-2 fuzzy logic and modular neural networks for face recognition applications. Appl. Soft Comput. J 9, 1377–1387 (2009)
15. Mendoza, O., Melin, P., Licea, G.: Interval type-2 fuzzy logic for edges detection in digital images. Int. J. Intell. Syst. 24, 1115–1133 (2009)
16. MIT-BIH Arrhythmia Database. PhysioBank, Physiologic Signal Archives for Biomedical Research. http://www.physionet.org/physiobank/database/mitdb/ (2012). Accessed 12 Nov 2012
17. Nikmam, T., Amiri, B.: An efficient hybrid approach based on pso, aco and k-means for cluster ananlysis. Appl. Soft Comput. 10(1), 183–197 (2010)
18. Osowski, S., Siwek, K., Siroic, R.: Neural system for heartbeats recognition using genetically integrated ensemble of classifiers. Comput. Biol. Med. 41(3), 173–180 (2011)
19. Sepulveda, R., Castillo, O., Melin, P., Rodriguez-Diaz, A., Montiel, O.: Experimental study of intelligent controllers under uncertainty using type-1 and type-2 fuzzy logic. Inf. Sci. 177(10), 2023–2048 (2007)
20. Sepulveda, R., Montiel, O., Lizarraga, G., Castillo, O.: Modeling and simulation of the defuzzification stage of a type-2 fuzzy controller using the xilinx system generator and simulink. Stud. Comput. Intell. 257, 309–325 (2009)
21. Sepulveda, R., Montiel, O., Castillo, O., Melin, P.: Optimizing the mfs in type-2 fuzzy logic controllers, using the human evolutionary model. Int. Rev. Autom. Control 3(1), 1–10 (2011)
22. Torrecilla, J.S., Rojo, E., Oliet, M., Domínguez, J.C., Rodríguez, F.: Self-organizing maps and learning vector quantization networks as tools to identify vegetable oils and detect adulterations of extra virgin olive oil. Comput. Aided Chem. Eng. 28, 313–318 (2010)
23. Valdez, F., Melin, P., Castillo, O.: Evolutionary method combining particle swarm optimisation and genetic algorithms using fuzzy logic for parameter adaptation and aggregation: the case neural network optimisation for face recognition. IJAISC 2(1/2), 77–102 (2010)
24. Valdez, F., Melin, P., Licea, G.: Modular neural networks architecture optimization with a new evolutionary method using a fuzzy combination particle swarm optimization and genetic algorithms. In: Bio-inspired Hybrid Intelligent Systems for Image Analysis and Pattern Recognition, pp. 199–213. Springer, Berlin (2009)
25. Vázquez, J.C., Valdez F., Melin P.: Comparative study of particle swarm optimization variants in complex mathematics functions. Recent Adv. Hybrid. Intell. Syst. 223–235 (2013)
26. Wu, K.L., Yang, M.S.: Alternative learning vector quantization. Pattern Recogn. 39(3), 351–362 (2006)

A DTCNN Approach on Video Analysis: Dynamic and Static Object Segmentation

Mario I. Chacon-Murguia and David Urias-Zavala

Abstract This paper presents a DTCNN model for dynamic and static object segmentation in videos. The proposed method involves three main stages in the dynamic stage; dynamic background registration, dynamic objects detection and object segmentation improvement. Two DTCNNs are used, one to achieved object detection and other for morphologic operations in order to improve object segmentation. The static segmentation stage is composed of a clustering module and a DTCNN module. The clustering module is in charge of detecting the possible regions and the DTCNN generate the regions. Visual and quantitative results indicate acceptable results compared with existing methods.

1 Introduction

Human visual perception is a very important activity that allows humans to interact with their environment. Although it is daily used by humans, it is not a simple process. On the contrary, human visual perception involves a series of complex processes that include segmentation, detection and tracking of moving objects among other important activities. Detection and tracking of moving objects are two vital tasks that allow scene analysis at high perception levels. Detection of moving objects by computer vision methods is a paramount area for applications like; intelligent surveillance systems [1], mobile robotics, traffic control systems [2], driver awareness systems [3], face recognition [4], etc. Research on the

M. I. Chacon-Murguia (✉) · D. Urias-Zavala
Visual Perception Applications on Robotic Lab, Chihuahua Institute of Technology,
Chihuahua, Mexico
e-mail: mchacon@itchihuahua.edu.mx; mchacon@ieee.org

D. Urias-Zavala
e-mail: jduriaa@itchinuanua.edu.mx

O. Castillo et al. (eds.), *Recent Advances on Hybrid Approaches for Designing*   315
Intelligent Systems, Studies in Computational Intelligence 547,
DOI: 10.1007/978-3-319-05170-3_22, © Springer International Publishing Switzerland 2014

developing of object detection algorithms have been reported in the literature using different approaches and with different performances, [1, 5–7]. Despite these efforts, object detection in dynamics environments is still an open research area because of its complexity and importance. Several factors still need to be improved like; performance, computational burden, scenario variability, number of parameters to be tuned, solution of illumination changes and shadows, etc.

Among the different neural networks paradigms used for image and video processing the model of the cellular neural network, CNN, represents an important contribution. This model has been reported in works related to; thresholding [8], edge detection [9], segmentation [10] and robot navigation [11]. The CNN model has provided important contributions to these image processing tasks and it is still considered as a prominent model for new image processing solutions.

In a previous work, one of the authors reported a method to detect dynamic objects in videos [12] with acceptable results based on a variation of the SOM architecture. After that work and aware of the advantages documented about the Cellular Neural Network architecture, robust configuration, parallel computing capacity and nonlinear behavior [13–15], a new dynamic and static object detection methodology is proposed in this paper.

The organization of the paper is as follows. Section 2 presents the dynamic object segmentation system. Section 3 describes the details of the proposed method for static object segmentation. Finally Sects. 4 and 5 highlight the results and conclusions of the work.

2 Dynamic Object Segmentation System

2.1 Discrete Time CNN

The method presented in this work is based on the discrete time CNN, DTCNN. The model of the DTCN used is the one defined in [16] and corresponds to

$$x_{ij}(n+1) = \sum_{k=-r}^{r} \sum_{l=-r}^{r} A_{kl} y_{i+k,j+1}(n) + \sum_{k=-r}^{r} \sum_{l=-r}^{r} B_{kl} u_{i+k,j+l} + z_{ij} \tag{1}$$

$$y_{kl}(n+1) = f(x_{kl}(n+1)) \tag{2}$$

$$\begin{aligned} |x_{ij}(0)| &\leq 1, \quad 1 \leq i \leq M; 1 \leq j \leq N \\ |u_{ij}| &\leq 1, \quad 1 \leq i \leq M; 1 \leq j \leq N \end{aligned} \tag{3}$$

where $x_{ij} \in R$, $y_{ij} \in R$, $u_{ij} \in R$, and $z_{ij} \in R$ correspond to the state, output, input, and threshold of a cell $C(i, j)$ respectively. A_{kl} and B_{kl} are the feedback and input synaptic kernels.

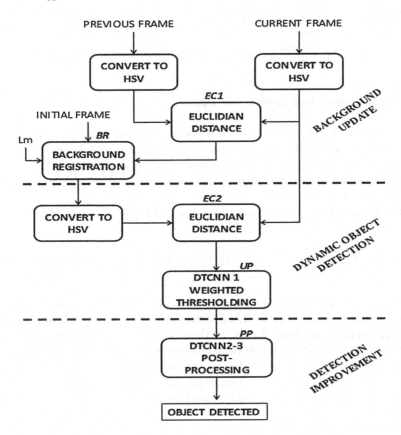

Fig. 1 General scheme of the dynamic object segmentation system

2.2 Dynamic Object Detection Module

The proposed method takes some ideas from the works documented in [7] and [12]. However, this work incorporates some changes with respect the type of neuron used and modifications with respect the DTCNN configuration and models. The proposed algorithm involves three main stages, dynamic background registration, dynamic object detection and object segmentation improvement, Fig. 1. The background update is achieved by frame subtraction and historical pixel changes. Once the dynamic background is updated, dynamic objects are detected by DTCNN1. A post-processing stage to improve the segmentation of the dynamics objects is performed by DTCNN2 and DTCNN3.

The method works on the HSV color space. This color space was selected because of its advantages reported in [17]. Therefore, the videos under analysis are converted to this space

$$p(x,y,t)^{RGB} \Rightarrow p(x,y,t)^{HSV} = [h,s,v]^T \tag{4}$$

assuming that $p(x, y, t)^{RGB}$ is a pixel of a frame in the RGB color space.

2.2.1 Background Registration and Update

In this section the process to determine the background and its update is described. The process considers the first frame as the current background. From this point, a dynamic update process starts to incorporate small variations into the background such that they do not yield false positives detections. The update process is explained next.

The Euclidian distance between the previous $I(x, y, t - 1)$ and current frame $I(x, y, t)$, $Ed_1(x, y, t)$, is as follows

$$Ed_1(x,y,t) = \|(v_{t-1}s_{t-1}\cos(h_{t-1}), v_{t-1}s_{t-1} \text{ s } en(h_{t-1}), v_{t-1} \\ -(v_t s_t \cos(h_t), v_t s_t \text{ s } en(h_t), v_t)\| \tag{5}$$

$Ed_1(x, y, t)$ contains relevant information to decide which pixels changed from the previous frame to the current one. The first step consists on assuming the first frame of the video as the background, $I_B(x, y, t)$. Then, information provided by $Ed_1(x, y, t)$ is used to detect significant pixels value changes between $p(x, y, t - 1)^{HSV} \in I(x, y, t - 1)$ and $p(x, y, t)^{HSV} \in I(x, y, t)$. A historical change of pixel values is registered in an accumulator $\mathbf{L}(t)$. Each element of this accumulator, matrix, indicates the number of frames the corresponding pixel has remained unchanged. This computation is performed by

$$\begin{aligned} If \quad &Ed_1(x,y,t) < Th_1 \quad then \quad \mathbf{L}(x,y,t) = \mathbf{L}(x,y,t) + 1 \\ else \quad &\mathbf{L}(x,y,t) = 0 \end{aligned} \tag{6}$$

where th_1 is a change level threshold and its value, 0.1, was determined by analysis of a set of videos in order to discard small illumination variations, video camera noise as well as to reduce false positives. This analysis is for a frame pixel and is not related yet to a DTCNN processing. The background update is achieved by the following rule

$$If \quad L(x,y,t) \geq Lm \quad then \quad p_B(x,y,t)^{HSV} = p(x,y,t)^{HSV} \tag{7}$$

where $p_B(x, y, t)^{HSV} \in I_B(x, y, t)$. That is, the pixel background is changed by the current frame pixel. In this way, the background is dynamical update to capture small variations of the scene that do not correspond to moving objects in the scene. Here Lm, a memory parameter, is the minimum number of consecutive frames a pixel must remain under the threshold th_1 to be considered static information that belongs to the background. A value of 60 for Lm is used as suggested in [7] except in Video 4 where a value of 80 was used instead.

2.2.2 DTCNN Dynamic Object Detection

Once the background has been update, the next step is to find possible moving objects in the scene. The pixels belonging to possible moving objects are detected by the difference between the current background $I_B(x, y, t)$ and the current frame $I(x, y, t)$ through the Euclidean distance, $Ed_1(x, y, t)$, computed by

$$
\begin{aligned}
Ed_2(x, y, t) = \| & (v_B s_B \cos(h_B), v_B s_B \, s \, en(h_B), v_B \\
& -(v_t s_t \cos(h_t), v_t s_t \, s \, en(h_t), v_t) \|
\end{aligned}
\tag{8}
$$

The $Ed_2(x, y, t)$ values are mapped to the range $[-1, 1]$ so they can be used as inputs to the DTCNN1 model as stated in Eq. (3) according to

$$
p_{ij} = \frac{0.7 - Ed_2(x, y)}{0.7} \quad \begin{aligned} i &= x = 1, 2, \ldots, M \\ j &= y = 1, 2, \ldots, N \end{aligned}
\tag{9}
$$

This transformation is determined by considering the possible values of $Ed_2(x, y, t)$, $[0, \sqrt{2}]$ and truncating the value of $\sqrt{2}-1.4$. The DTCNN1 uses as a weighted average thresholding processor to determine which candidate pixels are considered as moving objects. The result of the weighted average thresholding operation yields two classes of pixels that correspond to static or dynamic objects, O_S and O_D. The classification is determined by the threshold value specified in z_{ij}.

$$
\overline{Ed_2}(x, y, t) : \text{DTCNN1} \Rightarrow \begin{cases} O_S & \text{if} \quad \overline{Ed_2}(x, y, t) < z_{ij} \\ O_D & \text{if} \quad \overline{Ed_2}(x, y, t) \geq z_{ij} \end{cases}
\tag{10}
$$

where $\overline{Ed_2}(x, y, t)$ is the weighted average value of the 8-neighbor pixels of $Ed_2(x, y, t)$. The weighted average thresholding is a proposal in this work that differ from the traditional threshold operator for DTCNN used in the literature. In the traditional operator only the central cell is considered to take a decision. Considering neighbor cell information instead just the central cell information avoids the classification of isolated pixels as dynamic objects and in consequence reduced the false positive cases.

The definition of the DTCNN1 model for the weighted average thresholding is given in Eqs. (11)–(16)

$$
\left\langle A_{kl} = \begin{bmatrix} 1/16 & 1/16 & 1/16 \\ 1/16 & 1/2 & 1/16 \\ 1/16 & 1/16 & 1/16 \end{bmatrix} \right\rangle, \left\langle B_{kl} = \begin{bmatrix} 0 & 0 & 0 \\ 0 & 0 & 0 \\ 0 & 0 & 0 \end{bmatrix} \right\rangle, \langle z_{ij} = -0.7 \rangle
\tag{11}
$$

$$
x_{ij}(n+1) = \sum_{k=-r}^{r} \sum_{l=-r}^{r} A_{kl} y_{i+k,j+l}(n) + \sum_{k=-r}^{r} \sum_{l=-r}^{r} B_{kl} u_{i+k,j+l} + z_{ij}
\tag{12}
$$

$$x_{ij}(n+1) = \cfrac{y_{ij}(n) + \frac{1}{8} \sum\limits_{k=-1}^{1} \sum\limits_{\substack{l=-1 \\ i-k,j-l\neq i,j}}^{1} y_{i-k,j-l}(n)}{2} - 0.7 \qquad (13)$$

$$y_{ij}(0) = f(x_{ij}(0)) = \frac{1}{2}|x_{ij}(0)+1| - \frac{1}{2}|x_{ij}(0)-1|$$

$$y_{ij}(1) = f(x_{ij}(1)) = sgn(x_{ij}(1)) = \begin{cases} 1, & si \quad x_{ij}(1) > 0 \\ 0, & si \quad x_{ij}(1) = 0 \\ -1, & si \quad x_{ij}(1) < 0 \end{cases} \qquad (14)$$

$$x_{ij}(0) = p_{ij} \qquad (15)$$

$$[\mathbf{U}] = [\mathbf{Y}] = [0] \qquad (16)$$

In this model the coefficients A_{kl} determine the weights for the central and neighbor cells. In this case the central cell weight corresponds to 1/2 and the other coefficients half of the average of the neighbor cells 1/16. With this value isolated pixels as well as small regions are discarded as dynamic objects. The initial conditions for this model correspond to the Euclidean distances p_{ij} adjusted to the interval $[-1, 1]$ according to the traditional threshold configuration. The model involves two different output functions defined in (14). The first output function is a piecewise linear function. This function is used to compute $y_{ij}(0)$ from the value of $x_{ij}(0)$ because using a hard-limit function $y_{ij}(0)$ will be saturated before it is processed by the threshold. The second output function is a sign function and it is used to compute $y_{ij}(1)$. In this way, any value greater than z_{ij} is classified as a dynamic object, otherwise it is classified as a static object. The boundary conditions $\mathbf{U}$ and $\mathbf{Y}$ of the model are set to zero. The threshold value was selected experimentally and it was set to 0.7.

Another difference of the weighted average thresholding with respect the traditional threshold model is that it is not necessary to iterate the model because using the hard-limit function it warranties a stable output in one iteration.

The output of DTCNN1 is a binary matrix $\mathbf{Y}_{UP}$ where

$$\mathbf{Y}_{UP} = \begin{cases} -1 & for \ O_D \\ 1 & for \ O_S \end{cases} \qquad (17)$$

Thus, the matrix $\mathbf{Y}_{UP}$ represents the segmentation of the current frame for static and dynamic objects. However, it is common to find some false positives, small regions, and false negatives errors, small holes in dynamic objects. Therefore a post-processing stage is achieved over $\mathbf{Y}_{UP}$ to reduce these types of errors. The post-processing stage involves the morphologic operations of dilation and erosion performed with the CNNs, DTCNN2 and DTCNN3 respectively.

The definitions of the DTCNN2 and DTCNN3 models taken from [18] are

$$DTCNN2 \Rightarrow \left\langle A_{kl} = \begin{bmatrix} 0 & 0 & 0 \\ 0 & 0 & 0 \\ 0 & 0 & 0 \end{bmatrix} \right\rangle, \left\langle B_{kl} = \begin{bmatrix} 0 & 1 & 0 \\ 1 & 1 & 1 \\ 0 & 1 & 0 \end{bmatrix} \right\rangle, \langle z_{ij} = -4 \rangle \quad (18)$$

$$DTCNN3 \Rightarrow \left\langle A_{kl} = \begin{bmatrix} 0 & 0 & 0 \\ 0 & 0 & 0 \\ 0 & 0 & 0 \end{bmatrix} \right\rangle, \left\langle B_{kl} = \begin{bmatrix} 0 & 1 & 0 \\ 1 & 1 & 1 \\ 0 & 1 & 0 \end{bmatrix} \right\rangle, \langle z_{ij} = 4 \rangle \quad (19)$$

$$y_{ij}(0) = f(x_{ij}(0)) = \frac{1}{2}|x_{ij}(0) + 1| - \frac{1}{2}|x_{ij}(0) - 1|$$

$$y_{ij}(1) = f(x_{ij}(1)) = \text{sgn}(x_{ij}(1)) = \begin{cases} 1, & si \quad x_{ij}(1) > 0 \\ 0, & si \quad x_{ij}(1) = 0 \\ -1, & si \quad x_{ij}(1) < 0 \end{cases} \quad (20)$$

$$\mathbf{U} = \mathbf{Y}_{UP} \quad (21)$$

$$\mathbf{X}(0) = 0 \quad (22)$$

$$[\mathbf{U}] = [\mathbf{Y}] = [0] \quad (23)$$

After this morphologic operations, the outputs $y_{ij}(n)$ of DTCNN3 are mapped into the interval [0, 1] to visualize the result using

$$p_y(x, y, t) = \frac{(1 - y_{ij}(n))}{2} \quad (24)$$

3 Static Object Segmentation System

This section proposes a method to segment static objects from the background. The number of regions are established as Nr. The method is based on a DTCNN and the *K-Means* clustering algorithm. The input to this system is the output of the dynamic segmentation system. The weighted thresholdding DTCNN is also used by this systems and other DTCNN is employed to performed an AND operation.

Figure 2 shows a diagram of the static segmentation system. The input to this system is the current background $I_B(x, y, t)$. The first module of the system converts the HSV values of $I_B(x, y, t)$ to Cartesian coordinates, $I_B(x, y, t)^{XYZ}$, in order to use that information in the clustering algorithm, using

$$p_B(x, y, t)^{HSV} = [h_B, s_B, v_B]^T \Rightarrow p_B(x, y, t)^{xyz} = [x, y, z]^T \quad (25)$$

where

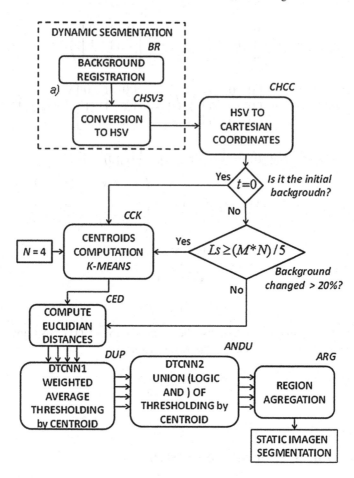

Fig. 2 General static segmentation scheme

$$x = s_B * v_B * \cos(h_B)$$
$$y = s_B * v_B * sen(h_B) \qquad (26)$$
$$z = v_B$$

The second module computes the centroids of possible background regions and third one performs the segmentation based on the centroids and DTCNN schemes.

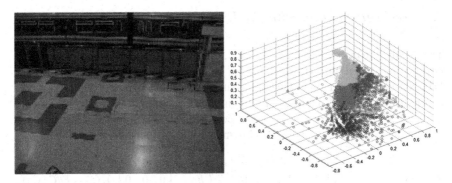

Fig. 3 Example of a background and its clusters generated by the *K-means*

3.1 Clustering Module

The second module of the system computes the centroids of the information in, $I_B(x, y, t)^{XYZ}$. This computation is achieved if at least one of the following conditions is true

If $I_B(x, y, t)$ is the first background /* compute centroids for the first time
If $Ls \geq 20\%$ /* the background has a significant change.

where Ls is a memory that keeps track of how many background pixels have changed. If more than 20 % of the pixels changed it is necessary to compute the centroids again to obtain a new background segmentation. *If $Ls \geq 20\%$* is true, then Ls is initialized to zero. The new centroids, $Ce_1(x, y, z)$, $Ce_2(x, y, z)$, $Ce_3(x, y, z)$ and $Ce_4(x, y, z)$ are compute with the *K-means* starting from the previous centroids trying to preserve segmentation coherence among the segmented regions. Figure 3 illustrates and example of a scenario and its clusters.

Once the centroids have been obtained the Euclidean distance between the centroids and pixels $p_b(x, y, z)^{XYZ} \in I_B(x, y, t)$ is computed by

$$\mathbf{EdC_1}(x,y) = \sqrt{[Ce_1(x) - p_B(x)]^2 + [Ce_1(y) - p_B(y)]^2 + [Ce_1(z) - p_B(z)]^2}$$

$$\mathbf{EdC_2}(x,y) = \sqrt{[Ce_2(x) - p_B(x)]^2 + [Ce_2(y) - p_B(y)]^2 + [Ce_2(z) - p_B(z)]^2}$$

$$\mathbf{EdC_3}(x,y) = \sqrt{[Ce_3(x) - p_B(x)]^2 + [Ce_3(y) - p_B(y)]^2 + [Ce_3(z) - p_B(z)]^2}$$

$$\mathbf{EdC_4}(x,y) = \sqrt{[Ce_4(x) - p_B(x)]^2 + [Ce_4(y) - p_B(y)]^2 + [Ce_4(z) - p_B(z)]^2}$$

$$(27)$$

where $\mathbf{EdC_1}$, $\mathbf{EdC_2}$, $\mathbf{EdC_3}$ and $\mathbf{EdC_4}$ are Euclidean matrix distances. These distances will be used by a weighted thresholding DTCNN to assign pixels to regions in the next module.

3.2 Pre-Segmentation with the DTCNN1

The distance matrices $\mathbf{EDC_i}$ $i = 1, \ldots, 4$, are used by the DTCNN1 to perform a threshoding operation. DTCNN1 has an architecture similar to the thresholding DTCNN used for dynamic detection, but DTCNN1 changes its threshold z_{ij} accordig to the distance matrices implied in the thresholding. Also, the initial conditions $\mathbf{X(0)}$ in each thresholding corresponds to $\mathbf{EdC_i}$ such that $\mathbf{EdC_i}$ is different than the one used as z_{ij}. The DTCNN1 model is thus specified as

$$DTCNN1 \Rightarrow \left\langle A_{kl} = \begin{bmatrix} 1/16 & 1/16 & 1/16 \\ 1/16 & 1/2 & 1/16 \\ 1/16 & 1/16 & 1/16 \end{bmatrix} \right\rangle, \left\langle B_{kl} = \begin{bmatrix} 0 & 0 & 0 \\ 0 & 0 & 0 \\ 0 & 0 & 0 \end{bmatrix} \right\rangle, \langle z_{ij} = EdC_i(x,y) \rangle$$

$$i = 1, 2, 3, 4$$

$$y_{ij}(0) = f(x_{ij}(0)) = \frac{1}{2}|x_{ij}(0) + 1| - \frac{1}{2}|x_{ij}(0) - 1|$$

$$y_{ij}(1) = f(x_{ij}(1)) = \text{sgn}(x_{ij}(1)) = \begin{cases} 1, & si \ x_{ij}(1) > 0 \\ 0, & si \ x_{ij}(1) = 0 \\ -1, & si \ x_{ij}(1) < 0 \end{cases}$$

$$\mathbf{U} = 0$$

$$\mathbf{X}(0) = \mathbf{EdC_j} \quad \text{where } j=1,2,3,4 \text{ and } \neq i$$

$$[\mathbf{U}] = [\mathbf{Y}] = [0]$$

$$(28)$$

The outputs of the DTCNN1 are the matrices

$$\begin{aligned}
\mathbf{Yu}_{ij} &= DTCNN1(\mathbf{X}(0), \mathbf{z}) & i &= j = 1, 2, 3, 4 \\
\mathbf{Yu}_{12} &= DTCNN1(\mathbf{EdC_2, EdC_1}) & \mathbf{Yu}_{21} &= DTCNN1(\mathbf{EdC_1, EdC_2}) \\
\mathbf{Yu}_{13} &= DTCNN1(\mathbf{EdC_3, EdC_1}) & \mathbf{Yu}_{23} &= DTCNN1(\mathbf{EdC_3, EdC_2}) \\
\mathbf{Yu}_{14} &= DTCNN1(\mathbf{EdC_4, EdC_1}) & \mathbf{Yu}_{24} &= DTCNN1(\mathbf{EdC_4, EdC_2}) \\
\mathbf{Yu}_{31} &= DTCNN1(\mathbf{EdC_1, EdC_3}) & \mathbf{Yu}_{41} &= DTCNN1(\mathbf{EdC_1, EdC_4}) \\
\mathbf{Yu}_{32} &= DTCNN1(\mathbf{EdC_2, EdC_3}) & \mathbf{Yu}_{42} &= DTCNN1(\mathbf{EdC_2, EdC_4}) \\
\mathbf{Yu}_{34} &= DTCNN1(\mathbf{EdC_4, EdC_3}) & \mathbf{Yu}_{43} &= DTCNN1(\mathbf{EdC_3, EdC_4})
\end{aligned} \quad (29)$$

The aim of the thresholding operation is to determine which distances of each centroid are closer to form regions in regard the other 3 centroids. The pixel with closest distance is thus assigned to the corresponding region represented by that centroid.

Later the thresholding results will be combined with an AND operation implemented with DTCNN2 to obtain the background segmentation.

3.3 Region Segmentation with the DTCNN2

The AND operation with the DTCNN2 model is defined as

$$DTCNN2 \Rightarrow \left\langle A_{kl} = \begin{bmatrix} 0 & 0 & 0 \\ 0 & 1 & 0 \\ 0 & 0 & 0 \end{bmatrix} \right\rangle, \left\langle B_{kl} = \begin{bmatrix} 0 & 0 & 0 \\ 0 & 1 & 0 \\ 0 & 0 & 0 \end{bmatrix} \right\rangle, \langle z_{ij} = 1.1 \rangle$$

$$y_{ij}(0) = f(x_{ij}(0)) = \frac{1}{2}|x_{ij}(0) + 1| - \frac{1}{2}|x_{ij}(0) - 1| \tag{30}$$

$$y_{ij}(1) = f(x_{ij}(1)) = \mathrm{sgn}(x_{ij}(1)) = \begin{cases} 1, & si \ x_{ij}(1) > 0 \\ 0, & si \ x_{ij}(1) = 0 \\ -1, & si \ x_{ij}(1) < 0 \end{cases}$$

U and $X(0) = [0]$ and the AND operation is computed as

$$ACe_{pi} = DTCNN2(U, X(0))$$
$$ACe_i = DTCNN2(U, X(0))$$
where $i = 1, 2, 3, 4$ to obtain:

$$\begin{aligned} &ACe_{p1} = DTCNN2(Yu_{12}, Yu_{13}) &&ACe_{p2} = DTCNN2(Yu_{21}, Yu_{23}) \\ &ACe_1 = DTCNN2(ACe_{p1}, Yu_{14}) &&ACe_2 = DTCNN2(ACe_{p2}, Yu_{24}) \end{aligned} \tag{31}$$

$$\begin{aligned} &ACe_{p3} = DTCNN2(Yu_{31}, Yu_{32}) &&ACe_{p4} = DTCNN2(Yu_{41}, Yu_{42}) \\ &ACe_3 = DTCNN2(ACe_{p3}, Yu_{34}) &&ACe_4 = DTCNN2(ACe_{p4}, Yu_{43}) \end{aligned}$$

The AND is performed over the three Yu of each centroid. It is observed that DTCNN2 is used twice in order to process each Yu. ACe_{pi} is a preliminary result and AC_{ei} is the final result. The outputs AC_{e1}, AC_{e2}, AC_{e3}, and AC_{e4} represent the pixels closer to the centroids Ce_1, Ce_2, Ce_3 and Ce_4 respectively. The matrices ACe_1, ACe_2, ACe_3 a and ACe_4 with values in $[-1, 1]$ are mapped into $[0, 1]$ to represent information that correspond to the regions R_1, R_2, R_3 and R_4 generated.

The final step is a weighted aggregation operation of R_1, R_2, R_3 and R_4 to yield the background segmentation

$$Is = ACe_1 + 86 * ACe_2 + 170 * ACe_3 + 256 * ACe_4 \tag{32}$$

Fig. 4 Video 1, interior scenario with non uniform illumination. **a–d** Original frames, **e–h** dynamic objects detected

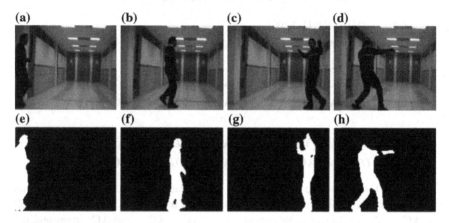

Fig. 5 Video 2, interior scenario with good illumination. **a–d** Original frames, **e–h** dynamic objects detected

4 Results

4.1 Dynamic Object Segmentation Results

The proposed method was tested in scenarios with different conditions and complexities related to illumination and dynamic objects speed among other aspects. The next figures illustrate visual results of the proposed method.

Video 1 is an interior scenario with non-uniform illumination. Findings show in general acceptable segmentation of the moving persons except in the right side of the scenario were the illumination is lower causing in some cases blob partition, Fig. 4. A better scenario is presented in Video 2, Fig. 5. This is also an interior scenario with good illumination and a better contrast between the color characteristics of the

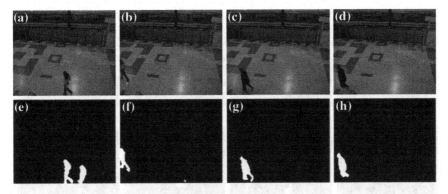

Fig. 6 Video 3, initial background with dynamic object. **a–d** Original frames, **e–h** dynamic objects detected

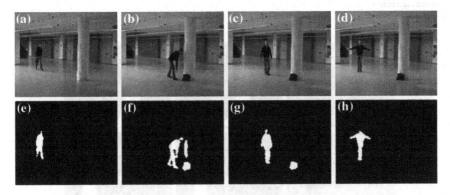

Fig. 7 Video 4, weak shadows and abandoned object. **a–d** Original frames, **e–h** dynamic objects detected

dynamic object and the background. In most of the frames the segmentation defines completely the dynamic object. However, in some frames part of the dynamic object is lost because the color similitude with part of the upper windows.

A different case is illustrated in Video 3, Fig. 6. Video 3 is an interior scenario with regular illumination, shadows and reflections. Besides, in the first frame, considered as the initial background, appears a dynamic object which is considered as background initially and then it is gradually removed from the background after 60 frames according to the memory parameter.

Video 4 is an interior scenario with good illumination, no reflections, a weak shadow, and one abandoned object, Fig. 7. The abandoned object is incorporated into the background in 80 frames. It can be noticed that weak shadows are removed from the dynamic object.

The last video, Video 5, is an exterior scenario with good illumination conditions and no shadows or reflections, Fig. 8. The region of the dynamic object is

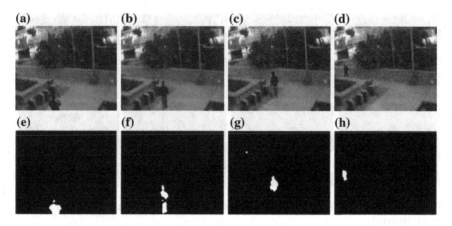

Fig. 8 Video 5 exterior scenario with good illumination. **a–d** Original frames, **e–h** dynamic objects detected

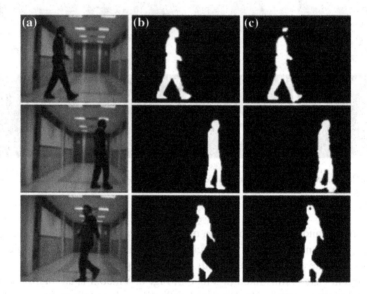

Fig. 9 Frames 45, 60 and 135 of Video 2. **a** Original. **b** Ground truth. **c** DTCNN method

Table 1 Performance metrics for frames of video 2

Video 2	DTCNN		
Metric	Frame 45	Frame 60	Frame 135
F1	0.9160	0.9331	0.9368
Precision	0.9266	0.8820	0.9308
Recall	0.9057	0.9905	0.9427
Similarity	0.8451	0.8747	0.8810

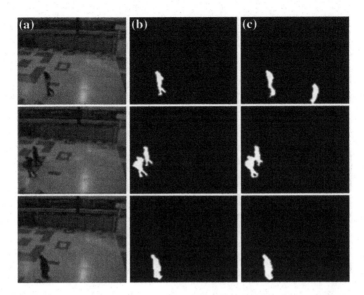

Fig. 10 Frames 40, 220 and 550 of Video 3. **a** Original. **b** Ground truth. **c** DTCNN method

Table 2 Performance metrics for frames of video 3

Video 3	DTCNN		
Metric	Frame 40	Frame 220	Frame 550
F1	0.5119	0.8595	0.8927
Precision	0.3935	0.8258	0.8548
Recall	0.7320	0.8960	0.9341
Similarity	0.3440	0.7536	0.8062

well segmented except in some frames where the morphologic operations reduce its region.

The visual results shown on the previous figures, indicates acceptable segmentation of the dynamic objects. Next, quantitative results are shown. The performance of the proposed method was computed using different metrics used in [19]. The frames used for comparison were selected such that most of the dynamics objects appear on them.

Figure 9 illustrates the segmentation for frames 45, 60 and 135 of Video 2 and Table 1 shows the results. The results include 4 performance metrics, *F1*, *Precision*, *Recall*, *Similarity* as well as visual results of the segmentation process for the DTCNN.

The performance on Video 3 is shown in Fig. 10 and Table 2. The method has low performance in frame 40 because a dynamic object is present in the initial foreground, which is not eliminated until the frame 60.

Figure 11 and Table 3 illustrate the performance for Video 5. The method is able to detect a small object in frame 80.

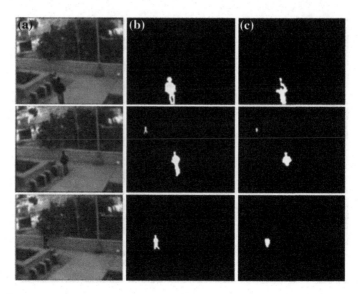

Fig. 11 Frames 45, 80 and 125 of Video 5. **a** Original. **b** Ground truth. **c** DTCNN method

Table 3 Performance metrics for frames of video 5

Video 5	DTCNN		
Metric	Frame 45	Frame 80	Frame 125
F1	0.7822	0.7186	0.7662
Precision	0.9263	0.9121	0.9390
Recall	0.6769	0.5929	0.6471
Similarity	0.6423	0.5608	0.6210

The Video 6 performances are indicated in Fig. 12 and Table 4. In this video, the CNN method could be benefited to produce better results because the average thresholding operator and the morphologic operations included in the method which reduce the amount of holes in the dynamic objects.

Metrics for Video 7 are given in Fig. 13 and Table 5. As in Video 6 the DTCNN method shows good performance in the frames 60, 120 and 190. As aforementioned, the improvement may be because of the thresholding operator and the morphologic operations included in the DTCNN model.

4.2 Static Object Segmentation Results

Different scenarios were selected to test the static segmentation system. The scenarios include indoor and outdoor conditions and different illumination conditions.

Figure 14 illustrates Video 2, an indoor scenario. The background includes distinct gray levels with low contrast, including the ceiling, floor and walls.

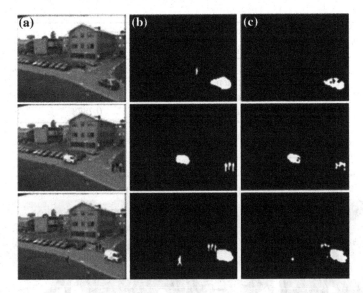

Fig. 12 Frames 520, 820 and 940 of Video 6. **a** Original. **b** Ground truth. **c** DTCNN method

Table 4 Performance metrics for frames of video 6

Video 6	DTCNN		
Metric	Frame 520	Frame 820	Frame 940
F1	0.7822	0.8433	0.8231
Precision	0.9192	0.8972	0.8860
Recall	0.6808	0.7956	0.7685
Similarity	0.6424	0.7291	0.6994

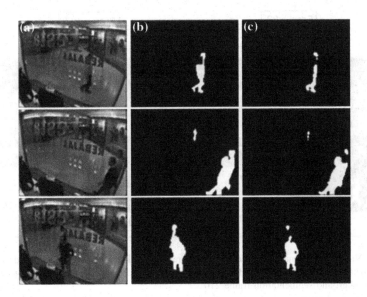

Fig. 13 Frames 60, 120, 190 Video 7. **a** Original. **b** Ground truth. **c** DTCNN method

Table 5 Performance
metrics for frames of video 7

Video 7	DTCNN		
Metric	Frame 60	Frame 120	Frame 190
F1	0.7512	0.8813	0.8434
Precision	0.9132	0.9411	0.9748
Recall	0.6381	0.8287	0.7432
Similarity	0.6016	0.7878	0.7292

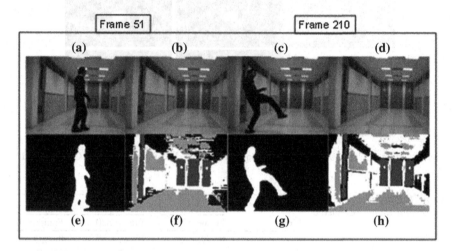

Fig. 14 Video 2. **a** Frame 51. **b** Background of frame 51. **c** Frame 210. **d** Background of frame 210. **e** Dynamic object frame 51. **f** Background segmentation frame 51. **g** Dynamic object frame 210. **h** Background segmentation frame 210

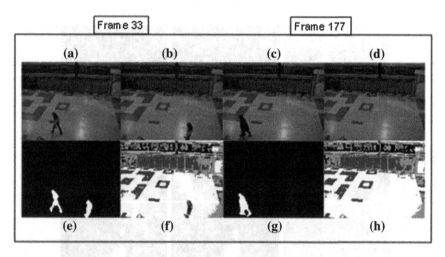

Fig. 15 Video 3. **a** Frame 33. **b** Background of frame 33. **c** Frame 177. **d** Background of frame 177. **e** Dynamic object frame 33. **f** Background segmentation frame 33. **g** Dynamic object frame 177. **h** Background segmentation frame 210

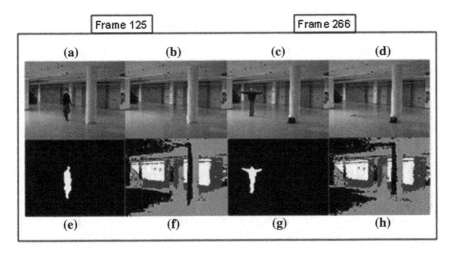

Fig. 16 Video 4. **a** Frame 125. **b** Background of frame 125. **c** Frame 266. **d** Background of frame 266. **e** Dynamic object frame 125. **f** Background segmentation frame 125. **g** Dynamic object frame 266. **h** Background segmentation frame 266

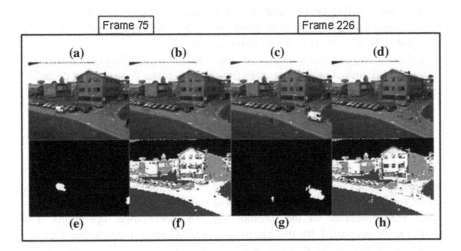

Fig. 17 **a** Frame 75. **b** Background of frame 75. **c** Frame 226. **d** Background of frame 226. **e** Dynamic object frame 75. **f** Background segmentation frame 75. **g** Dynamic object frame 226. **h** Background segmentation frame 226

Figure 15 presents Video 3 another indoor scenario. The results are better because there is less changes during the sequence. It can be observed that region segmentation is acceptable except in some parts where shine and shadows are present.

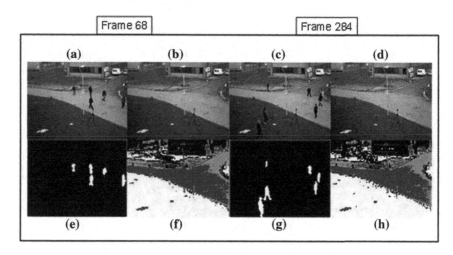

Fig. 18 **a** Frame 68. **b** Background of frame 68. **c** Frame 284. **d** Background of frame 284.
e Dynamic object frame 68. **f** Background segmentation fame 68. **g** Dynamic object frame 284.
h Background segmentation fame 284

The results of frames 125 and 266 of Video 4 are shown in Fig. 16. High
contrast regions are correctly segmented, meanwhile the ceiling and floor present
some problems because they have similar colors.

Results for two exterior scenarios are illustrated in Figs. 17 and 18. Because of
good natural illumination, the two scenarios present better contrast than de indoor
scenarios. This situation results on better region definition.

5 Conclusions

Considering the experiments reported, it can be said that the dynamic segmentation system shows acceptable results, where the DTCNN method presents some
improvements because of the thresholding and morphologic operations. This
conclusion is supported by the quantitative performance metrics and they are
comparable with existing dynamic object detection systems. The static segmentation system also produces promising results especially in high contrast scenarios,
although more work is needed to extend the number of possible regions. Both
systems, dynamic and static object present an interesting alternative for parallel
computing because of their CNN architecture.

Acknowledgments The authors thanks to DGEST, by the support of this research under grant
CHI-MCIET-2012-105.

References

1. Chen, Y.: Study of moving object detection in intelligent video surveillance system. In: Proceedings of International Conference on Computer Engineering and Technology, pp. V7–62–V7–66 (2010)
2. Faro, A., Giordano, D., Spampinato, C.: Evaluation of the traffic parameters in a metropolitan area by fusing visual perceptions and CNN processing of webcam images. IEEE Trans. Neural Netw. **19**(6), pp. 1108–1129 (2008)
3. Su, MC., A., Hsiung, C.Y., Huang, D.Y.: A simple approach to implementing a system for monitoring driver inattention. In: Proceedings of Systems, Man and Cybernetics, 2006, SMC '06, vol. 1, pp. 429–433, Oct 2006
4. Li, P., Ai, H., Li, Y., Huang, C.: Video parsing based on head tracking and face recognition. In: Proceedings of the 6th ACM International Conference on Image and Video retrieval, pp. 57–64 (2007)
5. Chen, L., Zhu, P., Zhu, G.: Moving objects detection based on background subtraction combined with consecutive frames subtraction. In: Proceedings of International Conference on Future Information Technology and Management Engineering, pp. 545–548, Dec 2010
6. Liu, Y., Bin, Z.: The improved moving object detection and shadow removing algorithms for video surveillance. In: Proceedings of International Conference on Computational Intelligence and Software Engineering, pp. 1–5, Dec 2010
7. Rodriguez, D., Vilarino, D., Pardo, X.: CNN implementation of a moving object segmentation approach for real-time video surveillance. In: Proceedings of 11th International Workshop on Cellular Neural Network and their Application, pp. 129–134, July 2008
8. Kang, J.: Image thresholding using cellular neural networks combined with fuzzy c-means. In: Proceeding of 2nd International Symposium on Intelligent Information Technology Application, pp. 319–322, Dec 2008
9. Babatunde, H., Folorunso, O., Akinwale, A.: A cellular neural-based model for edge detection. J. Inform. Comput. Sci. **5**(1), pp. 3–10 (2010)
10. Ketout, H., Gu, J., Horne, G.: MVN_CNN and FCNN for FCNN for endocardial edge detection. In: Proceedings of 1st Middle East Conference on Biomedical Engineering, pp. 208–212, Feb 2011
11. Arena, P., De Fiore, S., Fortuna, L., Lombardo, D., Patane, L.: Implementation of CNN-based perceptual framework on a roving robot. In: Proceedigns IEEE International Symposium on Circuits and System, pp. 1588–1591, May 2008
12. Chacon, M., González, S., Vega, J.: Simplified SOM-neural model for video segmentation of moving objects. In: Proceedings of International Joint Conference on Neural Network, pp. 474–480, June 2009
13. Gacsadi, A., Szolgay, P.: Variational computing based segmentation methods for medical imaging by using CNN. In: Proceedings of 12th International Workshop on Cellular Nanoscale Network and their Application, pp. 1–6, Feb 2010
14. Tang, M.: Edge detection and image segmentation based on cellular neural network. In: Proceedings of 3rd International Conference on Bioinformatics and Biomedical Engineering, pp. 1–4, June 2009
15. Kang, J., Zhang, W.: An approach for image thresholding using CNN associated with histogram analysis. In: Proceedings of International Conference on Measuring Technology and Mechatronics Automation, pp. 421–424, Apr 2009
16. Chua, L.O., Roska, T.: Cellular Neural Networks and Visual Computing. Cambridge University Press, Cambridge (2002)
17. Chen, B., Lei, Y.: Indoor and outdoor people detection and shadow suppression by exploiting HSV color information. In: Proceedings of IEEE International Conference Computer Information Technology, pp. 137–142, Sept 2004

18. Cellular Sensory and Wave Computing Laboratory of the Computer and Automation Research Inst., Hungarian Academy of Sciences and Jedlik Laboratory of the Pázmány P. Catholic Univ. Budapest.: Software Library for Cellular Wave Computing Engines (Online). Available: http://cnn-technology.itk.ppke.hu/Template_library_v3.1.pdf (2010)
19. Maddalena, L., Petrosino, A.: A self-organization approach to background substraction for visual surveillance applications. IEEE Trans. Image Process. 17, pp. 1168–1177 (2008)

Identification of Epilepsy Seizures Using Multi-resolution Analysis and Artificial Neural Networks

Pilar Gómez-Gil, Ever Juárez-Guerra, Vicente Alarcón-Aquino, Manuel Ramírez-Cortés and José Rangel-Magdaleno

Abstract Finding efficient and effective automatic methods for the identification and prediction of epileptic seizures is highly desired, due to the relevance of this brain disorder. Despite the large amount of research going on in identification and prediction solutions, still it is required to find confident methods suitable to be used in real applications. In this paper, we discuss the principal challenges found in epilepsy identification, when it is carried on offline analyzing electro-encephalograms (EEG) recordings. Indeed, we present the results obtained so far in our research group, with a system based on multi-resolution analysis and feed-forward neural networks, which focus on tackling three important challenges found in this type of problems: noise reduction, feature extraction and pertinence of the classifier. A *3-fold* validation of our strategy reported an accuracy of 99.26 ± 0.26 %, a sensitive of 98.93 % and a specificity of 99.59 %, using data provided by the University of Bonn. Several combinations of filters and wavelet transforms were tested, found that the best results occurs when a Chebyshev II filter was used to eliminate noise, 5 characteristics were obtained using a Discrete Wavelet Transform (DWT) with a Haar wavelet and a feed-forward neural network with 18 hidden nodes was used for classification.

P. Gómez-Gil (✉)
Department of Computer Science, Instituto Nacional de Astrofísica, Óptica y Electrónica, Tonantzintla, PUE, México
e-mail: pgomezexttu@gmail.com

E. Juárez-Guerra · V. Alarcón-Aquino
Department of Electronics, Universidad de las Américas, Cholula, PUE, México
e-mail: ever.juarezga@udlap.mx

M. Ramírez-Cortés · J. Rangel-Magdaleno
Department of Electronics, Instituto Nacional de Astrofísica, Óptica y Electrónica, Tonantzintla, PUE, México

O. Castillo et al. (eds.), *Recent Advances on Hybrid Approaches for Designing Intelligent Systems*, Studies in Computational Intelligence 547, DOI: 10.1007/978-3-319-05170-3_23, © Springer International Publishing Switzerland 2014

1 Introduction

It is estimated that around 1 % of the world population suffers from epilepsy. This disease is the result of sudden changes in the dynamics of the brain, producing abnormal synchronization in its networks, which is called a "seizure" [16]. In some cases, this disease cannot be controlled, producing serious physical and social disturbances in the affected patients. Seizures occur unexpectedly, lasting from few seconds to few minutes and showing symptoms as loss of consciousness, convolutions, lip smacking, blank staring or jerking movements of arms or legs [8]. Due to the negative effects of their occurrence, the medical community is looking for ways to predict seizures in real-time, so that patients can be warned of incoming episodes. However, still this is an open problem, but many research efforts are being dedicated to this. Among several strategies, the monitoring of the electroencephalographic activity (EEG) of the brain is considered one of the most promising options for building epilepsy predictors. However, this is very difficult problem, because the brain is a very complex system, and an EEG time series may be generated by dynamics among neurons nearby or far away from the electrodes where it was acquired, which makes difficult to model the dynamics involved.

The electrical activity of the brain was first mentioned in 1875, when Galvani and Volta performed their famous experiments [7, 23]. The first EEG was recorded from the surface of a human skull by Borger in 1929 [5] and 1935 is considered the year of birth of today's clinical electroencephalography. Since then, EEGs are part of the required medical tools for diagnosis and monitoring of patients. Figure 1 shows a portion of a EEG of a healthy patient.

Experts have identified several morphological characteristics in an EEG, related to the type of patients and brain stages when the EEG is recorded. The EEG of a healthy, relaxed patient with closed eyes presents a predominant physiological rhythm, known as the "alpha rhythm", with frequencies ranging from 8 to 13 Hz; if patient's eyes are open, broader frequencies are presented. Seizure-free intervals, acquired within the epileptogenic zone of a brain, that is, the zone where the abnormality occurs, present rhythmic and high amplitude patterns, known as "interictal epileptiform activities"; these activities are fewer and less pronounced when intervals are sensed in places distant from the epileptogenic zones. Contrary to the expected, an EEG taken during an epileptic event (ictal activity) is almost periodic and presents high amplitudes [16].

Several stages may be identified in an EEG related to epilepsy: the ictal stage, which corresponds to the occurrence a seizure, the interictal stage, which occurs in between epileptic events, the pre-ictal state, occurring few minutes prior to a seizure and the postictal stage [26]. The term "healthy stage" is used here to identify the behavior of a EEG obtained from relaxed, healthy patients with closed eyes.

With the possibility of using fast and accurate information technologies ubiquitously, the promise of trustily devices able to monitor and predict health conditions in real time is growing, and a fact in many cases. Based on this, a huge

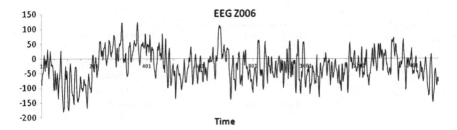

Fig. 1 An example of a EEG of a healthy patient (taken form set Z of the database created by the University of Bonn)

amount of research has been published, with the hope of finding better uses of signal processing techniques to support clinical diagnosis. Related to the use of EEG for diagnosis, the health community has agreed on some nomenclature related to time–frequency analysis. Table 1 lists the identifications given by most authors to frequency sub-bands found in an EEG spectra, and its relation to specific conditions of the subjects being analyzed. These sub-bands have been used to characterize different brain conditions as Alzheimer disease or epilepsy and as a tool for building brain computer interfaces [7]. In this respect, some researchers have reported that delta and alpha sub-bands are suitable for the identification of epileptic episodes [28, 31]. Signal processing has also been widely involved in the use of time–frequency transformations as an aid for feature extraction in classification and prediction in medical applications and as pre-processing techniques for biomedical signals [34]. In particular, wavelet analysis has been widely used for classification and prediction of epilepsy (see for example [1, 10, 15, 19, 29]).

In this chapter we review some important concepts and research work related with the automatic identification of epilepsy using EEG, and we analyze the use of filtering, wavelet analysis and artificial neural networks to tackle the main problems associated to this identification: noise reduction, definition of a suitable feature vector and design of a classifier. The behavior of several classifiers is presented, which were built using combinations of: Infinite Impulse Response (IIR) filters, Finite Impulse Response (FIR) filters, discrete wavelet transforms (DWT), Maximal Overlap Discrete Wavelet Transform (MODWT) and Feed-Forward Artificial Neural Networks (FF-ANN). The chapter is organized as follows: Sect. 2 presents some of the recent work related to this research; Sect. 3 comments on the basic concept used for identification of epilepsy; the main steps for identifying epilepsy using EEG are detailed in Sect. 4; Sect. 5 shows the results obtained with an experiment using the just mentioned characteristics. Section 6 presents some conclusions and future work.

Table 1 Frequency sub-bands identified in EEG (adapted from)

Name	Frequency range (Hz)	Subject's conditions
Delta	0–4	Deep sleep, Infancy, Brain diseases
Tetha	4–8	Parental an temporal regions in children, emotion stress in some adults,
Alpha	8–12 or 8–13 (Ihle 2013)	Awaked, resting healthy subjects
Beta	12–30	Intense mental activity
Gamma	>30	Information processing Onset of voluntary movements

2 Related Works

The amount of published work related to the automatic identification of epilepsy in EEGs is amazing. For this reason, a complete state-of the art review is out of the scope of this chapter. Therefore, in this section, we limit our analysis to works similar to ours. We present only recent works that have tested using a free-available EEG database, originally presented in [2] from the University of Bonn and available at [33]. This collection contains unfiltered EEGs of five subjects, including different channels per patient, recorded with a sampling rate of 173 Hz. The database is divided in 5 sets, identified as (A, B, C, D, E) or (Z, O, N, F, and S). Each set contains 100 single-channel segments of EEG with the characteristics described in Table 2, lasting 23.6 s each. Even though larger databases exist (see [16] for a review), this database has the advantages of being free, highly popular and very easy to use. For this reason, we decided to use this database to test our algorithms, referred herein as the "Bonn database." In addition, we focus our review in works based on the use of wavelet analysis for feature extraction and feed-forward neural networks (FF-NN) for identification of 2 classes (healthy and ictal stage) and 3 classes (healthy, ictal and inter-ictal stages). Table 3 summaries this review. A detailed review of other works can be found in [32].

3 Filtering and Wavelet Analysis

This section describes some basic concepts related to filtering and wavelet analysis, which are used in this paper. It is important to point that that these paragraphs include just basic definitions and brief descriptions of techniques; it does not pretend to be a detailed review of them. The interested reader is encouraged to review [1, 27].

In this context, filtering refers to a process that removes unwanted frequencies in a signal. This is required for some tasks in order to eliminate noise, even though

Table 2 A brief description of the EEG data collection provided by the Univerisity of Bonn [33]. For a detailed description see [2]

Set	Description
Z	Extra-cranial EEG recording of healthy subjects with open eyes
O	Extra-cranial EEG recording of healthy subjects with closed eyes
N	Intra-cranial EEG recording into the hippocampal formation zone of interictal stages
F	Intra-cranial EEG recording into the epileptogenic zone of interictal stages
S	Intra-cranial EEG recording of seizure stages

Table 3 Comparison of published works in epilepsy identification. Two or three classes are identified using data provided by [2, 33]

Author(s)	Feature extraction method	Accuracy	Number of classes and observations on the database
Tzallas et al. [32]	Based on power spectrum density (PSD) calculated using fast-fourier transform	100 % (using sets Z and S)	2 classes (healthy and ictal) using sets Z and S;
		100 % (using sets Z, F and S)	3 classes (healthy, interictal and ictal)
		89 % (using all sets)	using sets Z, F and S and using all sets
Anusha et al. [3]	Sliding windows	95.5 % (ictal) 93.37 % (healthy)	2 classes using sets A and E
Ghosh-Dastidar et al. [10]	Standard deviation, correlation dimension and largest Lyapunov exponent of delta, theta, alpha beta and gamma sub-bands obtained by wavelet analysis	96.7 %	3 classes using sets Z, F and S
Husain & Rao [15]	Energy, covariance inter-quartile range and Median Absolute Deviation, applied to sub-bands obtained from segments of 1 s	98.3 %	2 classes using all sets
Juarez et al. [17]	Maximal overlap discrete wavelet transform over delta and alpha bands of segments of 23.6 s, previously filtered using a 10-order Butterworth low pass filter	90 %	2 classes using sets Z and S
Juarez et al. [18]	Chebyshev II filtering, DWT with Haar wavelet over segments of 1 s	99.26 %	2 classes, sets Z and S

it may create some distortion in the features represented in the frequency-time domain of the signal. In spite of that, filtering in epilepsy identification is highly recommended; according to Mirzaei, EEG frequencies above 60 Hz can be neglected [25]. There are different types of filters: linear and non-linear, time-invariant or time-variant, digital or analog, etc.

Digital filters are divided as infinite impulse response (IIR) filters and finite-impulse response filters (FIR). A FIR filter is one whose response to any finite length input is of finite duration, because it settles to zero in finite time. On the other hand, an IIR filter has internal feedback and may continue to respond indefinitely, usually decaying. The classical IIR filters approximate the ideal "brick wall" filter. Examples of IIR filters are: Butterworth, Chebyshev type I and II, Elliptic and Bessel; of special interest are Chebyshev type II and Elliptic. The Chebyshev type II filter minimizes the absolute difference between the ideal and actual frequency response over the entire stop band. Elliptic filters are equiripple in both the pass and stop band and they generally meet filter requirements with the lowest order of any supported filter type. FIR filters have useful properties, for example, they require no feedback and they are inherently stable. The main disadvantage of FIR filters is that they require more computational power that IIR filters, especially when low frequency cutoffs are needed [24].

To design a filter means to select their coefficients in a way that the system covers specific characteristics. Most of the time these filter specifications refer to the frequency response of the filter. There are different methods to find the coefficients suitable from frequency specifications, for example: window design, frequency sampling method, weighted least squares design and equiripple [6]. In the experiments presented in this chapter, the methods "weighted least squares design" and "Equiripple" were applied, as recommended by [4]. Figure 2 shows a filtered EEG and its frequency spectrum of an ictal stage. The filtering was obtained using a Least Squares FIR filter.

Wavelet analysis is a very popular signal processing technique, widely applied in non-periodic and noisy signals [1]. It consists on calculating a correlation among a signal and a basis function, known as the wavelet function; this similarity is calculated for different time steps and frequencies. A wavelet function $\varphi(.)$ is a small wave, which means that it oscillates in short periods of time and holds the following conditions:

1. Its energy is finite, that is:

$$E = \int\limits_{-\infty}^{\infty} |\varphi(t)|^2 dt < \infty \tag{1}$$

2. Its admissibility constant C_φ is finite, that is:

$$C_\varphi = \int\limits_{0}^{\infty} \frac{|\Phi(u)|^2}{u} du < \infty \tag{2}$$

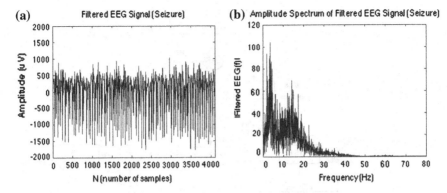

Fig. 2 A filtered EEG of an ictal stage (**a**), its corresponding frequency spectra (**b**). Signal was filtered using a least-square FIR filter. Data taken from [33]

where:

$$\Phi(u) = \int\limits_{-\infty}^{\infty} \varphi(t)e^{-i2\pi u}dt \qquad (3)$$

is the Fourier transform of $\varphi(.)$.

Any function $f(t)$ can be expressed as a linear decomposition:

$$f(t) = \sum_{\ell} a_{\ell}\varphi_{\ell}(t) \qquad (4)$$

where a_{ℓ} are real-valued expansion coefficients and $\varphi_{\ell}(t)$ are a set of real functions, known as the expansion set. Using wavelets as an expansion set, a series expansion of a signal is defined as:

$$f(t) = \sum_{k}\sum_{j} a_{j,k}\varphi_{j,k}(t) \qquad (5)$$

$a_{j,k}$ are known as the Discrete Wavelet Transform (DWT) of $f(t)$.

In addition to DWT, there are several wavelet transforms, which may be applied using different types of wavelets. Two transforms frequently used are the Discrete Wavelet Transforms (DWT) and Maximal Overlap Discrete Wavelet Transform (MODWT). DWT can be estimated using a bank of low-pass and high-pas filters. Filters are combined with down-sampling operations, to create coefficients that represent different frequencies in different resolution levels. Two types of coefficients are obtained: approximation coefficients and detailed coefficients. The output of each filter contains a signal with half the frequencies of the input

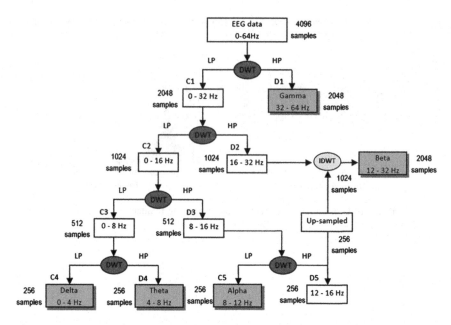

Fig. 3 Wavelet analysis of an EEG using a DWT. The *yellow blocks* correspond to identifiers of sub-bands found in a EEG (see Table 1). *LP* Low-pass filter, *HP* High-pass filter, *C#* Approximation coefficient number #, *D#* Detailed coefficient number#

signal, but double the points, requiring then to be down-sampled. Figure 3 shows a tree representing the decomposition of an EEG using a DWT algorithm. Notice that the EEG sub-bands, described in Table 1, are obtained in different levels of the decomposition (marked as yellow blocks in the figure). Figure 4 shows the delta sub-band of an EEG signal obtained by a MODWT with Daubechies degree 2 and its corresponding spectra. Figure 5 shows the same for Alpha sub-band.

The selection of the right wavelet to be used in a specific problem is related with the characteristics found in the signal being analyzed. In this paper, we used Haar, second order Daubechies (Db2) and fourth order Daubechies (Db4) wavelets. For a complete review of wavelet analysis, see [1].

4 Identifying Epilepsy

Off-line identification of epilepsy stages using EEGs is composed of 3 main steps, which are described next:

1. *Preprocessing of the input signal.* In this step the EEG stream is divided in segments and each segment is clean of unwanted frequencies that may represent noise or other artifacts. Section 3 describes some methods for filtering. In

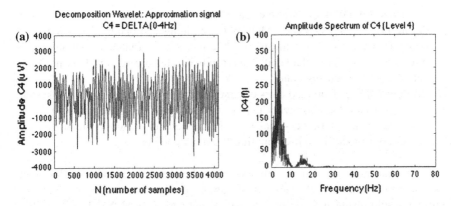

Fig. 4 Delta sub-band of an EEG signal obtained by a MODWT with Daubechies degree 2 (**a**) and its corresponding frequency spectrum (**b**)

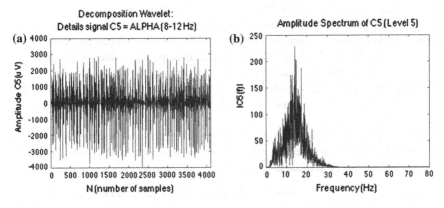

Fig. 5 Alpha sub-band of an EEG signal obtained by a MODWT with Daubechies degree 2 (**a**) and its corresponding frequency spectrum (**b**)

this chapter we compare the performance obtained for classifiers using different types of filters: Chebyshev II, Elliptic, Equiripple and Least Squares. For de results showed in this paper, we used segments lasting 23.6 s, 1 s, and 0.7375 segments each.

2. *Feature extraction.* The design of this part of a classifier is one of the most important challenges in this and other identification problems. Using signal processing, statistics, or other math techniques, each filtered EEG segment (sample) has to be represented with appropriate values that will allow the recognizer to separate the different classes. In Table 3, we list some examples of feature extraction methods, used for some works related to epilepsy identification, but there are many more. Here we present the results obtained using

DWT and MODWT as feature extractors. First, each segment of one second is decomposed using DWT or MODWT, obtaining sub-bands alpha and delta (among others). Next, the mean, absolute mean and variance of the amplitude of each sub-band are calculated. This results in a feature vector of six positions characterizing each segment.

3. *Recognition.* The feature vectors obtained in previous step are input to a system that will decide the class with highest probability of being the sample's class. There are many classifiers that have been tried for epilepsy identifications, being those based on soft computing the most popular.

In special, artificial neural networks have showed to be good modelers with excellent generalization abilities, when compared with other strategies. Here we present results using a feed forward neural network (FF-ANN). The activation function used in all neurons was a sigmoid. The network used here has 6 input neurons (one for each feature) and two output neurons, each representing a class: ictal stage or healthy stage.

The best combination of filter method, features and recognizer parameters need to be found in order to get the best possible performance in a particular application. To find this, it is advised to execute several experiments, testing each model with some validation criteria. Several of these methods have been proposed in the literature, being *k-fold* validation one of the most popular, due to its ability to provide a good statistical estimation of the performance of the classifier [20]. The selection of a value for k depends upon the number of samples available for training and testing the system. We used a *3-fold* validation to test all the combinations reported here.

With respect to the way of evaluating the performance of epilepsy identifiers, it is a common practice to use 3 metrics: accuracy, sensitivity and specificity. When two classes are involved (healthy and ictal stages), these metrics are defined as [26]:

$$Accuracy = \frac{number\ of\ correct\ classified\ segments}{total\ of\ segments} \tag{6}$$

$$Sensitivity = \frac{number\ of\ correct\ classified\ ictal\ segments}{total\ of\ ictal\ segments} \tag{7}$$

$$Specificity = \frac{number\ of\ correct\ classified\ healthy\ segments}{total\ of\ healthy\ segments} \tag{8}$$

Accuracy gives the proportion of correctly identified samples. Sensitivity, also known as the recall rate, measures a proportion of sick cases correctly identified as such. Specificity gives the proportion of healthy cases correctly identified as such [13].

In the next section, we present the results of an experiment that we executed for identification of ictal and healthy stages in EEG, which were originally reported in [18].

Table 4 Parameters for the FF-NN, applied for both experimental cases

Characteristic	Value(s)
Number of input nodes	6
Number of hidden nodes	6, 9, 12, 15, 16, 18, 21 and 24
Number of output nodes	2
Software	Based on Matlab 2010 and Neural Toolbox V 6.0.3, available in [11]
Training algorithm	Leverage at a learning rate of 0.5, with a max. of 1,000 epochs
Activation functions	sigmoid

5 An Experiment for Ictal Identification

Our research group is working with the design of new models for identification and prediction of epilepsy [17]. In our way to do so, we have experimented with some models currently proved to obtain good results in this task. Here, as an example of this process, we detail the results obtained for a FF-NN when trained with filtered segments of EEG [18]. As described in Sect. 4, we tested different combinations of filters, wavelet transforms and number of hidden nodes, to obtain the most suitable architecture for the data provided by sets Z and S from the Bonn database. We tested two cases, the first using segments of 23.6 s and the second using segments of one second or of 0.7375 s. For the second case, segments of one second were used when a DWT transform were applied and segments of 0.7375 were used when MODWT was applied. Table 4 presents the parameters used for setting the FF-NN classifier. As we stated before, three-fold validation was applied to obtain all performance measures.

Performances obtained in the first case are summarized in Table 5, and the performances obtained for second case are summarized in Table 6. Measures are calculated according to Eqs. (6), (7) and (8). These values are the average obtained by a *3-fold* validation procedure. The best results in each case are bolded. Notice that the best results are obtained using segments of one second (second case) cleaned using a Chebyshev II filter, features are obtained using a DWT with a Haar wavelet. In this case an accuracy of 99.26 ± 0.26 is obtained, with a sensitivity of 98.93 % and a specificity of 99.59 %. This results are better than most of the results shown in Table 3, except for the work of [32].

6 Conclusions and Ongoing Work

In this paper, we present basic ideas related to the identification of different stages identified in a EEG, related to epilepsy. Given the importance of this brain disorder, it is mandatory to find better ways to identify and predict seizures in real time. However, this is still an open problem, given the complexity inherent to the brain and the amount of information provided by EEGs. Some recent works using

Table 5 Performance of models using segments of 23.6 s

Filter	Wavelet	Hidden nodes	Accuracy (%)	Acc. of standard deviation	Sensitivity (%)	Specificity (%)
Chebyshev II	DWT—Haar	9	82.82	21.10	82.79	74.64
Chebyshev II	DWT—Db2	15	83.73	21.52	82.56	82.04
Chebyshev II	DWT—Db4	6	91.11	14.59	91.72	88.37
Chebyshev II	MODWT—Haar	9	83.33	16.58	80.78	86.42
Chebyshev II	MODWT—Db2	6	84.94	16.40	86.06	84.52
Chebyshev II	MODWT—Db4	18	84.44	21.33	86.44	85.74
Elliptic	DWT—Haar	21	88.38	16.76	86.46	91.58
Elliptic	DWT—Db2	6	80.30	21.73	79.87	76.85
Elliptic	DWT—Db4	9	82.82	20.20	80.19	87.61
Elliptic	MODWT—Haar	21	85.45	18.21	84.13	82.21
Elliptic	MODWT—Db2	6	90.00	12.91	88.17	96.32
Elliptic	MODWT—Db4	24	87.17	14.85	87.22	85.61
Equiripple	DWT—Haar	18	87.17	19.09	85.47	90.96
Equiripple	DWT—Db2	12	83.03	20.57	81.48	85.36
Equiripple	DWT—Db4	18	86.56	17.05	85.90	91.14
Equiripple	MODWT—Haar	6	87.07	17.66	84.83	92.02
Equiripple	MODWT—Db2	6	88.88	18.52	89.17	82.80
Equiripple	MODWT—Db4	6	85.52	19.04	83.71	84.74
Least squares	DWT—Haar	6	84.44	21.32	80.00	82.82
Least squares	**DWT—Db2**	**6**	**93.23**	**14.85**	**93.87**	**90.07**
Least squares	DWT—Db4	18	82.72	20.13	91.24	81.42
Least squares	MODWT—Haar	9	83.73	18.32	82.71	90.97
Least squares	MODWT—Db2	21	84.14	15.96	82.76	87.16
Least squares	MODWT—Db4	21	87.37	15.82	85.88	90.89

Table 6 Performance of the models using segments of 1 or 0.7375 s

Filter	Wavelet	Hidden nodes	Accuracy (%)	Acc. of Standard deviation	Sensitivity (%)	Specificity (%)
Chebyshev II	**DWT—Haar**	**18**	**99.26**	**0.26**	**98.93**	**99.59**
Chebyshev II	DWT—Db2	18	99.03	0.27	98.75	99.32
Chebyshev II	DWT—Db4	18	96.57	5.87	95.38	98.91
Chebyshev II	MODWT—Haar	15	99.24	0.32	98.86	99.64
Chebyshev II	MODWT—Db2	24	95.80	12.84	94.77	96.19
Chebyshev II	MODWT—Db4	24	97.72	4.55	96.76	99.13
Elliptic	DWT—Haar	21	95.49	11.10	97.48	95.30
Elliptic	DWT—Db2	21	95.96	12.33	96.65	96.23
Elliptic	DWT—Db4	9	98.44	1.26	97.91	99.06
Elliptic	MODWT—Haar	18	95.98	12.52	95.49	99.74
Elliptic	MODWT—Db2	6	99.12	0.40	98.73	99.51
Elliptic	MODWT—Db4	24	95.34	12.62	91.27	96.14

filtering, FF-NN and wavelet decomposition were analyzed. We also presented the design of a identifier of ictal and healthy stages, based on filters, wavelet analysis and FF-NN, which obtained an average of 99.26 ± 0.26 % of accuracy.

Here we present just the first steps of an ongoing research, and still the most important ideas are being explored. A critical issue to be considered is that this problem is highly related to temporal classification. Some temporal classification problems may be solved using static classification strategies (as the one presented here), provided that the information about time is represented in some way in a feature vector. However, it is difficult to identify accurately the time lag required to build the feature vector, which in this case corresponds to the right size of the segment to be characterized. It has been showed that recurrent neural networks are a better option than FF-NN for time-dependent problems where chaos is present [12]. Particularly, some studies have outlined the advantage of recurrent models over feed-forward models for EEG classification [13]. Indeed, there is a strong evidence of the chaotic behavior of EEG during seizures [2] and that recurrent neural networks have presented good results modeling chaotic systems [12]. Therefore, the next step to be explored in our research is the use of recurrent neural networks for temporal identification. Encourage for the results obtained here, we will explore wavelet-recurrent neural networks, as the ones presented in [30] and [9], modeled with Haar wavelets as activation functions.

A search for better features extractor methods has to be performed. A method that has reported with good results in this context is Empirical Mode Decomposition (see for example [22]) will be also analyzed. EMD, introduced by Huang in 1971 [34], has become very popular in biomedicine in the last few years [35]. This is a spontaneous multi-resolution method that represents nonlinear and non-stationary data as a sum of oscillatory modes inherent in the data, called Intrinsic Mode Functions (IMFs) [21].

Acknowledgments The first author gratefully acknowledges the financial support from the "Universidad Autónoma de Tlaxcala" and PROMEP by scholarship No. UATLX-244. This research has been partially supported by CONACYT, project grant No. CB-2010-155250.

References

1. Addison, P.S.: The Illustrated Wavelet Transform Handbook: Introductory Theory and Applications in Science, Engineering Medicine and Finance. IOP Publishing, England (2002)
2. Andrzejak, R.G., Lehnertz, K., Mormann F., Rieke D., David P., Elger, C.: Indications of nonlinear deterministic and finite dimensional structures in time series of brain electrical activity: dependence on recording region and brain state. Phys. Rev. E. **64**(6), 061907-1, 061907-8 (2001). doi:10.1103/PhysRevE.64.061907
3. Anusha, K.S., Mathew, T.M., Subha, D.P.: Classification of normal and epileptic EEG signal using time & frequency domain features through artificial neural network. In: International Conference on Advances in Computing and Communications. IEEE (2012)
4. Bashashati, M., Fatourechi, R., Ward, K., Birch, G.E.: A survey of signal processing algorithms in brain-computer interfaces based on electrical brain signals. J. Neural Eng. **4**(2), R32–R57 (2007)
5. Berger, H.: Über das elektrenkephalogramm des menschen. Arch. F. Psichiat. **87**, 527–570 (1929)
6. Cetin, E., Gerek, O.N., Yardimci, Y.: Equiripple FIR filter design by the FFT algorithm. IEEE Signal Process. Mag. **14**, 60–64 (1997)
7. Durka, P.: Matching Pursuit and Unification in EEG Analysis. Artech House Norwood, Boston (2007)
8. EU FTP 7. ICT -2007 5. Epilepsy and seizures. EPILEPSIAE project (Online).: Advanced ICT for risk assessment and patient safety grant 211713. Available: http://www.epilepsiae.eu/about_project/epilepsy_and_seizures/, (2013). Accessed in 27 Nov 2013 (2913)
9. García-González, Y.: Modelos y algoritmos para redes Neuronales recurrentes basadas en wavelets aplicados a la detección de intrusos. Master thesis, Department of Computing, Electronics and Mechatronics, Universidad de las Américas, Puebla (2011)
10. Ghosh-Dastidar, S., Adeli, H., Dadmehr, N.: Mixed-band wavelet-chaos neural network methodology for epilepsy and epileptic seizure detection. IEEE Trans. Biomed. Eng. **54**(9), 1545–1551 (2007)
11. Gómez-Gil, P.: Tutorial: an introduction to the use of artificial neural networks. Available at: http://ccc.inaoep.mx/ ~ pgomez/tutorials/ATutorialOnANN2012.zip
12. Gómez-Gil, P., Ramírez-Cortés, J.M., Pomares Hernández, S.E., Alarcón-Aquino, V.: A neural network scheme for long-term forecasting of chaotic time series. Neural Process. Lett. **33**(3), 215–233 (2011)
13. Güler, N.F., Übeylib, E.D., Güler, I.: Recurrent neural networks employing Lyapunov exponents for EEG signals classification. Expert Syst. Appl. **29**, 506–514 (2005)
14. Huang, N.E., Shen, Z., Long, S.R., Wu, M.L.C., Shih, H.H., Zheng, Q.N., et al.: The empirical mode decomposition and the Hilbert spectrum for nonlinear and non-stationary time series analysis. P Roy Soc Lond a Mat **1998**(454), 903–995 (1971)
15. Husain, S.J., Rao, K.S.: Epileptic seizures classification from EEG signals using neural networks. In: International Conference on Information and Network Technology, (37) (2012)
16. Ihle, M., Feldwisch-Drentrupa, H., Teixeirae, C., Witonf, A., Schelter, B., Timmerb, J., Schulze-Bonhagea, A.: EPILEPSIAE—a European epilepsy database. Comput. Methods Programs Biomed. **106**, 127–138 (2012)
17. Juárez-Guerra, E., Gómez-Gil, P., Alarcon-Aquino, V.: Biomedical signal processing using wavelet-based neural networks. In: Special Issue: Advances in Pattern Recognition, Research in Computing Science, vol. 61, pp. 23–32 (2013a)

18. Juárez-Guerra, E., Alarcón-Aquino, V., Gómez-Gil P.: Epilepsy seizure detection in eeg signals using wavelet transforms and neural networks. To be published in: Proceedings of the Virtual International Joint Conference on Computer, Information and Systems Sciences and Engineering (CISSE 2013), 12–14 Dec 2013 (2013b)

19. Kaur, S.: Detection of epilepsy disorder by EEG using discrete wavelet transforms. Thesis in Master of Engineering in Electronic Instrumentation and Control, Thapar University, July 2012

20. Kohavi, R.: A study of cross-validation and bootstrap for accuracy estimation and model selection. In: Proceedings of the International Joint Conference on Artificial Intelligence, vol. 14, pp. 1137–1145. Lawrence Erlbaum Associates Ltd (1995)

21. Mandic, D., Rehman, N., Wu, Z., Huang, N.: Empirical mode decomposition-based time-frequency analysis of multivariate signals. IEEE Signal Process. Mag. 30(6), 74–86 (2013)

22. Martis, J.R., Acharya, U.R., Tan, J.H., Petznick, A., Yanti, R., Chua, C.K., Ng, E.K.Y., Tong, L.: Application of empirical mode decomposition (EMD) for automated detection of epilepsy using EEG signals. Int. J. Neural Syst. 22(6), 1250027-1, 16 (2012)

23. Martosini, A.N.: Animal electricity, CA 2+ and muscle contraction. A brief history of muscle research. Acta Biochim. Pol. 47(3), 493–516 (2000)

24. MathWorks Incorporation: Documentation, Signal Processing Toolbox, Analog and Digital Filters. Matlab R2013b (2013)

25. Mirzaei, A., Ayatollahi, A., Vavadi, H.: Statistical analysis of epileptic activities based on histogram and wavelet-spectral entropy. J. Biomed. Sci. Eng. 4, 207–213 (2011)

26. Petrosian, A., Prokhorov, D., Homan, R., Dasheiff, R., Wunsch II, D.: Recurrent neural network based prediction of epileptic seizures in intra- and extracranial EEG. Neurocomputing 30, 201–218 (2000)

27. Proakis, J.G., Manolakis, D.G.: Digital signal processing. Principles, algorithms and applications, 3rd edn. Englewood Prentice Hall, Cliffs (1996)

28. Ravish, D.K., Devi, S.S.: Automated seizure detection and spectral analysis of EEG seizure time series. Eur. J. Sci. Res. 68(1), 72–82 (2012)

29. Subasi, A., Ercelebi, E.: Classification of EEG signals using neural network and logistic regression. J Comput. Methods Programs Biomed. 78, 87–99 (2005)

30. Sung, J.Y., Jin, B.P., Yoon, H.C.: Direct adaptive control using self recurrent wavelet neural network via adaptive learning rates for stable path tracking of mobile robots. In: Proceedings of the 2005 American Control Conference, pp. 288–293. Portland (2005)

31. Sunhaya, S., Manimegalai, P.: Detection of epilepsy disorder in EEG signal. Int. J. Emerg Dev. 2(2), 473–479 (2012)

32. Tzallas, A.T., Tsipouras, M.T., Fotiadis, D.I.: Epileptic seizure detection in EEGs using time-frequency analysis. IEEE Trans. Inf. Technol. Biomed. 13(5), 703–710 (2009)

33. Universitat Bonn, Kinik für Epitepbologie: EEG time series download page. URL: http://epileptologie-bonn.de/cms/front_content.php?idcat=193, Last Accessed at 12 Dec 2013

34. Wacker, M., Witte, H.: Time-frequency techniques in biomedical signal analysis: a tutorial review of similarities and differences. Methods Inf. Med. 52(4), 297–307 (2013)

35. Wang, Y. et al.: Comparison of ictal and interictal EEG signals using fractal features. Int. J. Neural Syst. 23(6), 1350028 (11 pages) (2013)

Temporal Validated Meta-Learning for Long-Term Forecasting of Chaotic Time Series Using Monte Carlo Cross-Validation

Rigoberto Fonseca and Pilar Gómez-Gil

Abstract Forecasting long-term values of chaotic time series is a difficult task, but it is required in several domains such as economy, medicine and astronomy. State-of-the-art works agree that the best accuracy can be obtained combining forecasting models. However, selecting the appropriate models and the best way to combine them is an open problem. Some researchers have been focusing on using prior knowledge of the performance of the models for combining them. A way to do so is by meta-learning, which is the process of automatically learning from tasks and models showing the best performances. Nevertheless, meta-learning in time series impose no trivial challenges; some requirements are to search the best model, to validate estimations, and even to develop new meta-learning methods. The new methods would consider performance variances of the models over time. This research addresses the meta-learning problem of how to select and combine models using different parts of the prediction horizon. Our strategy, called "Temporal Validated Combination" (TVC), consists of splitting the prediction horizon into three parts: short, medium and long-term windows. Next, for each window, we extract knowledge about what model has the best performance. This knowledge extraction uses a Monte Carlo cross-validation process. Using this, we are able to improve the long-term prediction using different models for each prediction window. The results reported in this chapter show that TVC obtained an average improvement of 1 % in the prediction of 56 points of the NN5 time series, when compared to a combination of best models based on a simple average. NARX neural networks and ARIMA models were used for building the predictors and the SMAPE metric was used for measuring performances.

R. Fonseca · P. Gómez-Gil (✉)
National Institute of Astrophysics, Optics and Electronics, Tonantzintla, Mexico
e-mail: pgomez@acm.org; pgomezexttu@gmail.com

R. Fonseca
e-mail: rfonseca@inaoep.mx

O. Castillo et al. (eds.), *Recent Advances on Hybrid Approaches for Designing Intelligent Systems*, Studies in Computational Intelligence 547,
DOI: 10.1007/978-3-319-05170-3_24, © Springer International Publishing Switzerland 2014

1 Introduction

Improving the performance in the forecasting of chaotic time series is a complex task. Chaotic time series are cataloged as unpredictable, due their high sensibility to initial conditions [17]. Despite that, many applications deal with chaotic systems and require a reasonable estimation of long-term future values. Many domains are looking for an improvement of the accuracy obtained by current prediction models, for example in financial applications, load forecasting or wind speed [9]. The problem of predicting multi-steps-ahead, based on data captured from the chaotic system, is still an open problem [9].

Several authors have studied how to improve the performance of long-term forecasting. The state of the art of time series forecasting shows three main ideas: combinations of models obtained the best results [8, 23]; some models have better performance than other models depending on the number of steps to predict [8, 9], that is, the size of the prediction horizon; data features determine the relative performance of different models [20, 23, 32]. A method that considers these ideas could improve long-term forecasting.

Considering that combinations obtain the best results and data features determine the performance, one viable way of joining these ideas is the stacking method. The aim of the stacking method is to minimize the generalization error rate of one or more classifiers [33]. The stacking method allows combining models based on meta-features. In general, this method uses different learning algorithms for generating models to combine. Then, the models are validated with a cross validation method, generally, leave-one-out cross validation LOOCV [12]. The combinations obtained with the stacking method outperform the individual accuracy of the involved models [35].

Considering that the prediction horizon determines the performance of the model, a desired prediction horizon can be divided into three parts, every single part with the same number of elements, named short-term, medium-term and long-term; these are the prediction windows. Our goal is to extend the bases of the stacking method for combining N-models with the best performance in short, medium and long term window. Indeed, in our strategy the selection process validates the best models using a Monte Carlo cross validation (MCCV) instead of an LOOCV. MCCV was introduced by Picard and Cook [25]; Shao [29] proved that MCCV is asymptotically consistent and has a greater chance than LOOCV of selecting the best model with more accurate prediction ability. The main contribution of our proposed work, named Temporal Validated Combination (TVC), is in the use of MCCV for the selection of the best models for combining in different prediction windows.

Following the philosophy of stacking, the used models have a different base. The base models are two: "Autoregressive Integrated Moving Average" (ARIMA) [4] and "Non-linear Autoregressive with eXogenous inputs" (NARX) [21]. The involved models were built changing the main parameters of the ARIMA and NARX base models.

In order to asses TVC reliability, we compared TVC with the combination of the N-best models. The time series set used for this assessment is the reduced set of the benchmark provided by the international forecasting tournament NN5 [7]. The maximum Lyapunov exponent [28] of every single time series of the NN5 reduced set is positive, showing that these are chaotic time series. In addition, the experiments were executed with three synthetic time series, generated using an ARMA model [4], a Mackey–Glass equation [22] and a sine function.

Our results showed that the proposed combination has on average a better accuracy than the N-best combination using few models; it was also found that both methods tend to have an equal accuracy when the number of models increases. This is because with several models the two methods combine the same models.

This chapter is organized as follows: Sect. 2 describes fundamental concepts; Sect. 3 describes the base models ARIMA and NARX. The "Temporal Validated Combination" is shown in Sect. 4. Section 5 presents experiments comparing TVC with the N-best combination. Finally, Sect. 6 presents some conclusions and future work.

2 Fundamental Concepts

2.1 Time Series Forecasting

A time series is a set of n 1 observations $y_t \in Y$, each one being recorded at a specified time t [5]. Observations are made at fixed time intervals. A sequence of h future values can be estimated based on a set of observations of Y, where h is the size of the prediction horizon. A general expression of multi-step ahead prediction may be described as:

$$\{y_{t+1}, y_{t+2}, \ldots, y_{t+h}\} = F(y_t, y_{t-1}, \ldots, y_{t-m+1}) \tag{1}$$

where F predicts h future values using m values of the past. There are some forms to calculate h values of Y; a strategy used to forecast many real-world time series, which was implemented in this work, is the recursive (also called iterated or multi-stage) strategy [31]. In this strategy, a single model f is trained to perform a one-step ahead forecast as:

$$y_{t+1} = f(y_t, y_{t-1}, \ldots, y_{t-m+1}) \tag{2}$$

For forecasting many steps ahead, the model f forecasts the first step y_{t+1}. Subsequently, the value just forecasted is used as part of the input of f for estimating the next step value:

$$y_{t+2} = f(y_{t+1}, y_t, y_{t-1}, \ldots, y_{t-m+2}) \tag{3}$$

The process continues until the last value of the horizon reached.

$$y_{t+h} = f(y_{t+h-1}, y_{t+h-2}, \ldots, y_{t-m+h}) \tag{4}$$

Uncertainty increases the difficulty of estimating several future values in a time series. A reason for the potential inaccuracy when recursive prediction is used is the accumulation of errors within the forecasting horizon. Errors present in intermediate forecasts will propagate forward as these forecasts are used to determine subsequent values [31]. However, a small improvement in the long-term forecasting can imply a big improvement in the benefits in some domains like planning [24].

2.2 Chaotic Time Series

The theory of chaos deals with complex nonlinear systems; this theory had its breakthrough in the late 1800s, when Poincaré [26] addressed the stability of the solar system and the position of the planets. Abarbanel and Gollub [1] proposed that: "chaos is the deterministic evolution of a nonlinear system which is between regular behavior and stochastic behavior or 'noise'". Dhanya and Kumar [10] summarize the features of a chaotic system as: (1) they are deterministic, i.e., there are some determining equations ruling their behavior; (2) they are sensitive to initial conditions, that is, a slight change in the starting point can lead to significantly different outcomes; (3) they are neither random nor disorderly. Chaotic systems do have a sense of order and pattern, even though they do not repeat. In this work, a chaotic time series is the outcome of a chaotic system.

Chaos theory is useful for analyzing time series. It is well known that a time series has a chaotic behavior if it has a maximum Lyapunov exponent greater than zero [28]. On the other hand, a critical factor in predicting time series is to determine the number of past values m required on the forecasting. Chaos theory offers interesting ideas for finding suitable values for this regard. For example, the Cao method [6], which is an extension of the false nearest neighborhoods method [18], can help to define ranges of these m values.

2.3 Symmetric Mean Absolute Percentage Error (SMAPE)

In this chapter, the quality of a model is evaluated according to its ability of prediction. There are several metrics for measuring the expected ability of prediction. The international tournaments of forecasting [7] and many other works (for example [20]) use the SMAPE function. SMAPE generates values from 0 to 200; value 0 means that the obtained prediction matches exactly with the expected output; the worst possible prediction implies a value of 200. The SMAPE function measures the expected accuracy of a model and it is given by [2]:

$$SMAPE(Y, \hat{Y}) = \frac{1}{h} \sum_{t=1}^{h} \frac{|y_t - \hat{y}_t|}{\frac{1}{2}(|y_t| + |\hat{y}_t|)} 100 \qquad (5)$$

where Y is the expected time series, $y_t \in Y$ is the expected output at time t, $\hat{Y}$ is the sequence obtained by the model, $\hat{y}_t \in \hat{Y}$ is the model prediction at time t, and h is the prediction horizon.

2.4 Monte Carlo Cross Validation (MCCV)

Cross-validation (CV) is a method commonly used to validate models. It is more attractive than other methods, since it gives a statistical estimation of the expected prediction ability [19]. CV requires splitting the samples into two parts: a training set and a validation set. Notice that this division must maintain the order of observations when CV is applied in a time series prediction model. In other words, training and validating sets contain consecutive values. The CV follows these steps: first, fitting the model with the training set; second, obtaining a prediction of h steps ahead; and third, evaluating the prediction obtained $\hat{Y}$, comparing it with the expected values of Y which are known.

Monte Carlo cross validation (MCCV) [25] is a simple and effective procedure [34]. As it occurs with all cross-validation procedures, a sequence of observations is required for training and validating a model. In this work, the training process uses a set r plus a random number of observations. The maximum number of random observations is $v = n - r - h$, where n is the total number of observations. After training, the model predicts h values. SMAPE evaluates the results of next h values predicted comparing them with the expected values. This process is repeated k times, as shown in Fig. 1. Finally, the SMAPE average of all iterations composes the MCCV. In a previous work [13], we found that the MCCV obtained good results in the task of selecting the best NARX model for time series forecasting.

2.5 Stacking Method

The aim of the stacking method is to minimize the generalization error rate of one or more classifiers [33]. The N used models are generated with different learning algorithms $L_1,\ldots,L_N$ in a unique data set S, which consists of n samples $s_i = (x_i, y_i)$ that are pairs of feature vectors x_i and their classifications y_i. In the first phase, a set of base-level classifiers $C_1, C_2,\ldots,C_N$ is generated, where $C_i = L_i(S)$. In the second

Fig. 1 Monte Carlo cross
validation that takes a
window of random size for
training. The process iterates
k times [13]

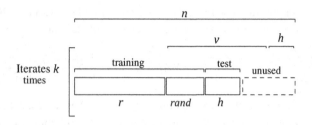

phase, a meta-level classifier is trained for combining the outputs of the base-level
classifiers [12]; for this reason, this is a meta-learning method.

A leave-one-out cross validation (LOOCV) generates a training set of meta-
level. The LOCCV applies each of the base-level learning algorithms to almost the
entire dataset, leaving one example for testing:

$$C_k^i = L_k(S - s_i) \mid \forall i = 1, \ldots, n, \ \forall k = 1, \ldots, N \tag{6}$$

Then, the learned classifiers generate predictions for s_i; the output of the
k-classifier with the i sample is represented as $\hat{y}_i^k = C_i^k(x_i)$. The meta-level dataset
consists of examples of the form $((\hat{y}_i^1, \ldots, \hat{y}_i^n), y_i)$, where the features are the
predictions of the base-level classifiers and the class is the correct class of the
example at hand. Finally, a meta-model is trained for combining the base-level
models.

3 Base Models

The base models trained with different parameters generate several new models.
For the experiments reported here, we used the ARIMA model and the NARX
neural networks as base models. These methods are described in the next sections.

3.1 ARIMA Model

An autoregressive moving average model (ARMA) expresses the conditional mean
of y_t as a function of both past observations $y_{t-1}, \ldots, y_{t-p}$ and past errors
$\varepsilon_{t-1}, \ldots, \varepsilon_{t-q}$. The number of past observations p required is the AR degree. The
number of past error q is the MA degree. In general, these models are denoted by
ARMA (p, q) [4]. The general form of the ARMA (p, q) is:

$$y_t = \phi_1 y_{t-1} + \ldots + \phi_p y_{t-p} + \varepsilon_t + \theta_1 \varepsilon_{t-1} + \ldots + \theta_q \varepsilon_{t-q} \tag{7}$$

where ε_t is an uncorrelated innovation process with mean zero. The values of
constants ϕ_i and θ_i are determined from the data.

The autoregressive integrated moving average (ARIMA) model can be generated adding or integrating the ARMA model d times. In this work the ARIMA models are used due to the good results that they have shown to obtain for forecasting. A detailed description of ARIMA model can be found in [4].

3.2 NARX Networks

An important class of discrete-time non-linear systems is the Non-linear Auto-Regressive with eXogenous Inputs (NARX) model [21]. This may be defined as [30]:

$$y_t = f\left(u_t, u_{t-1}, \ldots, u_{t-n_u}, y_{t-1}, y_{t-2}, \ldots, y_{t-n_y}\right) \tag{8}$$

where u_t and y_t represent the input and output of the network at time t, n_u and n_y are the input and output order, and the function f is non-linear. When the function f can be approximated by a feed-forward neural network [15], the resulting system is called a NARX network.

In this work, we deal with prediction using only past values of the time series, without any exogenous variables involved. Figure 2 shows the NARX network structure used in this research. The first component is a tapped delay line (TDL) [3] where the input time series enters from the left and passes through m delays. The output of the TDL is an m-dimensional vector, made up of the input time series at the current time, and m previous observations [3]. The next block is the hidden layer with several neurons. The right block is the output layer with only one neuron. The weights associated with the neurons are learned in the training phase. After the NARX network is trained, the output is fed back to the input of the feed-forward neural network as part of the standard NARX architecture. The model iterates until it reaches h predictions.

NARX had been evaluated in a theoretical and empirical way, with good results that guarantee the effectiveness of NARX models in time series forecasting [8, 11]. A key component is the learning algorithm used in the NARX model [14]. The most favorable behavior of NARX networks depends on its parameters: the dimension of input and output layers and the number of neurons in the hidden layer. The determination of these architectural elements, which is a selection problem, is a critical and difficult task for using NARX networks [11].

4 Temporal Validated Combination (TVC)

The desired prediction horizon can be divided into three prediction windows, each one with the same number of elements, named short-term, medium-term and long-term. The basic idea is to combine models with the best performance in each

Fig. 2 NARX network structure with tapped delay line (*TDL*) of *m* neurons, hidden layer with several neurons, and one neuron in the output layer. Image based on [3]

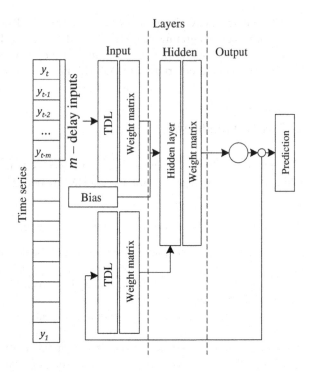

prediction window. This idea was tested in a previous work [14] where only one model was used by each prediction window.

TVC follows some stacking principles. First, the models have different bases and a cross-validation process generates meta-features. Our goal in this work is to select the *N*-best models in each prediction window from performance features extracted with a Monte Carlo cross validation. Later, in a second phase a forecasting process (FP) combines the predictions of the selected models.

The selection process (SP) of the *N*-best models in each prediction window requires some preprocessing. The SP divides every test set of the MCCV iterations into three windows. For each model, the MCCV returns an expected SMAPE for every prediction window. The *N*-selected models for the short-term windows will have the smallest short-term values of expected SMAPEs from all models. A similar procedure is followed to select models of the medium and long-term windows. Finally, the SP results are three sets of models *S* (*N*-best in short-term), *M* (*N*-best in medium-term), *L* (*N*-best in long-term).

The FP starts training the selected models with the entire training set. Then, each selected prediction model (*j*) forecasts the entire horizon of size *h*; predictions are represented by $\{\hat{y}_i^j \mid 1 \leq i \leq h\}$. The windows short, medium and long term have the same size *b*. The FP takes a segment of size *b* from each prediction by window and model. Complete TVC prediction C_{TVC} is obtained by joining the predictions' averages of each prediction window. The binding is expressed in

Eq. 9 and Fig. 3 resumes the FP process. The prediction C_{TVC} is assessed by calculating SMAPE of the entire prediction horizon.

$$
C_{TVC} = \left\{ \begin{array}{l} \left\{ \dfrac{1}{N} \sum_{j \in S} \hat{y}_i^j, 1 \leq i < b \right\} \cup \left\{ \dfrac{1}{N} \sum_{j \in M} \hat{y}_i^j, b \leq i < 2b \right\} \cup \\[2ex] \left\{ \dfrac{1}{N} \sum_{j \in L} \hat{y}_i^j, 2b \leq i < 3b \right\} \end{array} \right\} \tag{9}
$$

5 Experimental Analysis

In this section, we compare the TVC with the combination of the N-best models. The N-best combination consists of selecting N-models with the better SMAPE of forecasting the entire prediction horizon. The combination of N-best models obtained good results with real datasets as the NN3 benchmark [14]. In this experiment, a set of time series was modeled using several prediction models. The base models were NARX neural networks and ARIMA models.

5.1 Data Description

Four types of time series were used to test our strategy. One of them, the NN5, comes from real samplings of drawbacks from teller machines in London; the reminding three are synthetic values generated by an ARMA model, a Mackey–Glass Equation [22] and a sine function. Next we explain in detail each one.

The first subset of time series comes from the NN5 prediction tournament, which can be downloaded from: http://www.neural-forecasting-competition.com/NN5/datasets.htm. We used the available reduced set, which consists of 11 time series. Each training sequence contains 735 observations, while the prediction horizon is composed of 56 future values for all series. The set of values to predict is called the test set. In order to assess the non-linearity of NN5 dataset, we calculated the maximum Lyapunov exponent of each time series. The exponent estimator used was the function defined by Rosenstein et al. [27] and provided by the TISEAN package [16], using the default parameters. Table 1 shows the maximum Lyapunov exponent of each time series. The first column shows the time series identifier; the second column presents its maximum Lyapunov exponent. All the time series from the NN5 reduced dataset have a Lyapunov exponent greater than 0, this means that these time series have a chaotic behavior. Therefore, traditional techniques of forecasting could not be applied.

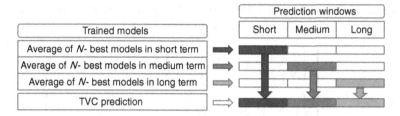

Fig. 3 The temporal validate combination of prediction models previously trained. The result is composed by the average of models' predictions in their different prediction windows

Table 1 Maximum Lyapunov exponents evaluated from the NN5 reduced set

Id	Max. Lyapunov exponent
1	0.15
2	0.14
3	0.14
4	0.15
5	0.14
6	0.14
7	0.14
8	0.15
9	0.14
10	0.16
11	0.14

For testing our model for big prediction horizons, we generated three synthetic time series using Mackey–Glass [22], an ARMA model [4] and a sine function. The Mackey–Glass values came from integrating the differential equation:

$$\frac{dx(t)}{dt} = \frac{ax(t - \tau)}{1 + x(t - \tau)^{10}} - bx(t) \tag{10}$$

This function has a chaotic behavior when: $a = 0.2$, $b = 0.1$, $\tau = 17$, $x_0 = 1.2$ and the time step size is 0.1. We generated 750 training values and 250 testing values. An ARMA(2, 1) model defined as:

$$y_t = 0.5y_{t-1} - 0.3y_{t-2} + \varepsilon_t + 0.2\varepsilon_{t-1} \tag{11}$$

where ε_t follows a Gaussian distribution with mean 0 and variance 0.1. This ARMA model was used to generate 550 values of a time series; the first 500 values form the training set and the last 50 the test set. A third series was generated using a sine function with a time step size of $2\pi/64$. This series contains 750 observations for training and 250 values for testing. All the time series were normalized in a magnitude between 0 and 1.

5.1.1 Prediction Models

We built several models with NARX and ARIMA base forms. First, different models with the same NARX base are trained with different parameters. Notice that, if a NARX is trained using different algorithms, their weight values will be different and consequently its performance may change. For that reason, this experiment considers the training algorithm as a parameter.

There are three parameters used to generate the NARX models: the number of delay neurons m (25, 30 and 35), the number of neurons in the hidden layer (20, 25 and 30) and the training algorithm. The training algorithms used were:

- Bayesian regulation back propagation (BP),
- Conjugate gradient BP with Fletcher-Reeves updates, and
- Leveberg-Marquardt (BP).

In total, we generated 27 models with NARX form.

The ARIMA models have four parameters: the number of autoregressive terms p (1, 2, and 3), the number of non-seasonal difference d (1, 2, and 3), the number of lagged forecast errors (1, 2, and 3) and the seasonality (7 and 12). In total, we generated 54 models with the ARIMA form, which comes from a total of 81 models used in these experiments.

5.2 Experiments Setup and Analysis of Results

We designed two kinds of experiments for evaluating the performance of TVC. The first experiment used a fixed number of models to combine, N equal to 3. This first experiment shows the forecasting accuracy. The second experiment presents the behavior of TVC when N varies. The experiments and their results are presented in the next subsections.

5.2.1 Evaluation with N Equal to 3

The objective of our first experiment was to test whether a TVC could perform better than the combination of the N-best models (N-best, for simplicity). The N-best selected the models with the best SMAPE in the complete prediction horizon. Later, in the prediction phase, N-best averaged the outputs of the selected models. We evaluated the two models' combinations with the test set. The error estimation function was SMAPE. The results are shown in Table 2. The first column indicates the identification (ID) of the series; the second column shows the SMAPE of the TVC and the third column is the SMAPE of the N-best strategy. The TVC used $N = 3$ for the number of models to combine by prediction window.

From the SMAPEs obtained for each time series, it can be seen that both combination methods have a good performance for a specific time series. The average of

Table 2 Comparison of the TVC and N-best combinations using a $N = 3$	ID	TVC SMAPE	N-best SMAPE
	NN5 1	17.88	18.43
	NN5 2	31.34	22.18
	NN5 3	25.78	46.93
	NN5 4	21.36	20.57
	NN5 5	23.99	23.03
	NN5 6	18.43	14.13
	NN5 7	36.31	26.06
	NN5 8	28.90	52.69
	NN5 9	12.76	12.13
	NN5 10	42.09	33.75
	NN5 11	19.82	17.99
	Mackey–Glass	2.47E−03	2.47E−03
	ARMA (2,1)	144.74	148.19
	Sine	4.37E−05	3.83E−07
	Average	*30.24 ± 33.77*	*31.15 ± 35.49*

SMAPEs indicates that TVC has better accuracy and less standard deviation than N-best method. On the other hand, the two models obtained the same SMAPE for the Mackey–Glass series. The Mackey–Glass time step is small, which turns on windows containing smooth portions of the time series, or with few peaks. This may be a reason why TVC selects the same models in the three prediction windows. Notice that both strategies obtained bad performances for the ARMA series. This is because this time series presents many peaks, even in some cases each value is a peak, which makes very difficult to predict. Finally, the sine function is non-lineal but this is not a chaotic time series. However, the prediction models had problems estimating several values even with this time series.

5.2.2 Evaluation Varying the Size of N

In this experiment we executed TVC and N-best varying the number of models N from 1 to 6, with the entire dataset. Figure 4 shows the average of SMAPEs obtained by both methods when N varies.

Notice that when the N is small, the SMAPE average value of TVC is shorter than the corresponding value of N-best. In order to analyze the behavior of both methods with large values of N, we repeated the experiment varying N from 1 to 25. Figure 5 shows the SMAPEs average of both methods for these values. Here we can observe that, when N has small values, TVC has a better SMAPE than N-best. On the other hand, when N has large values, TVC and N-best tend to have the same SMAPE average; this is because the models combined in the prediction windows tend to be the same. Additionally, we can observe that the best possible result is obtained by N-best using 16 models. The TVC has its best performance

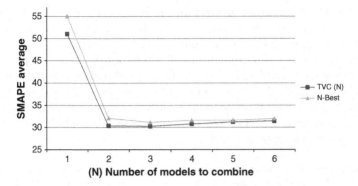

Fig. 4 SMAPEs average of TVC and N-best when N varies from 1 to 6

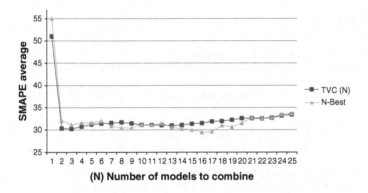

Fig. 5 SMAPEs average of TVC and N-best when N varies from 1 to 6

combining 3 models. In addition, TVC tends to have a more regular behavior than N-best when N varies. N-best has an irregular behavior because it does not use a validation process.

6 Conclusions and Future Work

This work presents a method, called Temporal Validate Combination (TVC), for selecting and combining prediction models based on meta-features. For the experiments presented here, TVC has on average better accuracy than the combination of the N-best models when the number of models is small. For large values of N, both combination methods tend to have the same performances. TVC shows a regular behavior when N varies, because this method estimates each model performance with a Monte Carlo cross validation. Therefore, we can use TVC with small values of N; this has less computational cost that search several values of N. On the other hand, when the prediction horizon is long, the models would have lower accuracy. A long prediction horizon can be determined by the time step of the time series and the number of future values.

For future work, we will explore if the TVC performance would improve if the Monte Carlo cross validation would estimate the combination accuracy, instead of the independent models' accuracy. Other field to be analyzed next is the use of chaotic theory for the construction of the models. Several real work time series like NN5 have a chaotic behavior, and then chaotic theory could help to develop forecasting methods more accurately.

Acknowledgments R. Fonseca thanks the National Council of Science and Technology (CONACYT), México, for a scholarship granted to him, No. 234540. This research has been partially supported by CONACYT, project grant No. CB-2010-155250.

References

1. Abarbanel, H.D.I., Gollub, J.P.: Analysis of observed chaotic data. Phys. Today **49**, 86 (1996)
2. Armstrong, J.S.: Long-Range Forecasting from Crystal Ball to Computer, 2nd edn. Wiley, New York (1985)
3. Beale, M.H., Hagan, M.T., Demuth, H.B.: Neural Network Toolbox User's Guide R2012b: MathWorks (2012)
4. Box, G.E.P., Jenkins, G.M., Reinsel, G.C.: Time Series Analysis Forecasting and Control, 3rd edn. In: Jerome Grant, (ed.) Prentice-Hall International, Upper Saddle River (1994)
5. Brockwell, P.J., Davis, R.A.: Time Series: Theory and Methods, 2nd edn. Springer, New York (2006)
6. Cao, L.: Practical method for determining the minimum embedding dimension of a scalar time series. Phys. D: Nonlinear Phenom. **110**(1–2), 43–50 (1997)
7. Crone, S.F.: Competition instructions (Online). http://www.neural-forecasting-competition.com/instructions.htm (2008, Feb)
8. Crone, S.F., Hibon, M., Nikolopoulos, K.: Advances in forecasting with neural networks? Empirical evidence from the NN3 competition on time series prediction. Int. J. Forecast. **27**(3), 635–660 (2011)
9. De Gooijer, J.G., Hyndman, R.J.: 25 years of time series forecasting. Int. J. Forecast. **22**(3), 443–473 (2006). Twenty five years of forecasting
10. Dhanya, C.T., Nagesh Kumar, D.: Nonlinear ensemble prediction of chaotic daily rainfall. Adv. Water Res. **33**(3), 327–347 (2010)
11. Diaconescu, E.: The use of NARX neural networks to predict chaotic time series. WSEAS Trans. Comp. Res. **3**(3), 182–191 (2008)
12. Dzeroski, S., Zenko, B.: Is combining classifiers with stacking better than selecting the best one? Mach. Learn. **54**, 255–273 (2004)
13. Fonseca-Delgado, F., Gómez-Gil, P.: An assessment of ten-fold and Monte Carlo cross validations for time series forecasting. In: 10th International Conference on Electrical Engineering, Computing Science and Automatic Control. Mexico (2013)
14. Fonseca-Delgado, R., Gómez-Gil, P.: Temporal self-organized meta-learning for predicting chaotic time series. In: 5th Mexican Conference on Pattern Recognition. Queretaro (2013)
15. Haykin, S.: Neural Networks A Comprehensive Foundation, 2nd edn. Pearson Prentice Hall, New York (1999)
16. Hegger, R., Kantz, H., Schreiber, T.: Practical implementation of nonlinear time series methods: The TISEAN package. Chaos: Interdiscip. J. Nonlinear Sci. **9**(2), 413–435 (1999)

17. Kantz, H., Schreiber, T.: Nonlinear Time Series Analysis. Cambridge University Press, Cambridge (2003)
18. Kennel, M.B., Brown, R., Abarbanel, H.D.I.: Determining embedding dimension for phase-space reconstruction using a geometrical construction. Phys. Rev. A **45**, 3403–3411 (1992)
19. Kohavi, R., John, G.H.: Wrappers for feature subset selection. Artif. Intell. **97**(1–2), 273–324 (1997). Relevance
20. Lemke, C., Gabrys, B.: Meta-learning for time series forecasting and forecast combination. Neurocomputing **73**(10–12), 2006–2016 (2010)
21. Leontaritis, I.J., Billings, S.A.: Input-output parametric models for non-linear systems Part II: Stochastic non-linear systems. Int. J. Control **41**(2), 329–344 (1985)
22. Mackey, M.C., Glass, L.: Oscillation and chaos in physiological control systems. Science **197**(4300), 287–289 (1977)
23. Makridakis, S., Hibon, M.: The M3-competition: Results, conclusions and implications. Int. J. Forecast. **16**(4), 451–476 (2000). The M3-Competition
24. Matijas, M., Suykens, J.A.K., Krajcar, S.: Load forecasting using a multivariate meta-learning system. Expert Syst. Appl. **40**(11), 4427–4437 (2013)
25. Picard, R.R., Cook, R.D.: Cross-validation of regression models. J. Am. Stat. Assoc. **79**(387), 575–583 (1984)
26. Poincare, H.: Memoire sur les courbes definies par une equation differentielle. Resal J. **3**, VII. 375–422 (1881)
27. Rosenstein, M.T., Collins, J.J., De Luca, C.J.: A practical method for calculating largest Lyapunov exponents from small data sets. Phys. D **65**(1–2), 117–134 (1993)
28. Sandri, M.: Numerical calculation of Lyapunov exponents. Math. J. **6**(3), 78–84 (1996)
29. Shao, J.: Linear model selection by cross-validation. J. Am. Stat. Assoc. **88**(422), 486–494 (1993)
30. Siegelmann, H.T., Horne, B.G., Giles, C.L.: Computational capabilities of recurrent NARX neural networks. Syst. Man Cybern. B Cybern. IEEE Trans. **27**(2), 208–215 (1997)
31. Taieb, S.B., Bontempi, G., Atiya, A.F., Sorjamaa, A.: A review and comparison of strategies for multi-step ahead time series forecasting based on the NN5 forecasting competition. Expert Syst. Appl. **39**(8), 7067–7083 (2012)
32. Wang, X., Smith-Miles, K., Hyndman, R.: Rule induction for forecasting method selection: Meta-learning the characteristics of univariate time series. Neurocomputing **72**(10–12), 2581–2594 (2009)
33. Wolpert, D.H.: Stacked generalization. Neural Netw. **5**(2), 241–259 (1992)
34. Xu, Q.S., Liang, Y.Z., Du, Y.P.: Monte Carlo cross-validation for selecting a model and estimating the prediction error in multivariate calibration. J. Chemom. **18**(2), 112–120 (2004)
35. Yang, Z.R., Lu, W., Harrison, R.G.: Evolving stacked time series predictors with multiple window scales and sampling gaps. Neural Process. Lett. **13**(3), 203–211 (2001)

MLP for Electroencephalographic Signals Classification Using Different Adaptive Learning Algorithm

Roberto Sepúlveda, Oscar Montiel, Daniel Gutiérrez, Gerardo Díaz
and Oscar Castillo

Abstract For the identification of muscular pain caused by a puncture in the right arm and eye blink, electroencephalographic (EEG) signals are analyzed in the frequency and temporal domain. EEG activity was recorded from 15 subjects in range of 23–25 years of age, while pain is induced and during blinking. On the other hand, EEG was converted from time to frequency domain using the Fast Fourier Transform (FFT) for being classified by an Artificial Neural Network (ANN). Experimental results in the frequency and time domain using five adaptation algorithms show that both neural network architecture proposals for classification produce successful results.

1 Introduction

The human brain is a complex network of synaptic connections between neurons, which generate the electric impulses necessaries to develop human functions like movements, communication, language, feelings, memory, reasoning, etc. These

R. Sepúlveda (✉) · O. Montiel · D. Gutiérrez · G. Díaz
Instituto Politécnico Nacional, Centro de Investigación y Desarrollo de Tecnología Digital
(CITEDI-IPN), Av. Del Parque No. 1310, Mesa de Otay 22510, Tijuana, B.C., Mexico
e-mail: rsepulveda@ipn.mx; rsepulve@citedi.mx

O. Montiel
e-mail: oross@ipn.mx

D. Gutiérrez
e-mail: dgutierrez@citedi.mx

G. Díaz
e-mail: gdiaz@citedi.mx

O. Castillo
Tijuana Institute of Technology, Calzada Tecnológico s/n,
Fracc. Tomas Aquino CP 22379, Tijuana, B.C., Mexico
e-mail: ocastillo@tectijuana.mx

O. Castillo et al. (eds.), *Recent Advances on Hybrid Approaches for Designing*
Intelligent Systems, Studies in Computational Intelligence 547,
DOI: 10.1007/978-3-319-05170-3_25, © Springer International Publishing Switzerland 2014

functions are represented by EEG signals while it is working [1]. Since the technology has allowed reading the EEG signals in humans, it was thought for interpreting and using them as communication channels with auxiliary devices that can help people with mental and physical problems [2].

For reading and interpret the EEG signals, devices like BCIs are used; these electrical signals generated by the brain are generated by a stimulus, a physical action (motion) or mental status as feeling, imagination, memory, etc.

In the last 15 years, the field of research and generated products obtained from BCI systems has increased considerably. In 1995, there were no more than six research groups in this kind of systems; nowadays, thanks to the reduction of costs of neuro head sets using external electrodes to acquire EEG signals by a non-invasive way, the existence of open-sources of software and databases focalized to make the experiment and research of EEG signals easier, the number of research groups and also the independent people doing research has increased considerably. In consequence, cognitive neuroscience and brain imaging technologies have expanded the ability to interact directly with the human brain. Considering the research field of these areas, it is becoming more important to focus on technological applications for the human health care [3–5]. In the same way, the use of engineering tools in medical research allows to extend the life of people, falling this to detect a variety of common diseases related to the human brain [1]. In fact, research centers such as "The Laboratory of Neural Injury and Repair" from Dr. Jon Wolpaw at the Wadsworth Center in New York, the "Advanced Technology Research Institute", in Japan, the "Schalklab" from Dr. Gerwin Schalk at the Wadsworth Center in New York, and more have invested in the BCI technologies [6–8].

The use of BCI interfaces and its applications is a reachable way of stability for people with motor and sensory disabilities. These applications obtained from development of neural interfaces are the base of artificial systems for motor controllers, proprioception, neurogenesis and nervous system repair when are combined with disciplines as tissue engineering and gene therapy.

Nowadays, few classification techniques used in the study and research of EEG signals are the statistical techniques, ANNs, Adaptive Neuro-Fuzzy Inference System, Hidden Markov Models and Support Vector Machines [9–13].

The EEG classification developed in this chapter is based on an ANN with backpropagation algorithm for training, it is used to search patterns of two external stimulus, muscle pain and blinking, in EEG signals. The ANN is one of the most common machine learning techniques used to recognition and classification of patterns; its functioning is based on communication between brain neurons through synaptic connections.

2 Overview

2.1 EEG Concepts

The EEG is one of the most common techniques used to analyze the activity in the human brain. The brain is a source of electrical signals that can be acquired and analyzed by an array of electrodes positioned specifically in certain sections of the scalp according to the international 10–20 system of electrode position; these signals detected by the electrodes are transformed in patterns, which are the base of the datasets of study [14].

The electrode position is identified by two parameters shown in Fig. 1, the letter and number which describe the lobule and the cerebral hemisphere, respectively. Where the F, T, C, P and O letters, corresponds to frontal, temporal, central, parietal and occipital zone. It is noteworthy that the central lobe identified with the letter "C" does not exist; however, it is used as a reference point; and the letter "Z" is used to identify an electrode located in an intermediate line. To identify the right hemisphere, the even numbers 2, 4, 6, 8 are used, whereas the odd numbers 1, 3, 5, and 7 identifies the left hemisphere [14].

It is important to know that the brain is divided by sections, two hemispheres (left and right) and four lobules (frontal, parietal, temporal and occipital), where each section of it is responsible on a specific part of the human body. For example, if we want to analyze a stimulus on the left side of the body, then the right hemisphere has to be analyzed; and viceversa, Fig. 2.

From another point of view, EEG technique is used to quantify the neuronal activity by placing electrodes on the scalp. Within the necessary characteristics for EEG are frequency, voltage, location, morphology, polarity, state, reactivity, symmetry and artifacts involved [14]. The frequency of brain activity oscillates between 30 Hz and is divided in four principals bands of frequency; delta (0.4 to 4 Hz), theta (4 to 8 Hz), alpha (8 to 13 Hz) and beta (13 to 30 Hz); being considered as an artifact any signal over 30 Hz of frequency.

The human–computer interfaces (HCI) are of great help in everyday life; some of these examples are: computer mouse, keyboard, digital pen, etc.; however, there is an area where you need to make a direct communication between the brain and the computer, this technique is known as brain computer interface (BCI) which is a communication method based in the neuronal activity generated by the brain and independent of the output pathways of peripheral nerves and muscles.

Figure 3 describes in general form, the basic functional scheme of a BCI system, we show on the left side an acquisition system that senses EEG signals through electrodes connected to the scalp, cortical surface or inside the brain; the acquired data are processed to a step of feature extraction needed to identify emotions, sensations or user intentions. The characteristics of the EEG signals are converted to interpretable commands by a computer to perform some action. Neuronal activity used in the BCI can be registered by invasive or noninvasive techniques [15]. One advantage of BCI's is that provide noninvasive communication channels

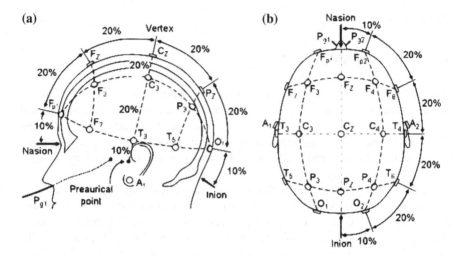

Fig. 1 International 10–20 system standard for EEG electrodes positions on the scalp **a** left side, and **b** above the head [6]

Fig. 2 Interpretation brain action

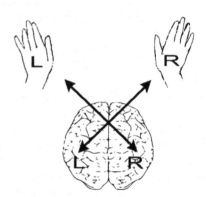

independent from peripheral nerves and muscles for people with motor disabilities [2].

A BCI system is divided by sections, in the first section the electric activity generated by the brain and the interface to acquire that brain activity is shown. The brain activity has specific characteristics like the time, frequency, amplitude, magnitude and kurtosis; some of these characteristics can be observed in time domain, other ones have to be converted to frequency domain using different methods as Fourier and Wavelet transform. In the second section, the signal processing is achieved using a computer which is responsible of processing data using methods as correlation, ANN, Adaptive Neuro-Fuzzy Inference Systems, etc. The third section consists of the implementation, where a variety of applications like the wheel chair controlled by mind, health assistant, rehabilitation, epilepsy detection, prosthesis of body extremities can be developed [13, 16–18].

2.2 Artificial Neural Networks Theory

The ANN algorithm is a method commonly used to classify patterns of data, has the peculiarity to adapt itself according to the information provided by the user. The knowledge is created with the first set of information and new input patterns modified with different adaptive algorithms. The method has the ability to focus on the characteristics of an arbitrary input that resembles other previously seen [9].

Algorithm 1. Pseudocode for implementing the backpropagation algorithm

Initialize all weights with small random numbers, typically between -1 and 1

repeat

for every pattern in the training set

Present the pattern to the network

//   Propagated the input forward through the network:
   for each layer in the network
     for every node in the layer
       1. Calculate the weight sum of the inputs to the node
       2. Add the threshold to the sum
       3. Calculate the activation for the node
     end
   end

//   Propagate the errors backward through the network
   for every node in the output layer
     calculate the error signal
   end

   for all hidden layers
     for every node in the layer
       1. Calculate the node's signal error
  2. Update each node's weight in the network
     end
   end

//   Calculate Global Error
   Calculate the Error Function

end

while ((maximum number of iterations < than specified) AND
    (Error Function is > than specified))

Backpropagation Algorithm could be used in two modes of training; the first one, called *supervised*, is defined as a type of network that learns with a teacher cataloged as the expert of the problem to solve and a pair of inputs and outputs used as dataset for training phase. The second mode called *no supervised*, learns with its own knowledge, and it does not need an expert (teacher), which means that the neural network establishes an association between neurons and its synaptic connections with new input data [19, 20].

Fig. 3 Scheme of a brain
computer interface (*BCI*)
system

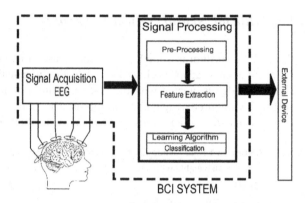

Supervised training mode was used in the experiments developed in this chapter. The methodology consists in a teacher presenting the training dataset as input vectors to the network, using pattern by pattern training. This learning is performed by adjusting the parameters of the network through the influence of the training dataset and the error of the output signal. The error in the output signal is the difference between the actual network output and the expected output.

The pseudocode for implementing Backpropagation Algorithm is shown in **Algorithm 1** [9].

3 Proposed Methodology

The experimental platform used is based on the basic block diagram described in Fig. 3.

1. Signal Acquisition. It is done by "Emotiv EEG neuroheadset" which is a multi-channel high resolution acquisition of EEG signals, has 14-bit resolution (1 LSB = 0.51 µV), signals are digitized at 128 samples/s, the bandwidth of the headband is 0.2–45 Hz, it has a digital band reject filter with high-quality factor for frequencies between 50 and 60 Hz. The neuroheadset has 14 channels that are set of international standard 10–20, which are: AF3, F7, F3, FC5, T7, P7, O1, O2, P8, T8, FC6, F4, F8, and AF4. It also has the CMS/DRL references, usually they are set in P3/P4 positions. Epoc neuroheadset generates a file with ".Edf" format, where the readings taken are saved. Communication to the computer is done using wireless communication. The Edf format is compatible with the free distribution of EEGLAB toolbox.
2. Signal Processing. At this stage, the interpretation of EEG signal is realized through Matlab EEGLAB Toolbox, which is used for analysis of EEG signals in OFF-LINE mode. EEGLAB is a simulator that allows to observe

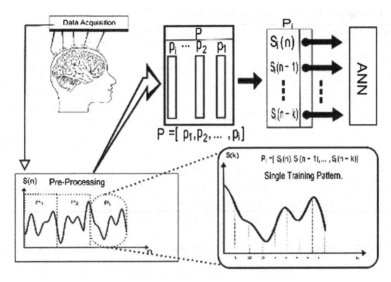

Fig. 4 Data input management for ANN training EEG signals classification

some properties of the EEG signals, like FFT, ICA, dipolar sources, ERP, etc. Once the interpretations of the signals were done, we proceed to pass EEG information to matrix format. This matrix contains the EEG signal sampled by the neuroheadset and is treated in Excel format.

3. Classification. Finally, the matrix representation of readings is performed to identifying patterns; this is done by preprocessing the signal and feature extraction, and then patterns found are classified by ANN using the Matlab toolbox "Neural Networks".

The EEG classification is achieved on Matlab, where matrix data read from Epoc neuroheadset is imported. Once the data is ready, the next step is to select patterns based on teacher heuristics. All EEG signals are divided into patterns that make up the batch of data to be presented to the ANN. The batch of patterns used to train the network is created as is shown in Fig. 4.

Two cases of study of classification of EEG signals are presented in this work; we are using ANN backpropagation multilayer perceptron type with hyperbolic tangent activation functions and supervised learning [19]. The output of the ANN returns values between 1 and −1, to make the classification possible a threshold of 0.9 is applied. If the output is 1 means that the ANN classifies it correctly as a stimulus, in contrast a −1 means not classified.

Table 1 shows the acronyms used to identify the adaptive algorithms.

In section of ANN training for each stimulus, the parameters for the two types of architectures are standardized, setting the same training characteristics except for inputs, the reason is the sample window established. While blinking signal

Table 1 Adaptive algorithms used for experiments

Acronym	Algorithm
LM	Levenberg–Marquardt
OSS	One-step secant
BFGS	Quasi-Newton
RP	Resilient backpropagation
GD	Gradient descendent

Table 2 Training characteristics for stimulus classification

Stimulus	ANN architectures	Error training	Epoch training	Adaptive algorithm	Activation function
Blinking	120:20:10:5:1	1e−03	100,000	LM, OSS, BFGS, RP, GD	Tansig
MusclePainArm	1280:20:10:5:1	1e−03	100,000	LM, OSS, BFGS, RP, GD	Tansig

needs 120 samples, the arm pain signal has 1,280 samples. Table 2 shows the established parameters.

3.1 Blinking Classification

The first stimulus to identify is the blinking activity, although a blinking signal is considered an artifact, its classification is important because it may interfere with the EEG results and their interpretation, also it is a natural movement of the body. In Fig. 5, a scheme of blinking dataset acquisition, processing and classification is shown.

The experiment consists of recording the blinking of a person each second, a simple clock alerts the user when to blink. The EEG dataset was taken from AF3 channel.

3.2 Pain Muscle

The second stimulus to classify is the pain induced by an external agent, this stimulus is induced by a prick in the right arm; the subject is in a relax status by 2 min approximately and then pain is induced, the EEG signal gotten after the prick is the most important to use in the ANN training and testing.

The EEG signal obtained from the pain does not present appreciable variations in the time domain, so the signal has to be converted to frequency domain using

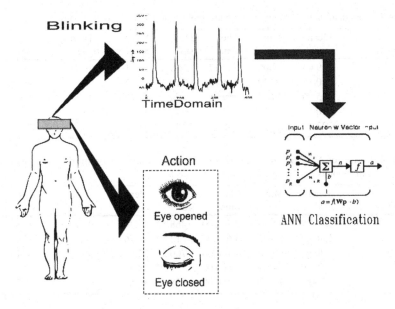

Fig. 5 Implementing blinking signal processing and classification

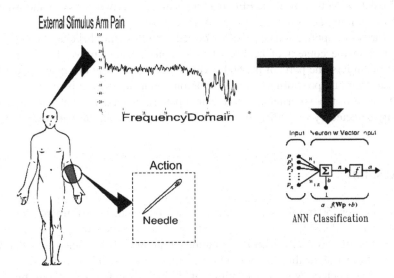

Fig. 6 Scheme process of muscle main for implementing EEG signal processing and classification

Table 3 Statistical values of time in seconds about training ANN for Blinking

Adaptive algorithm	Mean time	Min time	Max time	Desv. std	Mean epoch	Train set (%)	Test set (%)
LM	9.43	2.05	28.17	5.89	11	98.7	93.4
OSS	0.83	0.49	1.70	0.26	27	99.3	94.1
BFGS	145.96	59.12	269.35	60.32	19	99.3	92.9
RP	0.50	0.36	2.79	0.43	27	99.5	96.4
GD	22.43	10.17	55.14	10.19	3,825	99.7	88.7

Table 4 Results of ANN training statistics of adaptive learning algorithm of pain in right arm, the time is in seconds

Adaptive algorithm	Mean time	Min time	Max time	Desv. std	Mean epoch	Train set (%)	Test set (%)
LM	1,167.00	420.00	2,700.00	468.52	33,344	99.7	96.8
OSS	1,864.43	840.00	5,687.00	1,012.52	36,462	96.5	95.4
BFGS	1,403.60	600.00	4,140.00	706.33	33,677	97.3	93.1
RP	1,568.00	840.00	4,800.00	771.39	48,982	95.3	92.9
GD	1,290	720.00	4,140.00	728.86	39,677	98.7	96.7

the Fast Fourier Transform, and then a filter is applied to eliminate the signal noise, as it is seen on Fig. 6.

The dataset gotten from the EEG reading is divided in two parts of patterns, one used for training and other to test the ANN created.

A variety of applications can be performed with this stimulus classification, for example, in the context of health, when a subject cannot move, but he is still feeling pain in some parts of his body. This is the reason why this chapter focuses on analyzing and providing more information about the muscle pain.

The goal of the experiment is to create a pattern or model of behavior in brain activity detected by EEG that could be used to identify a similar pain in a subject.

4 Results

In the first step, 50 trainings were conducted to observe their performance with the adaptive algorithms. In Table 3, the results of blink classification corresponding to 50 patterns are shown. As seen RP got a meantime of 0.50 s which means that RP was the fastest algorithm to train. On the other hand, BFGS obtained the worst training parameters with a meantime of 145.96 s, and a maximum time of 269.35 s. It is important to note that GD performed 3,825 epochs to train, but the time is really good if it is compared it with BFGS.

For arm pain classification presented in Table 4, the LM algorithm shows good results with 96.8 % contrary to the RP that got 92.9 % of reliability; in time terms also the LM was the best with a mean time of 1167 s, not so good was the OSS with 1,864.43 s. The results of the blinking stimulus presented in Table 3 shows that the RP obtain a classification of 96.4 % from the 50 EEG readings used to test the ANN, in contrast, the GD obtained an 88.7 % of classification. This algorithm behavior is probably due to the difference in number of samples used for analyzing each stimulus; for muscle pain were 1,280 and 120 samples for blinking.

5 Conclusions

It is known that the use of ANN is a powerful and an efficient tool for patterns classification, so in this chapter, it was used to classify EEG signals. There is not enough research work focused to efficient detection pain from external agents, hence this work provides information in this regard, a methodology for classification is presented. Moreover, ANN architectures multilayer perceptron types are used to achieve classification of muscle pain and ocular artifact known as Blink.

For Blinking detection, the adaptive learning algorithm Gradient Descendent was the best in recognizing; however, it needs to iterate more training epochs. On the other hand, the RP adaptive learning algorithm demonstrates more power to achieve generalization, and average training epochs is low. It can be concluded that RP is convenient to perform this classification.

For muscle pain detection, Levenberg–Marquardt shows the best qualities to classify training and the testing data, with success percentage of 99.8 %. In general, the obtained results can be improved positively simply by moving the threshold 0.9 to a lower value, which requires a clinic study and represents a possible future work.

References

1. Arias, G., Felipe, H.: Detección y clasificación de artefactos en señales EEG. In: Memorias de STSIVA'09. Universidad Tecnológica De Pereira (2009)
2. Chen, A., Rappelsberger, P.: Brain and human pain: topograph EEG amplitude and coherence mapping. Brain Topogr. 7(2), 129–140 (1994)
3. Morchen, F.: Time Series Feature Extraction for Data Mining Using DWT and DFT. Philipps University Marburg, Marburg (2003)
4. Papadourakis, G., Vourkas, M., Micheloyannis, S., Jervis, B.: Use of artificial neural networks for clinical diagnosis. Math. Comput. Simul. 1996(40), 623–635 (1996)
5. Pera, D.L., Svensson, P., Valeriani, M., Watanabe, I., Arendt-Nielsen, L., Chen, A.C.: Long-lasting effect evoked by tonic muscle pain on parietal EEG activity in humans. Clin. Neurophysiol. 111(12), 2130–2137 (2000)
6. Sharbrough, F., Chatrian, G.-E., Lesser, R.P., Luders, H., Nuwer, M., Picton, T.W.: American electroencephalographic society guidelines for standard electrode position nomenclature. J. Clin. Neurophysiol 8(200), 2 (1991)
7. Sovierzoski, M., Argoud, F., De Azevedo, F.: Identifying eye blinks in EEG signal analysis. In: International Conference on Information Technology and Applications in Biomedicine (ITAB 2008), pp. 406–409 (2008)

8. Venkataramanan, S., Kalpakam, N.V.: Aiding the detection of Alzheimer's disease in clinical electroencephalogram recording by selective denoising of ocular artifacts. In: International Conference on Communications, Circuits and Systems (ICCCAS2004), vol. 2, pp. 965–968 (2004)
9. Melin, P., Castillo, O.: An intelligent hybrid approach for industrial quality control combining neural networks, fuzzy logic and fractal theory. Inf. Sci. **177**, 1543–1557 (2007)
10. Mendoza, O., Melin, P., Castillo, O., Licea, G.: Type-2 fuzzy logic for improving training data and response integration in modular neural networks for image recognition. Lect. Notes Artif. Intell. **4529**, 604–612 (2007)
11. Mendoza, O., Melin, P., Castillo, O.: Interval type-2 fuzzy logic and modular neural networks for face recognition applications. Appl. Soft Comput. J. **9**, 1377–1387 (2009)
12. Mendoza, O., Melin, P., Licea, G.: Interval type-2 fuzzy logic for edges detection in digital images. Int. J. Intell. Syst. **24**, 1115–1133 (2009)
13. Pérez, M., Luis, J.: Comunicación con Computador mediante Señales Cerebrales. Aplicación a la Tecnología de la Rehabilitación. Ph.D. thesis, Universidad Politécnica de Madrid (2009)
14. Hirsch, L., Richard, B.: EEG basics. In: Atlas of EEG in Critical Care, pp. 1–7. Wiley (2010)
15. Chang, P.F., Arendt-Nielsen, L., Graven-Nielsen, T., Svensson, P., Chen, A.C.: Comparative EEG activation to skin pain and muscle pain induced by capsaicin injection. Int. J. Psychophysiol. **51**(2), 117–126 (2004)
16. De la O Chavez, J.R.: BCI para el control de un cursor basada en ondas cerebrales. Master's thesis, Universidad Autónoma Metropolitana (2008)
17. Erfanian, A., Gerivany, M.: EEG signals can be used to detect the voluntary hand movements by using an enhanced resource-allocating neural network. In: Proceedings of the 23rd Annual International Conference of the IEEE Engineering in Medicine and Biology Society 2001, vol. 1, pp. 721, 724 (2001)
18. Lin, J.-S., Chen, K.-C., Yang, W.-C.: EEG and eye-blinking signals through a brain-computer interface based control for electric wheelchairs with wireless scheme. In: 4th International Conference on New Trends in Information Science and Service Science (NISS) 2010, pp. 731–734
19. Haykin, S.: Neural networks a comprehensive foundation. Pearson Prentice Hall, Delhi (1999)
20. Mitchell, T.M.: Machine Learning, Chap. 4. McGraw-Hill (1997)

Chemical Optimization Method for Modular Neural Networks Applied in Emotion Classification

Coral Sánchez, Patricia Melin and Leslie Astudillo

Abstract The goal of this chapter is the classification of the voice transmitted emotions, and in order to achieve it we worked with the Mel cepstrum coefficients for the pre-processing of audio. We also used a Modular Neural Network as the classification method and a new optimization algorithm was implemented: Chemical Reaction Algorithm which hadn't been used before on the optimization of neural networks to find the architecture of the neural network optimizing the number of layers in each module and the number of neurons by layer. The tests were executed on the Berlin Emotional Speech data base, which was recorded by actors in German language in six different emotional states of which they only considered anger, happiness and sadness.

1 Introduction

Speech is a complex signal that contains information about the message to be transmitted, about emotions and the physiological state of the speaker, therefore its production mechanisms have been studied and systems that are able to simulate it and recognize have been created, given that the voice is not nothing but a sound and this is why is characterized by a set of elements [1].

On the other hand, the use of Neural Networks have shown to provide good results in the recognition and classification, and in some studies they have

C. Sánchez · P. Melin (✉) · L. Astudillo
Tijuana Institute of Technology, Tijuana, Mexico
e-mail: pmelin@tectijuana.com

C. Sánchez
e-mail: es.coral.sa@gmail.com

L. Astudillo
e-mail: leslie.astudillo@suntroncorp.com

O. Castillo et al. (eds.), *Recent Advances on Hybrid Approaches for Designing Intelligent Systems*, Studies in Computational Intelligence 547,
DOI: 10.1007/978-3-319-05170-3_26, © Springer International Publishing Switzerland 2014

considered the application of paradigms such as genetic algorithms, ACO and PSO as optimization methods [2–4].

A new optimization approach is the Chemical Reactions Paradigm which has already been used in optimizing fuzzy control systems, thus it can also provide a good alternative for this application of neural networks, as well as it is innovative in this area.

The contribution of this chapter is the proposed method for Emotion Classification based on the Chemical Optimization method for the optimal design of Modular Neural Networks.

This chapter is composed of 6 sections described next: Sects. 2, 3, 4 describe the relationship between emotions and voice, description and operation of modular neural networks and finally the detailed description of the Chemical Reactions Algorithm. Section 5 describes the methodology used to complete the main objective of this study, starting with the database used, the applied pre-processing method, neural modular network's architecture and finally the emotions classification. Section 6 presents the obtained results until now.

2 The Voice and Emotions

According to the physiology of emotion production mechanisms, it has been found that the nervous system is stimulated by the expression of high arousal emotions like anger, happiness and fear. This phenomenon causes an increased heart beat rate, higher blood pressure, changes in respiration pattern, greater sub-glottal air pressure in the lungs and dryness of the mouth. The resulting speech is correspondingly louder, faster and characterized with strong high-frequency energy, higher average pitch, and wider pitch range. On the other hand, for the low arousal emotions like sadness, the nervous system is stimulated causing the decrease in heart beat rate, blood pressure, leading to increased salivation, slow and low-pitched speech with little high-frequency energy [1].

3 Modular Neural Networks

A modular or ensemble neural network uses several monolithic neural networks to solve a specific problem. The basic idea is that combining the results of several simple neural networks we can achieve a better overall result in terms of accuracy and also learning can be done faster.

In general, the basic concept resides in the idea that combined (or averaged) estimators may be able to exceed the limitation of a single estimator.

The idea also shares conceptual links with the "divide and conquer" methodology. Divide-and-conquer algorithms attack a complex problem by dividing it

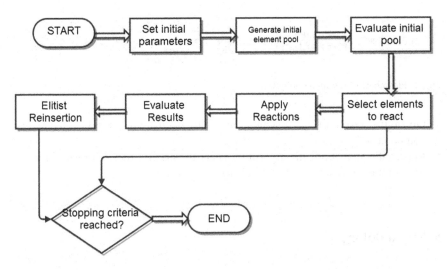

Fig. 1 General flowchart of the chemical reaction algorithm

into simpler problems whose solutions can be combined to yield a solution to the complex problem [5].

4 Chemical Reaction Algorithm (CRA)

The proposed chemical reaction algorithm is a metaheuristic strategy that performs a stochastic search for optimal solutions within a defined search space. In this optimization strategy, every solution is considered as an element (or compound), and the fitness or performance of the element is represented by the final energy (which is determined according to the objective function).

The main difference with other optimization techniques is that no external parameters are taken into account to evaluate the results, while other algorithms introduce additional parameters (kinetic/potential energies, mass conservation, thermodynamic characteristics, etc.), this is a very straight forward methodology that takes the characteristics of the chemical reactions (synthesis, decomposition, substitution and double substitution) to find the optimal solution [6–10].

The steps to consider in this optimization method are illustrated in Fig. 1 and given.

1. First, we need to generate an initial pool of elements/compounds.
2. Once we have the initial pool, we have to evaluate it.
3. Based on the previous evaluation, we will select some elements/compounds to "induce" a reaction.

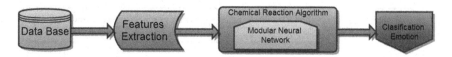

Fig. 2 Flowchart of the methodology

4. Given the result of the reaction, we will evaluate the obtained elements/ compounds.
5. Repeat the steps until the algorithm meets the criteria (desired result or maximum number of iterations is reached).

5 Methodology

The main goal of this work is to build a voice emotion classifier using a modular neural network optimized by a chemical reactions algorithm, tested with the Berlin Database of Emotional Speech.

The methodology is illustrated in Fig. 2 and it is described below.

5.1 Database

Berlin Emotional Speech: This database is in German. It was recorded by actors. The recordings took place in an anechoic chamber at the Technical University of Berlin. Contains 535 files, 10 audio players (5 men and 5 women between 21 and 35 years) in 7 emotions (anger, boredom, disgust, fear, happiness, sadness and neutral) with 10 different sentences for emotion [1].

5.2 Features Extraction

Not as in another conventional method the transformed time–frequency allows a combined representation in space of time and frequency, in other words, we can acknowledge the dynamic of the spectral contents in time course.

The main objectives are to achieve a reduction of dimensionality and enhance signal aspects that contribute significantly to perform further processing (recognition, segmentation or classification) [11].

Mel Frequency Cepstral Coefficients (MFCC)

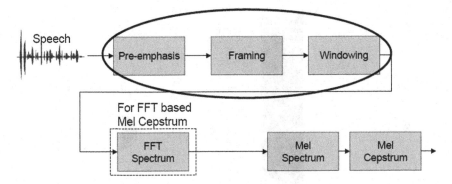

Fig. 3 Cepstral coefficients in the Mel scale

The MFCC feature extraction technique basically includes windowing the signal, applying the DFT, taking the log of the magnitude and then warping the frequencies on a Mel scale, followed by applying the inverse DCT [2].

Figure 3 shows the cepstral coefficients obtained from the speech signal.

1. Pre-emphasis
2. Frame blocking and windowing
3. DFT spectrum
4. Mel-spectrum

5.3 Modular Neural Network Architecture

This is the modular neural network architecture, it consists of three modules where each one is optimized by the chemical reaction algorithm individually and trained with only one emotion, with a simple integrator "winner takes all" as shown in Fig. 4.

6 Simulation Results

This section shows the results of some experiments made so far with different combinations in the parameters of the chemical reactions algorithm, with their corresponding percentage of recognition.

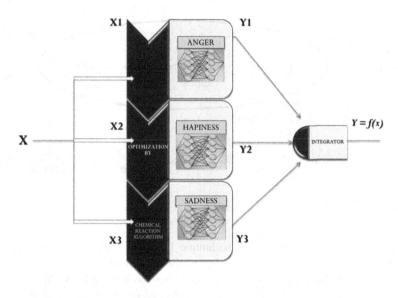

Fig. 4 Modular neural network architecture

Parameter	Value
Maximum iteration	12
Population size	10
Output data	5
Selection rate	0.2
Synthesis rate	0.2
Decomposition rate	0.2
Single substitution rate	0.2
Double substitution rate	0.2

Table 1 Values for the parameters of CRA for experiment 1

6.1 Experiment 1

In the first experiment the best obtained percentage of recognition was:

46.15 % for anger
81.81 % for happiness
68.42 % for sadness

Table 1 shows the parameters of the CRA and Fig. 5 shows the results.

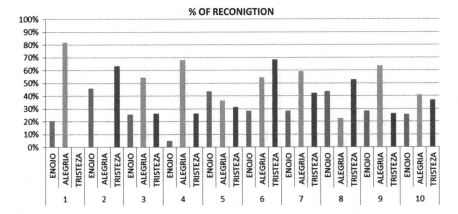

Fig. 5 Chart of recognition rates for Experiment 1

Table 2 Values for the parameters of CRA for experiment 2

Parameter	Value
Maximum iteration	12
Population size	10
Output data	5
Selection rate	0.3
Synthesis rate	0.4
Decomposition rate	0.5
Single substitution rate	N/A
Double substitution rate	N/A

6.2 Experiment 2

In the second experiment the best percentage of recognition was:

48.71 % for anger
59.09 % for happiness
89.47 % for sadness

Table 2 shows the parameters of the CRA and Fig. 6 the results.

6.3 Experiment 3

In the third experiment the best percentage of recognition was:

66.66 % for anger
86.36 % for happiness
78.94 % for sadness

Table 3 shows the parameters of the CRA and Fig. 7 the results.

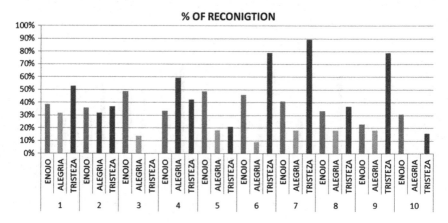

Fig. 6 Chart of recognition rates for experiment 2

Table 3 Values for the parameters of CRA for experiment 3

Parameter	Value
Maximum iteration	12
Population size	10
Output data	5
Selection rate	0.3
Synthesis rate	0.4
Decomposition rate	0.5
Single substitution rate	0.6
Double substitution rate	N/A

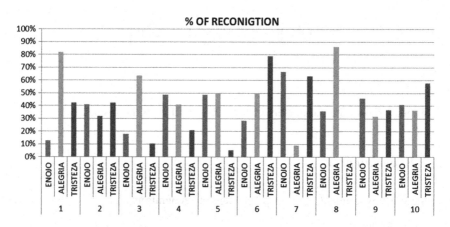

Fig. 7 Chart of recognition rates for experiment 3

7 Conclusions

Although we have been obtaining good results, there is a lot to work to do, like with the audio preprocessing, to detect which features are outstanding referred to the emotion received in the voice, and keep experimenting with different combinations in the parameters of the optimization algorithm to achieve an improvement in the network architecture and obtain an meaningful increase in the classification percentage. As future work we plan to incorporate the use of fuzzy logic in the proposed methodology [12–14].

Acknowledgments We would like to express our gratitude to CONACYT, and Tijuana Institute of Technology for the facilities and resources granted for the development of this research.

References

1. Burkhardt, F., Kienast, M., Paeschke, A., Weiss, B.: 1997–1999. http://pascal.kgw.tu-berlin.de/emodb/. Accessed Jan 2013
2. Sombra, A., Valdez, F., Melin, P., Castillo, O.: A new gravitational search algorithm using fuzzy logic to parameter adaptation. IEEE Congress on Evolutionary Computation 2013, pp. 1068–1074
3. Valdez, F., Melin, P., Castillo, O.: Evolutionary method combining particle swarm optimization and genetic algorithms using fuzzy logic for decision making. In: Proceedings of the IEEE International Conference on Fuzzy Systems, 2009, pp. 2114–2119
4. Valdez, F., Melin, P., Castillo, O.: Fuzzy logic for parameter tuning in evolutionary computation and bio-inspired methods. MICAI (2), pp. 465–474 (2010)
5. Melin, P., Castillo, O.: Hybrid Intelligent Systems for Pattern Recognition Using Soft Computing. Springer, Berlin (2005)
6. Astudillo, L., Melin, P., Castillo, O., Valdez, F., Garcia, M.: Optimal design of type-2 and type-1 fuzzy tracking controllers for autonomous mobile robots under perturbed torques using a new chemical optimization paradigm. Expert Syst Appl 40(8), 3185–3195 (2013)
7. Astudillo, L., Melin, P., Castillo, O.: Design of fuzzy systems using a new chemical optimization paradigm. Fuzzy information processing society (NAFIPS). In: 2011 Annual Meeting of the North American, pp. 1–6. 18–20 Mar 2011
8. Astudillo, L., Melin, P., Castillo, O.: A new optimization algorithm based on a paradigm inspired by nature. J. Autom. Mobile Rob. Intel. Syst. 5(1) (2011)
9. Astudillo, L., Melin., P, Castillo, O.: Chemical optimization paradigm applied to a fuzzy tracking controller for an autonomous mobile robot. Int. J. Innovative Comput. Inf. Control ICIC International 2013. 9(5), 2007–2018. ISSN 1349-4198 May 2013
10. Astudillo, L., Melin, P., Castillo, O.: Nature optimization applied to design a type-2 fuzzy controller for an autonomous mobile robot. In: 2012 fourth world congress on nature and biologically inspired computing (NaBIC), Mexico City, pp. 212–217
11. Melin, P., Olivas, F., Castillo, O., Valdez, F., Soria, J., García Valdez, J.: Optimal design of fuzzy classification systems using PSO with dynamic parameter adaptation through fuzzy logic. Expert Syst. Appl. 40(8), 3196–3206 (2013)
12. Montiel, O., Camacho, J., Sepúlveda, R., Castillo, O.: Fuzzy system to control the movement of a wheeled mobile robot. Soft computing for intelligent control and mobile robotics 2011, pp 445–463

13. Montiel, O., Sepulveda, R., Melin, P., Castillo, O., Porta, M., Meza, I.: Performance of a simple tuned fuzzy controller and a PID controller on a DC motor, FOCI 2007, pp 531–537
14. Porta, M., Montiel, O., Castillo, O., Sepúlveda, R., Melin, P.: Path planning for autonomous mobile robot navigation with ant colony optimization and fuzzy cost function evaluation. Appl. Soft Comput. 9(3), 1102–1110 (2009)

Comparing Metaheuristic Algorithms on the Training Process of Spiking Neural Networks

Andrés Espinal, Martín Carpio, Manuel Ornelas, Héctor Puga, Patricia Melin and Marco Sotelo-Figueroa

Abstract Spiking Neural Networks are considered as the third generation of Artificial Neural Networks. In these networks, spiking neurons receive/send the information by timing of events (spikes) instead by the spike rate; as their predecessors do. Spikeprop algorithm, based on gradient descent, was developed as learning rule for training SNNs to solve pattern recognition problems; however this algorithm trends to be trapped in local minima and has several limitations. For dealing with the supervised learning on Spiking Neural Networks without the drawbacks of Spikeprop, several metaheuristics such as: Evolutionary Strategy, Particle Swarm Optimization, have been used to tune the neural parameters. This work compares the performance and the impact of some metaheuristics used for training spiking neural networks.

A. Espinal · M. Carpio · M. Ornelas · H. Puga (✉) · M. Sotelo-Figueroa
Instituto Tecnológico de León, León Gto, Mexico
e-mail: pugahector@yaho.com

A. Espinal
e-mail: andres.espinal@itleon.edu.mx

M. Carpio
e-mail: jmcarpio61@hotmail.com

M. Ornelas
e-mail: mornelas67@yahoo.com.mx

M. Sotelo-Figueroa
e-mail: marco.sotelo@itleon.edu.mx

P. Melin
Instituto Tecnológico Tijuana, Baja California, Mexico
e-mail: pmelin@tectijuana.edu.mx

O. Castillo et al. (eds.), *Recent Advances on Hybrid Approaches for Designing Intelligent Systems*, Studies in Computational Intelligence 547,
DOI: 10.1007/978-3-319-05170-3_27, © Springer International Publishing Switzerland 2014

1 Introduction

Spiking Neural Networks (SNNs) are considered as the third generation of Artificial Neural Networks (ANNs) [1]. These networks are formed by spiking neurons; which deal with information encoded in timing of events (spikes), instead by the spike rate as their predecessors do. It have been proved that spiking neural networks are computationally more stronger than sigmoid neural networks [2]. In fact, there is evidence that fewer spiking neurons are required for solving some functions than neurons of previous generations [1].

As their predecessors, SNNs have been applied for dealing with pattern recognition problems. Spikeprop algorithm is a learning rule for training SNNs, which is based on gradient descent [3]; it is capable of training SNNs to solve complex classification problems. However, this algorithm trends to be trapped in local minima and it has several limitations i.e. it does not allow both, positive and negative synaptic weights. The following list explains some drawbacks about using Spikeprop on the supervised training of SNNs [4]:

- The convergence of the algorithm is vulnerable to being caught in local minima.
- The convergence of the algorithm is not guaranteed as it depends on fine tuning of several parameters before start of the algorithm.
- The structure of the synapse which consists of a fixed number of sub-connections, each of which has a fixed synaptic delay, leads to a drastic increase in the number of connecting synapses and therefore the number of adjustable weights.
- The huge number of synaptic connections makes it difficult for such algorithms to scale up when processing high-dimensional dataset is considered.
- Also having a fixed number of sub-connections with fixed delay vales is not necessity and yields a lot of redundant connections.
- The algorithm entails the problem of 'silent neurons'; i.e. if the outputs do not fire then the algorithm halts, since no error can be calculated.

For overcoming the drawbacks of Spikeprop on the supervised learning of SNNs, some works have proposed to use metaheuristic for tuning the neural parameters (weights and delays) [4–7]. Two similar proposals of the state of the art for training SNNs by using metaheuristics are compared in this work. Both methodologies are tested with classical test benchmarks: XOR Logic Gate and Iris plant dataset (UCI Machine Learning Repository), with different architectures over several experiments.

This chapter is organized as follows: Sect. 2 gives fundamentals for simulating SNNs, including neural network structure, codifier and spiking neuron model. In Sect. 3, the methodology used for training SNNs is explained. The experimental design and results are showed in Sect. 4. Finally, in Sect. 5 conclusions about the work and future work is presented.

Fig. 1 *Top* Spiking neural
network with three layers:
input, hidden and output.
Bottom Synapse connection
(weights and delays) between
presynaptic neurons and the
postsynaptic neuron *j* (figure
taken from [6])

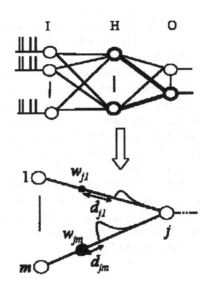

2 Spiking Neural Networks

A neural network can be defined as an interconnection of neurons, such that
neuron outputs are connected, through weights, to all other neurons including
themselves; both lag-free and delay connections are allowed [8]. There are several
models or topologies of ANNs, which are defined around three aspects: computing
nodes, communication links and message types [9].

For this work were used feed-forward SNNs. They were formed by 3 layers:
input, hidden and output (see Fig. 1), their topology was set as follows: the
computing nodes (neuron models) are spiking neurons, specifically neurons of the
Spike Response Model (SRM) [10]; this model is explained later. The communication links (synaptic connections, each synaptic connection is defined by a
synaptic weight and a synaptic delay) were set to form a fully interlayer connected
neural network; where all possible interlayer connections are present and the
network contains no intralayer (including self-connections), or supralayer connections [11], this kind of pattern connection is also know it as fully connected
neural network. The message types are based on the time-to-first-spike encoding
scheme; it means that each neuron can produce at most one spike in all the
simulation time for each stimuli.

2.1 Gaussian Receptive Fields

The traditional form of patterns as multidimensional raw data, which consist of
real values can't be used to feed a SNNs in a simulation process. The patterns need
to be transformed into temporal patterns (a set of events in time or spikes as known
as spike trains) before being processed by the SNN.

In [12], was proposed an encoding scheme to generate firing times from real values; the Gaussian Receptive Fields (GRFs). This scheme enables the representation of continuously valued input variables by a population of neurons with graded and overlapping sensitivity profiles, such as Gaussian activation functions (the receptive fields). To encode values into a temporal pattern, it is sufficient to associate highly stimulated neurons with early firing times and less stimulated neurons with later (or no) firing times. Each input variable is encoded independently, it means that each input dimension is encoded by an array of one-dimensional receptive fields; the GRFs constitute a biologically plausible and well studied method for representing real-valued parameters [3].

In [13] is given a detailed definition about the construction and use of the GRFs. Each input datum is fed to all the conversion neurons covering the whole data range. For a range $[n_{min}, n_{max}]$ of a variable n, m neurons were used with GRF. For the i-th neuron coding for variable n, the center of the Gaussian function is set to C_i according to Eq. (1).

$$C_i = n_{min} + \frac{2i - 3}{w} \times \frac{n_{max} - n_{min}}{m - 2}, \quad m > 2 \tag{1}$$

The width w of each Gaussian function is set according to Eq. 2. Where $1 \leq \gamma \leq 2$ is a coefficient of multiplication which is inversely proportional to the width of the Gaussian functions.

$$w = \frac{n_{max} - n_{min}}{\gamma(m - 2)} \tag{2}$$

Using the centers of each neuron/Gaussian function and their width, the magnitude of firing $f(x)$ for each sample points of the input x is calculated using Eq. 3.

$$f_i(x) = e^{-\frac{(x - c_i)^2}{2w^2}} \tag{3}$$

The magnitude values are then converted to time delay values by associating the highest magnitude value to a value close to zero milliseconds and neurons with lower firing magnitude are associated with a time delay value close to the limit maximum of time of simulation. Conversion from magnitude to time delay values is done using Eq. (4). Where τ is the total time of simulation.

$$T_i = (1 - f_i) \times \tau, \quad i = 1, 2, \ldots, m \tag{4}$$

Time delay values greater than some value $\tau_{threshold} < \tau$ are coded no to fire as they are consider being insufficiently excited. Therefore, neurons with time delay value between zero and $\tau_{threshold}$ milliseconds carry all the encoded information of the input data.

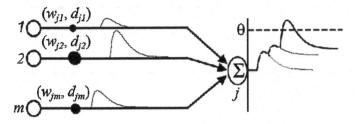

Fig. 2 Incoming spikes transformed into postsynaptic potentials, they are delayed and weighted by the synapse and finally they are integrated by the postsynaptic neuron (figure taken from [5])

2.2 Spike Response Model

The SRM [10, 14] is an approximation of the dynamics of the integrate-and-fire neurclose to the limit maximum of time of simulation.ons. The neuron status is updated through a linear summation of the postsynaptic potentials resulting from the impinging spike trains at the connecting synapses. A neuron fires whenever its accumulated potential reaches a threshold from below (see Fig. 2) [6].

Due to spiking neurons in this work use the time-to-first-spike encoding scheme for sending/receiving messages, a reduced version of the SRM is implemented; this has been used in [3–7, 12].

The reduced SRM is defined according [6] as follows. Let us consider that a neuron j has a set Γ_j of immediate predecessors called presynaptic neurons and receives a set of spikes with firing times $t_i, i \in \Gamma_j$. Neurons fire when their state variable $x(t)$, called membrane potential, reaches a certain threshold Θ. The internal state of a neurons is determined by (5), where w_{ji} is the synaptic weight to modulate $y_i(t)$, which is the unweighted postsynaptic potential of a single spike coming from neuron i and impinging on neuron j.

$$x_j(t) = \sum_{i \in \Gamma_j} w_{ji} y_i(t) \tag{5}$$

The unweighted contribution $y_i(t)$ is given by Eq. (6), where $\varepsilon(t)$ is a function that describes the form of the postsynaptic potential.

$$y_i(t) = \varepsilon\left(t - t_i - d_{ji}\right) \tag{6}$$

The form of the postsynaptic potential is given by Eq. (7), and it requires the next parameters: t is the current time, t_i is the firing time of the presynaptic neuron i and d_{ji} is the associated synaptic delay. Finally the function has a τ parameter, it is the membrane potential time constant defining the decay time of the postsynaptic potential.

$$\varepsilon(t) = \begin{cases} \frac{t}{\tau} e^{1-\frac{t}{\tau}} & if f > 0 \\ 0 & else \end{cases} \tag{7}$$

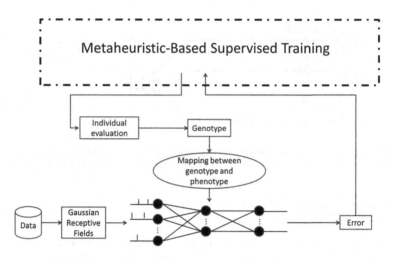

Fig. 3 Generic scheme for training SNNs with metaheuristic algorithms

The firing time t_j of neuron j is determined as the first time the state variable crosses the threshold from below. The threshold Θ and τ are constants and equal for all neurons in the network.

3 Metaheuristic-Based Supervised Learning

Learning is a process by which the free parameters of a neural network are adapted through a process of stimulation by the environment in which the network is embedded. The type of learning is determined by the manner in which the parameter changes take place [15]. In this case, the learning is driven by some metaheuristic algorithm (see Fig. 3 the free parameters of the SNN are all weights and delays of each synapse.

Several works have used metaheuristics for training SNN [4–7]; each of them defines a particular methodology, next is presented the methodology followed based on similar aspects of the works revised for this work.

In Metaheuristic-Based Supervised Learning, each individual contains all the free parameters of a previously structured SNN. Every individual is evaluated by means of a fitness function, the first step of the fitness function makes a mapping process; this sets the individuals parameter as weights and delays in the SNN. The second step of the fitness function uses the batch training as learning protocol, where all patterns are presented to the network before the learning takes place [16]. The third step of the fitness function is calculated an error (to be minimized) according Eq. (8) (equation taken from [4]); where T are all training patterns, O are all output spiking neurons, $t_O^a(t)$ is the current timing output of the SNN and $t_O^t(t)$ is the desired timing output. The error calculated in the fitness function determines the fitness value of each individual and drives the supervised learning based on metaheuristic algorithms.

$$E = \sum_{t}^{T} \sum_{o}^{O} \left(t_0^a(t) - t_o'(t) \right)^2 \tag{8}$$

Next are presented two metaheuristic used for training SNNs in the state of the art.

3.1 Evolutionary Strategy

The Evolutionary Strategies (ES) [17], deal natively with problems in real domain. In [4] was designed a Self-Adaptive ES for training SNNs, in this ES each population member consists of n-dimensional vectors, where n is the total number of tuneable network parameters within input, hidden and output layer. The population at any given generation g is denoted as $P(g)$. Each individual is taken as a pair of real-valued vectors, (x_i, η_i), where x_i's are objective variables representing the synaptic free parameters, and η_i's are standard deviations for mutations. Each individual generates a single offspring (x'_i, η'_i), where each variable $x'_i(j)$ of the offspring can be defined by either Eq. 9 (local search) or Eq. 10 (global search) and the standard deviation is defined by Eq. (11).

$$x'_i(j) = x_i(j) + \eta_i(j)N_j(0, 1) \tag{9}$$

$$x'_i(j) = x_i(j) + \eta_i(j)\delta_j \tag{10}$$

$$\eta'_i(j) = \eta_i(j) \exp\left(\tau'N(0, 1) + \tau N_j(0, 1)\right) \tag{11}$$

where:

- $N(0, 1)$ denotes a normally distributed one dimensional random number with $\mu = 0$ and $\sigma = 1$.
- $N_j(0, 1)$ indicates that the random number is generated anew for each value of j.
- δ_j is a Cauchy random variable, and it is generated anew for each value of j (Scale $= 1$).
- Factor $\tau = \frac{1}{\sqrt{2\sqrt{n}}}$
- Factor $\tau' = \frac{1}{\sqrt{2n}}$

The Self-adaptive ES is presented in the Algorithm 1.

Algorithm 1 Self-adaptive ES 1

1: Generate the initial population of μ individuals.
2: Evaluate the fitness score for each individual (x_i, η_i), $i = 1, \ldots, \mu$ of the population based on
$E = \sum_t^T \sum_o^O (t_o^a(t) - t_o^t(t))^2$
3: **while** the maximum iteration is not reached **do**
4: Each parent (x_i, η_i) generates a single offspring (x_i', η_i')
5: Calculate the fitness of each offspring (x_i', η_i'), $i = 1, \ldots, \mu$
6: Generate a new population $P(g)$ using tournament selection and elitism to keep track of the best individual at each generation
7: **end while**

3.2 Particle Swarm Optimization

The Particle Swarm Optimization (PSO) [18] is an algorithm originally used for continuous optimization problems, it is inspired in the behavior of flocks of birds or schools of fish. A version of PSO called Cooperative PSO (CPSO) has been used in [7] for training SNNs. In this metaheuristic, each individual is defined by the vector velocity v_i and the vector position x_i.

The vector v_i is updated by Eq. (12) where ρ_1 and ρ_2 are acceleration coefficients. χ is a constriction factor, it can be defined according Eq. (14). γ_1 and γ_2 are two independent uniform numbers randomly distributed in the range [0,1]. And ω is the inertia weight, which controls the impact of the previous velocity; it is defined by Eq. (13) where ω_0 is the maximum of ω, $\omega_0 - \omega_1$ is the minimum of ω, t is the current iteration and T_{max} is the maximal iteration of PSO.

The vector x_i is updated by Eq. (12).

$$v_{id} = \chi(\omega v_{id} + \rho_1 \gamma_1 (p_{id} - x_{id})) \tag{12}$$

$$\omega = \omega_0 - \omega_1 \frac{t}{T_{max}} \tag{13}$$

$$\chi = \frac{2}{\left| 2 - \varphi - \sqrt{\varphi^2 - 4\varphi} \right|} ; \varphi = \rho_1 + \rho_2, \varphi > 4 \tag{14}$$

$$x_{id} = x_{id} + v_{id} \tag{15}$$

The CPSO is presented in the Algorithm 2.

Algorithm 2 CPSO-based supervised learning

1: Generate the initial swarm.
2: **for all** particle i **do**
3: $x_{id} \leftarrow rand(x^{min}, x^{max})$

(continued)

(continued)

Algorithm 2 CPSO-based supervised learning

4: $v_{id} \leftarrow rand(v^{min}, v^{max})$

5: Its past best solution can be set to the same as x_i initially.

6: **end for**

7: **while** the maximum iteration is not reached **do**

8: Compute the fitness of each particle for its current position

9: Compute for each input pattern by Spiking neural network, and obtain the spike times of the output neurons

10: Calculate the square error usign $E = \sum_t^T \sum_o^O \left(t_o^a(t) - t_o^t(t)\right)^2$ as the fitness of the particle

11: Find the global best particle g with the minimal fitness from the swarm of the current iteration

12: $v_{id} \leftarrow \chi\left(\omega v_{id} + p_1\gamma_1\left(p_{id} - x_{id}\right) + p_2\gamma_2\left(p_{gd} - x_{id}\right)\right)$

13: $x_{id} \leftarrow x_{id} + v_{id}$

14: **end while**

---

4 Experiments and Results

Classical benchmarks of pattern recognition were used for experimentation: XOR Logic Gate and Iris Plant dataset. For each benchmark, 5 SNNs with different configurations on their layers were used; each of them was trained (using both metaheuristics) 35 times. For each benchmark, the training by using ES and CPSO of each SNN configuration is contrasted using the fitness values by calculating the minimum, the maximum, median, mean and standard deviation.

The common parameters of the metaheuristics through all experiments and trainings are: 30 individuals and 15,000 function calls.

4.1 XOR Logic Gate

For the XOR logic gate, it was not necessary to use GRFs. Inputs values were manually encoded, the encoded inputs and the desired timing output are showed in Table 1.

The values of CPSO parameters were empirically set, they were set as $\rho_1 = 0.8$ and $\rho_2 = 3.3$. The results for the training of each configuration are showed in Table 2. For this benchmark the testing phase was not necessary, due that any fitness value greater than 0 produces a poor performance of the SNN. Taking the median value as representative error of training for each metaheuristic over all configurations, can be inferred that ES outperforms to CPSO for the training phase in this problem.

Table 1 Configuration used for solving XOR

Original XOR				Encoded XOR	
x_1	x_2	y	x_1'	x_2'	Output firing time (ms)
0	0	0	0	0	20
0	1	1	0	6	10
1	0	1	6	0	10
1	1	0	6	6	20

Table 2 Results of fitness values for training SNNs by using ES and CPSO to solve XOR Logic Gate

Network	ES Optimization				CPSO Optimization			
	Minimum	Median	Maximum	$\mu \pm \sigma$	Minimum	Median	Maximum	$\mu \pm \sigma$
2-3-1	0.0	8.0	51.0	15.4 ± 16.8	1.0	22.0	52.0	25.8 ± 15.6
2-5-1	0.0	1.0	50.0	5.7 ± 9.9	0.0	5.0	43.0	10.3 ± 12.0
2-7-1	0.0	0.0	18.0	1.3 ± 3.3	0.0	2.0	24.0	5.9 ± 7.0

Table 3 Results of fitness values for training SNNs by using ES and CPSO to solve Iris plant dataset (1st fold)

Network	ES Optimization (Fitness)				CPSO optimization (Fitness)			
	Minimum	Median	Maximum	$\mu \pm \sigma$	Minimum	Median	Maximum	$\mu \pm \sigma$
16-1-1	35.0	99.0	218.0	101.4 ± 40.1	83.0	182.0	267.0	181.2 ± 47.8
16-5-1	10.0	67.0	167.0	69.7 ± 30.1	33.0	117.0	192.0	121.2 ± 40.6
16-10-1	16.0	62.0	207.0	64.9 ± 34.0	55.0	111.0	202.0	115.6 ± 35.5
16-15-1	20.0	64.0	123.0	65.9 ± 23.1	13.0	86.0	215.0	97.4 ± 42.8
16-20-1	29.0	66.0	111.0	66.3 ± 23.5	16.0	88.0	204.0	92.6 ± 41.4

4.2 Iris Plant Dataset

The Iris plant dataset contains 3 classes of which 2 are not linearly separable, each class is formed by 50 patterns each of them described by 4 features. For this benchmark, each feature was encoded by GRFs using 4 encoding. The desired timing outputs for setosa, versicolor and virginica classes are respectively 6,10 and 14 ms. The dataset was divided into 2 equal parts (2-Folds Cross Validation). Then 2 tests were performed.

The values of CPSO parameters were empirically set, they were set as $\rho_1 = 2.1$ and $\rho_2 = 2.1$. For the first test instance Tables 3 and 4 show the optimization phase and classification phase respectively. And for the second test instance Tables 5 and 6 show the optimization phase and classification phase respectively. The classification phase was performed by using the configuration given by the

Table 4 Results of performance of SNN trained with ES and CPSO (1st fold)

Network	Classification (ES Training)				Classification (CPSO Training)			
	Training		Testing		Training		Testing	
	Minimum (%)	Maximum (%)	Minimum (%)	Maximum (%)	Minimum (%)	Maximum (%)	Minimum (%)	Maximum (%)
16-1-1	4.0	69.3	13.3	81.3	9.3	22.7	12.0	34.7
16-5-1	41.3	85.3	38.7	90.7	20.0	68.0	21.3	80.0
16-10-1	4.0	89.3	8.0	98.7	1.3	32.0	2.7	37.3
16-15-1	41.3	92.0	41.3	97.3	36.0	66.7	32.0	82.7
16-20-1	54.7	86.7	58.7	86.7	34.7	78.7	33.3	89.3

Table 5 Results of fitness values for training SNNs by using ES and CPSO to solve Iris plant dataset (2nd fold)

Network	ES Optimization (Fitness)				CPSO Optimization (Fitness)			
	Minimum	Median	Maximum	$\mu \pm \sigma$	Minimum	Median	Maximum	$\mu \pm \sigma$
16-1-1	16.0	67.0	218.0	65.5 ± 16.3	69.0	133.0	312.0	134.7 ± 32.2
16-5-1	10.0	39.0	167.0	37.3 ± 11.4	33.0	95.0	220.0	86.7 ± 23.1
16-10-1	16.0	32.0	207.0	34.5 ± 11.0	37.0	83.0	202.0	79.3 ± 20.2
16-15-1	7.0	39.0	123.0	37.6 ± 11.3	13.0	62.0	215.0	62.3 ± 18.4
16-20-1	16.0	39.0	111.0	38.0 ± 8.8	16.0	52.0	204.0	53.5 ± 13.8

Table 6 Results of performance of SNN trained with ES and CPSO (2nd fold)

Network	Classification (ES Training)				Classification (CPSO Training)			
	Training		Testing		Training		Testing	
	Minimum (%)	Maximum (%)	Minimum (%)	Maximum (%)	Minimum (%)	Maximum (%)	Minimum (%)	Maximum (%)
16-1-1	4.0	89.3	13.3	80.0	1.3	53.3	4.0	52.0
16-5-1	41.3	85.3	38.7	90.7	33.3	68.0	36.0	80.0
16-10-1	4.0	89.3	8.0	98.7	1.3	78.7	2.7	68.0
16-15-1	41.3	90.7	41.3	85.3	36.0	66.7	32.0	82.7
16-20-1	54.7	98.7	58.7	93.3	34.7	78.7	33.3	89.3

minimum and maximum fitness values on the training phase, the best performances correspond to trainings with low fitness and worst performances correspond to trainings with high fitness. Same as the XOR logic gate, taking the median value as representative error of training for each metaheuristic over all configurations, can be inferred that ES outperforms to CPSO for the training phase in this problem. In general the best performances on classification tasks, for both: known patterns and unknown patters, can be observed in the trainings made by ES.

5 Conclusions and Future Work

This work compares two metaheuristics on the training of SNNs. The few ES's parameters made easier to implement. It as training algorithm than CPSO. Moreover, in general ES gets better fitness values than CPSO and the SNNs trained with ES get better classification performance too.

The results obtained show evidence that there is a relationship between the neural architecture and the capability of the metaheuristic to train it. Due that good results in phases of training and testing can be achieved by different architectures, but some neural configurations are easier to train and show more stability through several experiments. Even when a metaheuristic can achieved low fitness values, it doesn't ensure the good performance of the SNN for classifying unseen data; this is an important aspect that needs to be analyzed when using metaheuristic-based supervised learning on SNNs.

As future work, authors propose try to improve the CPSO performance on the training of SNNs by tuning its parameters. It is interesting to use other metaheuristic algorithms as learning rules to analyze their behavior. Finally it is necessary to use statistical tests for a best comparison of these algorithms in this task.

Acknowledgments Authors thanks the support received from Consejo Nacional de Ciencia y Tecnologia (CONACyT).The authors want to thank to *Instituto Tecnológico de León* (ITL) for the support to this research. Additionally they want to aknowledge the generous support from the *Mexican National Council for Science and Technology* (CONACyT) for this research project.

References

1. Maass, W.: Networks of spiking neurons: the third generation of neural network models. Neural Networks **10**(9), 1659–1671 (1997)
2. Maass, W.: Noisy Spiking Neurons with Temporal Coding Have More Computational Power Than Sigmoidal Neurons, pp. 211–217. MIT Press, Cambridge (1996)
3. Bohte, S.M., Kok, J.N., LaPoutre, H.: Error-backpropagation in temporally encoded networks of spiking neurons. Neurocomputing **48**, 17–37 (2002)
4. Belatreche, A.: Biologically Inspired Neural Networks: Models, Learning, and Applications. VDM Verlag Dr. Müller, Saarbrücken (2010)
5. Belatreche, A., Maguire, L.P., McGinnity, M., Wu, Q.X.: An evolutionary strategy for supervised training of biologically plausible neural networks. In: The Sixth International Conference on Computational Intelligence and Natural Computing (CINC), Proceedings of the 7th Joint Conference on Information Sciences. pp. 1524–1527 (2003)
6. Belatreche, A., Maguire, L.P., McGinnity, T.M.: Advances in design and application of spiking neural networks. Soft. Comput. **11**(3), 239–248 (2007)
7. Shen, H., Liu, N., Li, X., Wang, Q.: A cooperative method for supervised learning in spiking neural networks. In: CSCWD. pp. 22–26. IEEE (2010)
8. Zurada, J.M.: Introduction to Artificial Neural Systems. West (1992)
9. Judd, J.S.: Neural Network Design and the Complexity of Learning. Neural Network Modeling and Connectionism Series, Massachusetts Institute Technol (1990)

10. Gerstner, W.: Time structure of the activity in neural network models. Phys. Rev. E **51**(1), 738–758 (1995)
11. Elizondo, D., Fiesler, E.: A survey of partially connected neural networks. Int. J. Neural Syst. **8**(5–6), 535–558 (1997)
12. Bohte, S.M., La Poutre, H., Kok, J.N.: Unsupervised clustering with spiking neurons by sparse temporal coding and multilayer RBF networks. Neural Networks IEEE Trans. **13**, 426–435 (2002)
13. Johnson, C., Roychowdhury, S., Venayagamoorthy, G.K.: A reversibility analysis of encoding methods for spiking neural networks. In: IJCNN. pp. 1802–1809 (2011)
14. Gerstner, W., Kistler, W.: Spiking Neuron Models: Single Neurons, Populations, Plasticity. Cambridge University Press, Cambridge (2002)
15. Haykin, S.: Neural Networks: Comprehensive Foundation. Prentice Hall (1999)
16. Duda, R.O., Hart, P.E., Stork, D.G.: Pattern Classification. Wiley, New York (2012)
17. Rechenberg, I.: Evolutions Strategie: optimierung technischer systeme nach prinzipien der biologischen evolution. Frommann-Holzboog (1973)
18. Kennedy, J., Eberhart, R.C.: Particle swarm optimization. IEEE Int. Conf. Neural Netw. **4**, 1942–1948 (1995)

A Hybrid Method Combining Modular Neural Networks with Fuzzy Integration for Human Identification Based on Hand Geometric Information

José Luis Sánchez and Patricia Melin

Abstract In this chapter a hybrid approach for human identification based on the hand geometric information is presented. The hybrid approach is based on using modular neural networks and fuzzy logic in a synergetic fashion to achieve effective and accurate hand biometric identification. Modular neural networks are used to recognize hand images based on the geometric information. Fuzzy logic is used to combine the performed recognition from several modular neural networks and the same time handle the uncertainty in the data. The proposed hybrid approach was implemented and tested with a benchmark database and experimental results show competitive identification rates when compared with the best methods proposed by other authors.

1 Introduction

In recent decades an ever increasing number of automatic recognition methods have been developed, all of which are intended to help manage and control access to many different goods, places and services. Computational Intelligence paradigms such as artificial neural networks (ANNs) and Fuzzy systems (based on fuzzy logic) have proven invaluably useful when applied to pattern recognition problems, as well as being able to perform at very good levels when dealing with such problems, these being the reason they are used in this work [1–4].

The features to be found in a human's hands are attractive as a means to build upon for the construction of methods for recognition; not the least of qualities associated with them are their permanence and uniqueness.

Among these methods only two will be mentioned: palmprint recognition and hand geometry recognition [5].

J. L. Sánchez · P. Melin (✉)
Tijuana Institute of Technology, Tijuana, Mexico
e-mail: pmelin@tectijuana.mx

O. Castillo et al. (eds.), *Recent Advances on Hybrid Approaches for Designing Intelligent Systems*, Studies in Computational Intelligence 547,
DOI: 10.1007/978-3-319-05170-3_28, © Springer International Publishing Switzerland 2014

Palmprint recognition is an amplification of fingerprint recognition methods, and may or may not build upon the presence of three principal lines on most everyone's hands.

This kind of methods tends to be very precise, yet spend sizable amounts of computing power.

Hand geometry methods, in their original implementation, look to make several measurements from images of the outline of the hand (these images are taken while the hand is on a level surface and placed between unmovable pegs, to improve measurements).

The measurements commonly include: length of fingers, palm width, knuckle size.

Commonly, a very reduced amount of memory is needed for a single individual, but identification tasks aren't as accurately resolved through such methods, more so if the database is really big; for verification purposes, performance can be good [5].

The rest of the chapter is organized as follows:

Section 2 describes the tools used for this work and what they are about, Sect. 3 describes the methodology applied for the resolution of the stated problem, Sect. 4 describes the results so far obtained, and Sect. 5 gives the conclusions drawn so far, along with ideas for further works.

2 Computational Intelligence Tools

There is an enormous amount of Computational Intelligence tools that are useful for pattern recognition purposes. Mentioned below are those that are important to this work.

2.1 Fuzzy Logic

Fuzzy logic follows a path unlike that of traditional logic, allowing not only two possible states for a situation, but an infinite gamut between those two traditional states, which in a sense approaches more closely the way in which humans use language and what has traditionally been considered a byproduct of language: reasoning.

From this particular standpoint, fuzzy logic builds until it reshapes every tool of logic, including deductive inference [6].

2.2 Fuzzy Systems

A fuzzy inference system (FIS) is a system that takes advantage of fuzzy logic to create and control models of complex systems and situations without making use of ground-up specifications, which are many times unavailable and relying instead on descriptions more or less at a human language level [3].

A much used tool for working with fuzzy systems is that included within the Matlab programming language [6].

2.3 Artificial Neural Networks

Artificial neural networks are much simplified models of generic nervous systems, studied with the idea of applying them for the same tasks organisms excel at, including pattern recognition.

Nowadays there are many kinds of ANNs, but all build upon using lots of simple elements (neurons) capable of being interconnected and of maintaining weighted responses to "stimuli" between those connections; the similarities end there for many types of ANNs.

The afore mentioned networks are typically divided into layers, and for those nets needing training to get the desired response to a given input, such training is accomplished through a training algorithm, like the much used backpropagation algorithm [7].

There would be a last stumbling stone to consider when in need to have a certain output derive from a given input: not all network architectures are equally successful at attaining them [8].

2.4 Modular Neural Networks (MNNs)

It is possible to divide a problem so that several networks (modules or experts) work separately to solve it. Much needed improvements can be obtained with these MNNs, such as a lower training time or an efficiency boost.

The final step in getting an answer to a given problem with an MNN involves the integration of individual results, to mention just a few (not all applicable to every MNN): gating networks, voting, weighted average, winner-takes-all.

Winner takes all is possibly the simplest method and works by considering the highest ranking results in each module and comparing all such results, allowing only the very highest ranking to remain and be part of the final result [9].

It is also possible to have a fuzzy system integrate responses to several modules or networks, according to the system's rules and the compared modules' results.

3 Methodology

The following items were necessary aspects for this work's problem solving methodology:

1. Having a database.
2. Having a general architecture for the system.
3. Having an architecture for each of the used modular neural networks.
4. Having an architecture for each of the modules within a modular neural network.
5. Having preprocessing modules for each of the MNNs.
6. Having an architecture for each of the used fuzzy integrators.
7. Having fuzzy rules for each of the fuzzy integrators.
8. Train all the modules of every modular neural network and integrate results.

3.1 Utilized Database

The database used with this work is the Polytechnic University of Hong Kong's multispectral palmprint database (MS PolyU database) [10].

The images contained in the database are very nearly fingerless, and the middle and ring fingers are opened wide due to the presence of a central metallic stud; each image has a size of 352 by 288 pixels.

The database has four directories with a total of 6,000 images each, taken from 250 volunteers.

The database gets its name from the volunteers being taken an image of every hand four times, under red, green, blue or infrared illumination.

The only directory used for this work is the "Red" one.

3.2 The General System's Architecture

The system comprises two modular neural networks, one dealing with the images and the other the geometric measurements, with each of them having a different preprocessing for the whole dataset. At the end, the results of both MNNs are integrated by a FIS, as shown in Fig. 1:

3.3 Modular Neural Network 1s Architecture

This MNN is comprised of ten modules, which equally divide their dataset, along with an integrator (winner takes all).

The layout for this MNN is shown in Fig. 2:

3.3.1 MNN 1s Parameters by Module

The parameters for this MNN are as follows (all further networks share these parameters, except the number of neurons):

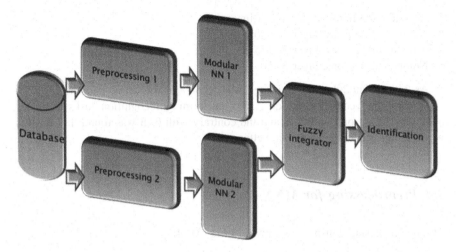

Fig. 1 Complete system's architecture

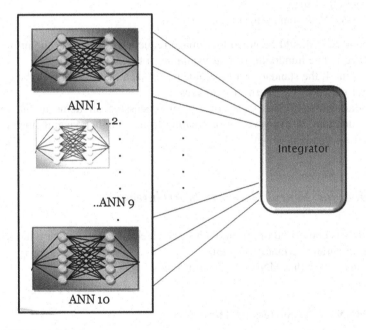

Fig. 2 MNN 1s architecture

1. Number of layers: 3
2. Training function: scaled conjugate gradient (though in the initial trainings for this MNN, it was Fletcher's conjugate gradient)
3. Transference function, layer 1: Tangent sigmoid
4. Transference function, layer 2: Logarithmic sigmoid

5. Transference function, layer 3: Linear
6. Number of neurons, layer 1: 350
7. Number of neurons, layer 2: 175
8. Number of neurons, layer 3: 50

As with all modules, the training set is 58.3 % of the whole dataset.

A common vector target for ANNs performing recognition tasks indicates belonging to a group with a 1, and the contrary with 0. It was found that for this work, a number higher than 1 is better.

3.4 Preprocessing for MNN 1

The preprocessing consists (as shown in Fig. 3) of:

1. Equalization of histogram.
2. Edge detection.
3. Reduction of images.
4. Binarizing and standardization of images.

In point 3, it should be noted that image reduction is such that the information to manage is one hundredth of that available in the previous stage.

For point 4, the standardization alluded to is such that it improves the capacity of neural networks to "learn" the images.

Standardization in this case consists in processing all items in the training dataset such that for every vector originating from a single image, its mean equals zero and its standard deviation, 1.

3.5 Modular Neural Network 2s Architecture

This MNN is comprised of four modules, which equally divide their dataset, along with an integrator (winner takes all).

The layout for this MNN is shown in Fig. 4.

3.5.1 MNN 2s Parameters by Module

1. Number of neurons, layer 1: 875
2. Number of neurons, layer 2: for successive series, 60, 654, 1,200
3. Number of neurons, layer 3: 125

(At the beginning, inputs to these modules were formatted to Gray code, but it proved unfruitful).

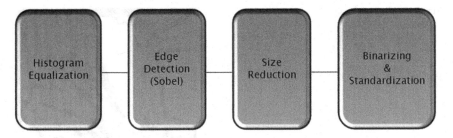

Fig. 3 Block diagram for MNN 1s preprocessing

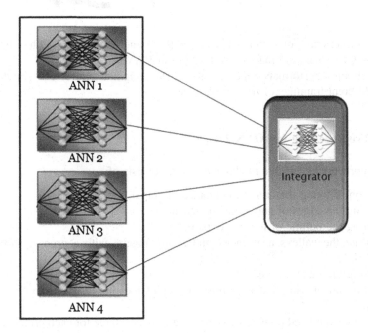

Fig. 4 MNN 2s architecture

3.5.2 MNN 2s Integrator's Network

When it was evident that results of the modules in this MNN were ill-scaled for integration, a NN was developed that could tell which module should be carrying an individual's identity, by first placing each one in the appropriate fourth of the dataset, and therefore the right module.

This network has the following parameters:

1. Number of neurons, layer 1: 3,500
2. Number of neurons, layer 2: 34
3. Number of neurons, layer 3: 4

Fig. 5 Finding the outer
valleys

The success rate for such a net is around 95 % and it took some 20 h of training in a 2.7 GHz dual core machine with 2 Gb of RAM.

To obtain a performance close to 98 %, the weighted sum of two simulations from different trainings of the net was used.

3.5.3 Preprocessing for MNN 2

The preprocessing for this modular network consists of:

1. Custom filtering the image to enhance principal lines.
2. Getting the edges of above mentioned image.
3. Getting the outline of the whole palm.
4. Finding the valleys near index and small fingers, with a process shown in Fig. 5.
5. Getting the finger baseline.
6. Getting finger width aided by prior steps plus location of the central reference (stud).
7. Starting from fixed points, the same for all images, three for each principal line, the nearest points to them, presumably belonging to the principal lines are found and three splines are traced through them. This is necessary because edge detection leaves segments too fragmented.
8. The center of each principal line is located with aid from its spline.
9. The total length of each spline is found.
10. The distance from each "center" to the central reference is found.
11. Wrist width is found.
12. Distances from each spline's extremes to the central reference are found.

For point 2, edge detection is performed with Canny's method.

For point 3, outline of the palm is acquired by properly flood filling the resulting image of point 2 and edge detecting the filled image; the resulting image is subtracted with the filled one to obtain the outline.

Fig. 6 A *central line*
touching all half segments

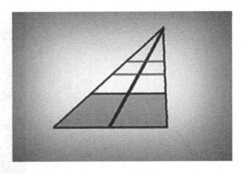

Fig. 7 Image of origin

For point 4, the image obtained in point 3 is used; the motivating idea for the
expected approximateness of the points found to the real ones is a simple fact from
projective geometry: in a square on a plane, with an edge that's away from oneself,
but having it and its opposed edge completely in front of us, a line passed through
each segments' half (or other congruent partition) will go on to touch the point at
"infinity" formed at the crossing of the square's sides not considered so far, and
halfing all the properly aligned squares [11]. It is to be expected that curves that
are not overly bent behave in a similar manner, that being the reason that in
locating the valleys, two points are proposed that one can be sure won't be touched
by any hand and that will be within a valley's curve; then parallels through those
points are traced until they touch the curve; then through both segments' half, a
line is traced until it touches something, which will be the point sought for.

Figure 6 illustrates the mentioned idea. Notice the point at infinity (the upper
blue corner).

For point 6, location of the central reference, found by inspection, is the point in
the stud, at its right, and on the stud's symmetry axis. Knowing the finger baseline
(particularly, its extremes) allows, along with the valleys' location, to estimate
finger widths by calculating distances in this order: extreme—valley—central
reference—valley—extreme.

Figures 7 and 8 show an image prior to spline tracing and another, with its
spline overlayed (notice central reference):

Fig. 8 Overlayed image

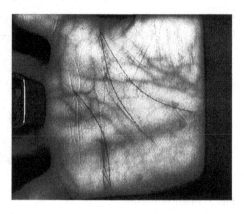

All hands that were left were reflected to become right. The preprocessing of the whole dataset took 1.91 h.

3.6 Fuzzy Integrator

The output from both MNNs are used as input to a fuzzy system to give the final system's response. The integrator defuzzifies with the centroid method and is of Mamdani type (Fig. 9):

The input variables are "imagen" and "geom", with three membership functions (MFs) each (all triangular), as shown plotted in Figs. 10 and 11:

Where the MFs "bajo", "medio", "alto" mean "low", "medium", "high". The output variable is "modular" (Fig. 12):

3.6.1 Fuzzy Rules

There's a total of nine fuzzy rules:

1. If (imagen is bajo) and (geom is bajo) then (modular is imagen)
2. If (imagen is bajo) and (geom is medio) then (modular is imagen)
3. If (imagen is bajo) and (geom is alto) then (modular is geom)
4. If (imagen is medio) and (geom is bajo) then (modular is imagen)
5. If (imagen is medio) and (geom is medio) then (modular is imagen)
6. If (imagen is medio) and (geom is alto) then (modular is geom)
7. If (imagen is alto) and (geom is bajo) then (modular is imagen)
8. If (imagen is alto) and (geom is medio) then (modular is imagen)
9. If (imagen is alto) and (geom is alto) then (modular is imagen)

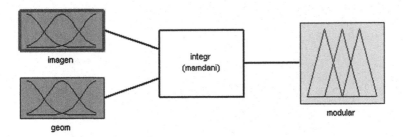

Fig. 9 Fuzzy integrator's architecture

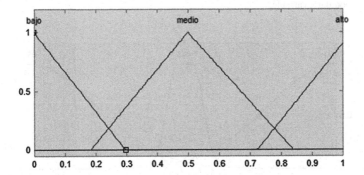

Fig. 10 "Imagen" variable

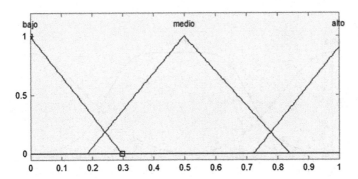

Fig. 11 "Geom" variable

3.6.2 Fuzzy System # 2

Another fuzzy system was created starting from the first one, with the same rules, variables and membership functions, the difference being that instead of triangular membership functions, Gaussian functions and generalized bells were used.

Figures 13 and 14 graphically show the input variables.

Figure 15 shows a plotting of the output variable.

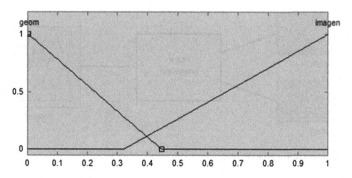

Fig. 12 "Modular" variable

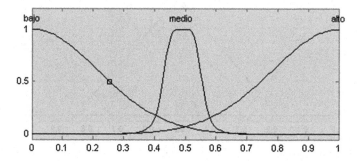

Fig. 13 "Imagen" variable

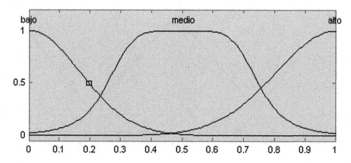

Fig. 14 "Geom" variable

4 Results

The following are several tables for series of ten trainings each with the best identification performer marked in yellow. The tables for MNN 1 have this difference, in incremental order: for the first series, neurons in the hidden layer

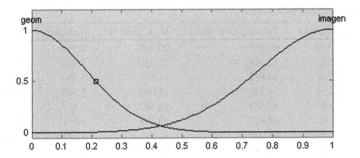

Fig. 15 "Modular" variable

numbered 875; the others, 5 more than its predecessor series. Moreover, the first series didn't use a modified target (in the same series, the same is true for MNN 2).

For MNN 2, for its ordered series, 60, 654 and 1,200 neurons were used in its hidden layer; for the fourth series, 1,200 were also used, but here and in the third series its integrator used the weighted result mentioned in Sect. 3.4; in the fourth series the error goal was 0.005 points lower than all others.

4.1 Results for MNN 1

Results for the first series are shown in Table 1; the best success rate (SR) was 93.74 %.

Results for the second series are shown in Table 2; the best SR was 97.73 %.
Results for the third series are shown in Table 3; the best SR was 97.82 %.
Results for the fourth series are shown in Table 4; the best SR was 97.80 %.

4.2 Results for MNN 2

Results for the first series are shown in Table 5; the best SR was 77.92 %.
Results for the second series are shown in Table 6; the best SR was 80.45 %.
Results for the third series are shown in Table 7; the best SR was 83.03 %.
Results for the fourth series are shown in Table 8; the best SR was 82.96 %.

4.3 Complete System Results

The results in these series were produced with the first FIS, those in the next subsection, with the second.

Results for the first series are shown in Table 9; the best SR was 93.74 %.

Table 1 Results for MNN 1, series 1

Training	Σ Epochs	Training time (min)	Ident. time (s)	Identified	Success rate
1	4,220	40.919	20.465	5,558	92.63
2	4,005	39.028	13.930	5,528	92.13
3	3,798	36.832	14.520	5,556	92.60
4	3,954	38.229	15.065	5,589	93.15
5	*4,143*	*40.055*	*16.562*	*5,624*	*93.74*
6	4,170	38.452	19.728	5,511	91.85
7	4,229	41.021	14.591	5,545	92.42
8	4,143	40.145	14.978	5,524	92.07
9	4,072	39.490	14.076	5,526	92.09
10	4,193	40.461	15.180	5,562	92.70

Table 2 Results for MNN 1, series 2

Training	Σ Epochs	Training time (min)	Ident. time (s)	Identified	Success rate
1	2,151	29.1508	4.5072	5,860	97.67
2	2,177	29.1985	4.3679	5,864	97.73
3	2,169	28.8461	4.6182	5,861	97.68
4	2,181	28.8524	4.8882	5,861	97.68
5	*2,151*	*28.5450*	*5.1730*	*5,864*	*97.73*
6	2,178	28.8729	4.3149	5,856	97.60
7	2,180	28.9235	4.4670	5,858	97.63
8	2,185	28.9521	4.3540	5,860	97.67
9	2,174	28.8521	4.3386	5,860	97.67
10	2,113	28.0448	4.4143	5,851	97.52

Table 3 Results for MNN 1, series 3

Training	Σ Epochs	Training time (min)	Ident. time (s)	Identified	Success rate
1	2,187	35.658	4.6441	5,858	97.63
2	2,203	35.571	4.9544	5,860	97.67
3	2,146	34.855	4.6527	5,867	97.78
4	*2,169*	*29.509*	*4.4374*	*5,869*	*97.82*
5	2,163	28.254	4.7263	5,852	97.53
6	2,129	27.866	4.6170	5,858	97.63
7	2,225	29.200	4.7011	5,866	97.77
8	2,174	28.549	4.2187	5,867	97.78
9	2,160	28.240	4.7767	5,868	97.80
10	2,208	28.880	4.5861	5,868	97.80

Results for the second series are shown in Table 10; the best success rate SR was of 97.75 %.

Results for the third series are shown in Table 11; the best SR was 97.82 %.

Results for the fourth series are shown in Table 12; the best SR was 97.80 %.

Table 4 Results for MNN 1, series 4

Training	Σ Epochs	Training time (min)	Ident. time (s)	Identified	Success rate
1	2,154	28.099	4.8450	5,859	97.65
2	2,175	28.478	4.3345	5,868	97.80
3	2,118	27.695	4.6245	5,859	97.65
4	2,102	27.499	4.7626	5,865	97.75
5	2,145	28.016	4.6701	5,859	97.65
6	2,136	28.012	4.6923	5,861	97.68
7	2,161	28.318	4.1485	5,855	97.58
8	2,454	32.272	4.7369	5,851	97.52
9	2,182	28.696	4.5687	5,867	97.78
10	2,139	28.001	4.7217	5,861	97.68

Table 5 Results for MNN 2, series 1

Training	Σ Epochs	Training time (min)	Ident. time (s)	Identified	Success rate
1	5,874	86.820	31.779	4,653	77.55
2	6,289	87.635	31.785	4,670	77.83
3	4,811	85.611	31.924	4,655	77.58
4	5,487	90.340	31.932	4,668	77.80
5	5,879	89.005	32.041	4,661	77.68
6	4,906	102.528	32.178	4,677	77.95
7	3,818	71.473	31.917	4,662	77.70
8	3,494	57.437	31.595	4,673	77.88
9	3,627	79.091	31.682	4,669	77.82
10	3,851	92.051	31.974	4,675	77.92

Table 6 Results for MNN 2, series 2

Training	Σ Epochs	Training time (min)	Ident. time (s)	Identified	Success rate
1	4,990	250.0487	14.6676	4,813	80.22
2	4,857	232.5927	14.5411	4,812	80.20
3	4,978	236.7514	14.5536	4,817	80.28
4	4,926	240.5805	14.6846	4,809	80.15
5	4,970	238.6703	14.8786	4,807	80.12
6	5,049	218.5406	14.6734	4,813	80.22
7	4,992	222.8399	14.6320	4,810	80.17
8	4,979	227.9744	14.6869	4,816	80.27
9	5,010	223.3908	14.6582	4,827	80.45
10	4,991	224.7490	14.9844	4,814	80.23

4.4 Complete System Results (2)

Results for the first series are shown in Table 13; the best SR was 93.74 %.
Results for the second series are shown in Table 14; the best SR was 97.77 %.

Table 7 Results for MNN 2, series 3

Training	Σ Epochs	Training time (min)	Ident. time (s)	Identified	Success rate
1	5,956	192.942	14.9475	4,938	82.30
2	5,942	192.900	15.2534	4,974	82.90
3	5,996	193.613	14.6771	4,971	82.85
4	*6,033*	*195.942*	*15.6747*	*4,982*	*83.03*
5	5,957	192.291	14.6878	4,963	82.72
6	6,004	193.695	14.9864	4,965	82.75
7	5,963	192.273	14.4913	4,960	82.66
8	6,024	194.540	15.1295	4,967	82.78
9	5,917	190.888	14.3270	4,977	82.95
10	5,926	191.293	15.2687	4,971	82.85

Table 8 Results for MNN 2, series 4

Training	Σ Epochs	Training time (min)	Ident. time (s)	Identified	Success rate
1	6,019	227.865	14.6875	4,976	82.93
2	5,928	227.100	14.8267	4,959	82.65
3	5,908	221.800	14.9905	4,977	82.95
4	5,977	227.459	15.3694	4,935	82.25
5	5,916	225.239	15.2899	4,950	82.50
6	6,081	214.206	15.0198	4,973	82.88
7	6,020	220.839	15.0429	4,978	82.96
8	6,065	220.730	15.7641	4,963	82.72
9	*5,981*	*227.295*	*15.6019*	*4,978*	*82.96*
10	5,935	222.037	15.5677	4,976	82.93

Table 9 Complete system results, series 1

Training	Σ Epochs	Training time (min)	Ident. time (s)	Identified	Success rate
1	10,094	86.820	20.465	5,558	92.63
2	10,294	87.635	13.930	5,528	92.13
3	8,609	85.611	14.520	5,556	92.60
4	9,441	90.340	15.065	5,589	93.15
5	*10,022*	*89.005*	*16.562*	*5,624*	*93.74*
6	9,076	102.528	19.728	5,511	91.85
7	8,047	71.473	14.591	5,545	92.42
8	7,637	57.437	14.978	5,524	92.07
9	7,699	79.091	14.076	5,526	92.09
10	8,044	92.051	15.180	5,562	92.70

Results for the third series are shown in Table 15; the best SR was 97.82 %.
Results for the fourth series are shown in Table 16; the best SR was 97.80 %.
In all previously shown tables, identification time leaves out preprocessing
time.

Table 10 Complete system results, series 2

Training	Σ Epochs	Training time (min)	Ident. time (s)	Identified	Success rate
1	7,141	279.1995	23.4003	5,859	97.65
2	*7,034*	*261.7912*	*23.1624*	*5,865*	*97.75*
3	7,147	265.5975	23.2537	5,860	97.67
4	7,107	269.4329	23.2042	5,861	97.68
5	7,121	267.2153	23.1404	5,864	97.73
6	7,227	247.4135	22.9807	5,856	97.60
7	7,172	251.7634	22.9578	5,858	97.63
8	7,164	256.9265	23.9409	5,860	97.67
9	7,184	252.2429	23.1345	5,860	97.67
10	7,104	252.7938	23.1556	5,851	97.52

Table 11 Complete system results, series 3

Training	Σ Epochs	Training time (min)	Ident. time (s)	Identified	Success rate
1	8,143	228.60	23.9001	5,858	97.63
2	8,145	228.47	23.5287	5,860	97.67
3	8,142	228.47	23.8567	5,867	97.78
4	*8,202*	*225.45*	*23.5640*	*5,869*	*97.82*
5	8,120	220.55	23.5114	5,852	97.53
6	8,133	221.56	23.9038	5,859	97.65
7	8,188	221.47	23.6061	5,866	97.77
8	8,198	223.09	23.3990	5,867	97.78
9	8,077	219.13	23.6577	5,868	97.80
10	8,134	220.17	23.1945	5,868	97.80

Table 12 Complete system results, series 4

Training	Σ Epochs	Training time (min)	Ident. time (s)	Identified	Success rate
1	8,157	295.04	23.1908	5,859	97.65
2	*8,045*	*277.96*	*23.5887*	*5,868*	*97.80*
3	8,109	281.34	23.5317	5,859	97.65
4	8,041	284.97	23.7930	5,865	97.75
5	8,128	283.58	23.0712	5,859	97.65
6	8,198	263.44	23.6783	5,862	97.70
7	8,166	268.05	23.5027	5,854	97.57
8	8,446	277.14	23.3098	5,851	97.52
9	8,205	268.98	23.6990	5,867	97.78
10	8,143	269.64	23.9236	5,861	97.68

The tables in this section show that the greatest boost in performance when integrating both of the MNNs occurred in series 3 of either complete system results, but it has nothing to do with integration, and that MNN 1 always carries most of the weight in finding the correct identifications, and it happens to be the non-geometric MNN.

Table 13 Complete system results (2), series 1

Training	Σ Epochs	Training time (min)	Ident. time (s)	Identified	Success rate
1	10,094	86.820	21.5115	5,558	92.63
2	10,294	87.635	15.3230	5,528	92.13
3	8,609	85.611	14.9720	5,556	92.60
4	9,441	90.340	16.5715	5,589	93.15
5	*10,022*	*89.005*	*18.2182*	*5,624*	*93.74*
6	9,076	102.528	20.7008	5,511	91.85
7	8,047	71.473	16.0501	5,545	92.42
8	7,637	57.437	16.4758	5,525	92.08
9	7,699	79.091	15.4836	5,526	92.09
10	8,044	92.051	15.6980	5,562	92.70

Table 14 Complete system results (2), series 2

Training	Σ Epochs	Training time (min)	Ident. time (s)	Identified	Success rate
1	7,141	279.1995	23.5901	5,860	97.67
2	*7,034*	*261.7912*	*23.3187*	*5,866*	*97.77*
3	7,147	265.5975	23.5999	5,861	97.68
4	7,107	269.4329	24.0719	5,860	97.67
5	7,121	267.2153	24.4985	5,864	97.73
6	7,227	247.4135	23.3928	5,858	97.63
7	7,172	251.7634	23.5359	5,858	97.63
8	7,164	256.9265	23.4708	5,860	97.67
9	7,184	252.2429	23.3946	5,860	97.67
10	7,104	252.7938	23.9661	5,851	97.52

Table 15 Complete system results (2), series 3

Training	Σ Epochs	Training time (min)	Ident. time (s)	Identified	Success rate
1	8,143	228.60	23.6137	5,858	97.63
2	8,145	228.47	23.3012	5,860	97.67
3	8,142	228.47	23.5831	5,867	97.78
4	*8,202*	*225.45*	*23.2797*	*5,869*	*97.82*
5	8,120	220.55	23.7590	5,852	97.53
6	8,133	221.56	23.2175	5,858	97.63
7	8,188	221.47	23.1074	5,866	97.77
8	8,198	223.09	23.5439	5,867	97.78
9	8,077	219.13	23.5984	5,868	97.80
10	8,134	220.17	23.3198	5,868	97.80

Prior experiences with lower yielding trainings of the MNNs showed that an improvement of about 1.9 % could be obtained from the complete system over MNN 1, for a 95 % total.

Table 16 Complete system results (2), series 4

Training	Σ Epochs	Training time (min)	Ident. time (s)	Identified	Success rate
1	8,157	295.04	23.3043	5,859	97.65
2	*8,045*	*277.96*	*23.5547*	*5,868*	*97.80*
3	8,109	281.34	23.7833	5,859	97.65
4	8,041	284.97	23.8182	5,865	97.75
5	8,128	283.58	24.3388	5,859	97.65
6	8,198	263.44	23.3681	5,861	97.68
7	8,166	268.05	23.9501	5,855	97.58
8	8,446	277.14	23.4867	5,850	97.50
9	8,205	268.98	23.2219	5,867	97.78
10	8,143	269.64	23.7547	5,861	97.68

The best performance in these series that has to do with integration is in series 2 with the second FIS, for a 97.77 % total.

What happens in these series is easy to explain, even though both MNNs get to perform better: since MNN 2 doesn't improve as much as MNN 1 (over initial experiences), there is close to no net gain in using MNN 2, and in fact, sometimes there is a loss; this is only marginally better with FIS #1, which gave the best overall performance, at 97.82 % (though without a FIS direct relationship).

A few comparisons with prior works in the same (or close) area of research can be very illustrative:

Jiansheng Cheng's group uses the PolyU palmprint database (different to the one used with this work), applying their own 2D extension of the SAX standard to encode regions of interest of every image, and uses a similarity comparison to match the templates, obtaining a success rate of 98.901 % for identifications [12].

Guo [13] uses the same database we do, but uses all spectra at the same time; he uses a fusion algorithm based on Haar wavelets and PCA, and on one of many configurations tried, for identification tasks obtained 97.877 % as a success rate.

Kumar and Zhang [14] use entropy based discretization for a database smaller than ours and use Support vector machines and neural networks as classifiers for a respective accuracy of 95 and 94 % identifying, for a hand geometry method.

Öden's et al. [15] claims that their method, using implicit polynomials, is capable of achieving a 95 % accuracy in identification tasks.

Lu et al. [16] process palm images to extract their eigenpalms, intended and supposed to convey the principal components of the palms, they then work with them considering they're embedded in an eigenspace, for matching purposes; they report a 99.149 % accuracy for identification tasks, with a database smaller than ours.

Wu et al. [17] using a database similar in size to the aforementioned work's, applied to their images the Fisher's linear discriminant transform; a Fisherspace exists in this case too. In some instances identification rates are better than verification rates; the method obtained an accuracy of 99.75 % for identifications.

Table 17 Results of methods

Author	Year	Method	Success rate
Cheng et al. [12]	2010	2D SAX coding	98.901
Guo [13]	2009	PCA fusion	97.877
Kumar and Zhang [14]	2007	H. Geometry—Entropy b.d.—SVM	95
Kumar and Zhang [14]	2007	H. Geometry—Entropy b.d.—NN	94
Lu et al. [16]	2003	Eigenpalms	99.149
Wu et al. [17]	2003	Fisherpalms	99.750
Öden et al. [15]	2001	Implicit polynomials	95
Gayathri—Ramamoorthy [14]	2012	wavelet-based fusion—PCA	98.82
Gayathri—Ramamoorthy [14]	2012	wavelet-based fusion	95
Sánchez, Melín (proposed method)	2012	H. Geometry—Modular NN—Fuzzy integrator	97.820

It is noteworthy that, when dealing with hand geometry systems in verification mode, Ingersoll Rand's ID3D system is reported by Sandia Labs as performing as low as 0.2 % of total error, this as far back as 1991 [18].

Table 17 shows a summary of the identification results of works considered in this section.

5 Conclusions and Future Work

As seen by the comparisons exposed in the previous section, the system's performance, compared to some prior works does not seem too bad, but as a whole, comparisons show that there is still much to be improved and open a few lines of future work. First, trying to elevate performance of the net within MNN 2s integrator; that would gain no more than a few tens of well identified individuals over MNN 1 alone. Second, performance problems might reside in the arrangement of modules (their number) in MNN 2. Third, comparing with other systems including verification, would be aided by performing complete statistical evaluations of the system. As a final remark, a true verification comparison should take into account that MNN 1s modules on their own have a success rate at least two percentage points higher than MNN 1 itself.

References

1. Melin, P., Castillo, O.: Hybrid Intelligent Systems for Pattern Recognition Using Soft Computing. Springer, Berlin, Heidelberg (2005)
2. Melin, P., Olivas, F., Castillo, O., Valdez, F., Soria, J., García Valdez, J.: Optimal design of fuzzy classification systems using PSO with dynamic parameter adaptation through fuzzy logic. Expert Syst. Appl. **40**(8), 3196–3206 (2013)

3. Montiel, O., Camacho, J., Sepúlveda, R., Castillo O.: Fuzzy system to control the movement of a wheeled mobile robot. In: Soft Computing for Intelligent Control and Mobile Robotics, pp. 445–463. Springer, Berlin, Heidelberg (2011)
4. Montiel, O., Sepulveda, R., Melin, P., Castillo, O., Porta, M., Meza, I.: Performance of a simple tuned fuzzy controller and a PID controller on a DC motor. In: FOCI 2007, pp. 531–537
5. Kong, A., Zhang, D., Kamel, M.: A survey of palmprint recognition. Pattern Recogn. **42**, 1408–1418 (2009)
6. Jang, R., Tsai-Sun, C., Mizutani, E.: Neuro-Fuzzy and Soft Computing, 1st edn. Prentice Hall, Upper Saddle River (1996)
7. Hagan, M., et al.: Neural Network Design. Thomson Learning, Pacific Grove (1996)
8. Luger, G., Stubblefield, W.: Artificial Intelligence. Structures and Strategies for Complex Problem Solving, 3rd edn. Addison Wesley, Reading (1998)
9. Rojas, R.: Neural Networks: A Systematic Introduction. Springer, New York (1996)
10. PolyU Multispectral Palmprint Database. http://www4.comp.polyu.edu.hk/Ebiometrics/MultispectralPalmprint/MSP_files. Accessed October 2011
11. Coxeter, H.: Projective Geometry, 2nd edn. Springer, New York (2003)
12. Chen, J., Moon, Y., et al.: Palmprint authentication using a symbolic representation of images. Image Vis. Comput. **28**, 343–351 (2010)
13. Guo, Z.: Online multispectral palmprint recognition. Thesis, Hong Kong Polytechnic University (2009)
14. Kumar, A., Zhang, D.: Biometric recognition using entropy-based discretization. In: Proceedings of 2007 IEEE International Conference on Acoustics Speech and Signal Processing, pp. 125–128 (2007)
15. Öden, C., Yildiz, T., et al.: Hand recognition using implicit polynomials and geometric features. In: Lecture Notes in Computer Science, vol. 2091, pp. 336–341. Springer, New York (2001)
16. Lu, G., Zhang, D., et al.: Palmprint recognition using eigenpalms features. Pattern Recogn. Lett. **24**, 1463–1467 (2003)
17. Wu, X., Zhang, D., et al.: Fisherpalms based palmprint recognition. Pattern Recogn. Lett. **24**, 2829–2838 (2003)
18. Zunkel, R.: Hand geometry based authentication. In: Jain, A.K., Bolle, R., Pankanti, S. (eds.) Biometrics: Personal Identification in Networked Society. Kluwer Academic Publishers, Boston (1998)

Echocardiogram Image Recognition Using Neural Networks

Beatriz González, Fevrier Valdez, Patricia Melin
and German Prado-Arechiga

Abstract In this chapter we present a neural network architecture to recognize if the echocardiogram image corresponds to a person with a heart disease or is an image of a person with a normal heart, so that it can facilitate the medical diagnosis of the person that may hold an illness. One of the most used methods for the detection and analysis of diseases in the human body by doctors and specialists is the use of medical imaging. These images become one of the possible means to achieve a safe estimate of the severity of the injuries and thus to initiate treatment for the benefit of the patient.

1 Introduction

Research on medical imaging has been growing over the past few years as it is normally a non-invasive method of diagnosis and so its results may bring benefits to people's health. These research works cover many aspects of image processing and medicine, such as disease predicting and accurate diagnostics [12]. For this reason, we are considering in this chapter the application of neural networks for recognition of medical images.

Normally there are many kinds of noise in the ultrasound images. So the resulting images are contaminated with this noise that corrodes the borders of the cardiac structures [11]. This characteristic turns difficult the automatic image processing, and specially the pattern recognition. Besides this kind of noise, other factors influence the outcome of an ultrasound image. For instance in pregnancy

B. González · F. Valdez · P. Melin (✉)
Tijuana Institute of Technology, Calzada Tecnologico s/n, Tijuana, Mexico
e-mail: pmelin@tectijuana.mx

G. Prado-Arechiga
Excel Medical Center, Paseo de los Heroes No. 2507, Zona Rio, Tijuana, Mexico

O. Castillo et al. (eds.), *Recent Advances on Hybrid Approaches for Designing Intelligent Systems*, Studies in Computational Intelligence 547,
DOI: 10.1007/978-3-319-05170-3_29, © Springer International Publishing Switzerland 2014

Normal   Hypokinesia

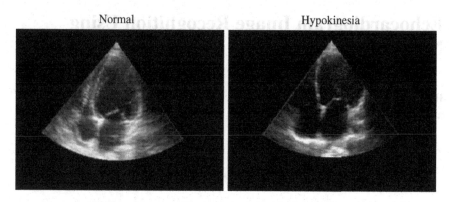

Fig. 1 (*left*) Echocardiographic frames from normal (apical 4 chamber view); (*right*) Motions of the heart for Hypokinesia patients [6]

tests, the transducer and the fetus position, the rotation and the scale variations in images of different patients and the composition of the tissue separating the fetus heart are issues that must be taken into account when dealing with heart images [23, 24].

However, the disease recognition problem is complicated by the heart's non-rigid motion (see Fig. 1). Furthermore, the poor imaging quality of 2D echo videos due to low contrast, speckle noise, and signal dropouts also cause problems in image interpretation [6, 7].

The rest of the chapter describes this approach in detail and is organized as follows. In Sect. 2, we describe some of the works related to image recognition echocardiography and basic concepts. In Sect. 3 we describe the methodology used for pattern recognition imaging echocardiography, and the database for this study, the pre-processing that is used, and the neural network that was designed. In Sect. 4 experimental results are presented. In Sect. 5 the conclusions are offered.

2 Background and Basic Concepts

2.1 Previous Work

In the first automatic cardiac view recognition system, Ebadollahi et al. [8] proposed a constellation-of-parts based method. They used a generic heart chamber detector [1] to locate heart chambers, and they represented the spatial arrangement of the chambers using a Markov Random Field (MRF) based relational graph. Final classification of a test image was performed using a Support Vector Machine on the MRF network output. This method suffers from sensitivity of the chamber detection method to frequently present noise in the echocardiogram images while demonstrating limited robustness to basic image transformations.

In another work of Ebadollani et al. an automatic identification of the views of the heart from the content of the echocardiogram videos is presented. In this approach the structure of the heart is represented by the constellation of its parts (chambers) under the different views. The statistical variations of the parts in the constellation and their spatial relationships are modeled using Markov Random Field models. A discriminative method is then used for view recognition which fuses the assessments of a test image by all the view-models [14].

Jacob et al. [2] and Sugioka et al. [19] developed research using patterns to detect cardiac structures using active contours (snakes) in echocardiographic images. In another work, Comaniciu [9] proposed a methodology to tracking cardiac edges in echocardiographic images using several information extracts of the images.

2.2 Echocardiography

Echocardiography is an important diagnostic aid in cardiology for the morphological and functional assessment of the heart. During an echocardiogram exam, sonographer images of the heart using ultrasound by placing a transducer against the patient's chest. Reflected sound waves reveal the inner structure of the heart walls and the velocities of blood flows. Since these measurements are typically made using 2D slices of the heart, the transducer position is varied during an echo exam to capture different anatomical sections of the heart from different viewpoints [4] (Fig. 2).

Echocardiography is often used to diagnose cardiac diseases related to regional and wall motion as well as valvular motion abnormalities. It provides images of cardiac structures and their movements giving detailed anatomical and functional information about the heart [5].

2.3 Noise in Ultrasound

Speckle is a characteristic phenomenon in laser, synthetic aperture radar images, or ultrasound images. Its effect is a granular aspect in the image. Speckle is caused by interference between coherent waves that, backscattered by natural surfaces, arrive out of phase at the sensor [3, 10]. Speckle can be described as random multiplicative noise. It hampers the perception and extraction of fine details in the image. Speckle reduction techniques can be applied to ultrasound images in order to reduce the noise level and improve the visual quality for better diagnoses [11].

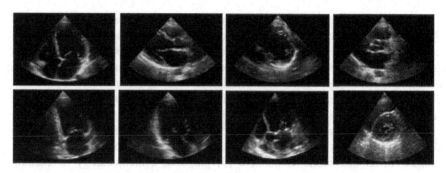

Fig. 2 Images from the eight echocardiogram viewpoints in this chapter. In clockwise order from top left, apical four chamber (a4c), parasternal long axis (pla), parasternal short axis—basal (psab), parasternal short axis—papillary (psap), apical two chambers (a2c), apical three chambers (a3c), apical five chambers (a5c), parasternal short axis—mitral (psam) [4]

3 Methodology

The methodology used for Echocardiography recognition in this chapter is described as follows:

- Database Acquisition
- Application of Preprocessing
- Neural Network Design
- Neural Network Training

    - Scaled Conjugate Gradient (SCG)
    - Gradient descent with adaptive learning rate backpropagation (GDA)
    - Gradient Descent with Momentum and Adaptive Learning Rate (GDX)

3.1 Data Acquisition

To create the database, we acquired videos of echocardiograms, of 30 patients, of which 18 echocardiograms are some heart in-disease patients and 12 echocardiograms from patients without any disease, in where each video is captured with 10 images. The size of the images is 200 × 125 pixels.

3.2 Preprocessing

For image preprocessing, we reduced the image size from 400 × 325 to 200 × 125 pixels taken the region of interest (ROI) to eliminate as much as possible the noise. The first thing we did was to clean up the image, then we apply a filter to the image

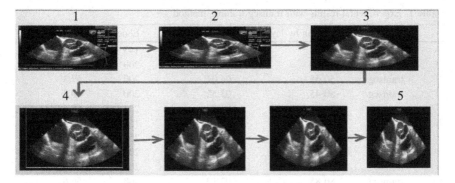

Fig. 3 In the above images *1* captured image, *2* clean image, *3* image filter to reduce the noise, *4* crop image, *5* color image

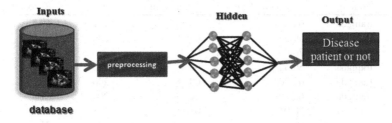

Fig. 4 Monolithic neural network architecture

to reduce noise, then we cut the image so that only the part of the image that is of interest remains, and finally we apply color to these images (Fig. 3).

3.3 Neural Network

We are considering a monolithic neural network in this chapter. Figure 4 shows the architecture used for this work and is described as follows: We have a data base, then we apply preprocessing to the images, then in the architecture of the neural network the output is if the image corresponds to a person with a cardiac illness or not.

4 Experimental Results

For the 30 patients in the study, we trained the neural network with the 2 available images of each patient. The size of the images is of 200 × 125 pixels.

Table 1 shows the experimental results with the "trainscg" training method, with 150 and 200 epochs and a goal error of 0.001.

Table 1 Experimental results with trainscg training method

	Training	Neuron by layer	Training time	Epoch	Recognition %
1	trainscg	60 55	01:26	143	59.1
2	trainscg	60 50	01:55	200	64.7
3	trainscg	50 55	04:04	200	57.3
4	trainscg	50 45	03:52	200	60.3
5	trainscg	42 40	03:27	200	58.5
6	trainscg	40 50	02:25	150	69.4
7	trainscg	70 60	07:13	250	58.6
8	trainscg	50 50	04:11	200	64.6
9	trainscg	65 65	01:52	200	70.3
10	trainscg	80 85	02:12	200	56.6

Table 2 Experimental results with traingda training method

	Training	Neuron by layer	Training time	Epoch	Recognition %
1	traingda	62 64	00:54	191	69.5
2	traingda	50 55	00:45	200	77.6
3	traingda	70 60	01:01	200	73.6
4	traingda	50 45	00:40	200	61.2
5	traingda	65 70	00:58	200	63.8
6	traingda	60 60	00:54	200	81.8
7	traingda	90 95	01:14	198	86.7
8	traingda	60 55	00:51	200	75
9	traingda	56 54	00:49	200	82.5
10	traingda	53 50	00:48	200	83.3

Table 3 Experimental results with traingdx training method

	Training	Neuron by layer	Training time	Epoch	Recognition %
1	traingdx	90 90	01:13	199	68.2
2	traingdx	60 60	00:47	200	60.1
3	traingdx	50 45	00:44	200	80.9
4	traingdx	88 72	02:41	200	59.1
5	traingdx	53 50	00:44	200	77.6
6	traingdx	56 54	00:56	200	60.4
7	traingdx	65 65	02:58	247	65.4
8	traingdx	75 75	03:29	200	60.7
9	traingdx	70 60	01:49	146	64.1
10	traingdx	64 64	01:27	144	60.5

Table 2 shows the experimental results with the "traingda" training method, with 200 epochs and a goal error of 0.001.

Table 3 shows the experimental results with the "traingd" training method, with 200 epochs and a goal error of 0.001.

Table 4 Learning algorithm comparison

Training	Neuron by layer	Training time	Epoch	Recognition %
trainscg	65 65	01:52	200	70.3
traingda	90 95	01:14	198	86.7
traingdx	53 50	00:44	200	77.6

Table 4 shows the learning algorithm comparison. The best learning algorithm in this case was Gradient descent with adaptive learning rate backpropagation (GDA). Comparing with scaled Conjugate Gradient (SCG) and gradient Descent with Momentum and Adaptive Learning Rate (GDX), GDA gets the best training time and recognition.

5 Conclusions

In this chapter we presented a monolithic neural network architecture to recognize if the image is of a person with a heart disease or is an image of a person with normal heart. Results obtained are good; the best learning algorithm in this case was gradient descent with adaptive learning rate backpropagation (GDA) with 86.7 % recognition rate. Comparing with scaled conjugate gradient (SCG) and gradient descent with momentum and adaptive learning rate (GDX), GDA gets the best training time and recognition.

Results obtained are good, however, other methods could be used to improve results, for example just to compare a modular neural network architecture can be used [18, 19, 20, 21, 22], or apply another kind of preprocessing, like in [13, 14, 15, 16, 17]. Other future work could include applying an optimization method to the design the neural network, like in [25, 26, 27].

Acknowledgments We would like to express our gratitude to CONACYT and Tijuana Institute of Technology for the facilities and resources granted for the development of this research.

References

1. Bailes, R.: The use of the gray level sat to find the salient cavities in echocardiograms. J. Vis. Commun. Image Represent. **7**(2), 169–195 (1996)
2. Banning, A.P., Behrenbruch, C., Kelion, A.D., Jacob, G., Noble, J.A.: A shape-space-based approach to tracking myocardial borders and quantifying regional left-ventricular function applied in echocardiography. Medical Imaging, IEEE, pp. 226–238 (2002)
3. Barillot, C., Greenleaf, J.F., Kinter, T.M., McCann, H.A, McEwan, C.N., Sharp, J.C., Multidimensional ultrasonic imaging for cardiology. In: Proceedings of the IEEE, pp. 1063–1071 (1988)

4. Beymer, D., Kumar, R, Tanveer, S., Wang, F.: Echocardiogram view classification using edge filtered scale-invariant motion feactures. Computer Vision and Pattern Recognition. pp. 723–730 (2009)

5. Beymer, D., Tanveer, S., Wang, F.: Exploiting spatio-temporal information for view recognition in cardiac echo videos. Computer Vision and Pattern Recognition Workshops. Conference on IEEE Computer Society (2008)

6. Beymer, D., Kumair, R., Tanveer, S.M., Wang, F.: Cardiac disease detection from echocardiogram using edge filtered scale-invariant motion feactures. Computer Vision and Pattern Recognition Workshops (CVPRW). Conference on IEEE Computer Society (2010)

7. Cerrolaza, M., Gavidia, G., Landrove, M., Soudah, E.: Generación de modelos discretos de tejidos del ser humano a través del procesamiento y segmentación de imágenes médicas. Revista Internacional de Métodos Numéricos para Cálculo y Diseño en Ingeniería, pp. 200–226 (2011)

8. Chang, S.F., Ebadollahi, S., Wu, H.: Automatic view recognition in echocardiogram videos us ing parts-based representation. Computer Vision and Pattern Recognition CVPR. Conference on IEEE Computer Society pp.2–9, (2004)

9. Comaniciu, D., Krishnan, S., Zhou, X.S.: Robust real-time myocardial border tracking for echocardiography. Med. Imaging IEEE Trans 23, 849–860 (2004)

10. Ealy, M., Lu, D., Weaver, J.B.: Contrast enhancement of medical images using multiscale edge representation. Opt. Eng. 33(7), 1251–1261 (1994)

11. Kang, S.C., Hong, S. H.: A speckle reduction filter using wavelet-based methods for medical imaging application. International Conference on Digital Signal Processing (DSP), pp. 1169–1172 (2002)

12. Lemos, M., Navaux, A., Nehme, D., Olvier, P., Cardiac structure recognition ultrasound images. Conference focused on Speech and Image Processing, Multimedia Communications and Services, pp. 463–466 (2007)

13. Melin, P., Castillo, O.: Hybrid Intelligent Systems for Pattern Recognition. Springer, Heidelberg (2005)

14. Melin, P., Castillo, O.: An intelligent hybrid approach for industrial quality control combining neural networks, fuzzy logic and fractal theory. Inf. Sci. 177, 1543–1557 (2007)

15. Mendoza, O., Melin, P., Castillo, O., Licea, G.,: Type-2 fuzzy logic for improving training data and response integration in modular neural networks for image recognition. Lecture Notes in Artificial Intelligence, vol. 4529, pp. 604–612 (2007)

16. Mendoza, O., Melin, P., Castillo, O.: Interval type-2 fuzzy logic and modular neural networks for face recognition applications. Appl. Soft Comput. J. 9, 1377–1387 (2009)

17. Mendoza, O., Melin, P., Licea, G.: Interval type-2 fuzzy logic for edges detection in digital images. Int. J. of Intell. Syst. 24, 1115–1133 (2009)

18. Sánchez, D., Melin, P.: Modular neural network with fuzzy integration and its optimization using genetic algorithms for human recognition based on iris, ear and voice biometrics. In: Melin, P., et al. (eds.) Soft Computing for Recognition Based on Biometrics, pp. 85–102. (2010)

19. Sánchez, D., Melin, P., Castillo, O., Valdez, F.: Modular neural networks optimization with hierarchical genetic algorithms with fuzzy response integration for pattern recognition. MICAI 2, 247–258 (2012)

20. Sánchez, D., Melin, P.: Multi-objective hierarchical genetic algorithm for modular granular neural network optimization. In: Batyrshin, I., Gonzalez Mendoza, M. (eds.) Soft Computing Applications in Optimization, Control, and Recognition, pp. 157–185 (2013)

21. Sánchez D., Melin P., Oscar Castillo, Fevrier Valdez: Modular granular neural networks optimization with Multi-Objective Hierarchical Genetic Algorithm for human recognition based on iris biometric. IEEE Congress on Evolutionary Computation pp. 772–778 (2013)

22. Sánchez, D., Melin, P.: Optimization of modular granular neural networks using hierarchical genetic algorithms for human recognition using the ear biometric measure. Eng. Appl. AI. 27, 41–56 (2014)

23. Shahram, E., Shih-Fu, C., Wu, H.: Automatic view recognition in echocardiogram videos using parts-based representation. Computer Vision and Pattern Recognition (2004)
24. Sugioka, K., Automated quantification of left ventricular function by the automated contour tracking method. J. Cardiovasc. Ultrasound Allied Tech. 20(4), 313–318 (2003)
25. Valdez, F., Melin, P., Castillo, O., Evolutionary method combining particle swarm optimization and genetic algorithms using fuzzy logic for decision making. In: Proceedings of the IEEE International Conference on Fuzzy Systems, pp. 2114–2119 (2009)
26. Valdez, F., Melin, P., Castillo, O.: Parallel particle swarm optimization with parameters adaptation using fuzzy logic. MICAI 2, 374–385 (2012)
27. Vázquez, J., Valdez, F., Melin, P.: Comparative study of particle swarm optimization variants in complex mathematics functions. In: Castillo, O. (eds.) Recent Advances on Hybrid Intelligent Systems, pp. 223–235. Springer, Heidelberg (2013)

Face Recognition with Choquet Integral in Modular Neural Networks

Gabriela E. Martínez, Patricia Melin, Olivia D. Mendoza
and Oscar Castillo

Abstract In this chapter a new method for response integration, based on Choquet Integral is presented. A type-1 fuzzy system for edge detections based in Sobel and Morphological gradient is used, which is a pre-processing applied to the training data for better performance in the modular neural network. The Choquet integral is used how method to integrate the outputs of the modules of the modular neural networks (MNN). A database of faces was used to perform the pre-processing, the training, and the combination of information sources of the MNN.

1 Introduction

An integration method is a mechanism which takes as input a number n of data and combines them to result in a value representative of the information, methods exist which combine information from different sources which can be aggregation operators as arithmetic mean, geometric mean, OWA [1], and so on., In a modular neural network (MNN) is common to use some methods like fuzzy logic Type 1 and Type 2 [2–4], the fuzzy Sugeno integral [5], Interval Type-2 Fuzzy Logic Sugeno Integral [6], a probabilistic sum integrator [7], A Bayesian learning method [8], among others.

The Choquet integral is an aggregation operator which has been successfully used in various applications [9–11], but has not been used as a method of integrating a modular neural network, and in this chapter is proposed for achieving this goal 12.

G. E. Martínez · O. D. Mendoza
University of Baja California, Tijuana, Mexico

P. Melin (✉) · O. Castillo
Tijuana Institute of Technology, Tijuana, Mexico
e-mail: pmelin@tectijuana.mx

O. Castillo et al. (eds.), *Recent Advances on Hybrid Approaches for Designing Intelligent Systems*, Studies in Computational Intelligence 547,
DOI: 10.1007/978-3-319-05170-3_30, © Springer International Publishing Switzerland 2014

This chapter is organized as follows: Sect. 2 shows the concepts of Fuzzy Measures and Choquet integral which is the technique that was applied for the combination of the several information sources. Section 3 presents Edge detection based in Sobel and Morphological gradient with type-1 fuzzy system. Section 4 shows the modular neural network proposed and in the Sect. 5 the simulation results are shown. Finally, Sect. 6 shows the conclusions.

2 Fuzzy Measures and Choquet Integral

Sugeno first defined the concept of "fuzzy measure and fuzzy integral" in 1974 [13]. A fuzzy measure is a nonnegative function monotone of values defined in "classical sets". Currently, when referring to this topic, the term "fuzzy measures" has been replaced by the term "monotonic measures", "non-additive measures" or "generalized measures" [14–16].When fuzzy measures are defined on fuzzy sets, we speak of fuzzified measures monotonous [16].

2.1 Fuzzy Measures

A fuzzy measure μ with respect to the dataset X, must satisfy the following conditions:

1.
$$\mu(X) = 1; \ \mu(\emptyset) = 0$$

2.
$$Si \ A \subset B, \ then \ \mu(A) \leq \mu(B)$$

In condition 2 A and B are subsets of X.

A fuzzy measure is a Sugeno measure or λ-fuzzy, if it satisfies the condition (1) of addition for some $\lambda > -1$.

$$\mu(A \cup B) = \mu(A) + \mu(B) + \lambda\mu(A) \ \mu(B) \tag{1}$$

where λ can be calculated with (2):

$$f(\lambda) = \left\{ \prod_{i=1}^{n} (1 + M_i(x_i)\lambda) \right\} - (1 + \lambda) \tag{2}$$

The value of the parameter λ is determined by the conditions of the theorem 1.

Theorem 1 *Let $\mu(\{x\}) < 1$ for each $x \in X$ and let $\mu(\{x\}) > 0$ for at least two elements of X. Then (2) determines a unique parameter λ in the following way:*
If $\sum_{x \in X} \mu(\{x\}) < 1$, then λ is in the interval $(0, \infty)$.
If $\sum_{x \in X} \mu(\{x\}) = 0$, then $\lambda = 0$; That is the unique root of the equation.
If $\sum_{x \in X} \mu(\{x\}) > 1$, then λ se is in the interval $(-1, 0)$.

The method to calculate Sugeno measures, carries out the calculation in a recursive way, using (3) and (4).

$$\mu(A_1) = \mu(M_i) \tag{3}$$

$$\mu(A_i) = \mu(A_{(i-1)}) + \mu(M_i) + (\lambda \mu(M_i) * \mu(A_{(i-1)})) \tag{4}$$

where $1 < M_i \leq \cdots \leq n$, and the values of $\mu(x_i)$ correspond to the fuzzy densities determined by an expert.

To perform this procedure $\mu(M_i)$ should be permuted with respect to the descending order of their respective $\mu(A_i)$.

There are 2 types of Integral, the integral of Sugeno and Choquet Integral.

2.2 Choquet Integral

The Choquet integral can be calculated using (5) or an equivalent expression (6)

$$Choquet = \sum_{i=1}^{n} \left\{ \left[A_i - A_{(i-1)} \right] * D_i \right\} \tag{5}$$

With $A_0 = 0$

Or also

$$Choquet = \sum_{i=1}^{n} A_i * \left\{ \left[D_i - D_{(i+1)} \right] \right\} \tag{6}$$

With $D_{(n+1)} = 0$

where A_i represents the fuzzy measurement associated with data D_i.

2.2.1 Pseudocode of Choquet Integral

INPUT: Number of information sources n; information sources x_1, x_2, ..., x_n; fuzzy densities of information sources M_1, M_2, ..., $M_n \in (0,1)$.
OUTPUT: Choquet integral $(\sigma(x_1), \sigma(x_2), ..., \sigma(x_n))$.
STEP 1: Calculate λ finding the root of the function (2).
STEP 2: Fuzzify variable x_i.

$$D_i = \{x, \mu_{Di}(x) | x \in X\}, \ \mu D_i(x) \in [0, 1]$$

STEP 3: Reorder M_i with respect to D (x_i) in descending order
STEP 4: Calculate fuzzy measures for each data with (3), (4).
STEP 5: Calculate Choquet integral with (5) or (6).
STEP 6: OUTPUT Choquet.
STOP

3 Edge Detection

Edge detection can be defined as a method consisting of identifying changes that exist in the light intensity, which can be used to determine certain properties or characteristics of the objects in the image.

We used the database of the ORL [17] to perform the training of the modular neural network, which has images of 40 people with 10 samples of each individual. To each of the images was applied a pre-processing by making use of Sobel edge detector and morphological gradient with type 1 fuzzy logic system [18] in order to highlight features, some of the images can be displayed in Fig. 4a.

3.1 The Morphological Gradient

To perform the method of morphological gradient is calculated every one of the four gradients as commonly done in the traditional method using (7–11), see Fig. 1, however, the sum of the gradients is performed by a fuzzy system type 1 [17] as shown in Fig. 2, and the resulting image can be viewed in Fig. 4b (Fig. 3).

$$D1 = \sqrt{(z5 - z2)^2 + (z5 - z8)^2} \tag{7}$$

$$D2 = \sqrt{(z5 - z4)^2 + (z5 - z6)^2} \tag{8}$$

$$D3 = \sqrt{(z5 - z1)^2 + (z5 - z9)^2} \tag{9}$$

$$D4 = \sqrt{(z5 - z7)^2 + (z5 - z3)^2} \tag{10}$$

$$G = D1 + D2 + D3 + D4 \tag{11}$$

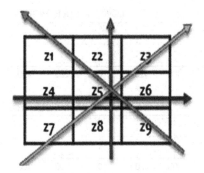

Fig. 1 Calculation of the gradient in the 4 directions

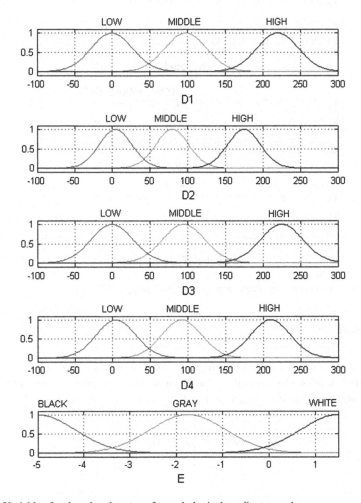

Fig. 2 Variables for the edge detector of morphological gradient type 1

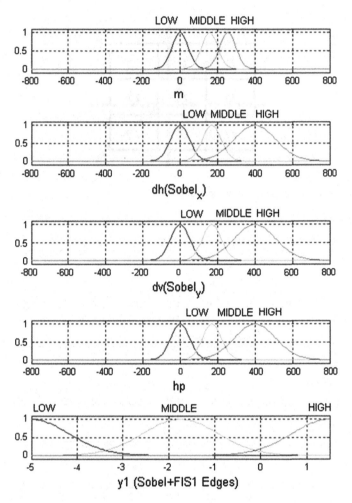

Fig. 3 Variables for the edge detector with the type-1 fuzzy Sobel

3.2 Sobel

The Sobel operator is applied to a digital image in gray scale, is a pair of 3×3 convolution masks, one estimating the gradient in the x-direction (columns) (12) and the other estimating the gradient in the y-direction (rows) (13) [19].

$$sobel_x = \begin{bmatrix} -1 & 0 & 1 \\ -2 & 0 & 2 \\ -1 & 0 & 1 \end{bmatrix} \qquad (12)$$

$$sobel_y = \begin{bmatrix} 1 & 2 & 1 \\ 0 & 0 & 0 \\ -1 & -2 & -1 \end{bmatrix} \tag{13}$$

If we have $I_{m,n}$ as a matrix of m rows and r columns where the original image is stored, then g_x and g_y are matrices having the same dimensions as I, which at each element contain the horizontal and vertical derivative approximations and are calculated by (14) and (15) [19].

$$g_x = \sum_{i=1}^{i=3} \sum_{j=1}^{j=4} Sobel_{x,ij} * I_{r+i-2,c+j-2} \quad \begin{matrix} for = 1,2,\ldots,m \\ for = 1,2,\ldots,n \end{matrix} \tag{14}$$

$$g_y = \sum_{i=1}^{i=3} \sum_{j=1}^{j=4} Sobel_{y,i.j} * I_{r+i-2,c+j-2} \quad \begin{matrix} for = 1,2,\ldots,m \\ for = 1,2,\ldots,n \end{matrix} \tag{15}$$

In the Sobel method the gradient magnitude g is calculated by (16).

$$g = \sqrt{g_x^2 + g_y^2} \tag{16}$$

For the type-1 fuzzy inference system, 3 inputs can be used, 2 of them are the gradients with respect to the x-axis and y-axis, calculated with (14) and (15), which we call DH and DV, respectively. The third variable M is the image after the application of a low-pass filter hMF in (17); this filter allows to detect image pixels belonging to regions of the input were the mean gray level is lower. These regions are proportionally affected more by noise, which is supposed to be uniformly distributed over the whole image [19].

$$hMF = \frac{1}{25} * \begin{bmatrix} 1 & 1 & 1 & 1 & 1 \\ 1 & 1 & 1 & 1 & 1 \\ 1 & 1 & 1 & 1 & 1 \\ 1 & 1 & 1 & 1 & 1 \\ 1 & 1 & 1 & 1 & 1 \end{bmatrix} \tag{17}$$

After applying the edge detector type 1 with Sobel, the resulting image can be viewed in Fig. 4c.

4 Modular Neural Networks

Were trained a MNN of 3 modules with 80 % of the data of ORL. Each image was divided into 3 sections horizontal and each of which was used as training data in each of the modules, as shown in Fig. 5.

The integration of the modules of the MNN was made with the Choquet integral.

(a)

(b)

(c)

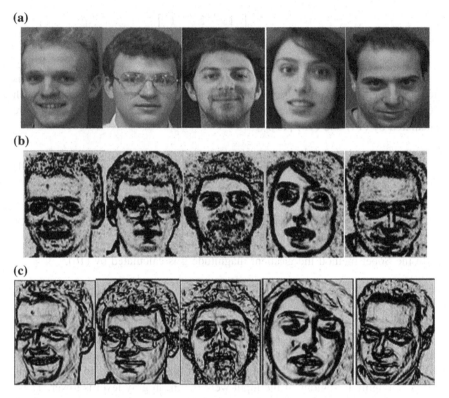

Fig. 4 **a** Original image, **b** image with morphological gradient, **c** image with Sobel

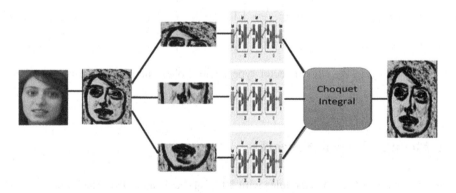

Fig. 5 Architecture proposal of the modular neural network

Table 1 Distribution of the training data

Training (%)	Validation (%)	Test (%)
70	15	15

Table 2 Procedure performed in the experiment

1.—Define the database of images
2.—Define the edge detector
3.—Detect the edges of each image
4.—Add the edges to the train set
5.—Divide the images in three parts
6.—Calculate the recognition rate using the k-fold cross validation method
   (a) Calculate the indices for k folds
   (b) Train the modular neural network k − 1 times for each training fold
   (c) Simulate the modular neural network with the k test old
7.—Calculate the mean of rate for all the k-folds using Choquet integral

Training parameters were

Training method: gradient descendent with momentum and adaptive learning rate back-propagation (Traingdx).
Each module with two hidden layers [200 200].
Error goal: 0.00001
Epochs: 500
In Table 1 the distribution of the training data is shown.

4.1 The Experiment with a Modular Neural Network Recognition System and Choquet Integral for the Modules Fusion

The experiment consist on applying each evaluated edge detector to obtain a data set of same well know benchmark data base of images like ORL database of faces and then train a neural network to compare the recognition rate using the k-fold cross validation method [20], see Table 2.

5 Simulation Results

In the experiments we performed 27 tests in simulation of the trainings with each edge detectors making variations in fuzzy densities and performing the calculation

Table 3 Results of training with morphological gradient edge detector

Fuzzy densities			lambda	Training data			Test data		
				Morphological gradient and sobel			Morphological gradient		
				Mean rate	Mean rate	Std rate	Max rate	Mean rate	Std rate
0.1	0.1	0.1	−1.40E−16	1	0	1	0.8825	0.0456	0.9375
0.1	0.1	0.5	2.80E−16	1	0	1	0.8825	0.0456	0.9375
0.1	0.1	0.9	−0.5401	1	0	1	0.885	0.0379	0.9375
0.1	0.5	0.1	2.76E−16	1	0	1	0.8825	0.0381	0.925
0.1	0.5	0.5	−0.2918	1	0	1	0.8825	0.0381	0.925
0.1	0.5	0.9	−0.9107	1	0	1	0.8825	0.0338	0.925
0.1	0.9	0.1	−5.40E−01	1	0	1	0.88	0.036	0.925
0.1	0.9	0.5	−0.9107	1	0	1	0.8825	0.0371	0.925
0.1	0.9	0.9	−0.9891	1	0	1	0.8875	0.0319	0.925
0.5	0.1	0.1	2.76E−16	1	0	1	0.88	0.0527	0.9375
0.5	0.1	0.5	−0.2918	1	0	1	0.88	0.0527	0.9375
0.5	0.1	0.9	−0.9107	1	0	1	0.88	0.0489	0.9375
0.5	0.5	0.1	−0.2918	1	0	1	0.88	0.0456	0.925
0.5	0.5	0.5	−0.7639	1	0	1	0.88	0.0456	0.925
0.5	0.5	0.9	−0.9647	1	0	1	0.8775	0.0454	0.925
0.5	0.9	0.1	−0.9107	1	0	1	0.88	0.042	0.925
0.5	0.9	0.5	−0.9647	1	0	1	0.8825	0.0429	0.925
0.5	0.9	0.9	−0.9945	1	0	1	0.885	0.0428	0.925
0.9	0.1	0.1	−0.5401	1	0	1	0.875	0.05	0.9375
0.9	0.1	0.5	−0.9107	1	0	1	0.875	0.05	0.9375
0.9	0.1	0.9	−0.9891	1	0	1	0.875	0.0476	0.9375
0.9	0.5	0.1	−0.9107	1	0	1	0.875	0.0476	0.925
0.9	0.5	0.5	−0.9647	1	0	1	0.875	0.0476	0.925
0.9	0.5	0.9	−0.9945	1	0	1	0.875	0.0442	0.925
0.9	0.9	0.1	−0.9891	1	0	1	0.8775	0.0428	0.925
0.9	0.9	0.5	−0.9945	1	0	1	0.88	0.0438	0.925
0.9	0.9	0.9	−0.99	1	0	1	0.88	0.042	0.925
				1	**0**	**1**	**0.8800**	**0.0436**	**0.9292**

of the parameter λ with the bisection method. The results obtained with the morphological gradient are shown in Table 3 and with Sobel in Table 4.

In Table 5 the percentages of recognition of the Choquet integral with each edge detector are displayed. It can be noted that when using the morphological gradient it was obtained 94 % recognition of the MNN, while with Sobel a 93.125 % was obtained.

Table 4 Results of training with sobel edge detector

Fuzzy densities			lambda	Training data			Test data		
				Morphological gradient and Sobel			Sobel		
				Mean rate	Mean rate	Std rate	Max rate	Mean rate	Std rate
0.1	0.1	0.1	−1.40E−16	1	0	1	0.865	0.0056	0.875
0.1	0.1	0.5	2.80E−16	1	0	1	0.8625	0.0153	0.8875
0.1	0.1	0.9	−0.5401	1	0	1	0.8575	0.0143	0.875
0.1	0.5	0.1	2.76E−16	1	0	1	0.8675	0.0143	0.8875
0.1	0.5	0.5	−0.2918	1	0	1	0.8625	0.0234	0.9
0.1	0.5	0.9	−0.9107	1	0	1	0.86	0.0163	0.8875
0.1	0.9	0.1	−5.40E−01	1	0	1	0.8675	0.0143	0.8875
0.1	0.9	0.5	−0.9107	1	0	1	0.8675	0.0209	0.9
0.1	0.9	0.9	−0.9891	1	0	1	0.865	0.0137	0.8875
0.5	0.1	0.1	2.76E−16	1	0	1	0.86	0.0105	0.875
0.5	0.1	0.5	−0.2918	1	0	1	0.855	0.0143	0.875
0.5	0.1	0.9	−0.9107	1	0	1	0.8525	0.0105	0.8625
0.5	0.5	0.1	−0.2918	1	0	1	0.865	0.0105	0.875
0.5	0.5	0.5	−0.7639	1	0	1	0.8575	0.019	0.8875
0.5	0.5	0.9	−0.9647	1	0	1	0.8575	0.0112	0.875
0.5	0.9	0.1	−0.9107	1	0	1	0.87	0.0143	0.8875
0.5	0.9	0.5	−0.9647	1	0	1	0.87	0.0143	0.8875
0.5	0.9	0.9	−0.9945	1	0	1	0.865	0.0105	0.875
0.9	0.1	0.1	−0.5401	1	0	1	0.86	0.0105	0.875
0.9	0.1	0.5	−0.9107	1	0	1	0.8625	0.0088	0.875
0.9	0.1	0.9	−0.9891	1	0	1	0.86	0.0137	0.875
0.9	0.5	0.1	−0.9107	1	0	1	0.86	0.0056	0.8625
0.9	0.5	0.5	−0.9647	1	0	1	0.86	0.0105	0.875
0.9	0.5	0.9	−0.9945	1	0	1	0.86	0.0137	0.875
0.9	0.9	0.1	−0.9891	1	0	1	0.865	0.0137	0.8875
0.9	0.9	0.5	−0.9945	1	0	1	0.8675	0.0143	0.8875
0.9	0.9	0.9	−0.99	1	0	1	0.865	0.0105	0.875
				1	**0**	**1**	**0.8625**	**0.0131**	**0.8806**

Table 5 Results of training with Sobel edge detector

Method	% of recognition
Sobel	0.93125
Morphological gradient	0.94

6 Conclusions

The use of Choquet integral as a integration method answers of a modular neural network applied to face recognition has yielded favorable results when performing the aggregation process of the pre-processed images with the detectors of Sobel edges and morphological gradient, however it is still necessary to use a method that optimizes the value of the Sugeno measure assigned to each source of information because these were designated arbitrarily. Future work could be considering the optimization of the proposed method, as in [7, 21, 22].

Acknowledgments We thank the MyDCI program of the Division of Graduate Studies and Research, UABC, Tijuana Institute of Technology, and the financial support provided by our sponsor CONACYT contract grant number: 189350.

References

1. Zhou, L.-G., Chen b, H.-Y., Merigó, J.M., Anna, M.: Uncertain generalized aggregation operators. Expert Syst. Appl. **39**, 1105–1117 (2012)
2. Hidalgo, D., Castillo, O., Melin, P.: Type-1 and type-2 fuzzy inference systems as integration methods in modular neural networks for multimodal biometry and its optimization with genetic algorithms. Inf. Sci. **179**(13), 2123–2145 (2009)
3. Sánchez, D., Melin, P.: Modular neural network with fuzzy integration and its optimization using genetic algorithms for human recognition based on iris, ear and voice biometrics. In: Soft Computing for Recognition Based on Biometrics, pp. 85–102 (2010)
4. Sánchez, D., Melin, P., Castillo, O., Valdez, F.: Modular neural networks optimization with hierarchical genetic algorithms with fuzzy response integration for pattern recognition. MICAI **2**, 247–258 (2012)
5. Melin, P., Gonzalez C., Bravo, D., Gonzalez F., Martínez G.: Modular neural networks and fuzzy Sugeno integral for pattern recognition: the case of human face and fingerprint. In: Hybrid Intelligent Systems: Design and Analysis. Springer, Heidelberg, Germany (2007)
6. Melin, P., Mendoza, O., Castillo O.: Face recognition with an improved interval Type-2 fuzzy logic Sugeno integral and modular neural networks. IEEE Trans. Syst. Man Cybern. Part A Syst. Hum. **41**(5) (2011)
7. Sánchez, D., Melin, P.: Optimization of modular granular neural networks using hierarchical genetic algorithms for human recognition using the ear biometric measure. Eng. Appl. Artif. Intell. **27**, 41–56 (2014)
8. Wang, P., Xu, L., Zhou, S.M., Fan, Z., Li, Y., Feng, S.: A novel Bayesian learning method for information aggregation in modular neural networks. Expert Syst. Appl. **37**, 1071–1074 (2010) (Elsevier)
9. Kwak, K.-C., Pedrycz, W.: Face recognition: a study in information fusion using fuzzy integral. Pattern Recogn. Lett. **26**, 719–733 (2005)
10. Timonin, M.: Robust optimization of the Choquet integral. Fuzzy Sets Syst. **213**, 27–46 (2013)
11. Yang, W., Chen, Z.: New aggregation operators based on the Choquet integral and 2-tuple linguistic information. Expert Syst. Appl. **39**, 2662–2668 (2012)
12. Meena, Y.K., Arya, K.V., Kala, R.: Classification using redundant mapping in modular neural networks. In: Second World Congress on Nature and Biologically Inspired Computing, Kitakyushu, Fukuoka, Japan, 15–17 Dec 2010
13. Sugeno, M.: Theory of fuzzy integrals and its applications. Thesis Doctoral, Tokyo Institute of Technology, Tokyo, Japan (1974)
14. Murofushi, T., Sugeno, M.: Fuzzy measures and fuzzy integrals. Department of Computational Intelligence and Systems Science, Tokyo Institute of Technology, Yokohama, Japan (2000)
15. Song, J., Li, J.: Lebesgue theorems in non-additive measure theory. Fuzzy Sets Syst. **149**, 543–548 (2005)

16. Wang, Z., Klir, G.: Generalized Measure Theory. Springer, New York (2009)
17. Database ORL Face. Cambridge University Computer Laboratory. http://www.cl.cam.ac.uk/research/dtg/attarchive/facedatabase.html. Nov 2012
18. Mendoza, O., Melin, P., Castillo, O., Castro, J.: Comparison of fuzzy edge detectors based on the image recognition rate as performance index calculated with neural networks. In: Soft Computing for Recognition Based on Biometrics. Studies in Computational Intelligence, vol. 312, pp. 389–399 (2010)
19. Mendoza, O., Melin, P., Licea, G.: A hybrid approach for image recognition combining type-2 fuzzy logic, modular neural networks and the Sugeno integral. Inf. Sci. Int. J. **179**(13), 2078–2101 (2009)
20. Mendoza, O., Melin, P.: Quantitative evaluation of fuzzy edge detectors applied to neural networks or image recognition. Advances in Research and Developments in Digital Systems (2011)
21. Sánchez, D., Melin, P.: Multi-objective hierarchical genetic algorithm for modular granular neural network optimization. In: Soft Computing Applications in Optimization, Control, and Recognition, pp. 157–185 (2013)
22. Sánchez, D., Melin, P., Castillo, O., Valdez, F.: Modular granular neural networks optimization with Multi-Objective Hierarchical Genetic Algorithm for human recognition based on iris biometric. In: IEEE Congress on Evolutionary Computation, pp. 772–778 (2013)

Part IV
Optimization Methods and Applications

A Survey of Decomposition Methods for Multi-objective Optimization

Alejandro Santiago, Héctor Joaquín Fraire Huacuja,
Bernabé Dorronsoro, Johnatan E. Pecero, Claudia Gómez Santillan,
Juan Javier González Barbosa and José Carlos Soto Monterrubio

Abstract The multi-objective optimization methods are traditionally based on Pareto dominance or relaxed forms of dominance in order to achieve a representation of the Pareto front. However, the performance of traditional optimization methods decreases for those problems with more than three objectives to optimize. The decomposition of a multi-objective problem is an approach that transforms a multi-objective problem into many single-objective optimization problems, avoiding the need of any dominance form. This chapter provides a short review of the general framework, current research trends and future research topics on decomposition methods.

A. Santiago · H. J. F. Huacuja (✉) · C. G. Santillan · J. J. González. Barbosa ·
J. C. S. Monterrubio
Instituto Tecnológico de Ciudad Madero, Ciudad Madero, Mexico
e-mail: automatas2002@yahoo.com.mx

A. Santiago
e-mail: alx.santiago@gmail.com

C. G. Santillan
e-mail: cggs71@hotmail.com

J. J. González. Barbosa
e-mail: jjgonzalezbarbosa@hotmail.com

J. C. S. Monterrubio
e-mail: soto190@gmail.com

B. Dorronsoro
University of Lille, Lille, France
e-mail: bernabe.dorronsoro_diaz@inria.fr

J. E. Pecero
University of Luxembourg, Luxembourg, Luxembourg
e-mail: johnatan.pecero@uni.lu

O. Castillo et al. (eds.), *Recent Advances on Hybrid Approaches for Designing*   453
Intelligent Systems, Studies in Computational Intelligence 547,
DOI: 10.1007/978-3-319-05170-3_31, © Springer International Publishing Switzerland 2014

1 Introduction

The decomposition of an optimization problem is an old idea that appears in optimization works [1–4]. This approach transforms the original problem into smaller ones and solves each problem separately. The decomposition could be at different levels: decision variables, functions and objectives. The decomposition in decision variables split in sub groups the original set of decision variables, and each group is optimized as a subproblem. Decomposition in decision variables is a novel technique in multi-objective optimization, works using this approach are found in Dorronsoro et al. [5] and Liu et al. [6]. The decomposition in functions can be done when the objective function can be decomposed into two or more objective functions. In (1) an example of a decomposable function is showed.

$$Minimize\ g(\vec{x}) = f_1(x_1, y) + f_2(x_2, y) \tag{1}$$

In this case, it is possible to optimize f_1 and f_2 in separately optimization problems [7–10]. One example in multi-objective optimization is the large-scale multi-objective non-linear programming problems [11]. It is important to notice that most of the multi-objective optimization problems cannot be optimized as individual optimization problems because their objective functions are not composed from subproblems. The decomposition in objectives is more natural than decomposition in functions for multi-objective optimization, given that every objective could represent a subproblem to optimize. This is the approach followed by the Nash genetic algorithms [12], however this approach finds solutions in equilibrium, not the optimal set. The coevolutionary algorithms can decompose in objectives, but it requires a different population for every objective [13]. The approach to decompose a multi-objective problem reviewed in this chapter is the one used in Zhang and Li [14]. The advantage of this approach is that it is not affected by the problem structure (decision variables, functions, number of objectives) and it is able to obtain a representation of the optimal set. The main contributions of our work can be summarized as follows:

- We identify the trends to decompose a multi-objective problem into single-objective problems and their weaknesses.
- We identify the general framework for multi-objective optimization by decomposition.
- We identify the current and future research trends for multi-objective decomposition.

The rest of the chapter is organized as follows. In Sect. 2, we present the basic concepts of multi-objective optimization and the concept of aggregate objective function. In Sect. 3, we present how the decomposition framework works and its advantages over based on dominance. We present the current research on multi-objective decomposition in Sect. 4. Section 5 provides our perspective on new research directions to be addressed in multi-objective decomposition. We conclude this chapter in Sect. 6.

2 Definitions and Background Concepts

This section presents the basic concepts in multi-objective optimization and the concepts used in multi-objective decomposition methods.

Definition 1 *Multi-objective optimization problem (MOP)*

Given a vector function $\vec{f}(\vec{x}) = [f_1(\vec{x}), f_2(\vec{x}), \ldots, f_k(\vec{x})]$ and its feasible solution space Ω, the MOP consists in find a vector $\vec{x} \in \Omega$ that optimizes the vector function $\vec{f}(\vec{x})$. Without loss of generality we will assume only minimization functions.

Definition 2 *Pareto dominance*

A vector $\vec{x}$ dominates $\vec{x}'$ (denoted by $\vec{x} \prec \vec{x}'$) if $f_i(\vec{x}) \leq f_i(\vec{x}')$ for all i functions in $\vec{f}$ and there is at least one i such that $f_i(\vec{x}) < f_i(\vec{x}')$.

Definition 3 *Pareto optimal*

A vector $\vec{x}^*$ is Pareto optimal if not exists a vector $\vec{x}' \in \Omega$ such that $\vec{x}' \prec \vec{x}^*$.

Definition 4 *Pareto optimal set*

Given a MOP, the Pareto optimal set is defined as $P^* = \{\vec{x}^* \in \Omega\}$.

Definition 5 *Pareto front*

Given a MOP and its Pareto optimal set P^*, the Pareto front is defined as $PF^* = \{\vec{f}(\vec{x}) | \vec{x} \in P^*\}$.

Aggregate functions use the concept of aggregate value to represent a set of individual values in a single one. Uses of the aggregate functions are found in: probability, statistics, computer science, economics, operations research, etc. [15]. In multi-objective optimization the aggregate objective function (AOF) transforms the functions in $\vec{f}$ into a single objective function. The simplest aggregate objective function used in multi-objective optimization is the following global objective sum.

$$Z = \sum_{i=1}^{k} f_i(\vec{x}) \tag{2}$$

3 The Decomposition Framework

In this section we present the multi-objective decomposition approach, how to obtain Pareto optimal solutions, how to decompose a multi-objective optimization problem and the general decomposition framework.

Table 1 Common weighted aggregate objective functions

Name	Formulation	Designed for	Details in
Weighted sum	$\sum_{i=1}^{k} w_i f_i(\vec{x})$	Pareto optimality in convex solution spaces	Marler and Arora [17]
Weighted exponential sum	$\sum_{i=1}^{k} w_i [f_i(\vec{x})]^p$	Pareto optimality in non-convex solution spaces	Marler and Arora [17]
Weighted min–max	$MAX_{i=1}^{k} \{ w_i [f_i(\vec{x}) - f_i^o] \}$	Weak Pareto optimality	Marler and Arora [17]
Weighted product	$\prod_{i=1}^{k} [f_i(\vec{x})]^{w_i}$	Different magnitudes in objectives	Marler and Arora [17]

3.1 How to Achieve Different Pareto Optimal Solutions

With the global objective sum in (2), just one solution to the optimization problem is obtained. However in a multi-objective optimization problem many optimal solutions exist, therefore to get different optimal solutions it is necessary to find the minima in different regions of the Pareto front. To reach this goal the following weighted AOF is used.

$$Z = \sum_{i=1}^{k} w_i f_i(\vec{x}) \tag{3}$$

If all the weights in (3) are positive, the minimum is Pareto optimal [16]. Diverse weights may achieve different Pareto optimal solutions; this is the core of how the multi-objective decomposition approach works. The original optimization with aggregate objective functions (AOFs) was developed using mathematical programming and different AOFs have been proposed to overcome various difficulties. In linear programming problems with convex solution spaces the use of weighted sum method guarantees an optimal solution, however for other problems is not possible to obtain points in the non-convex portions of the Pareto optimal set [17]. Table 1 shows a summary of weighted AOFs used in multi-objective optimization.

In mathematical programming, it is mandatory to select the correct optimization method with the correct AOF to guide the search according to the feasible decision space. It is important to notice that not all the weighted AOFs ensure Pareto optimality as in the weighted min–max (also known as weighted Tchebycheff), which only ensures weak Pareto optimality (none solution in the feasible space is better for all objectives). On the other hand with metaheuristics the search is guided by different operators, for example differential evolution is a great idea for convex problems to exploit linear dependencies, or polynomial mutations to handle non-convex problems [18]. Hybridization of heuristic operators and weighted AOF is a powerful tool for a more generic multi-objective optimization approach, although it does not guarantee optimality.

3.2 Decomposition of a Multi-objective Optimization Problem

In previous section, we reviewed how to achieve different Pareto optimal solutions, the weighted AOFs, to get Pareto optimal solutions. Now it is necessary to decompose the original multi-objective optimization problem into multiple single-objective problems in order to achieve a good representation of the Pareto Front. Let $W = \{\vec{w}_1, \vec{w}_2, \ldots, \vec{w}_n\}$, where $\vec{w}_i \leftarrow [w_{i1}, w_{i2}, \ldots, w_{ik}]$ is a weight vector. Most authors suggest $\sum_{j=1}^{k} w_{ij} = 1$, there is no real reason for this, but in order to achieve uniformity between solutions found it is necessary that the sum be the same for all the weight vectors defined. The selection of weight vectors is not trivial, similar weights bias the search towards the same region and very diverse weights will not produce uniformity. The first uses of the decomposition method suggest systematically altering weights to yield a set of Pareto optimal solutions [19, 17]. Currently the state of the art manages three main ideas to generate the weight vectors: random vector generators, systematic vector generators and optimization vector generators.

Random vector generators are the common approach. In the work of Zhang et al. [20] a vectors generator of this kind is used with good results. The generator initiates with the corners of the objective space (the best solution for every objective when the others are with 0 weight of importance) in the initial set W, an example of a corner is given in (4).

$$\vec{w}_1 \leftarrow [1.0, 0.0, 0.0, \ldots, 0.0] \tag{4}$$

Then the vector with the biggest distance to W is added from a set of 5,000 vectors uniformly randomly generated. The process is repeated until the desired size of W is reached.

Systematic vector generators use patterns to generate the weight vectors. In Messac and Mattson [21], the proposed systematic vector generator, weights are generated by increments of size $1/(n - 1)$ such that the sum of weights is equal to one. The permutations formed with the weights of size k that satisfy $\sum_{j=1}^{k} w_{ij} = 1$ are used as in Fig. 1 from [21].

It is important to notice that when $k = n$ and the objectives have the same magnitude these weight vectors are similar to different orders in the lexicographic method. Another work with a systematic vector generator is in [22], it uses the Simplex-lattice design principle, where the components are formed with the equation $x_i = 0, \frac{1}{n}, \frac{2}{n}, \ldots, \frac{n}{n}$, the permutations formed with the components are used as weight vectors to achieve a uniform Pareto front.

It is possible to formulate a vectors generator as an optimization problem. The work of Hughes [23] proposed generate the weight vectors using the following optimization problem.

$$Minimize\ Z = MAX_{i=1}^{n} \left\{ MAX_{j=1, j\neq i}^{n} \left(\vec{w}_i \cdot \vec{w}_j \right) \right\} \tag{5}$$

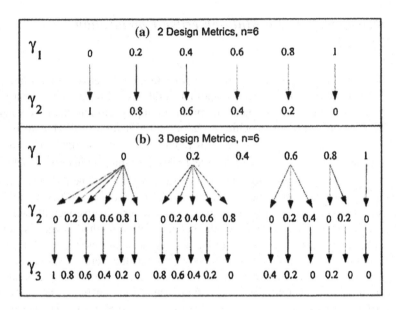

Fig. 1 Permutations of weights

The inner maximum operator finds the nearest vectors in angle, because the dot product provides the cosine of the angle between $\vec{w}_i$ and $\vec{w}_j$ and the minimum angle provides the maximum value $\cos(0) = 1$. The outer maximization operator finds the smallest of the angles in the vector set. It is important to clarify this vectors generator works with one assumption: the real optimal solution is when all the vectors in W are the same. In order to avoid this condition, all the corners of the objective space have to be added to the set W and remain in the entire optimization process.

3.3 General Multi-objective Optimization Framework by Decomposition

The multi-objective decomposition algorithms are based on population search algorithms (evolutionary algorithms, ant colony, and particle swarm) or path search. Figure 2 shows the framework used by most of the multi-objective decomposition algorithms.

The framework requires every vector $\vec{w}_i$ be in a cluster of similar vectors from the set W. The framework works in four steps. The first step sets an initial solution $\vec{x}_i$ for every vector $\vec{w}_i$. The second step selects one $\vec{w}_i$ from the set W and optimizes it (with any optimization method) from the best-known solution $\vec{x}_i$ for that vector. The third step uses the new solution found $\vec{x}'$ to update the best-known solutions for the weight vectors clustered with $\vec{w}_i$. The fourth step verifies if the stop criteria

Require: Every vector $\vec{w}_i$ be in a cluster C_z
for $\forall \vec{w}_i \in W$ **do**
    $\vec{x}_i \leftarrow$ an initial solution for the vector $\vec{w}_i$
end for
repeat
    select one vector $\vec{w}_i \in W$
    $\vec{x}' \leftarrow Optimize(\vec{x}_i)$;
    for $\forall \vec{w}_j \in C_z$ **do** $\triangleright\ C_z$ are the clustered vectors to $\vec{w}_i$
        if $\vec{x}' < \vec{x}_j$ **then**
            $\vec{x}_j \leftarrow \vec{x}'$
        end if
    end for
until stop criterion

Fig. 2 Multi-objective optimization framework by decomposition

have been reached; if not, return to the second step. Overlapping in the clusters is also allowed. For example, sometimes the optimization of a vector updates the entire population or sometimes a smaller portion. In the first multi-objective decomposition proposals of Hughes, the structure is different from Fig. 2, MSOPS [24] and MSOPS-II [23]. Later Hughes changed to the structure of Fig. 2 in his last decomposition approach MODELS [25] with better performance results.

3.4 Multi-objective Optimization Based on Decomposition Versus Dominance Based

The multi-objective decomposition methods have benefits over based dominance optimization methods. With dominance and relaxed dominance forms the non-dominance verification is a very repetitive operation of order $O(kn)$, where n is the set of non-dominated solutions and k the number of objective functions. While in decomposition methods the comparisons are directly between solutions in $O(k)$. When the number of objectives increases, the number of non-dominated solutions grows [26], which leads to stagnation in the search processes based on dominance, because for the optimizer most solutions are equally good.

4 Current Research on Decomposition Methods

This section reviews the current research trends on multi-objective decomposition methods: heuristic approaches, performance, efficiency, and scalability.

The current heuristic approaches are based on well-known metaheuristic optimization algorithms methods: differential evolution, memetic algorithms, ant colony and particle swarm optimization.

Differential evolution is an approach that emphasizes mutation using a proportional difference of two random vector solutions to mutate a target vector.

In multi-objective decomposition the first implementation found is from Li and Zhang [27], an evolution of the first MOEA/D in Zhang and Li [14], called MOEA/D-DE. The original version of MOEA/D keeps an external archive with all the non-dominated solutions found. MOEA/D-DE does not use an external archive because the final solutions for the set W are usually non-dominated. Another modification is a parameter of maximal number of solutions updated by a new one found. This algorithm has been tested with very hard Pareto shapes.

Memetic algorithms are based on the synergy of 2 ideas, population search and local improvements a very robust idea for optimization. MOGLS in Ishibuchi and Murata [28] was the first memetic proposal in multi-objective decomposition, but later an evolutionary algorithm MOEA/D in Zhang and Li [14] obtains a very similar performance without local improvements. In the work of Tan et al. [29] MOEA/D-SQA, the population search is the MOEA/D, and the local improvements are made by the simplified quadratic approximation (SQA), outperforming MOEA/D. Another memetic algorithm is in Mei et al. [30] for the capacitated arc routing problem (CARP), where the population search is a modification of the MOEA/D structure. The solutions are preserved using the preservation mechanism from NSGA-II [31], and the preserved solutions are assigned to the weight vector that solves best. The crossover and local search operators are from the memetic algorithm with extended neighborhood search (MAENS).

Ant colony optimization emulates the behavior of ants looking for a path between food and their colony; ants with promissory paths leave a pheromone path to be followed for other ants. The ants (solutions) are constructed stochastically between the best path and others. This optimization method uses 2 kinds of information, pheromone and heuristic. The pheromones represent the paths leading to promissory solutions, while heuristic information is how the solutions are constructed (for example restrictions). The work of Ke et al. [32] proposed MOEA/D-ACO where each ant is a weight vector to optimize and has its own heuristic information. The pheromone information is shared with a group of solutions (not necessarily the clustered weight vectors) this group tries to achieve a region of the Pareto front. Every ant is constructed with its heuristic information, pheromone path, and best local solution known; this information is combined to guide the ant.

Particle Swarm optimization (PSO) is inspired by the social foraging behavior of some animals, such as flocking behavior of birds and the schooling behavior of fish. Velocity guides the search of the particles, fluctuating between their own best past location (best local) and the best past location of the whole swarm (best global/leader). Peng and Zhang [33] proposed the first PSO based on multi-objective decomposition MOPSO/D. Every particle is assigned to a weight vector, MOPSO/D uses a crowding archive strategy to keep the non-dominated solutions found. Later Al Moubayed et al. [34] propose SDMOPSO, the main difference with MOPSO/D is that it uses Pareto dominance to update the best locations local and global; the leaders are selected from a random non-dominated particle from its clustered vectors. Martínez and Coello [35] proposed dMOPSO, which does not use turbulence (mutation) and the leader is a random solution from the best local particles in the swarm. In dMOPSO the concept of age is introduced, if a particle is not improved in a generation its age is increased by one; the particles that exceed certain age are reinitialized from scratch. In the work of Al Moubayed et al. [36] D^2MOPSO was proposed comparing it with SDMOPSO and dMOPSO, outperforming them. D^2MOPSO uses a crowding archive, when the archive is full the non-dominated particles are only added at the low dense regions replacing those ones at high dense regions. The selections of leaders are made from the archive using the solution that best solves the weight vector of the particle.

The performance is measured by the quality in the Pareto front representation found, a very competitive field in multi-objective optimization. There are different desirable characteristics: convergence, uniformity and a big dominated region. The inverted generational distance [37] measures the distance from the real front to the representation. It was the indicator used in the IEEE CEC 09 competition, where the winner for unconstrained problems was MOEA/D [20]; this version of MOEA/D uses a utility function that assigns less computational effort when the weight vectors are not improved in the generation.

Efficiency is an issue hard to improve in multi-objective optimization, the reason for this is the computational efficiency of non-dominance verification and sorting in fronts $O(kn^2)$ [38, 31]. The multi-objective decomposition methods do not use dominance, allowing lower execution times in multi-objective optimization. Bearing in mind Zhang and Li design MOEA/D in [14] with a lower complexity in space and time and similar performance than MOGLS in [28].

The scalability in objectives is a hard topic in multi-objective optimization because optimization methods based on dominance have difficulties. It is well known that dominance based optimization methods lose their effectiveness with more than 3 objectives [39]. When the number of objectives grows it increases the number of non-dominated solutions [26], causing stagnation in the search. Hughes proposed an evolutionary algorithm MSOPS [24] and improved it in MSOPS-II [23], for a problem of 5 objectives outperforming NSGA-II. Later Hughes proposed a path search based on golden section search named MODELS in [25], for a problem of 20 objectives outperforming MSOPS-II. The work of Yan et al. [22] UMOEAD ensures a uniform Pareto front, using a systematic vectors generator, tested with a 5 objectives problem.

5 Future Research on Decomposition Methods

This section presents the authors perspective about what is still need to be researched in multi-objective decomposition methods.

5.1 Weight Vector Generators

In order to achieve a good representation of the Pareto front, the final solutions have to be diverse enough to cover most of the Pareto front and the extremes; also uniformity is important to not ignore regions in the Pareto front. The problem with the current vector generators is that they only focus in diversity or uniformity, but not both. Random vector generators are just focused in diversity while systematic vector generators are just targeting uniformity. Another issue is that when increasing the number of objectives usually the current vector generators use more vectors to ensure diversity or uniformity, implying more computational effort in the optimization. The multi-objective decomposition needs new vector generators not sensitive to the number of vectors taking account the diversity and the uniformity.

5.2 Problem Clustering and Communications

The current multi-objective decomposition methods gather the problems with the metric of Euclidian distance and use the same cluster technique, the T closest problems. It is possible to apply other clustering techniques like k-means or k-medoids, and metrics like Manhattan distance or Mahalanobis distance [40], to the multi-objective decomposition methods. There are no studies about different metrics or cluster techniques and the general structure of these optimization methods is affected directly by the clusters of weight vectors. Different clustering techniques and metrics could be used and studied. Also the communication between clusters of vectors is generally random, a more structured communication mechanism could be designed to exploit clusters information and improve another ones.

5.3 New Heuristic Approaches

The current multi-objective decomposition methods are based on metaheuristic population search or path search, there is a great variety of population search algorithms not studied yet and path search is almost not studied. A search based on clusters and its communications could be a new topic for heuristic approaches.

5.4 Scalability in Decision Variables

Although scalability in objectives is studied by some multi-objective decomposition methods, the scalability in the number of decision variables is barely studied in multi-objective optimization; this was studied in the work of Durillo et al. [41].

The scalability in decision variables is not studied yet in multi-objective decomposition. It is important to notice that when some problems grow in objectives it also grows in decision variables [42, 43].

6 Conclusion

The multi-objective decomposition methods solve problems related to methods based on dominance forms like stagnation for non-dominated solutions and the need of dominance verification $O(kn)$ or sorting $O(kn^2)$ [31]. More research is needed on multi-objective decomposition methods in: vector generators, clustering techniques and heuristic approaches. Research on multi-objective decomposition may lead to diverse and uniform Pareto front representations for a wide variety of problems.

Acknowledgments B. Dorronsoro acknowledges the support by the National Research Fund, Luxembourg (AFR contract no. 4017742). A. Santiago would like to thank CONACyT Mexico, for the support no. 360199.

References

1. Dantzig, G.B., Wolfe, P.: Decomposition principle for linear programs. Oper. Res. **8**(1), 101–111 (1960)
2. Dantzig, G.B., Harvey, R.P., Lansdowne, Zachary F., Robinson, D.W., Maier, S.F.: Formulating and solving the network design problem by decomposition. Transp. Res. Part B: Methodol. **13**(1), 5–17 (1979)
3. Sobieszczanski-Sobieski, J., James, B.B., Dovi, A.R.: Structural optimization by multi level decomposition. AIAA J. **23**(11), 1775–1782 (1985)
4. Ovacik, I.M., Uzsoy, R.: Decomposition Methods for Complex Factory Scheduling Problems. Kluwer Academic Publishers, Boston (1997)
5. Dorronsoro, B., Danoy, G., Nebro, A.J., Bouvry, P.: Achieving super-linear performance in parallel multi-objective evolutionary algorithms by means of cooperative coevolution. Comput. Oper. Res. **40**(6), 1552–1563(2013) (Emergent nature inspired algorithms for multi-objective optimization)
6. Liu, M., Zou, X., Chen, Y., Wu, Z.: Performance assessment of DMOEA-DD with CEC 2009 MOEA competition test instances. In: IEEE Congress on Evolutionary Computation (CEC 09), pp. 2913–2918 (2009)
7. Boyd, S., Xiao, L., Mutapcic, A., Mattingley, J.: Notes on decomposition methods. Notes for EE364B, Stanford University (2008)
8. Abdelouahed, H., Mishra, S.K.: Decomposition methods based on augmented lagrangians: a survey. In: Mishra, S.K. (ed.) Topics in Nonconvex Optimization, Springer Optimization and Its Applications, pp. 175–203. Springer, New York (2011)
9. Koko, J.: A survey on dual decomposition methods. SeMA J. **62**(1), 27–59 (2013)
10. Palomar, D.P., Chiang, Mung: A tutorial on decomposition methods for network utility maximization. IEEE J. Sel. Areas Commun. **24**(8), 1439–1451 (2006)

11. Sakawa, M., Yano, H.: A fuzzy decomposition method by right-hand-side allocation for large-scale multiobjective nonlinear programming problems. In: Clmaco, J. (ed.) Multicriteria Analysis, pp. 237–246. Springer, Berlin, Heidelberg (1997)
12. Sefrioui, M., Perlaux, J.: Nash genetic algorithms: examples and applications. In: Proceedings of the 2000 Congress on Evolutionary Computation, vol. 1, pp. 509–516 (2000)
13. Zhan, Z.H., Li, J., Cao, J., Zhang, J., Chung, H.S.-H., Shi, Y.-H.: Multiple populations for multiple objectives: a coevolutionary technique for solving multiobjective optimization problems. IEEE Trans. Cybern. **43**(2), 445–463 (2013)
14. Zhang, Q., Li, H.: MOEA/D: a multiobjective evolutionary algorithm based on decomposition. IEEE Trans. Evol. Comput. **11**(6), 712–731 (2007)
15. Grabisch, M., Marichal, J.-L., Mesiar, R., Pap, E.: Aggregation functions: means. Inf. Sci. **181**(1), 1–22 (2011)
16. Zadeh, L.: Optimality and non-scalar-valued performance criteria. IEEE Trans. Autom. Control **8**(1), 59–60 (1963)
17. Marler, R.T., Arora, J.S.: Survey of multi-objective optimization methods for engineering. Struct. Multi. Optim. **26**(6), 369–395 (2004)
18. Sindhya, K., Ruuska, S., Haanp, T., Miettinen, K.: A new hybrid mutation operator for multiobjective optimization with differential evolution. Soft. Comput. **15**(10), 2041–2055 (2011)
19. Athan, T.W., Pa-palambros, P.Y.: A note on weighted criteria methods for compromise solutions in multi-objective optimization. Eng. Optim. **27**, 155–176 (1996)
20. Zhang, Q., Liu, W., Li, H.: The performance of a new version of MOEA/D on CEC09 unconstrained mop test instances. In: IEEE Evolutionary Computation (CEC'09), pp. 203–208 (2009)
21. Messac, A., Mattson, C.A.: Generating well-distributed sets of Pareto points for engineering design using physical programming. Optim. Eng. **3**(4), 431–450 (2002)
22. Tan, Y.Y., Jiao, Y.C., Li, H., Wang, X.: MOEA/D uniform design: a new version of MOEA/D for optimization problems with many objectives. Comput. Oper. Res. **40**(6), 1648–1660 (2013) (emergent nature inspired algorithms for multi-objective optimization)
23. Hughes, E.J.: Msops-ii: a general-purpose many-objective optimiser. In: IEEE Congress on Evolutionary Computation (CEC 2007), pp. 3944–3951 (2007)
24. Hughes, E.J.: Multiple single objective Pareto sampling. In: The 2003 Congress on Evolutionary Computation (CEC'03), vol. 4, pp. 2678–2684 (2003)
25. Hughes, E.J.: Many-objective directed evolutionary line search. In: Proceedings of the 13th Annual Conference on Genetic and Evolutionary Computation (GECCO'11), pp. 761–768. ACM, New York (2011)
26. Fabre, M.G., Pulido, G.T., Coello, C.A.C.: Alternative fitness assignment methods for many-objective optimization problems. In: Collet, P., Monmarch, N., Legrand, P., Schoenauer, M., Lutton, E. (eds.): Artificial Evolution. Lecture Notes in Computer Science, vol. 5975, pp. 146–157. Springer, Berlin, Heidelberg (2010)
27. Li, H., Zhang, Q.: Multiobjective optimization problems with complicated pareto sets, MOEA/D and NSGA-II. IEEE Trans. Evol. Comput. **13**(2), 229–242 (2009)
28. Ishibuchi, H., Murata, T.: A multi-objective genetic local search algorithm and its application to flowshop scheduling. IEEE Trans. Syst. Man Cybern. Part C Appl. Rev. **28**(3), 392–403 (1998)
29. Tan, Y.-Y., Jiao, Y.-C., Li, H., Wang, X.-K.: MOEA/D-SQA: a multi-objective memetic algorithm based on decomposition. Eng. Opt. **44**(9), 1095–1115 (2012)
30. Mei, Y., Tang, K., Yao, X.: Decomposition-based memetic algorithm for multiobjective capacitated arc routing problem. IEEE Trans. Evol. Comput. **15**(2), 151–165 (2011)
31. Deb, K., Agrawal, S., Pratap, A., Meyarivan, T.: A fast elitist non-dominated sorting genetic algorithm for multi-objective optimization: Nsga-ii. In: Proceedings of the 6th International Conference on Parallel Problem Solving from Nature. Lecture Notes in Computer Science, vol. 1917. Springer, Berlin, Heidelberg (2000)

32. Ke, L., Zhang, Q., Battiti, R.: MOEA/D-ACO: a multiobjective evolutionary algorithm using decomposition and ant colony. IEEE Trans. Cybern. **43**(6), 1845–1859 (2013)
33. Peng, W., Zhang, Q.: A decomposition-based multi-objective particle swarm optimization algorithm for continuous optimization problems. In: IEEE International Conference on Granular Computing (GrC 2008), pp. 534–537 (2008)
34. Al Moubayed, N., Petrovski, A., McCall, J.: A novel smart multi-objective particle swarm optimisation using decomposition. In: Schaefer, R., Cotta, C., Koodziej, J., Rudolph, G. (eds.) Parallel Problem Solving from Nature, PPSN XI. Lecture Notes in Computer Science, vol. 6239, pp. 1–10. Springer, Berlin Heidelberg (2010)
35. Martínez, S.Z., Coello, C.A.C.: A multi-objective particle swarm optimizer based on decomposition. In: Proceedings of the 13th Annual Conference on Genetic and Evolutionary Computation (GECCO'11), pp. 69–76. ACM, New York (2011)
36. Al Moubayed, N, Petrovski, A., McCall, J.: D^{2}MOPSO: multi-objective particle swarm optimizer based on decomposition and dominance. In: Hao, J.-K., Middendorf, M. (eds.) Evolutionary Computation in Combinatorial Optimization. Lecture Notes in Computer Science, vol. 7245, pp. 75–86. Springer Berlin Heidelberg (2012)
37. Van Veldhuizen, D.A., Lamont, G.B.: Multiobjective Evolutionary Algorithm Research: A History and Analysis (1998)
38. Coello, C.A.C.: Evolutionary multi-objective optimization: some current research trends and topics that remain to be explored. Front. Comput. Sci. China **3**(1), 18–30 (2009)
39. Hughes, E.: Evolutionary many-objective optimisation: many once or one many? In: The 2005 IEEE Congress on Evolutionary Computation, 2005, vol. 1, pp. 222–227 (2005)
40. Berkhin, P.: A survey of clustering data mining techniques. In: Kogan, J., Nicholas, C., Teboulle, M. (eds.) Grouping Multi-dimensional Data, pp. 25–71. Springer, Berlin, Heidelberg (2006)
41. Durillo, J.J., Nebro, A.J., Coello, C.A.C., Garcia-Nieto, J., Luna, F., Alba, E.: A study of multi-objective metaheuristics when solving parameter scalable problems. IEEE Trans. Evol. Comput. **14**(4), 618–635 (2010)
42. Laumanns, M., Deb, K., Thiele, L., Zitzler, E.: Evolutionary Multiobjective Optimization. Theoretical Advances and Applications. Scalable Test Problems for Evolutionary Multi-Objective Optimization. 105–145 (2005)
43. Barone, L., Huband, S., Hingston, P., While, L.: A review of multi-objective test problems and a scalable test problem toolkit. IEEE Trans. Evol. Comput. **10**(5), 447–506 (2006)

A Decision Support System Framework for Public Project Portfolio Selection with Argumentation Theory

Laura Cruz-Reyes, César Medina Trejo, Fernando López Irrarragorri and Claudia G. Gómez Santillán

Abstract In this chapter, we propose a framework for a Decision Support System (DSS) to aid in the selection of public project portfolios. Organizations are investing continuously and simultaneously in projects, however, they face the problem of having more projects than resources to implement them. Public projects are designed to favor society. Researches have commonly addressed the public portfolio selection problem with multicriteria algorithms, due to its high dimensionality. These algorithms focus on identifying a set of solutions in the Pareto frontier. However, the selection of the solution depends on the changing criteria of the decision maker (DM). A framework to support the DM is presented; it is designed to help the DM by a dialogue game to select the best portfolio in an interactive way. The framework is based on the argumentation theory and a friendly user interface. The dialogue game allows the DM to ask for justification on the project portfolio selection.

L. Cruz-Reyes (✉) · C. Medina Trejo · C. G. Gómez Santillán
Instituto Tecnológico de Ciudad Madero, 1o. de Mayo y Sor Juana I. de la Cruz S/N C.P 89440 Ciudad Madero, Tamaulipas, Mexico
e-mail: lauracruzreyes@itcm.edu.mx

C. Medina Trejo
e-mail: rpgamer86@hotmail.com

C. G. Gómez Santillán
e-mail: cggs71@hotmail.com

F. López Irrarragorri
Universidad Autónoma de Nuevo León, San Nicolás de Los Garza, Nuevo León, Mexico
e-mail: flopez65@gmail.com

O. Castillo et al. (eds.), *Recent Advances on Hybrid Approaches for Designing Intelligent Systems*, Studies in Computational Intelligence 547,
DOI: 10.1007/978-3-319-05170-3_32, © Springer International Publishing Switzerland 2014

1 Introduction

In this chapter we propose a framework for generating recommendations for the problem of public project portfolio selection through an approach with argumentation theory. The selection of public project portfolio is a complex optimization problem, which involves a high dimensionality, multiple conflicting attributes.

Deciding on the best portfolio lies within one person or a group of people, called the decision maker (DM). There is a subjective problem, because each DM rests on its own base of preferences, then the problem lies in satisfying his preferences with the selected portfolio.

The DM must be convinced that the recommendation provided to him satisfies his base of criteria, making good decisions is critical since a bad decision would affect society given they are public projects, so explaining and justifying the recommendation is needed. To provide explications alongside with the recommended portfolio to the user is a major feature of decision support tools [5].

The argumentation theory is used in the process of constructing and evaluating arguments to review conclusions. In the decision support systems (DSS) the use of argumentation has increased. The goal of such systems is to assist people in making decisions. The need for argumentation in such systems arises from the demand to justify and explain the options and the recommendations shown to the user [8].

An analyst may find it useful to have the support of a tool that will provide an explicit explanation, justification, and possible responses that may occur in the course of interaction with the framework. In the case of an automatic tool, it can be satisfactory to permit some interaction with the client that allows him to refine recommendations.

Decision support involves understanding, interpreting, justifying, explaining, convincing, reviewing and updating the results of the decision support process. With the argumentation theory a framework for recommendation can be strengthened, having sustentation to justify recommendations.

2 The Public Portfolio Problem

A central and very frequent issue in public policy analysis is how to allocate funds to competing projects. Public resources are limited, particularly for financing social projects. Very often, the cumulative budget being requested ostensibly exceeds what can be supported [1].

Public projects are designed to favor society. The public projects have special characteristics; Fernandez mentions a generalization of these [1]:

- They are hard to measure and can be without question profitable, their effects are sometimes only visible on the long term and are not direct.

- They have economic contributions to society; there are benefits that must be taken into account to achieve a total social view.
- The fairness of the conditions of the society and their impact on the individuals should be considered.

Having good decisions is critical, the problem is selecting the set of public projects that provides more benefits to society. Due to their nature, public project portfolio has been commonly addressed by multicriteria algorithms focusing on identifying a set of solutions in the Pareto frontier. The selection of the solution depends on the criteria and preferences of the DM.

3 Theoretical Concepts of Argumentation Theory for Decision Aiding

The recommendations in decision support need to be explained and justified; the argumentation theory is a reasoning model for evaluating interacting arguments, such arguments support or refute a conclusion. This theory is used to address the non-monotonic reasoning, in the light of new information we may change our conclusions, with more information new arguments or counter arguments may arise and lead to new deductions.

In an overview of argumentation for making decisions [10], usually one will construct a set of arguments and counter arguments for each possible decision (alternative). With an argumentation-based approach in a decision problem, there are benefits: The user will be provided with a good choice and the reasons explaining and justifying this recommendation in an easy way. Furthermore, decision making based on argumentation is quite similar to the way humans make choices.

3.1 Argumentation Schemes and Critical Questions

Argument schemes are forms of templates to represent inference structures of the arguments that are used daily and in special contexts, such as scientific or legal argumentation. Argument schemes can play two roles [9]:

1. They supply a repertoire of templates of arguments to be used for the pieces of information that are needed,
2. They provide a set of specific critical questions to attack potential vulnerabilities in the opponents' arguments.

Argument schemes serve as templates to draw conclusions through premises. Different schemes of arguments are needed to build the evaluation model

Table 1 Walton's argument scheme for expert opinion

Principal premise	Source E is an expert in the subject domain S which contains proposition A
Secondary premise	E affirms that proposition A is true (or false)
Conclusion	A is true (or false)

completely, such as schemes for comparing alternatives, multicriteria or unicriteria, among others.

Each argument scheme has an associated set of critical questions, which serve to identify the weaknesses of the schemes, challenge assumptions that conform them to verify if there are sufficient support reasons to accept the conclusion.

To better describe the critical questions, the outline of the argument scheme for an expert opinion and their associated critical questions is shown in Table 1 [2].

Six critical questions have been identified for arguments from expert opinion [14]:

1. **Expertise question**: How believable is E as an expertise source?
2. **Field question**: Is E an expert in the field that A is in?
3. **Opinion question**: What did E declares that implies A?
4. **Trustworthiness question**: Is E personally credible as a source?
5. **Consistency question**: Is A consistent with other experts' opinion?
6. **Backup evidence question**: Is E's asseveration based on solid evidence?

If a critical question is answered wrong, the conclusion of the argument scheme is questioned.

3.2 Argumentation Frameworks

To our knowledge, there is no argumentation framework that addresses the public project portfolio selection problem, the frameworks are focused on decision problems, and most of these frameworks are based on the principles of dialogue games.

3.2.1 Dialogue Systems

One way to define the argumentative logic is in the dialectical form of dialogue games (or dialogue systems). Such games model the interaction between two or more players, where the arguments for and against a proposition are exchanged according to certain rules and conditions.

Ouerdane [7] mentions that the dialogue systems define essentially the principle of coherent dialogue and the condition under which a declaration made by an individual is appropriate. There are different formal dialogues considering various information such as: the participants, the language of communication, roles of the participants, the purpose and the context of the dialogue.

Ouerdane focuses on the role of the dialogue system in a DSS. She identifies a set of rules necessary for a dialogue system; these can be divided into four groups:

1. **Locution Rules**: Rules that indicate which statements are allowed.
2. **Dialogue Rules**: Rules that specify the set of locution rules that are allowed at a given moment in the dialogue, and its possible answers.
3. **Commitment Rules**: Rules which indicate the commitment of each player when they make a move.
4. **Termination Rules**: Rules that define when the dialogue must end.

3.2.2 Argumentation System

Here is an overview of a framework for automating decision trough argumentation [8]. This framework uses an agent theory, whose components are divided in three levels:

1. In the first level the object-level decision rules need to be defined, these rules refer to the subject domain. The policy of preferences is divided into two parts.
2. In Level 2 the preference policy under normal circumstances is captured.
3. In Level 3 the part of the preference policy that is referred to the exceptional circumstances of specific contexts is addressed.

The Levels 2 and 3 describe the priority rules, on which the object-level decision rules of the first level are associated with the preference policy of normal and specific situations for the general decision making of the agent. The argumentation-based decision making constructed on the preference policy will then be perceptive to context changes due the changing situations.

Visser [13] presents a Qualitative Preference System (QPS), which is a formal framework to represent qualitative multicriteria preferences where the criterion's preference is based on a lexicographic or cardinality manner.

The principal goal of a QPS is to establish which alternative is preferred from all the outcomes via a value assignment to the variables that are important. There is a knowledge base on which all the constraints of the variables are stored.

The preferences of the alternatives are based on multicriteria, where every criterion has its own weight on the preference from one particular point of view. QPS works with compound criteria, multiple criteria combined to calculate a total preference value, the compound criteria is divided in two groups:

1. **Cardinality criteria**: In this group all subcriteria have the same weight, the preference is calculated by the sum of all the subcriteria that supports the compound criteria.
2. **Lexicographic criteria**: In this category the subcriteria is ordered by priority and the preference is determined with the value of the main subcriteria.

Visser proposes an argumentation framework based on a QPS, this framework provides a logical language to show events of the alternatives, the preference, and a way to construct an argument scheme to deduct preferences from certain input, normal or exceptional context.

3.2.3 Argumentation Framework for Groups

The DM can be a group of people who take the decision; Karacapilidis [4] proposes a decision support system with argumentation focused on either cooperative or non-cooperative groups.

In this framework the argumentation is made by an argumentative discourse, which is composed by diverse discourse acts that have specific roles, in this framework those acts are classified into three categories:

1. **Agent acts**: These acts are made from the user interface and refer to the actions made by the user, such as adding alternatives, opening an issue and correspond to functions directly supported by the user interface. Such functions include the opening of an issue, adding constraints and alternatives, inserting the pro and con arguments, etc.
2. **Internal acts**: As the name implies these are the acts that are performed internally by the system to verify the consistency of the argumentation, update the status of the discussion and to recommend solutions. These functions are called with the user interface via the agent acts and are hidden from him.
3. **Moderator acts**: These acts have a dynamic structure, they are activated by the mediator of the discussion, they may assign the importance of weights to agents, change of the proof standard, a proof that determines the acceptability of the arguments, among others.

The framework proposed by Karacapilidis combines aspects from a variety of areas such as decision theory, argumentation theory, cognitive modeling, this framework is intended to be an advisor for the recommendations of the solutions, but leaves the final execution of the decision to agents.

3.3 Argument Diagramming

There has been a recent growth in argument diagramming tools for visualizing the arguments to understand more efficiently the conclusions. Walton [12] presents a survey on the advances of argument diagramming covering several fields such as informal logic, argumentation theory, legal reasoning, artificial intelligence and computer science.

The diagramming technique is used to represent using the form of a graph the reasoning structure of an argument. Walton mentions a formal definition of an argument diagram made by Freeman, which consists of two components:

1. The first one is a set of numbers in circles; one component is a set of circled numbers organized as points. Each circle is the representation of a premise or conclusions of the argument going to be diagrammed.
2. The second component is a set of arrows that will link the points; each one represents an inference.

The two sets represent the argument graph, for reasoning in the given argument, with this diagram the premises leading to conclusions can be shown in a simple way to facilitate the cognitive capacity of the user in understanding the inference reasoning.

3.3.1 Araucaria Diagramming Software

Reed [11] implemented a model into a software called Araucaria, this system provides a user interface for diagramming arguments, it handles argumentation concepts such as argument schemes for inferences templates. Reed notes that by recognizing the argument structure and all the relationship associated to this, it becomes more efficiently to evaluate all the parts of the argument to draw conclusions. He mentions that argumentation theories are being used widely in artificial intelligence in areas such as natural language, user modelling, and decision support, among others.

Araucaria diagrams the arguments as trees, identifying the parts of the premises, the con and pro arguments and their relationship with the conclusion; here any analysis is debatable. The goal of these trees of arguments is to aid an analyst to dissect the arguments and have a more critical decision.

Araucaria consists of three main features [11]:

1. **Marking up arguments**: Here the user can start constructing a diagram from simple text. Portions of text are created as nodes and can be connected in pairs, one node being the premise and the other one the conclusion.
2. **Ownership and evaluation**: Each node can have one or more owners, which serves as a grouper of nodes of the same argument, each node can have an associated evaluation as well, and this represents a value of certainty on the premises.
3. **Scheme and schemesets**: The argument schemes are defined here; the premises, conclusions and critical questions of the scheme can be defined. The schemesets are sets of defined argument schemes to be used afterwards. With the schemeset, a scheme can be assigned to the argument diagram of the previous parts.

Araucaria is a flexible tool for diagramming arguments, and allows the user to manipulate the structure of the arguments trees.

3.3.2 Carneades Model

Carneades is a computational model proposed by Walton [3], is based on the argumentation theory used in philosophy. Walton mentions that the Carneades model has an advantage over other argument diagramming tools based on informal models, in such form that Carneades informs the user of the acceptability of the premises and conclusion in an efficient way, showing if these satisfy a proof standard with the available arguments of the concerned parties, showing in the diagram.

The argumentation framework of Carneades uses three different types of premises and four types of dialectical status of the statements made by the parties for the modelling of critical questions. Carneades can identify and analyze the structures of the premises and conclusions using the argument schemes, Carneades also evaluates the arguments, it verify if the premises of the arguments hold depending on the state of the dialogue (undisputed, at issue, accepted or reject).

The structure of the argument in Carneades uses the conventional structure of argumentation theory; the arguments contain premises leading to a conclusion, which can be accepted as true or false.

The evaluation of the argument is to decide if a premise is acceptable in the argument graph, it depends on the proof standard, which in turn depends on the defensibility of the pro and con arguments against the statement. The defensibility of the argument depends if the premises hold. This definition of acceptability of the statements is recursive [3]. So, to evaluate the set of arguments in the graph additional information is required: The dialectical status of each statement in the dialogue (if the status is at issue, undisputed, rejected or accepted), the proof standard assigned to each of the statements in the dialogue and the weight of the statements.

There is no approach to public project portfolio selection using argumentation theory, we have seen that with argumentation the conclusion of arguments is very well supported, so it can be used to justify recommendations.

4 Decision Support Systems

The proposed line of work proposes to address the development of a decision support system focused on the recommendation subsystem for the public project portfolio selection, with the purpose of integrating theoretical aspects with emerging technologies in an efficient way. Its goal is to develop a robust tool that integrates formal methods to solve a real problem.

The researches oriented at linking theory with the practice addresses particular aspects of the overall process of decision making, there are few works that follow a holistic approach.

There is no single approach to solving all the problems, some approaches work better than others for particular problems. It is important to choose the right

Table 2 General framework for solving portfolio decision

Phases	Steps	Activities
Preparation activities	1. Pre-evaluate alternatives	Verify accordance with the general objectives
	2. Individual analysis	Set individual criteria weights
Portfolio selection	1. Evaluate alternatives	Verify accordance with individual constraints
	2. Selection of optimal portfolio	Use the correct multicriteria method to select the portfolio
	3. Portfolio refinement	Verify the given portfolio with the general objectives, if not satisfied make adjustments to it

approach to solve different evaluation and selection problems. It is intended that using a decision support system, the problem is analyzed to facilitate choosing the right approach to the problem, choosing this approach can be very complex due to the subjective nature of the problem and the different preferences between the decision makers.

Weistroffer [15] presents a general scheme of a framework to solve problems of project portfolio, in Table 2 an overview based on his scheme is presented.

We propose for Step 5 of the process to use the argumentation theory, we believe it is a good approach for this step through an interaction with the DM to refine the portfolio that is proposed to him as optimal.

5 The Proposed Framework

A decision support system for public project portfolio is proposed focused mainly on a subsystem for generating recommendations in an interactive manner.

A definition from the Oxford dictionary of the term "interactive" is: allowing a two-way flow of information between a computer and a computer-user. Moreover, in the context of computational science, it is defined as the interaction with a human user, often in a conversational way, to obtain data or commands and to give immediate or updated results [6].

In decision support, the term "interactive" is associated with a particular type of methods that are well known and documented in the literature. Each one is based on several principles or philosophical assumptions.

Figure 1 shows the general outline of a framework to decision support, it gives a feedback between all subsystems, where a strong interaction between the analyst (system) and the DM is displayed.

In the organizational subsystem the information of the DM is obtained, such as the preferences, with this information the parameters of the evaluation model are assigned. The output of this subsystem is taken as the input for the exploration subsystem, which is supported on methods of multicriteria decision analysis for

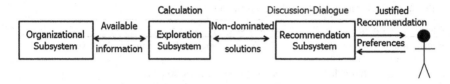

Fig. 1 General scheme of a MCDA process with interaction

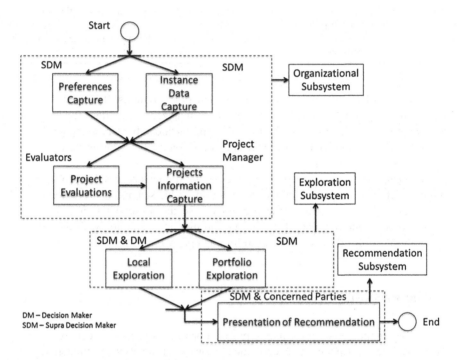

Fig. 2 Subsystems involved in a DSS

running instances of problems, obtaining solutions to these problems, to finally go into the recommendation subsystem.

A deep interaction with the DM is made to help him choose the solution that is more consistent to his model of preferences among all possible solutions, these recommendations should be justified, formal, and significant, the DM could change his preferences and go steps backwards in the process, this process is focused to support a self-learning of preferences of the DM, the interaction plays an important role in its evolution and development.

Elements of the interacting subsystems of the decision support system are detailed more thoroughly in Fig. 2. The proposed framework is to specifically address the recommendation subsystem in how to present the portfolio in a simple way to the DM and concerned parties such as the Supra Decision Maker (SDM), a

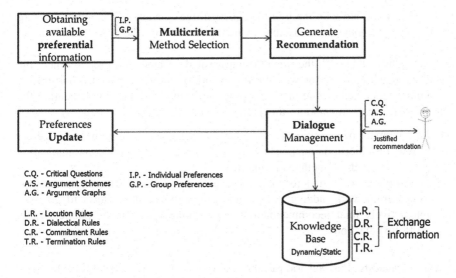

Fig. 3 Proposed framework for public project portfolio with argumentation theory

special player with authority to establish a preferences priority; with the framework they can make a decision that suits their preferences.

A more specific scheme is shown in Fig. 3, specifying in more detail the recommendation subsystem for public project portfolio selection complemented with argumentation theory.

First we have to get the preferential information available from the DM to select the appropriate multicriteria method to the information we gathered, this method provides us with a recommendation of a solution, this is where argumentation theory comes into play, in a dialogue manager between the DM and the framework to justify that the solution satisfies the preferences of the DM.

Throughout the dialogue, with new available information it is possibly for the DM to update his preferences and the process repeats. The theory of argumentation will be supported by a knowledge base consisting of a set of rules (locution, dialectical, commitment, termination) for the exchange of valid arguments and to explain the solution. This dialogue will be supported by a static and dynamic knowledge base; the static part is made up of elements such as sets of rules, argument schemes and initial preferences; the dynamic part consists of the argument graphs and the updated preferences.

6 Conclusions and Future Work

The argumentation theory has a solid base for justifying arguments; it serves as a complement to the recommendation of a portfolio obtained with a multicriteria method, allowing to explain why this option suits the DM's preference model.

This theory can aid with the limit of the cognitive capacity of the mind in regard with the non-monotonic reasoning which is commonly the way humans make decisions.

We believe that with the implementation of the framework proposed here the DM can reach a consensus with the framework on the recommended portfolio.

Future work includes refining the proposed framework and implementing the proposal for experimentation, this is planned in two steps:

1. Development of a multi-criteria evaluation framework that allows to integrate different multicriteria methods of the state of the art, designing communication protocols for these methods to interact.
2. Development of a first recommendation framework based on the aforementioned framework, adding the argumentation theory that allows to generate interactively final recommendations with explanations for the public project portfolio selection.

Acknowledgments This work was partially financed by CONACYT, PROMEP and DGEST.

References

1. Fernández-González, E., Vega-López, I., Navarro-Castillo, J.: Public portfolio selection combining genetic algorithms and mathematical decision analysis. In: Bio-Inspired Computational Algorithms and Their Applications, pp. 139–160, INTECH (2012)
2. Godden, D.J., Walton, D.: Advances in the theory of argumentation schemes and critical questions. Informal Logic 27(3), 267–292 (2007)
3. Gordon, T. F., Walton, D.: The Carneades argumentation framework–using presumptions and exceptions to model critical questions. In: 6th Computational Models of Natural Argument Workshop (CMNA), European Conference on Artificial Intelligence (ECAI), Italy, pp. 5–13 (2006)
4. Karacapilidis, N., Papadias, D.: A computational approach for argumentative discourse in multi-agent decision making environments. AI Commun. 11(1), 21–33 (1997)
5. Labreuche, C., Maudet, N., Ouerdane, W.: Minimal and complete explanations for critical multi-attribute decisions. In: Algorithmic Decision Theory, pp. 121–134. Springer, Heidelberg (2011)
6. Mousseau, V., Stewart, T.: Progressive methods in multiple criteria decision analysis. PhD Thesis, Université Du Luxemburg (2007)
7. Ouerdane, W.: Multiple criteria decision aiding: a dialectical perspective. PhD Thesis, Université Paris Dauphine, Paris (2009)
8. Ouerdane, W., Dimopoulos, Y., Liapis, K., Moraitis, P.: Towards automating decision aiding through argumentation. J. Multi-Criteria Decis. Anal. 18(5–6), 289–309 (2011)
9. Ouerdane, W., Maudet, N., Tsoukiàs, A.: Argument schemes and critical questions for decision aiding process. In: Proceedings of the 2nd International Conference on Computational Models of Argument (COMMA'08), pp. 285–296 (2008)
10. Ouerdane, W., Maudet, N., Tsoukias, A.: Argumentation theory and decision aiding. In: Trends in Multiple Criteria Decision Analysis, pp. 177–208. Springer, US (2010)
11. Reed, C., Rowe, G.: Araucaria: Software for argument analysis, diagramming and representation. Int. J. Artif. Intell. Tools 13(04), 961–979 (2004)

12. Reed, C., Walton, D., Macagno, F.: Argument diagramming in logic, law and artificial intelligence. The Knowledge Engineering Review **22**(01), 87–109 (2007)
13. Visser, W., Hindriks, K.V., Jonker, C.M.: An argumentation framework for qualitative multi-criteria preferences. In: Theory and Applications of Formal Argumentation, pp. 85–98. Springer, Heidelberg (2012)
14. Walton, D., Gordon, T.F.: Critical questions in computational models of legal argument. Argumentation Artif. Intell. Law, 103–111 (2005)
15. Weistroffer, H.R., Smith, C.H.: Decision support for portfolio problems. Southern Association of Information Systems (SAIS), Savannah, Georgia (2005)

Generic Memetic Algorithm for Course Timetabling ITC2007

Soria-Alcaraz Jorge, Carpio Martin, Puga Hector, Melin Patricia,
Terashima-Marin Hugo, Cruz Laura and Sotelo-Figueroa Marco

Abstract Course timetabling is an important and recurring administrative activity
in most educational institutions. This chapter describes an automated configuration
of a generic memetic algorithm to solving this problem. This algorithm shows
competitive results on well-known instances compared against top participants of
the most recent International ITC2007 Timetabling Competition. Importantly, our
study illustrates a case where generic algorithms with increased autonomy and
generality achieve competitive performance against human designed problem-
specific algorithms.

1 Introduction

The *Course timetable problem* (CTTP), commonly seen in most educational institution, requires the assignment of a fixed number of subjects into a number of
timeslots. The main objective is to obtain a timetable that minimizes the number of
student's conflicts. Several types of conflicts can be identified, but they are mostly
time-related i.e. one student with two or more subjects assigned at the same time.
Like most timetabling problems, the *Course timetable* has been cataloged as a

S.-A. Jorge · C. Martin (✉) · P. Hector · S.-F. Marco
Division de Estudios de Posgrado e Investigacion, Leon Institute of Technology, León,
Guanajuato, Mexico
e-mail: jmcarpio61@hotmail.com

M. Patricia
Tijuana Institute of Technology, Tijuana B.C, Mexico

T.-M. Hugo
Instituto de Estudios Superiores de Monterrey ITESM, Monterrey N.L, Mexico

C. Laura
Instituto Tecnologico de Cd. Madero, Madero Tamaulipas, Mexico

O. Castillo et al. (eds.), *Recent Advances on Hybrid Approaches for Designing*   481
Intelligent Systems, Studies in Computational Intelligence 547,
DOI: 10.1007/978-3-319-05170-3_33, © Springer International Publishing Switzerland 2014

NP-Complete problem [1, 2]. Due its complexity and the fact that most course timetables are still often constructed by hand, it is necessary to automate this time-table construction to improve upon the solutions reached by the human expert [3].

This chapter describes the construction of a generic prototype based on a Memetic algorithm (MA). Memetic algorithms (MAs) [4] are population-based metaheuristic search approaches that have been receiving increasing attention in the recent years. They are inspired by Neo-Darwinian principles of natural evolution and a notion of a meme. This *meme* has been defined as a unit of cultural evolution that is capable of local refinements [5]. This MA approach is used along to a generic representation called *Methodology of Design* [6, 7] in order to achieve a greater degree of generality. Undoubtedly the most important part of the proposed algorithm is a *Local Search phase*. This phase is executed on each individual where a technique of autonomous search selects from a heuristic pool which is the best next heuristic to be implemented, increasing the chances to improve the current solution. The main contribution of this chapter is to present a novel MA generic approach to solve the CTTP using a set of generic structures as well as an autonomous search strategy as local search phase. This can be a useful approach for the interested researcher since it is designed to be easily adaptable to wide range of particular instances. The proposed MA prototype is tested over well-know CTTP benchmarks (ITC2007 track 2 and ITC2002 developed by PATAT where the proposed algorithm obtains competitive results against top competitors and in two cases improves the solutions achieved by the winning algorithm for ITC2007.

The CTTP has been intensively studied, from early solution approaches based on graph heuristics [8], linear programming [9] and logic programming [10, 11] to meta-heuristic schemes as tabu list [12], Genetic algorithms [13, 14], Ant Colony [15, 16], Variable Neighbourhood search [17], Nature inspired search techniques [18], Simulated Annealing [19] and so on. In the same way a great number of CSP solvers schemes have been reported as used to solve the CTTP problem [20, 21]. Also, in recent years selection Hyper-heuristics has been applied to this problem [22, 23] with encouraging results. In general all of these previous works used two important CTTP benchmarks: ITC2002 and ITC2007 in order to gather data to measure its efficiency. In this context this chapter uses a Memetic Algorithm as solution method. The proposed approach also uses a layer of generality called *methodology of design* [7], this methodology responds to the necessity of generic solution schemes [3], applicable to real world instances [24, 25]. The main objective of this methodology is to make a real impact in the automated timet-abling practice, producing state of art algorithms that can be directly applicable to a wide range of real-work CTTP problems, without losing time and resources into building a new specific-instance algorithms.

This paper is organized as follows: Sect. 2 gives the background of Course Timetabling Problem, Sect. 3 introduces the solution methodology and its justi-fication. Section 3 contains the experimental set-up, results, and their analysis. Finally Sect. 4 shows some conclusions and future work.

2 The Course Timetabling Problem

2.1 Problem Formulation

The CTTP can be modeled as a Constraint Satisfaction Problem (CSP) where the main variables are events and the most common type of constraints are time-related [17]. A concise definition of the CTTP can be found in [26]: A set of events (courses or subjects) $E = \{e_1, e_2,..., e_n\}$ are the basic element of a CTTP. Also there are a set of time-periods $T = \{t_1, t_2, ..., t_s\}$, a set of places (classrooms) $P = \{p_1, p_2, ..., p_m\}$, and a set of agents (students registered in the courses) $A = \{a_1, a_2, ..., a_O\}$. Each member $e \in E$ is a unique event that requires the assignment of a period of time $t \in T$, a place $p \in P$ and a set of students $S \subseteq A$, so that an assignment is a quadruple (e, t, p, S). A timetabling solution is a complete set of n assignments, one for each event, which satisfies the set of hard constraints defined by each university of college. This problem is known to be at least NP-hard [1, 2].

2.2 Problem Model

In order to manage the previously seen generic model problem as well as specific constraints the *methodology of design* is used. This methodology explicitly described in Soria-Alcaraz Jorge et al. [6, 7] presents a way to work with the CTTP using an extra layer of generality that converts time and space related constraints into a single constraint type: Student Conflict. The *methodology of design* integrates all constraints into three basic structures *MMA matrix, LPH list and LPA list*.

- **MMA matrix**: This matrix contains the number of common students between two subjects i.e. the number of conflicts if two subjects are assigned in the same timeslot. This is the principal structure for this methodology. An example of this matrix can be seen in the Fig. 1.
- **LPH list**: This structure have in its rows the subjects offered. In its columns have the offered timeslots. This list gives information about allowed timeslots per subject. An example of this list can be seen on Table 1.
- **LPA list**: This list shows in the classrooms available to be assigned to each event without conflict. An example of this list can be seen on Table 2.

Once we obtain these structures by means of the original inputs of our CTTP problem, we ensures *by design* the non-existence of violations by the selection of any values shown in LPH and LPA [6]. This is an important characteristic of this methodology: *each time that a solution is constructed with the information of these structures it can be guarantee that the solution is feasible*. Then, only is necessary

Fig. 1 MMA matrix

	ACM0403	SCM0414	SCB0421	SCE0418	ACH0408	ACM0401	ACM0404
ACM0403	4	1	1				
SCM0414	1	10	3	3			6
SCB0421	1	3	3	2			2
SCE0418		3	2	3			3
ACH0408							
ACM0401						4	1

Table 1 LPH list

Events	Timeslots
a_0	<7am> or <8am>
a_1	<9am>
a_2	<7am>
⋮	⋮
a_3	<7am> or <8am > or <9am> or <10am>

Table 2 LPA list

event	Classrooms
a_0	$<p_A, p_B>$
a_1	$<p_A, p_C>$
a_2	$<p_A, p_B, p_C>$
⋮	⋮
a_3	$<p_A>$

to optimize the current timetabling according the next equations in order to achieve a *perfect solution* (a feasible timetabling with 0 student conflicts):

$$min(FA) = \sum_{i=1}^{|V|} FA_{V_i} \qquad (1)$$

$$FA_{V_i} = \sum_{s=1}^{|V_i|} \sum_{l=s+1}^{|V_i|} (S(s)S(l)) \qquad (2)$$

where: FA = Student conflicts in current timetable. V_i = Student conflicts from 'Vector' i in the current timetable. $S(s) \cap S(l)$ = set of students that simultaneously demand subjects s and l inside the 'Vector' i. S means a function that given a event s in a timetable i return the number or enrolled students on it.

The term *Vector* is used to describe a collection of events like the timeslot offered by the specific university to be solved. It is fair to say the number of

Fig. 2 Representation used

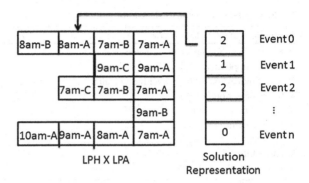

LPH X LPA   Solution Representation

timeslots offered by the university is the same number of vectors used in the methodology of design, this methodology simply describes the CTTP in general terms as the addition of events to vectors.

Finally the representation used by any search algorithm is a single integer array with a length equal to the number of variables (events) and its integer elements represent each pair (LPH_i, LPA_i), where LPH_i is a valid temporal value for the event i and LPA_i is a valid space value for the event i. This array constructs a complete timetable assignment. The objective is then to search in this Cartesian product space $LPH_i \times LPA_i$, looking for minimum student conflicts by Eqs. (1) and (2) and the *MMAmatrix*. One example of this representation can be seen on Fig. 2.

2.3 Memetic Algorithms

A Memetic Algorithm is a population-based search process that mimics culture evolution [5]. This kind of algorithm can be understood as a GA or generic algorithm with a more explicit *Improvement* phase, in the cannon MA algorithm, each individual has a local search phase just at the start of a new iteration a prior of any crossover, mutation or elitism operator. During this phase each individual is optimized by means of the application of heuristic strategies with the objective to accelerate the convergence of the algorithm.

The nature of the local search operator is still in discussion but it is commonly designed according the representation used, for example to a continuous-natured problem the use of a gradient descent operator as local search would be a good idea; On the another hand in the discrete space a hill-climbing method between several *s*wap operators would achieve good results. A canonical Memetic Algorithmpseudo code can be seen on Table 3.

Table 3 Canonical memetic algorithm

Requires: PopSize, StopCriteria
1: S_t = GenerateInitialPopulation(PopSize)
2: **while** !StopCriteria() **do**
3: S^* = ImprovementStage_EachIndividual(S_t)
4: $S^{*'}$ = Selection(S^*)
5: S^* = Crossover($S^{*'}$)
6: S^* = Mutation($S^{*'}$)
7: S_t = S^*
8: **end while**
9: **return** BestIndividual(S_t)

2.4 Combining MA with Methodology of Design for the Course Timetabling Problem

As seen on previous sections the MA algorithm uses the concept of Population, This population can be seen as a set of incoming feasible solutions or timetablings that can be measured by the application of Eqs. 1 and 2. In the proposed MA, the representation used will be the same one described on Fig. 2, Sect. 2.2. This means that a set of operators based in an integer representation are needed. First of all for *local search procedure* we proposed the simplest autonomous search method: *the Hill-Climbing method,* this strategy uses a set of Integer-based operators such as *swapTwo, Random Perturbation, Move to less Conflict and Brelaz heuristic.* The idea of this hill climbing is to apply all of these operators in a current solution and then analyses if the incoming neighbor solutions are better in terms of Fitness (Eqs. 1 and 2). If that happens the current solution is replaced by the best incoming solution. This process is iterated until no better solution is founded.

For *Selection* operator we use a roulette wheel where individuals with less fitness value have a higher probability to be chosen for further operators. For *Crossover* operator we use a two points crossover where two points inside the representation array are selected uniformly at random at each MA iteration, then the genetic material from two parents are interchanged between these points. For *Mutation* operator we simply chose the individual with the worse fitness value in order to regenerate it. Finally *Elitism* is applied, so a percentage defined by user of the best individuals is kept across all iterations. With these considerations we construct a Memetic algorithm with a generic timetabling representation noted as GMA.

3 Experiments

In this section the experiments executed as well as the parameters used for the GMA algorithm are detailed.

3.1 ITC 2007 Test Instances

The methodology of design allows the solution of several different instances, it is only necessary that these instances can be expressed in terms of the generic structures (MMA, LPH and LPA). This is an important advantage of the methodology of design approach. A well-known and referenced benchmark is used for comparison with our generic MA approach: PATAT 2007 Track 2 as employed in the second International Timetabling Competition sponsored by PATAT.

There are 24 instances in PATAT 2007 Track 2, these instances consist of:

- A set of n events that are scheduled into 45 timeslots.
- A set of r rooms, each which has a specific seating capacity.
- A set of *room-features* that are satisfied by rooms and required by events.
- A set of s students who attend various different combinations of events.

The hard constraints are:

- No student should be required to attend more than one event at the same time
- Each case the room should be big enough for all the attending students
- Only one event is put into each room in any timeslot.
- Events should only be assigned to timeslots that are pre-defined as available (2007 only).
- Where specified, events should be scheduled to occur in the correct order.

The Soft constraints are:

- Students should not be scheduled to attend an event in the last timeslot of a day.
- Students should not have to attend three or more events in successive timeslots.
- Student should not be required to attend only one event in particular day.

3.2 Parameter Configuration

The parameter configuration for any meta-heuristic is a complex problem by itself Hoos [27], for the GMA algorithm we have 4 main parameters (Population size, Elitism Percentage, Crossover Probability, Mutation Probability) to be configured, in this chapter we discretize the possible choices for each parameter at 10 elements in regular intervals by the next range per parameter: For Population size [25,275], Elitism [0,100], Crossover [0,100], Mutation [0,100]. For example the possible values to be used as population size will be: (25, 50, 75, 100, 125, 150, 175, 200, 225, 250, 275) these ranges were fixed by preliminary experiments. In order to search for the best possible parameter values we use ParamILS [28]. ParamILS is a framework for automated algorithm configuration based on performing local search in the parameters values space. The key idea behind ParamILS is to combine powerful stochastic local search algorithms (iterated local search) with

Table 4 Best parameter
configuration achieved by
ParamILS

Parameter	Value
Population size	75
Elitism	10 %
Crossover	85 %
Mutation	20 %

mechanisms for exploiting specific properties of algorithm configuration problems. The search process starts from a given configuration (which is generally the target algorithm's default configuration) as well as r additional configurations chosen uniformly at random from the given discretized ranges of the configuration parameters. These $r + 1$ initial configurations are evaluated, and the best performing is selected as the starting point of the ILS. To evaluate a configuration, the idea is to perform a fixed number of runs of the target algorithm with the given configuration on the set of training instances. The process starts with the iterated local search strategy, using one-exchange neighborhood (i.e. inducing an arbitrary change in a single target algorithm parameter) in the improvement stage, and a number of steps s of the same neighborhood as the perturbation stage. Small values of s (i.e. $s = 2$) have been found to be sufficient for obtaining good performance of the overall configuration procedure. In Hutter et al. [28], extensive evidence is presented that supports ParamILS can find substantially improved parameter configurations of complex and highly optimized algorithms. The parameter configuration achieved can be seen on Table 4.

3.3 Experiments with the ITC2007 Track 2 Benchmark

The proposed (GMA) with the best parameter configuration founded so far is tested in this experiment under the rules of the Second International Timetabling Competition ITC2007 Track 2 (Post-Enrolment Course Timetabling). For this test, a benchmark program were download from the webpage of ITC2007, this program indicates the run time allowed, this time is used for every instance of the ITC 2007 with the proposed algorithm. 10 independent tests were executed for each instance fallowing ITC rules. The Computer used for these experiments was an Intel core 7 CPU, 2.66 GHz × 8 with 8 Gb of Ram and Linux Ubuntu 12.10 64-bit with LXDE using OpenJDK7. In this test we made a comparison with the best results (achieved Timetables with lowest student conflict) obtained by the top entries (Winner Cambazard, Second Place Atsuna and Third Place Chiarandini) of the Patat ITC2007 track 2 competition. The final overall results can be seen on Table 5.

The GMA algorithm has shown a competitive performance over the ICT2007 benchmark. A box graph from the 10 independent results for all algorithms over the ITC2007-16 instance can be seen on Fig. 3.

Table 5 Comparative results (Best results ITC2007)

	Atsuna	Cambazard	Chiarandini	GMA
ITC2007-3	382	164	288	290
ITC2007-4	529	310	385	600
ITC2007-7	0	6	10	30
ITC2007-8	0	0	0	**0**
ITC2007-15	379	0	0	**0**
ITC2007-16	191	2	1	30
ITC2007-17	1	0	5	3
ITC2007-20	1215	445	596	670
ITC2007-21	0	0	602	**0**
ITC2007-24	720	21	822	130

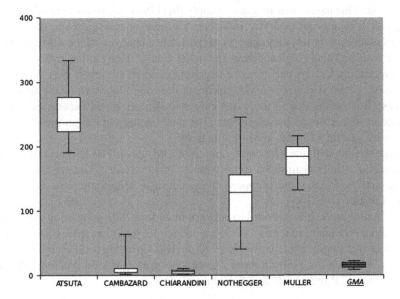

Fig. 3 Detailed results for 10 independent test over ITC2007-16

3.4 Discussion

As seen on Table 5 and Fig. 3 the GMA has shown competitive results against ad hoc algorithms designed explicit for the ITC2007 benchmark, the GMA algorithm is a generic scheme that allows the solution of a variety of CTTP instances when these instances can be transform into the a set of generic structures. The GMA algorithm achieves some best-know results in ITC2007-8, ITC2007-15 and ITC2007-21 These instances have different values in number of students, subjects, classrooms so this give us evidence about the adaptability of the GMA approach. The population used was 75 individuals with an elitism of 10 % so only

7 individuals were needed to store the best information across generations. Another important characteristic was the box graph in Fig. 3 where the GMA algorithm shows a very low dispersion for the ITC2007-16 instance, this means that all the independent executions achieved the best known result.

4 Conclusions

This chapter has discussed the use of a generic Memetic algorithm over one of the ITC2007 benchmark. The proposed GMA algorithm uses a methodology of design that presents a way to work with the CTTP employing an extra layer of generality that converts all the time and space constraints into a single student conflict constraint. This means that in order to apply this GMA algorithm to another set of instances it is only necessary to transform all these new instances to the generic representation, it is not necessary to make further changes in the GMA code to obtain results. This is a great advantage in the Course Timetabling area where ad-hoc algorithms lose its performance capabilities when are tested over real world instances, where it is necessary to adequate the used algorithm to the new conditions. This means that in the worst case re-codification and experimentation is needed to tuning the ad-hoc algorithm. The methodology of design presents more flexibility to deal with this kind of situations.

Experiments with the best configuration of GMA over ITC2007 instances give competitive results against the top 5 winners of the Second Timetabling competition, and in 5 cases the proposed hyper-heuristic algorithm achieved a best-known result.

For future work, it is intended to perform more tests with population algorithms with more complex schemes. We also intend to look for new CSP inspired operators that improve the performance of the Hill Climbing local search operator. Finally, we intend to perform more experiments with GMA on problems that are similar to CTTP (e.g. VRP, Job-Scheduling, etc.).

Acknowledgments Authors thanks the support received from the Consejo Nacional de Ciencia y Tecnologia (CONACYT) México and The University of Stirling UK.

References

1. Cooper, T.B., Kingston, J.H.: The complexity of timetable construction problems. Ph.D. Thesis, The University of Sydney (1995)
2. Willemen, R.J.: School timetable construction: algorithms and complexity. Ph.D. Thesis, Institute for Programming Research and Algorithms (2002)
3. Lewis, R.: Metaheuristics for University course timetabling. Ph.D. thesis, University of Nottingham (2006)

4. Ong, Y.S., et al.: Classification of adaptive memetic algorithms: a comparative study. Syst. Man Cybern. Part B: Cybern. IEEE Trans. 36(1), 141–152 (2006)
5. Radcliffe, et al. 1994. Formal memetic algorithms, vol. 85. Evolutionary Computing. Lecture Notes in Computer Science, pp. 1–16. Springer, Berlin
6. Soria-Alcaraz Jorge, A., et al.: Methodology of design: a novel generic approach applied to the course timetabling problem, vol. 451. Studies in Computational Intelligence. Springer Berlin (2013b)
7. Soria-Alcaraz Jorge, A., et al.: Comparison of Metaheuristic algorithms with a methodology of design for the evaluation of hard constraints over the course timetabling problem, vol. 451. Studies in Computational Intelligence. Springer, Berlin (2013a)
8. de Werra, D.: An introduction to timetabling. Eur. J. Oper. Res. 19(2), 151–162 (1985)
9. Carter, M.: A survey of practical applications of examination timetabling algorithms. Oper. Res. 34, 193–202 (1986)
10. Lajos, G.: Complete university modular timetabling using constraint logic programming. In E Burke and P Ross editors. Practice and Theory of Automated Timetabling (PATAT), vol. 1153, pp. 146–161. Springer, Berlin (1996)
11. Boizumault, P., et al.: Logic programming for examination timetabling. Logic Program 26, 217–233 (1996)
12. Lü, Z., Hao, J.-K.: Adaptive Tabu Search for course timetabling. Eur. J. Oper. Res. 200(1), 235–244 (2010)
13. Colorni, A., et al.: Metaheuristics for high-school timetabling. Comput. Optim. Appl. 9, 277–298 (1997)
14. Yu, E., Sung, K.S.: A genetic algorithm for a University Wekly courses timetabling problem. Int. Trans. Oper. Res. 9, 703–717 (2002)
15. Nothegger, C., Mayer, A., Chwatal, A., & Raidl, G. R.: Solving the post enrolment course timetabling problem by ant colony optimization. Annals of Operations Research. 194(1), 325–339 (2012)
16. Socha, K., et al.: A MAX-MIN Ant system for the University Course timetabling Problem. In: Dorigo, M., Caro, G.D., Samples, M. (eds.) Proceedings of Ants 2002—Third international workshop on Ant algorithms, Lecture Notes in Computer Science, pp. 1–13. Springer, Berlin (2002)
17. Burke, E., et al.: Hybrid variable neighbourhood approaches to university exam timetabling. Eur. J. Oper. Res. 206(1), 46–53 (2010)
18. Sabar, N.R., et al.: A honey-bee mating optimization algorithm for educational timetabling problems. Eur. J. Oper. Res. 216(3), 533–543 (2012)
19. Thompson, J.M., Dowsland, K.A.: A robust simulated annealing based examination timetabling system. Comput. Oper. Res. 25, 637–648 (1998)
20. Rudova, H., et al.: Complex university course timetabling. J. Sched. 14, 187–207 (2011). doi:10.1007/s10951-010-0171-3
21. Cambazard, H., et al.: Local search and constraint programming for the post enrolment-based course timetabling problem. Ann. Oper. Res. 194, 111–135 (2012)
22. Burke, E.K., et al.: A graph-based hyper-heuristic for educational timetabling problems. Eur. J. Oper. Res. 176(1), 177–192 (2007)
23. Soria-Alcaraz, J.A., et al.: Academic timetabling design using hyper-heuristics. Adv. Soft Comput. 1, 158–164 (2010). ITT Springer-Verlag
24. Causmaecker, P.D., et al.: A decomposed metaheuristic approach for a real-world university timetabling problem. Eur. J. Oper. Res. 195(1), 307–318 (2009)
25. Kahar, M., Kendall, G.: The examination timetabling problem at Universiti Malaysia Pahang: comparison of a constructive heuristic with an existing software solution. Eur. J. Oper. Res. 207(2), 557–565 (2010)
26. Conant-Pablos, S.E., et al.: Pipelining Memetic algorithms, constraint satisfaction, and local search for course timetabling. MICAI Mexican International Conference on Artificial Intelligence, vol. 1, pp 408–419. (2009)

27. Hoos, H.H.: Automated Algorithm Configuration and Parameter Tuning, Chap. 3, pp. 37–71. Springer, Berlin (2012)
28. Hutter, F., et al.: ParamILS: an automatic algorithm configuration framework. J. Artif. Intell. Res. **36**(1), 267–306 (2009)

Characterization of the Optimization Process

Marcela Quiroz, Laura Cruz-Reyes,
Jose Torres-Jimenez, Claudia Gómez Santillán,
Héctor J. Fraire Huacuja and Patricia Melin

Abstract Recent works in experimental analysis of algorithms have identified the need to explain the observed performance. To understand the behavior of an algorithm it is necessary to characterize and study the factors that affect it. This work provides a summary of the main works related to the characterization of heuristic algorithms, by comparing the works done in understanding how and why algorithms follow certain behavior. The main objective of this research is to promote the improvement of the existing characterization methods and contribute to the development of methodologies for robust analysis of heuristic algorithms performance. In particular, this work studies the characterization of the optimization process of the Bin Packing Problem, exploring existing results from the literature, showing the need for further performance analysis.

M. Quiroz · L. Cruz-Reyes (✉) · C. G. Santillán · H. J. F. Huacuja
Instituto Tecnológico de Ciudad Madero, Ciudad Madero, México
e-mail: lcruzreyes@prodigy.net.mx; lauracruzreyes@itcm.edu.mx

M. Quiroz
e-mail: qc.marcela@gmail.com

C. G. Santillán
e-mail: cggs71@hotmail.com

H. J. F. Huacuja
e-mail: hfraire@prodigy.net.mx

J. Torres-Jimenez
CINVESTAV-TAMAULIPAS, Cd. Victoria Tamps, México
e-mail: jtj@cinvestav.mx

P. Melin
Tijuana Institute of Technology, Tijuana, México
e-mail: pmelin@tectijuana.mx

O. Castillo et al. (eds.), *Recent Advances on Hybrid Approaches for Designing*   493
Intelligent Systems, Studies in Computational Intelligence 547,
DOI: 10.1007/978-3-319-05170-3_34, © Springer International Publishing Switzerland 2014

1 Introduction

Throughout the search for the best possible solutions for NP-hard problems, a wide variety of heuristic algorithms have been proposed. However, despite the efforts of the scientific community to develop new strategies, there is no efficient algorithm capable of finding the best solution for all possible situations [1, 2]. One of the main challenges of the experimental analysis of heuristic algorithms is to identify what strategies make them show an improved performance and under what conditions they get it.

Much of the recent progress in algorithm development has been facilitated by a better understanding of the properties of the problem instances and a better understanding of the performance of the heuristic algorithms; an example of this would be the extensive research on constraint satisfaction problems, gaining an improved understanding of the properties of the instances and the high-performance algorithms [2, 3]. However, there is still a strong lack of understanding of why heuristic algorithms follow certain behavior depending on the instances.

The main objective of this research is to promote the characterization and analysis of NP-hard problems and heuristic algorithms, for the purpose of designing better algorithms and benchmark instances for optimization problems. In particular, this work explores existing results in characterization for the Bin Packing Problem (BPP). The remainder of this chapter is structured as follows. Section 2 provides a brief introduction to NP-hard problems, heuristic algorithms, and experimental analysis of algorithms. Section 3 provides a summary of the main works related to the characterization of heuristic algorithms. Section 4 presents a review of the characterization of the optimization process of BPP. Finally, Sect. 5 concludes with final remarks, highlighting open problems in characterization of the optimization process and proposing future work in performance analysis.

2 Heuristic Algorithms for Hard Optimization Problems

The so-called NP-hard problems are of great interest in computer science. One feature of these problems is that the exact algorithms used to solve them require an exponential amount of time. In other words, these problems are very difficult to solve [4]. Under these circumstances, it is necessary to use heuristic algorithms which include strategies based on common sense to provide approximate solutions in a reasonable time, but do not guarantee the optimal solution. Heuristic algorithms are too complex to study analytically and become accessible via experimental analysis, experimentation can provide more accurate predictions of the performance of these algorithms in practice.

Most of the work on performance analysis of heuristic algorithms has been focused on comparative analysis of experimental results. However, recent works in the area have attempted to analyze heuristics algorithms to identify features that impact on their performance. To gain insight into the performance of a heuristic algorithm that solves a NP-hard problem, it is necessary to make a study of the entire solution process. The optimization process can be understood as the act of solving an instance of an optimization problem using an algorithm to obtain a final solution. The characterization of the instances of the problem and the solution process is an essential part in the performance analysis of algorithms, and allows identifying the factors that influence the algorithmic behavior.

The characterization of the optimization process seeks for relevant and measurement performance factors. These factors are characterized through indexes that provide useful information to describe the algorithmic performance. Then, the indexes are analyzed looking for relationships between features of the problem, features of the algorithm and the final performance. The main objective of the characterization is to understand how the performance of an algorithm is affected by a number of factors. The knowledge gained can lead to better predictions of performance in new situations, leading to the definition of improved algorithms.

The definition of a set of indexes that describe the optimization process is a difficult task that requires expert knowledge of the problem domain. The characterization indexes must be carefully chosen so as to facilitate a correct discrimination of the difficulty of the instances, as well as the behavior and performance of the algorithm. The success of any knowledge-discovery process depends on the quality of the data, and in this case, the indexes must use features that serve the purpose of differentiating algorithm performance [5].

3 Work in Characterization of the Optimization Process

Among related work on characterization and understanding of the optimization process, four general approaches have emerged: exploratory data analysis, analysis of the structure of the search space, study of the properties causing transitions in computational complexity of NP-hard instances, and study of the structure of optimal solutions.

3.1 Exploratory Data Analysis

The aim of the exploratory data analysis is to obtain knowledge from a set of characterization indexes and their underlying structure. It includes statistical methods, tabular comparisons, graphical analysis, causal inference and multivariate analysis that make possible the construction of a model that describes the set of relations between the factors under study.

There is a vast amount of work attempting to relate the characteristics of some classes of instances of NP-hard problems with the performance of algorithms [3, 5–11]. These works use tabular and graphical analysis as well as machine learning

Table 1 Works in characterization of the optimization process

Works	Input problem		Algorithm of solution			Final performance
	Instances	Optimal solutions	Parameters	Landscape	Structure	
Group 1	✔					✔
Group 2			✔			✔
Group 3				✔		✔
Group 4	✔			✔		✔
Group 5	✔[a]					✔
Group 6		✔				✔
Group 7		✔		✔		✔
Quiroz et al.	✔		✔	✔	✔	✔

Group 1 [3, 5–11]
Group 2 [12–18]
Group 3 [20–24]
Group 4 [3, 25, 26]
Group 5 [31–37] [a] Phase transition
Group 6 [38, 41, 42]
Group 7 [43–46]
Quiroz et al. [47]

techniques to predict which type of algorithm shows a better performance with some types of instances of the problem under study (see Table 1, Group 1). Other studies, attempt to find a relationship between various parameters and settings that control an algorithm and the final performance achieved when applied to a class of instances of a problem [12–18]. Some of these researches apply parameter tuning methods based on experimental designs while other apply graphical analysis and machine learning trying to find settings and parameters for improving the performance of the heuristics (see Table 1, Group 2).

However, the exploratory data analysis will be successful only if the set of indexes used to characterize the optimization process is the right one. These features need to be carefully chosen in such a way that they can characterize instances and algorithm structure as well as differentiate algorithm performance [5].

3.2 Analysis of the Structure of the Search Space

The fitness landscape is the visualization of the search space, the set of all possible solutions that satisfy the problem's constraints related to their fitness; it is a surface formed by peaks, valleys, mountains and plains. Thus, high fitness points are located in "peaks" and low fitness points in "valleys" [19]. The behavior of heuristic algorithms is determined by the structure of the search space, i.e., the objective function induced by a specific problem instance, and properties of the landscapes are expected to affect the performance. The analysis of the structure of the search space of the problem allows studying some features of the solutions traveled by the heuristic algorithms. This analysis helps to increase the knowledge about the behavior of the algorithm under study, suggesting explanations that provide intuitive descriptions for important aspects of the search process [20].

Some authors have attempted to characterize the structure of the search space follow by an algorithm that solve a class of instances of a NP-hard problem, and associate it with the properties of the final performance [20–24] (see Table 1, Group 3). Other researchers have explored the structure of the solution space of NP-hard problems, covering the landscape with random walks, and trying to find more general measures of difficulty [3, 25, 26] (see Table 1, Group 4).

However, the analysis of the proposed measures has shown that currently there are no indicators that are able to provide accurate predictions of the difficulty of the search. So far, it has not been possible to relate successfully the properties of the structure of the search space, with the algorithmic behavior and the final performance [2, 21, 23, 27–29].

3.3 Study of the Properties Causing Phase Transitions

A phase transition can be defined as an abrupt change in the qualitative behavior of a system [6, 30]. In the case of the optimization problems, a phase transition can be defined as an abrupt change in the complexity of the instances that occurs when some parameter varies. Many experimental results have shown that for a class of NP-hard problems one or more "order parameters" can be defined, and hard instances occur around the critical values of these order parameters. One reason for the interest in such phase transitions is that the most difficult problems tend to occur in the transition region. The phase transition analysis allows characterizing the structure and complexity of NP-hard instances.

The knowledge gained in the study of the phase transition phenomena has been applied to identify test instances with a high degree of difficulty and to suggest new algorithms specially designed for these instances [31–37] (see Table 1, Group 5). However, there are many optimization problems, whose phase transition has not been identified. Moreover, there seems to be no study on the relationships between the parameters defining the phase transition in the complexity of the instances and the behavior of the algorithms.

3.4 Analysis of the Structure of Optimal Solutions

The study of the structure of the optimal solutions of NP-hard problems instances has allowed to identify relationships between the characteristics of the values that define the optimal solutions and the difficulty of the instances. In the process of solving optimization problems, it is possible to identify some features in the values of the optimal solutions; a backbone can be defined as a set of variables having fixed values in all optimal solutions of an instance of the problem. For some NP-hard problems, it has been observed that the size of these backbones have an impact on the performance of the solution algorithms [38].

For some optimization problems it has been possible to identify backbones in optimal solutions, and in several cases it has been shown that the magnitude of these sets is related to the degree of difficulty of the instances [38–40]. However, currently, few studies have directly quantified the relationship between the size of such sets and the difficulty of the instances [38, 41, 42] (see Table 1, Group 6). And a limited number of studies have directly used this property to improve the performance of heuristic algorithms [43–46] (see Table 1, Group 7).

3.5 Characterization of the Optimization Process

In a previous work [47], we have proposed an experimental approach for a comprehensive study of the optimization process. The proposed approach combines methods of exploratory data analysis and causal inference to identify inherent relationships among the factors that affect the algorithmic performance, with the aim of understanding the behavior of the algorithm under study and visualize possible improvements in its structure.

Each one of the previous approaches have contributed to the characterization the optimization process. However, the complexity of the NP-hard problems and the heuristic algorithms has shown that a single approach is not sufficient to understand the performance of the algorithms, showing the need for further analysis. Table 1 summarizes the characteristics of the studies presented in the main works on characterization of the optimization process, highlighting the stages of the optimization process that have been analyzed by the different authors. The works have been integrated into seven groups, taking into account the type of research conducted.

The analysis of the main works on characterization of the optimization process revealed that, so far, it has not been possible to establish a connection between the indexes that have been defined to characterize the three stages of the optimization process: input problem, algorithm of solution and final performance. Most of the research has focused on demonstrating that the proposed heuristics are effective or superior to others in solving a set of instances of the problem, without understanding the reason for their good or bad performance.

Likewise, a large number of NP-hard problems and heuristic algorithms have not yet been studied and there is little knowledge about the difficulty of the instances of these problems and the behavior of the algorithms [2, 23]. Recent works in this field have pointed out the need to combine all proposed characterization techniques for better explanations of the difficulty of the NP-hard instances and the performance of heuristics [5]. A study of this type is important because it can provide solid foundations for the analysis and design of high performance algorithms.

4 Case of Study: Bin Packing Optimization Process

Given an unlimited number of bins with a fixed capacity $c > 0$ and a set of n items, each with a particular weight $0 < w_i \leq c$, BPP consist in packing all the items in the minimum number of bins without violating the capacity of any bin. Over the last twenty years, throughout the search for the best possible solutions for BPP, researchers have designed elaborate procedures that incorporate various techniques. The most relevant results have been obtained by means of heuristic algorithms [48–51].

4.1 Benchmark Problem Instances

The performances of the algorithms for BPP have been evaluated with different benchmark instances considered as challenging. The trial benchmark had been elected by different authors to compare the effectiveness of their proposals with other algorithms of the state-of-the-art. They form a group of 1,615 standard instances in which the number of items n varies between [50, 1,000] and the intervals of weights are between (0, c]. Such instances can be found on recognized websites [52–55].

Until now, Gau 1 and Hard28 sets are the most difficult instances for the known algorithms [48, 50, 51, 56–59], standing out the Hard28 class, for which exists a mayor number of instances that cannot be solved optimally by the known algorithms.

4.2 BPP Characterization Indexes

It is known that factors like the number of items, the central tendency of the weights, and their distribution; produces an impact the degree of difficulty of a BPP instance. Different authors have proposed sets of indexes for the characterization of BPP:

- With the goal of modeling the structure of an instance, Pérez et al. [60] formulated a specific-purpose indexes set, formed by five indexes: instance size, occupied capacity, dispersion, factors, and bin usage.
- Cruz-Reyes et al. [61] proposed 21 indexes based on descriptive statistics, these indexes characterize the weight distribution of the items and were grouped into four categories: centralization, dispersion, position and form of the frequency distribution of the weights.
- Quiroz et al. [47] studied all previous indexes by analyzing if they allowed discriminating between different BPP instances and introduced a new set of indexes for BPP and the algorithm behavior characterization. Quiroz also

proposed the use of 14 indexes to characterize the features of the problem, three indexes to characterize the algorithm behavior, and three indexes to characterize final performance. The study of these indexes allowed understanding the behavior of a GA, and improved its performance.

4.3 Performance of the State-of-the-Art BPP Algorithms

Table 2 shows the details in the performance achieved by the state-of-the-art algorithms for BPP: HI_BP [49], WA [62], Perturbation-SAWMBS [50], and HGGA-BP [51]. We note the superiority of HGGA-BP heuristic considering the class Hard28 of instances. It can be observed that, for this set of instances, HGGA-BP shows a higher effectiveness than the best BPP algorithms.

For Hard28 set, Fleszar and Charalambous [50] point out that even increasing the number of iterations of his Perturbation-SAWMBS from 2,000 to 100,000 an improvement could not be achieved, emphasizing the difficulty of these instances. Similar results were obtained when increasing the iterations in the improvement phase of HI_BP from 4,000 to 100,000. The observed results suggest that these two algorithms do not seem to integrate appropriate strategies to solve instances with these features.

Table 3 presents the detailed results of the HGGA-BP algorithm on 11 instances of the Hard28 set [51]. For each instance, it is included: the number of bins used in the optimal solution; and the best, worst and average solution obtained by HGGA-BP when solving the instance 30 times with different seeds of random numbers. We observed that there are three classes of instances. The "very easy" class has five instances (highlighted in bold) which are optimally solved by most of the state-of-the-art algorithms (hBPP14, hBPP359, hBPP716, hBPP119, hBPP175). The "medium difficulty" class has three instances (highlighted in underlines) that HGGA-BP solves in an optimal way in most of the runs, out-performing the best algorithms (hBPP640, hBPP531, hBPP814). The "difficult" class has three instances (highlighted in italics) for which HGGA-BP does not show robust behavior (hBPP360, hBPP742 and hBPP47).

The remaining Hard28 instances are considered "very difficult" because HGGA-BP algorithm cannot obtain the optimal solution in any of the 30 runs. However, within this class, it seems that there is not any relation between the instance difficulty and the instance features, the characterization indexes proposed for BPP do not allow discriminating among this set of instances. It would be interesting to know the characteristics that make a difference in the degree of difficulty within this set of instances, to be able to define appropriate strategies that allow solving them in an optimal way.

Table 2 Results obtained by the best heuristic algorithms applied to BPP

Class	Inst.	HI_BP [49]		WA [62]		Pert-SAWMBS [50]		HGGA-BP [51]	
		Opt.	Time	Opt.	Time	Opt.	Time	Opt.	Time
U	80	80	0.03	71	0.28	79	0.00	79	1.67
T	80	80	0.98	0	0.15	80	0.00	80	5.14
Data set 1	720	720	0.19	703	0.25	720	0.01	718	2.67
Data set 2	480	480	0.01	468	0.04	480	0.00	480	0.90
Data set 3	10	10	4.60	9	0.08	10	0.16	9	8.10
Was 1	100	100	0.02	100	0.00	100	0.00	100	0.04
Was 2	100	100	0.02	100	0.02	100	0.01	100	0.99
Gau 1	17	12	0.60	13	0.02	16	0.04	15	1.92
Hard28	**28**	**5**	**0.48**	**5**	**0.59**	**5**	**0.24**	**8**	**6.75**
Total	1,615	1,587	0.77	1,469	0.16	1,590	0.05	1,589	3.14
		P4 1.7 GHz		Core2 2.33 GHz		Core2 2.33 GHz		Xenon 1.86 GHz	

Table 3 Results obtained by the HGGA-BP algorithm on the set Hard28 [51]

Instance	Number of bins for solution			
	Optimal	Best	Worst	Average
hBPP14	**62**	**62**	**62**	**62**
hBPP360	*62*	*62*	*63*	*63*
hBPP742	*64*	*64*	*65*	*65*
hBPP47	*71*	*71*	*72*	*72*
hBPP359	**76**	**76**	**76**	**76**
hBPP640	74	74	75	74
hBPP716	**76**	**76**	**76**	**76**
hBPP119	**77**	**77**	**77**	**77**
hBPP175	**84**	**84**	**84**	**84**
hBPP531	83	83	83	83
hBPP814	81	81	82	81

4.4 Further Analysis of the BPP Optimization Process

To contribute with the process of characterizing the difficult Hard28 set, we generated 9,999 instances with similar characteristics to this class [59]: each with 160 items, item weights equally distributed [1, 800] and bin capacity $c = 1,000$. The difficulty of this new set of instances was characterized with the results obtained by the HGGA_BP algorithm in 30 runs with different seeds. In this new set, it was possible to identify a group of 955 instances for which the solution algorithm does not match the theoretical optimum (lower limit L_2 [63]), so for now we cannot verify whether the algorithm fails in obtaining the optimal solution or the theoretical optimum does not match the real optimum. The other 9,044 instances were grouped into four sets for which the algorithm showed a different performance:

1. A set of 72 instances "very difficult" for which HGGA-BP algorithm does not show a robust performance.
2. A set of 108 instances of "medium difficulty" whose optimal solution is found by the algorithm in most of the runs.
3. A set of 1,724 instances "easy" that are solved optimally in all runs.
4. A set of 7,140 instances "very easy" that are solved optimally in the first iteration of the algorithm.

To check the performance of the state-of-the-art algorithms on instances similar to the Hard28 set; we selected randomly, 13 instances of each of the groups described in the previous paragraphs. The selected instances were solved with HI_BP algorithm [49], the results achieved by this procedure allowed us to assess the performance of HGGA-BP: which is superior in every "very difficult" and "medium difficulty" instances and in 9 of the 13 "easy" instances; in the other "easy" instances and in all "very easy" instances both procedures obtained the optimal solution.

The analysis carried out allowed us to prove the difficulty of the Hard28 set and the superiority of HGGA-BP algorithm by solving instances with similar characteristics to this class. It is important to identify which are the characteristics that distinguish these BPP instances. It is also necessary to understand the behavior of the algorithms and evaluate the strategies that enable them to reach their performance.

On the other hand, we created larger test instances with features similar to the uniform class [48] to analyze the performance of heuristic algorithms on large scale instances: item weights equally distributed [20, 100] and bin capacity $c = 150$. Figure 1 shows some of the performance graphs for well-known BPP heuristics based on random permutations of objects. Each graph shows, for every size of the problem, the number of bins obtained by solving 10 instances with the FF, BF and WF packing heuristics [64]. It can be observed that by increasing the size of the problem, the behavior of the algorithms is opposite, projecting far away the results obtained by WF. Contrary to the observed behavior for instances of small size, where FF and BF show better performance than WF strategy.

The results observed in the performance graphs allow us to identify different behaviors in the simple BPP heuristics, questioning the applicability of the proposed algorithms, as well as the test instances proposed in the literature. It appears that the benchmark instances used to evaluate the state-of-the-art algorithms are not representative of real difficult instances, and do not allow to assess the performance of the algorithms on a large scale. The study carried out revealed the importance of knowing the characteristics of BPP that impact the algorithmic performance in order to generate representative benchmarks for those characteristics.

The review of the results obtained by the best algorithms for solving BPP revealed that there are still instances of the literature that show a high degree of difficulty, and the strategies included in the procedures do not seem to lead to better solutions. When analyzing the literature, we observed that none of the

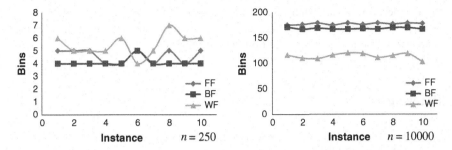

Fig. 1 Performance of simple BPP heuristics on instances similar to the Uclass

strategies of the state-of-the-art has been analyzed to explain the reason behind its good or bad performance. Furthermore, there is no study of the phase transition phenomena for BPP and no general discussion of the structure and complexity of the BPP instances.

5 Conclusions

The review of the works related to the characterization and study of the optimization process revealed the need for better methods of characterization and performance analysis. It also highlighted the importance of a study on the relationships between the measures proposed to describe the properties of the problem, the algorithmic behavior and the final performance. The characterization of the optimization process is a challenging problem, and there is still a long way to go before achieving a detailed understanding of the complexity of the optimization problems and the heuristic algorithms. There are currently a large number of open problems to research:

- Identification of new features of the structure of the problems and the search space that have a high impact on the algorithmic performance.
- Application of the characterization techniques to new NP-hard problems and heuristic algorithms.
- Generation of improved benchmark instances that represents real life problems, involving a higher degree of difficulty.
- Development of methodologies for robust analysis of heuristic algorithms with the goal of explaining the algorithmic optimization process.
- Incorporation of knowledge of the problem domain in algorithms to increase their effectiveness and improve their performance.

Several authors have agreed that currently there is still a strong lack of knowledge about the exact way in which the performance of different algorithms is related to the characteristics of the instances of a problem, highlighting the need for a guide for the description and analysis of performance of actual implementations of heuristics [2, 3, 29, 65]. It is expected that this work promotes the interest in the characterization and study of the optimization process.

References

1. Wolpert, D., Macready, W.: No free lunch theorems for optimizations. IEEE Trans. Evol. Comput. **1**, 67–82 (1997)
2. Smith-Miles, K., Lopes, L.: Measuring instance difficulty for combinatorial optimization problems. Comput. Oper. Res. **39**(5), 875–889 (2012)
3. Smyth, K.: Understanding stochastic local search algorithms: an empirical analysis of the relationship between search space structure and algorithm behaviour. The University of British Columbia, Ms. thesis (2004)
4. Garey, M., Jonson, D.: Computers and intractability: A guide to the theory of NP-completeness. W. H. Freeman and Company, a classic introduction to the field (1979)
5. Smith-Miles, K., James, R., Giffin, J., Tu, Y.: Understanding the relationship between scheduling problem structure and heuristic performance using knowledge discovery. In: Learning and Intelligent Optimization, LION 3 (2009)
6. Leyton-Brown, K., Nudelman, E., Shoham, Y.: Learning the empirical hardness of optimization problems: The case of combinatorial auctions. Principles Pract. Constraint Program. **2470**, 556–572 (2002)
7. Nudelman, E., Devkar, A., Shoham, Y., Leyton-Brown, K.: Understanding random SAT: Beyond the clauses-to-variables ratio. Principles Pract. Constraint Program. **3258**, 438–452 (2004)
8. Gagliolo, M., Schmidhuber, J.: Learning dynamic algorithm portfolios. Spec. Issue Ann. Math. Artif. Intell. **47**(3–4), 295–328 (2007)
9. Madani, O., Raghavan, H., Jones, R.: On the empirical complexity of text classification problems. SRI AI Center Technical Report (2009)
10. Messelis, T., Haspeslagh, S., Bilgin, B., De Causmaecker, P., Vanden, G.: Towards prediction of algorithm performance in real world optimization problems. In: Proceedings of the 21st Benelux Conference on Artificial Intelligence, pp. 177–183. BNAIC, Eindhoven (2009)
11. Hutter, F., Xu, L., Hoos, H.H., Leyton-Brown, K.: Algorithm runtime prediction: Methods and evaluation. Artif. Intell. **206**, 79–111 (2014)
12. McKay, R., Abbass, H.: Anti-correlation measures in genetic programming. In: Proceedings of the Australasia-Japan Workshop on Intelligent and Evolutionary Systems, pp. 45–51 (2001)
13. Burke, R., Gustafson, S., Kendall, G.: A survey and analysis of diversity measures in genetic programming. In: Proceedings of the Genetic and Evolutionary Computation Conference, pp. 716–723 (2002)
14. Thierens, D.: Predictive measures for problem representation and genetic operator design. Technical Report UU-CS-2002-055, Utrecht University: Information and Computing Sciences (2002)
15. Hutter, F., Hamadi, Y., Hoos, H.H., Leyton-Brown, K.: Performance prediction and automated tuning of randomized and parametric algorithms. In: Principles and Practice of Constraint Programming-CP, pp. 213–228 (2006)
16. Halim, S., Yap, R., Lau, H.: An integrated white + black box approach for designing and tuning stochastic local search. In: Principles and Practice of Constraint Programming-CP, pp. 332–347 (2007)

17. Birattari, M.: Tuning Metaheuristics: A machine learning perspective. SCI 197. Springer, Berlin (2009)
18. Akbaripour, H., Masehian, E.: Efficient and robust parameter tuning for heuristic algorithms. Int. J. Ind. Eng. **24**(2), 143–150 (2013)
19. Wright, S.: The roles of mutation, inbreeding, crossbreeding and selection in evolution. In: Proceedings of the Sixth International Genetics, vol. 1, pp. 356–366 (1932)
20. Jones, T.: Evolutionary algorithms, fitness landscapes and search. Ph.D. thesis, The University of New Mexico (1995)
21. Corne, D., Oates, M., Kell, D.: Landscape state machines: tools for evolutionary algorithm performance analyses and landscape/algorithm mapping. In: Applications of Evolutionary Computing, pp. 187–198 (2003)
22. Mitchell, B., Mancoridis, S.: Modeling the search landscape of metaheuristic software clustering algorithms. Lecture notes in computer science. In: Proceedings of the 2003 International Conference on Genetic and Evolutionary Computation, vol. 2, pp. 2499–2510 (2003)
23. Merz, P.: Advanced fitness landscape analysis and the performance of memetic algorithms. Evol. Comput. Spec. Issue Magn. Algorithms **12**(3), 303–325 (2004)
24. Ochoa, G., Qu, R., Burke, E.: Analyzing the landscape of a graph based hyper-heuristic for timetabling problems. In: Proceedings of the 11th Annual Conference on Genetic and Evolutionary Computation, pp. 341–348 (2009)
25. Czogalla, J., Fink, A.: Fitness landscape analysis for the resource constrained project scheduling problem. In: Lecture Notes in Computer Science, Learning and Intelligent Optimization, vol. 5851, pp. 104–118 (2009)
26. Verel, S., Liefooghe, A., Jourdan, L., Dhaenens, C.: On the structure of multiobjective combinatorial search space: MNK-landscapes with correlated objectives. Eur. J. Oper. Res. (2012)
27. Merz, P., Freisleben, B.: Fitness landscapes, memetic algorithms, and greedy operators for graph bipartitioning. Evol. Comput. **8**(1), 61–91 (2000)
28. Merz, P., Freisleben, B.: Fitness landscape analysis and memetic algorithms for the quadratic assignment problem. IEEE Trans. Evol. Comput. **4**(4), 337–352 (2000)
29. Borenstein, Y.: Information landscapes. In: Genetic and Evolutionary Computation Conference, pp. 1515–1522 (2005)
30. Borgs, C., Chayes, J., Pittel, B.: Phase transition and finite-size scaling for the integer partitioning problem. Random Struct. Algorithms **19**, 247–288 (2001)
31. Béjar, R., Vetsikas, I., Gomes, C., Kautz, H., Selman, B.: Structure and phase transition phenomena in the VTC problem. In TASK PI Meeting Workshop (2001)
32. Caramanis, C.: Survey propagation iterative solutions to constraint satisfaction problems. In: Expository Writing (2003)
33. Achlioptas, D., Naor, A., Peres, Y.: Rigorous location of phase transitions in hard optimization problems. Nature **435**(7043), 759–764 (2005)
34. Mertens, S.: The easiest hard problem: Number partitioning. Comput. Complex. Stat. Phys. **125**(2), 125–140 (2006)
35. Piñol, C.: CSP Problems as algorithmic benchmarks: Measures, methods and models. In: Universitat de Lleida, Departament d'Informàtica i Enginyeria Industrial (2008)
36. Rangel-Valdez, N., Torres-Jimenez, J.: Phase transition in the bandwidth minimization problem. In: Lecture Notes in Computer Science 5845, MICAI 2009: Advances in Artificial Intelligence, pp. 372–383 (2009)
37. Dewenter, T., Hartmann, A.: Phase transition for cutting-plane approach to vertex-cover problem. Phys. Rev. E **86**(4), 041128 (2012)
38. Slaney, J., Walsh, T.: Backbones in optimization and approximation. In: Proceedings the 17th International Joint Conference on Artificial Intelligence (IJCAI-01), pp. 254–259 (2001)
39. Monasson, R., Zecchina, R., Kirkpatrick, S., Selman, B., Troyansky, L.: Determining computational complexity from characteristic 'phase transitions'. Nature **400**(6740), 133–137 (1999)

40. Zeng, G., Lu, Y.: Survey on computational complexity with phase transitions and extremal optimization. In: Proceedings of 48th IEEE Conference on Decision and Control and 28th Chinese Control Conference Shanghai, People's Republic of China, 16–18 Dec 2009
41. Singer, J., Gent, I.P., Smaill, A.: Backbone fragility and the local search cost peak. J. Artif. Intell. Res. (JAIR) 12, 235–270 (2000)
42. Watson, J., Beck, J., Howe, A., Whitley, L.: Problem difficulty for tabu search in job-shop scheduling. Artif. Intell. 143, 189–217 (2003)
43. Schneider, J.: Searching for backbones—a high-performance parallel algorithm for solving combinatorial optimization problems. Future Gener. Comput. Syst. 19(1), 121–131 (2003)
44. Zhang, W.: Configuration landscape analysis and backbone guided local search: Part I: Satisfiability and maximum satisfiability. Artif. Intell. 158(1), 1–26 (2004)
45. Zhang, W., Looks, M.: A novel local search algorithm for the travelling salesman problem that exploits backbones. In: Proceedings of the 19th International Joint Conference on Artificial Intelligence, pp. 343–350 (2005)
46. Zeng, G., Lu, Y., Dai, Y., Wu, Z., Mao, W., Zhang, Z., Zheng, C.: Backbone guided extremal optimization for the hard maximum satisfiability problem. Int. J. Innovative Comput. Inf. Control 8(12), 8355–8366 (2012)
47. Quiroz, M., Cruz-Reyes, L., Torres-Jiménez, J., Melin, P.: Improving the performance of heuristic algorithms based on exploratory data analysis. In: Recent Advances on Hybrid Intelligent Systems, pp. 361–375 (2013)
48. Falkenauer, E.: A hybrid grouping genetic algorithm for bin packing. J. Heuristics 2, 5–30 (1996)
49. Alvim, A., Glover, F., Ribeiro, C., Aloise, D.: A hybrid improvement heuristic for the one-dimensional bin packing problem. J. Heuristics 10, 205–229 (2004)
50. Fleszar, K., Charalambous, C.: Average-weight-controlled bin-oriented heuristics for the one-dimensional bin-packing problem. Eur. J. Oper. Res. 210(2), 176–184 (2011)
51. Cruz-Reyes, L., Quiroz, M., Alvim, A., Fraire H., Gómez, C., Torres-Jiménez, J.: Heurísticas de agrupación híbridas eficientes para el problema. Computación y Sistemas 16(3) (2012)
52. Beasley, J.: OR-library: Distributing test problems by electronic mail. J. Oper. Res. Soc. 41(11), 1069–1072 (1990). http://people.brunel.ac.uk/~mastjjb/jeb/orlib/binpackinfo.html
53. Scholl, A., Klein, R.: Bin packing benchmark data sets (2014). http://www.wiwi.uni-jena.de/Entscheidung/binpp/. Accessed 24 Feb 2014
54. ESICUP.: Euro especial interest group on cutting and packing, one dimensional cutting and packing data sets (2014). http://paginas.fe.up.pt/~esicup/tiki-list_file_gallery.php?galleryId=1. Accessed 24 Feb 2014
55. CaPaD.: Cutting and packing at dresden university, benchmark data sets (2014). http://www.math.tu-dresden.de/~capad/cpd-ti.html#pmp. Accessed 24 Feb 2014
56. Scholl, A., Klein, R., Jürgens, C.: Bison: A fast hybrid procedure for exactly solving the one-dimensional bin packing problem. Comput. Oper. Res. 24(7), 627–645 (1997)
57. Wäscher, G., Gau, T.: Heuristics for the one-dimensional cutting stock problem: A computational study. OR Spektrum 18(3), 131–144 (1996)
58. Schwerin, P., Wäscher, G.: The bin-packing problem: A problem generator and some numerical experiments with FFD packing and MTP. Int. Trans. Oper. Res. 4(5–6), 337–389 (1997)
59. Schoenfield, J.E.: Fast exact solution of open bin packing problems without linear programming. In: Draft, US Army Space and Missile Defense Command, Huntsville, Alabama, USA (2002)
60. Pérez, J., Pazos, R.A., Frausto, J., Rodríguez, G., Romero, D., Cruz, L.: A statistical approach for algorithm selection. In: Ribeiro C.C., Matins, S.L. (eds.) WEA 2004. LNCS, vol. 3059, pp. 417–431 (2004)
61. Cruz-Reyes, L., Gómez-Santillán, C., Schaeffer, S.E., Quiroz-Castellanos, M., Alvarez-Hernández, V.M., Pérez-Rosas, V.: Enhancing accuracy of hybrid packing systems through general-purpose characterization. In: Hybrid Artificial Intelligent Systems, pp. 26–33 (2011)

62. Loh, K., Golden, B., Wasil, E.: Solving the one-dimensional bin packing problem with a weight annealing heuristic. Comput. Oper. Res. **35**(7), 2283–2291 (2008)
63. Martello, S., Toth, P.: Knapsack problems: Algorithms and computer implementations. Wiley, New York (1990)
64. Johnson, D.S.: Fast algorithms for bin packing. J. Comput. Syst. Sci. **8**(3), 272–314 (1974)
65. Chiarandini, M., Paquete, L., Preuss, M., Ridge, E.: Experiments on metaheuristics: Methodological overview and open issues. In: Technical Report DMF-2007-03-003, The Danish Mathematical Society (2007)

A New Integer Linear Programming Model for the Cutwidth Minimization Problem of a Connected Undirected Graph

Mario C. López-Locés, Norberto Castillo-García,
Héctor J. Fraire Huacuja, Pascal Bouvry, Johnatan E. Pecero,
Rodolfo A. Pazos Rangel, Juan J. G. Barbosa and Fevrier Valdez

Abstract In this chapter we propose a new integer linear programming model based on precedences for the cutwidth minimization problem (CWP). A review of the literature indicates that this model is the only one reported for this problem. The results of the experiments with standard instances shows that the solution of the problem with the proposed model outperforms in quality and efficiency to the one reported in the state of the art. Our model increases the number of optimal solutions by 38.46 % and the gap reduction by 45.56 %. Moreover, this quality

M. C. López-Locés (✉) · N. Castillo-García · H. J. F. Huacuja · R. A. Pazos Rangel · J. J. G. Barbosa
Instituto Tecnológico de Ciudad Madero, Av. 1o. de Mayo S/N Col. Los Mangos 89440, Cd. Madero, Tamaulipas, Mexico
e-mail: mariocesar@lopezloc.es

N. Castillo-García
e-mail: norberto_castillo15@hotmail.com

H. J. F. Huacuja
e-mail: automatas2002@yahoo.com.mx

R. A. Pazos Rangel
e-mail: r_pazos_r@yahoo.com.mx

J. J. G. Barbosa
e-mail: jjgonzalezbarbosa@hotmail.com

P. Bouvry · J. E. Pecero
Computer Science and Communications Research Unit, University of Luxembourg, 6 Rue Richard Coudenhove-Kalergi L-1359 Luxembourg City, Luxembourg
e-mail: pascal.bouvry@uni.lu

J. E. Pecero
e-mail: johnatan.pecero@uni.lu

F. Valdez
Instituto Tecnológico de Tijuana, Unidad Tomás Aquino S/N 22414 Tijuana, Baja California Norte, Mexico
e-mail: fevrier@tectijuana.mx

O. Castillo et al. (eds.), *Recent Advances on Hybrid Approaches for Designing Intelligent Systems*, Studies in Computational Intelligence 547,
DOI: 10.1007/978-3-319-05170-3_35, © Springer International Publishing Switzerland 2014

improvement is reached with a time solution reduction of 41.73 %. It is considered that the approach used in this work can be used in other linear ordering problems.

1 Introduction

The cutwidth minimization problem (CWP) is an NP-hard optimization problem that initially was employed for the modelling of VLSI circuits and more recently their applications includes automatic graph drawing, information retrieval and network migration [1–3]. To define CWP, we will consider an undirected connected graph $G = (V, E)$, with $n = |V|$ and $m = |E|$. A linear ordering of the vertices of G is a bijective function $\varphi : V \rightarrow \{1, 2, \ldots, n\}$. $\phi(G)$ is the set of all the linear orderings defined on G. $L(i, \varphi, G) = \{u \in V : \varphi(u) \leq i\}$ is the set of all the left side vertices with respect to the vertex cut (separation) i. $R(i, \varphi, G) = \{u \in V : \varphi(u) > i\}$ is the set of the vertices to the right of the vertex cut i. $\theta(i, \varphi, G) = |\{(u, v) \in E : u \in L(i, \varphi, G) \wedge v \in R(i, \varphi, G)\}|$ is the number of edges of the graph G that cross from L to R through the vertex cut i. The cutwidth of the graph G with the linear ordering φ is defined as $CW(\varphi, G) = \max_{i \in V}\{\theta(i, \varphi, G)\}$. The cutwidth minimization problem consists of determining the linear ordering $\varphi^* \in \phi(G)$ that minimizes the cutwidth of the graph G, that is $CW(\varphi^*, G) = \min_{\varphi \in \phi(G)}\{CW(\varphi, G)\}$ [2].

Given the NP-hardness of the problem, to solve hard instances of the cutwidth minimization problem is necessary the use of soft computing approximated algorithms. Previously an exact solution method is required to solve smaller size instances to the optimality, which results are used to assess the approximated algorithms behaviour. An approach frequently used consists in develop an integer linear programming (ILP) model, which is used to solve the smaller instances with a branch and bound enumerative exact algorithm.

To our knowledge, the only ILP model for CWP is the one reported by Luttamaguzi et al. [4]. A limitation of this model is that it represents the linear arrangement as a matrix, which generates a very costly formulation.

The main contribution of this work is a new ILP model for CWP based on the vertex precedences, which reduces the complexity of the generated models.

The remainder of the chapter is organized as follows. Section 2 introduces the ILP model proposed in this work. Section 3 presents the experimental results. We present the conclusions in Sect. 4.

2 Integer Linear Programming Models

2.1 Quadratic Model

Luttamaguzi et al. [4] described an approach to model linear ordering problems using integer linear programming, but in this formulation, a quadratic restriction that establishes the connectivity of the graph is generated. To model CWP, the following variables were defined:

The variable x_i^k is a binary variable used to represent the linear ordering and is defined as:

$$x_i^k = \begin{cases} 1, & k = \varphi(i) \\ \\ 0, & \text{otherwise} \end{cases}$$

The variable $y_{i,j}^{k,l}$ is a binary variable used to represent the graph connectivity and is defined as: $y_{i,j}^{k,l} = x_i^k \wedge x_j^l$.

The variable CW is an integer variable that represents the cutwidth value $CW(\varphi, G)$ of a given linear ordering.

Using the previously defined variables, the quadratic model is the following:

$$Min\ CW \tag{1}$$

$$s.t.$$

$$\sum_{k \in \{1,\dots,n\}} x_i^k = 1, \quad \forall i \in \{1,\dots,n\} \tag{2}$$

$$\sum_{k \in \{1,\dots,n\}} x_i^k = 1, \quad \forall k \in \{1,\dots,n\} \tag{3}$$

$$y_{i,j}^{k,l} = x_i^k x_j^l, \forall (i,j) \in E \vee (j,i) \in E, \quad \forall k,l \in \{1,\dots,n\} \tag{4}$$

$$\sum_{k \leq c < l} y_{i,j}^{k,l} \leq CW, \quad \forall c \in \{1,\dots,n-1\} \tag{5}$$

The constraints (2) and (3) ensure that the cutwidths are determined only for feasible linear orderings. Constraint (2) states that each vertex of the graph must be assigned to a single position. Constraint (3) specifies that only one vertex may be located at each position. Constraint (4) specifies the set of edges in terms of the relative positions of the vertices $i, j \in V$ in the linear ordering, such as $(i,j) \in E$, $x_i^k = 1$ and $x_j^l = 1$. Constraint (5) determines the cardinality of the sets defined in the previous constraints for each vertex cut and retains the largest of them. This value is the cutwidth of the graph G corresponding to the edges specified in the variable $y_{i,j}^{k,l}$ and to the linear ordering on the variable x_i^k.

2.2 State of the Art ILP Model

Because the model of Luttamaguzi et al. has a quadratic constraint (4), it must be linearized to transform the quadratic model into an integer linear programming model (ILP). To reach this goal, the authors applied the traditional linearization technique proposed by Fortet [5]. As a consequence, in the model, the quadratic constraint is transformed into three linear constraints, as shown in the following model (ILP$_2$):

$$Min \; CW \tag{6}$$

$$s.t.$$

$$\sum_{k \in \{1,...,n\}} x_i^k = 1, \forall i \in \{1, \ldots, n\} \tag{7}$$

$$\sum_{i \in \{1,...,n\}} x_i^k = 1, \; \forall k \in \{1, \ldots, n\} \tag{8}$$

$$y_{i,j}^{k,l} \leq x_i^k \tag{9}$$

$$y_{i,j}^{k,l} \leq x_j^l \tag{10}$$

$$x_i^k + x_j^l \leq y_{i,j}^{k,l} + 1 \tag{11}$$

$$\sum_{k \leq c < l} y_{i,j}^{k,l} \leq CW, \; \forall c \in \{1, \ldots, n-1\} \tag{12}$$

where the constraints (9), (10) and (11) replace the constraint (4) from the previous model.

This model generates $|V|^2|E| + 2|V|^2 + 1$ variables and $3|V|^2|E| + 2|V| + |V| - 1$ constraints.

2.3 ILP Model Based on Precedences

The precedence based model differs from the above formulation in that the node position in the linear array is specified in terms of their precedence over the others. A vertex $i \in V$ precedes the vertex $j \in V$ if and only if i is to the left of j in the linear ordering. Formally, the relation of precedence is defined as follows $i \prec j \leftrightarrow \varphi(i) < \varphi(j)$

This model uses the variables that will be described below:

2.3.1 Variable x_{ij}

This binary variable indicates whether the vertex i precedes the vertex j and is defined as:

$$x_{ij} = \begin{cases} 1, & \varphi(i) < \varphi(j) \\ 0, & \text{otherwise} \end{cases}, \quad \forall i, j \in V : i \neq j$$

Note that $x_{ii} = 0$, for any $i \in V$, and $x_{ij} = 1 \leftrightarrow x_{ji} = 0$. The linear constraints associated with this variable are:

$$x_{ij} + x_{ji} = 1, \quad \forall i, j \in V : i \neq j$$

$$x_{ij} + x_{jh} + x_{hi} \leq 2, \quad \forall i, j, h \in V : i \neq j \neq h$$

2.3.2 Variable y_{ij}

This dependent binary variable indicates whether there is an edge $(i, j) \in E$ such that $i \prec j$ and is defined as follows:

$$y_{ij} = \begin{cases} 1, & \varphi(i) < \varphi(j) \land (i, j) \in E \\ 0, & \text{otherwise} \end{cases}, \quad \forall i, j \in V : i \neq j$$

To establish the connections (edges) of the vertices, depending on the current linear ordering, the following constraints are defined:

$$y_{ij} = x_{ij}, \quad \forall i, j \in V : i \neq v : (i, j) \in E \lor (j, i) \in E$$

Additionally, to ensure that all those variables y_{ij} such that $(i, j) \notin E$ take the value of 0, the following restriction is established:

$$\sum_{i,j \in V : i \neq j : (i,j) \notin E} y_{ij} = 0$$

2.3.3 Variable p_i

This dependent integer variable indicates the position of vertex i, given a linear ordering. The p_i variable is defined by the following constraint:

$$p_i = n - \sum_{j \in V : i \neq j} x_{ij}, \forall i \in V$$

2.3.4 Variable R_{ip}

This dependent binary variable indicates whether the vertex i belongs to the right set with respect to the cutting point p and is defined as follows:

$$R_{ip} = \begin{cases} 1, & \varphi(i) > p \\ \\ 0, & \text{otherwise} \end{cases}, \qquad \forall i \in V, p = 1, \ldots, n-1$$

This definition is equivalent to the following double implication:

$$(R_{ip} = 1 \rightarrow p_i \geq p+1) \wedge (p_i \geq p+1 \rightarrow R_{ip} = 1).$$

Then this logical proposition is implemented by the following linear constraints:

$$p_i \geq (p+1) - n(1 - R_{ip}), \qquad \forall i \in V, p = 1, \ldots, n-1$$

$$1 + p_i - (p+1) \leq (n-1)R_{ip}, \qquad \forall i \in V, p = 1, \ldots, n-1$$

Because if a vertex $i \in V$ is on the right set of the cut point p, then it cannot be on the left set of the same cut point, and vice versa. The linear constraint that defines this variable is the negation of the variable R_{ip} as shown below:

$$1 - R_{ip}, \qquad \forall i \in V, p = 1, \ldots, n-1$$

2.3.5 Variable δ_{ijp}

This dependent binary variable indicates whether there is an edge $(i, j) \in E$ such that $i \prec j$, and the vertex $i \in V$ is on the left set and the vertex $j \in V$ on the right set of the cut point p. Its definition is:

$$\delta_{ijp} = y_{ij}(1 - R_{ip})R_{jp}, \qquad \forall i, j \in V : i \neq j, p = 1, \ldots, n-1$$

The linear constraints obtained on this variable are the result of applying the cubic linearization technique proposed by Chen et al. [6]:

$$y_{ij} + (1 - R_{ip}) + R_{jp} \geq 3\delta_{ijp} \qquad \forall i, j \in V : i \neq j, p = \{1, \ldots, n-1\}$$

$$y_{ij} + (1 - R_{ip}) + R_{jp} \leq \delta_{ijp} + 2 \qquad \forall i, j \in V : i \neq j, p = \{1, \ldots, n-1\}$$

2.3.6 Variable CW

This integer variable represents the objective value of a solution of the problem. It is defined as the maximum value of the partial objective values at each cutting point. This variable is defined in the following linear constraint:

Table 1 Theoretical complexity of the ILP models

	ILP$_2$	ILP$_3$
#Variables	$O(|V|^2|E|)$	$O(|V|^3)$
#Constraints	$O(|V|^2|E|)$	$O(|V|^3)$

$$\sum_{i,j \in V: i \neq j} \delta_{ijp} = CW, \qquad \forall p = \{1, \dots, n-1\}.$$

Below is shown the resulting new integer linear programming model based on precedences (ILP$_3$):

$$Min \; CW \tag{13}$$

$$s.t.$$

$$x_{ij} + x_{ji} = 1, \quad \forall i,j \in V : i \neq j \tag{14}$$

$$x_{ij} + x_{jk} + x_{ki} \leq 2, \quad \forall i,j,k \in V : i \neq j \neq k \tag{15}$$

$$y_{ij} = x_{ij}, \quad \forall i,j \in V : i \neq j : (ij) \in E \lor (ji) \in E \tag{16}$$

$$\sum_{ij \in V: i \neq j: (ij) \notin E} y_i^j = 0 \tag{17}$$

$$p_i = n - \sum_{i \in V: i \neq j} x_{ij}, \forall i \in V \tag{18}$$

$$p_i \geq (p+1) - n(1 - R_{ip}), \quad \forall i \in V, p = \{1, \dots, n-1\} \tag{19}$$

$$1 + p_i - (p+1) \leq (n-1)R_{ip}, \quad \forall i \in V, p = \{1, \dots, n-1\} \tag{20}$$

$$y_{ij} + (1 - R_{ip}) + R_{jp} \geq 3\delta_{ijp}, \quad \forall i,j \in V : i \neq j, p = \{1, \dots, n-1\} \tag{21}$$

$$y_{ij} + (1 - R_{ip}) + R_{j,p} \leq \delta_{ijp} + 2, \quad \forall i,j \in V : i \neq j, p = \{1, \dots, n-1\} \tag{22}$$

$$\sum_{i,j \in V: i \neq j} \delta_{ijp} = CW, \quad \forall p = \{1, \dots, n-1\} \tag{23}$$

This model generates $|V|(|V|-1)^2 + 3|V|(|V|-1) + |V| + 1$ variables and $|V|(|V|-1)(|V|-2) + 2|V|(|V|-1)^2 + 4|V|(|V|-1) + 2|V| - 1$ constraints.

Table 1, shows the theoretical complexity of the generated models with ILP$_2$ and ILP$_3$. The complexity of ILP$_2$ and ILP$_3$ is $O(|V|^2|E|)$ and $O(|V|^3)$, respectively. However, when |E| increases, the complexity of ILP$_2$ tends to $O(|V|^4)$. For bigger instances, this complexity level can significantly reduce the performance of branch and bound used to solve the smaller instances with the ILP$_2$ model.

Table 2 Experimental results obtained with the ILP models

		ILP$_2$	ILP$_3$
HB (4)	#opt	0	0
	%gap	100	89.85
	CPU Time	1800	1800
Grids (4)	#opt	1	2
	%gap	71.42	30.83
	CPU Time	1534.6	919.74
Small (5)	#opt	1	5
	%gap	50.27	0
	CPU Time	1534.7	117.19
Overall (13)	#opt	2	7
	%gap	73.89	40.22
	CPU Time	1623.1	945.64

In the next section, we will test this assertion by performing a set of computational experiments to compare the performance of both models.

3 Computational Experimentation

A set of computational experiments was conducted with standard CWP instances to assess the performance of the different ILP models. The implementation of the integer linear programming models was performed in the Java programming language, using the API of CPLEX optimization suite version 12.1 developed by IBM. The computer equipment in which the experiments were performed was an Intel i5 processor at 2.3 GHz with 8 GB of DDR3 RAM at 1333 GHz.

In the experiments with the models, thirteen instances were solved: four of the Harwell-Boeing group (ibm32, bcsstk01, bcspwr02 and bcspwr01), four of the Grids group (Grid9x3, Grid6x6, Grid6x3, and Grid3x3) and five of the Small group (p21_17_20, p20_16_18, p19_16_19, p18_16_21 and p17_16_24). These instances were chosen from the standard instance set used by Martí et al. [7] and are available on the Optsicom Project web page (http://heur.uv.es/optsicom/ cutwidth/). For the experiments, the smaller instances of each group were selected to obtain meaningful data on the time assigned to each one. The time limit given for the solution of an instance with the models assessed was set to 1800 s.

Table 2, shows the experimental results. The first column contains the identifier of the group and the number of instances used. In the second column, for each group and each ILP model, the number of optimal solutions found (#opt), the percentage of the average gap (%gap) and the average time to solve the instances are shown in CPU seconds.

In the last row of the table, is shown the overall performance of the ILP models: the accumulated number of the optimal solutions found, the average of the gap reduction and the average of the time solution.

As we can see, the ILP_3 model increases by 38.46 % the number of times the optimal solution is computed by the model compared to the ILP_2 state-of-the-art model. We can also observe that ILP_3 model increases the gap reduction by 45.56 %. Finally, as we can see this quality improvement is reached with a time solution reduction of 41.73 %.

4 Conclusions

In this work, we addressed the cutwidth minimization problem. We proposed a new ILP model based on precedences. To evaluate the performance of the proposed model, a set of computational experiments with standard instances were conducted.

The experimental results indicate that the ILP model based on precedences outperforms the state-of-the-art model.

Currently, we are working on exact solution methods based on the branch and bound methodology to solve the CWP.

Acknowledgments This research project has been partly financed by the CONACyT, the COTACyT, the DGEST and the University of Luxembourg. Additionally, we thank the IBM Academic Initiative for the support of this research project with the optimization software CPLEX, version 12.1.

References

1. Garey, M., Johnson, L.: Some simplified NP-complete graph problems. Theoret. Comput. Sci. **1**(3), 237–267 (1976)
2. Díaz, J., Petit, J., Serna, M.: A survey of graph layout problems. ACM. Comput. Surv. (CSUR) **34**(3), 313–356 (2002)
3. Andrade, D.V., Resende, M.G.C.: GRASP with path-relinking for network migration scheduling. In: Proceedings of the International Network Optimization Conference (INOC 2007), Citeseer, (2007)
4. Luttamaguzi, J., Pelsmajer, M., Shen, Z., Yang, B.: Integer programming solutions for several optimization problems in graph theory, In: 20th International Conference on Computers and Their Applications (CATA 2005), 2005
5. Fortet, R.: Applications de l'algébre de boole en recherche opérationnelle. Rev. Fr. de Rech. Opérationelle **4**, 17–26 (1960)
6. Chen, D.-S., Batson, R.G., Dang, Y.: Applied Integer Programming: Modeling and Solution. Wiley, (2010)
7. Martí, R., Pantrigo, J.J., Duarte, A., Pardo, E.G.: Branch and bound for the cutwidth minimization problem. Comput. Oper. Res. **40**(1), 137–149 (2013)

On the Exact Solution of VSP for General and Structured Graphs: Models and Algorithms

Norberto Castillo-García, Héctor Joaquín Fraire Huacuja,
Rodolfo A. Pazos Rangel, José A. Martínez Flores,
Juan Javier González Barbosa and Juan Martín Carpio Valadez

Abstract In this chapter the vertex separation problem (VSP) is approached. VSP is NP-hard with important applications in VLSI, computer language compiler design, and graph drawing, among others. In the literature there are several exact approaches to solve structured graphs and one work that proposes an integer linear programming (ILP) model for general graphs. Nevertheless, the model found in the literature generates a large number of variables and constraints, and the approaches for structured graphs assume that the structure of the graphs is known a priori. In this work we propose a new ILP model based on a precedence representation scheme, an algorithm to identify whether or not a graph has a Grid structure, and a new benchmark of scale-free instances. Experimental results show that our proposed ILP model improves the average computing time of the reference model in 79.38 %, and the algorithm that identifies Grid-structured graphs has an effectiveness of 100 %.

N. Castillo-García · H. J. F. Huacuja (✉) · R. A. P. Rangel · J. A. M. Flores ·
J. J. G. Barbosa
Instituto Tecnológico de Ciudad Madero, Madero, México
e-mail: automatas2002@yahoo.com.mx

N. Castillo-García
e-mail: norberto_castillo15@hotmail.com

R. A. P. Rangel
e-mail: r_pazos_r@yahoo.com.mx

J. A. M. Flores
e-mail: jose.mtz@itcm.edu.mx

J. J. G. Barbosa
e-mail: jjgonzalezbarbosa@hotmail.com

J. M. C. Valadez
Instituto Tecnológico de León, León, México
e-mail: jmcarpio61@hotmail.com

O. Castillo et al. (eds.), *Recent Advances on Hybrid Approaches for Designing Intelligent Systems*, Studies in Computational Intelligence 547,
DOI: 10.1007/978-3-319-05170-3_36, © Springer International Publishing Switzerland 2014

1 Introduction

The vertex separation problem (VSP) is an NP-hard combinatorial optimization problem [1], which consists of finding a minimal subset of vertices whose removal separates the graph into disconnected subgraphs [2]. VSP belongs to a family of graph layouts problems whose goal is to find a linear layout of an input graph such that a certain objective function is optimized [3]. A linear layout is a labeling of the vertices of a graph with different integers. This definition allows representing the solutions of the problem under a permutation scheme, in which the vertices correspond to the elements and their positions represent the labels.

Due to the strong relation with other graph problems, VSP has important applications in the context of very large scale integration (VLSI) design [4], computer language compiler design [5], natural language processing [6], and order processing of manufactured products [7].

In the literature exist several approaches for solving the problem to optimality, mostly focused on structured graphs, that is, special classes of graphs such as trees or grids, among others. However, these approaches assume that the structure of the given graph is known *a-priori*.

Additionally, there is an integer linear programming (ILP) model for general graphs [2]. This model is based on a *direct* permutation scheme, that is, one vertex is placed at a specific position in the permutation. The main drawback of this model is that it generates a large number of variables and constraints, which makes it impractical for relatively large instances. Specifically, the model generates $\mathcal{O}(mn^2)$ variables and $\mathcal{O}(mn^3)$ constraints, as reported in [2]. We use this formulation as reference from the literature. The main contributions of this work are the following:

- A new ILP model for VSP based on a different representation scheme of the candidate solution: the *precedence* scheme. In contrast to the reference model scheme, the precedence representation models the solution as a sequence, in which it is not known the exact position of a vertex u in the permutation; but it is known whether or not vertex u precedes vertex v in the sequence, instead.
- An algorithm to verify if an arbitrary graph has a grid structure. This algorithm takes $\mathcal{O}(n + m)$ steps to decide if the given graph is a grid.
- A new benchmark of small-sized scale-free instances to test both models. These instances characterize real-world complex networks and exhibit a power-law distribution.

The remainder of this chapter is organized as follows. In Sect. 2 we present a brief review of the related work found in the state-of-the-art. Section 3 presents the formal description of the approached problem. The ILP models are described in Sect. 4. Section 5 is devoted to describe the algorithm that identifies grid-structured graphs. In Sect. 6 we detailed the generation of the scale-free instances benchmark. The conducted experiments to test our proposed approaches are described and discussed in Sect. 7. Finally, in Sect. 8 we present the conclusions of this study.

2 Related Work

In 1994, Ellis et al. [8] proposed an exact algorithm for trees with complexity $\mathcal{O}(n \log n)$. In 2000, Skodinis in [9] proposed an improved algorithm that solves VSP for a tree in $\mathcal{O}(n)$. Bollobás et al. in [10] solved theoretically VSP for n-dimensional grid graphs. Later, Díaz et al. in [3] established a theorem that determines that the optimal vertex separation value of an m-sized square grid is m.

In 2012, Duarte et al. [2] proposed an ILP model for general graphs. As mentioned previously, the inconvenient of this approach is the large number of variables and constraints generated. Nevertheless, a relevant aspect of this work is the use of a standard benchmark of instances to evaluate the performance of their solution methods.

3 Formal Definition of VSP

Let $G = (V, E)$ be a connected undirected graph with $n = |V|$ vertices and $m = |E|$ edges. A linear layout φ is a bijective function $\varphi : V \to \{1, 2, \ldots, n\}$ which associates every vertex u with a unique label denoted by $\varphi(u)$. Let $\Phi(G)$ be the set of all linear orderings defined over G and $i = 1, 2, \ldots, n$ be an integer representing a *position*. We define $L(i, \varphi, G) = \{u \in V : \varphi(u) \le i\}$ as the set of vertices to the left with respect to position i in the linear layout φ. Similarly, $R(i, \varphi, G) = \{u \in V : \varphi(u) > i\}$ represents the set of vertices to the right with respect to i in φ. The vertex separators of G in the linear layout φ at position i are those vertices in $L(i, \varphi, G)$ with *at least* one adjacent vertex in $R(i, \varphi, G)$, for all $i = 1, \ldots, n$, and it is formally defined as follows:

$$\delta(i, \varphi, G) = \{u \in L(i, \varphi, G) : \exists v \in R(i, \varphi, G) : (u, v) \in E\} \qquad (1)$$

The number of vertex separators $|\delta(i, \varphi, G)|$ in a determined position i represents a value known as *cut-value*, and since there are n vertices in G, there are n cut-values for each φ. The objective value $vs(\varphi, G)$ of a linear layout φ on a graph G is given by the maximum value among all the cut-values, that is:

$$vs(\varphi, G) = \max_{i=1,\ldots,n} \{|\delta(i, \varphi, G)|\}. \qquad (2)$$

The goal of VSP is to find the linear layout $\varphi^* \in \Phi(G)$ such that $vs(\varphi^*, G) = \min_{\varphi \in \Phi(G)} \{vs(\varphi, G)\}$.

4 Integer Linear Programming Models for VSP

In this section we describe the mathematical formulations for the vertex separation problem. First, we detail the ILP model found in the literature and then we explain our proposed model based on a precedence scheme. For the rest of the chapter we will refer to the ILP model from the literature as ILP_1 and to our proposed ILP model as ILP_2.

4.1 ILP Model Based on Direct Permutations: ILP₁

This model uses three binary variables and one integer variable for computing the objective value. The main 0–1 variables are explained in the following paragraphs.

- Variable x_u^p is set to 1 if vertex u is placed at position p in a feasible linear layout, and 0 otherwise; for all $u, p = 1, 2, \ldots, n$. This variable models the candidate solution as a permutation.
- Variable $y_{u,v}^{p,q}$ is set to 1 if vertex u is placed at position p (i.e., $p = \varphi(u)$) and vertex v is placed at position q (i.e., $q = \varphi(v)$), and 0 otherwise; for all $(u, v) \in E$. It is easy to see that the number of y variables is mn^2. Nevertheless, $y_{u,v}^{p,q}$ avoids the product between x_u^p and x_v^q implicit in the logical and operation.
- Variable z_{pc} is set to 1 if the vertex placed at position p is adjacent to the vertex in position q, which is larger than position c (i.e., $q > c$), and 0 otherwise. Variable z_{pc} models the cut-values defined in Eq. (1).

In addition to the previous binary variables, ILP_1 uses an integer variable VS to model the objective value of VSP. The mathematical formulation for ILP_1 is presented below.

$$VS^* = \min VS \tag{3}$$

subject to:

$$\sum_{p \in \{1,2,\ldots,n\}} x_u^p = 1, \quad \forall u = 1, \ldots, n \tag{4}$$

$$\sum_{u \in \{1,2,\ldots,n\}} x_u^p = 1, \quad \forall p = 1, \ldots, n \tag{5}$$

$$y_{u,v}^{p,q} \leq x_u^p, \quad \forall (u, v) \in E \vee (v, u) \in E, p, q = 1, \ldots, n \tag{6}$$

$$y_{u,v}^{p,q} \leq x_v^q, \quad \forall (u, v) \in E \vee (v, u) \in E, p, q = 1, \ldots, n \tag{7}$$

$$x_u^p + x_v^q \leq y_{u,v}^{p,q} + 1, \quad \forall (u, v) \in E \vee (v, u) \in E, p, q = 1, \ldots, n \tag{8}$$

$$z_{pc} \leq \sum_{q=c+1}^{n} \sum_{u=1}^{n} \sum_{v=1}^{n} y_{u,v}^{p,q} \leq (n-1)z_{pc}, \quad \forall p,c = 1,\ldots n-1, p \leq c \qquad (9)$$

$$\sum_{p=1}^{c} z_{pc} \leq VS, \quad \forall c = 1,\ldots,n-1 \qquad (10)$$

$$x_u^p, y_{u,v}^{p,q}, z_{pc} \in \{0,1\}, \quad \forall u,v,p,q = 1,\ldots,n. \qquad (11)$$

Constraints (4) and (5) ensure that each vertex is only assigned to one position, and each position is only assigned to one vertex, respectively. Constraints (6–8) define the variable $y_{u,v}^{p,q}$ as the product of x_u^p and x_v^q, linearized in the traditional way. Constraints (9) compute the binary value of z_{pc} from the variable $y_{u,v}^{p,q}$. Constraints (10) compute the cut-values for each position c, and the variable VS (right hand side) keeps the maximum among all of these cut-values of the given feasible permutation. We refer the reader to [2] for a more detailed explanation of ILP_1.

4.2 ILP Model Based on a Precedence Scheme: ILP_2

ILP_2 models the candidate solution by using a precedence relation, that is, a vertex u precedes vertex v *iff* u appears before than v in the current feasible sequence. In mathematical terms, the precedence relation is defined in the following way.

$$u \prec v \leftrightarrow \varphi(u) < \varphi(v) \qquad (12)$$

Our proposed model uses five binary variables and two integer variables, which are described below.

- Variable x_{uv} is set to 1 if vertex u precedes vertex v in the feasible solution, and 0 otherwise; for all $u,v \in V : u \neq v$. This variable models the solution as a sequence. Notice that $x_{uu} = 0$ for any $u \in V$, that is, in a feasible solution any vertex cannot precede itself. Besides, if u precedes v, then v cannot precede u (i.e., $x_{uv} = 1 \leftrightarrow x_{vu} = 0$).
- Variable y_{uv} is set to 1 if there is an edge $(u,v) \in E$ such that $u \prec v$, and 0 otherwise; for all $u,v \in V : u \neq v$. This variable models the graph connectivity according to the current sequence.
- Variable z_{up} is set to 1 if vertex u must be quantified as part of the cut-value at position p (i.e., $u \in \delta(p,\varphi,G)$) in the current feasible sequence φ, and 0 otherwise; for all $u \in V, p = 1,\ldots,n-1$.
- Variable R_{up} is set to 1 if vertex u is in the set of right vertices at position p (i.e., $\varphi(u) > p$) in the sequence φ, and 0 otherwise; for all $u \in V, p = 1,\ldots,n-1$. This variable models the set of right vertices $R(p,\varphi,G)$. Notice that if $R_{up} = 0$, it means that vertex u is in the set of left vertices at position p, i.e., $u \in L(p,\varphi,G)$.

- Variable δ_{uvp} is set to 1 if there is an edge $(u, v) \in E$ such that $u \prec v$, $\varphi(u) \leq p$ and $\varphi(v) > p$; and 0 otherwise. In other words, this variable indicates whether or not there is and edge $(u, v) \in E$ with u in the set of left vertices and v in the set of the right vertices at position p.
- Integer variable p_u gives the relative position of vertex u in the current feasible sequence, i.e., $p_u = \varphi(u)$, for all $u \in V$.
- Integer variable VS allows computing the objective value of any feasible sequence, i.e., $VS = vs(\varphi, G)$. This value is minimized.

The formulation for ILP$_2$ is the following.

$$VS^* = \min VS \tag{13}$$

subject to:

$$x_{uv} + x_{vu} = 1, \quad \forall u, v \in V : u \neq v \tag{14}$$

$$x_{uv} + x_{vw} + x_{wu} \leq 2, \quad \forall u, v, w \in V : u \neq v \neq w \tag{15}$$

$$y_{uv} = x_{uv}, \quad \forall u, v \in V : u \neq v : (u, v) \in E \vee (v, u) \in E \tag{16}$$

$$\sum_{\substack{u, v \in V : u \neq v : \\ (u, v) \notin E \vee (v, u) \notin E}} y_{uv} = 0 \tag{17}$$

$$p_u = n - \sum_{v \in V : u \neq v} x_{uv}, \quad \forall u \in V \tag{18}$$

$$p_u \geq (p + 1) - n(1 - R_{up}), \quad \forall u \in V, p = 1, \ldots n - 1 \tag{19}$$

$$1 + p_u - (p + 1) \leq (n - 1)R_{up}, \quad \forall u \in V, p = 1, \ldots n - 1 \tag{20}$$

$$y_{uv} + (1 - R_{up}) + R_{vp} \geq 3\delta_{uvp},$$
$$\forall u, v \in V : u \neq v, p = 1, \ldots, n - 1 \tag{21}$$

$$y_{uv} + (1 - R_{up}) + R_{vp} \leq \delta_{uvp} + 2,$$
$$\forall u, v \in V : u \neq v, p = 1, \ldots, n - 1 \tag{22}$$

$$z_{up} \leq \sum_{u \in V : u \neq v} \delta_{uvp} \leq (n - 1)z_{up}, \quad \forall u \in V, p = 1, \ldots, n - 1 \tag{23}$$

$$\sum_{u \in V} z_{up} \leq VS, \quad \forall p = 1, \ldots, n - 1 \tag{24}$$

$$x_{uv}, y_{uv} \in \{0, 1\}, \quad \forall u, v \in V : u \neq v \tag{25}$$

$$z_{up}, R_{up} \in \{0, 1\}, \quad \forall u \in V, p = 1, \ldots, n - 1 \tag{26}$$

Table 1 Number of variables and constraints generated by ILP$_1$ and ILP$_2$

ILP model	Number of variables	Number of constraints
ILP$_1$	$n^2 + 2mn^2 + n(n-1)/2 + 1$	$n^2 + 6mn^2 + 2n - 1$
ILP$_2$	$n^3 + 2n^2 - 2n + 1$	$3n^3 - 2n^2 + n + 2m$

$$\delta_{uvp} \in \{0,1\}, \quad \forall u, v \in V : u \neq v, p = 1, \ldots, n-1 \qquad (27)$$

$$p_u \in \mathbb{Z}^+, \quad \forall u \in V \qquad (28)$$

$$VS \in \mathbb{Z}^+ \qquad (29)$$

Constraints (14) state that if vertex u precedes vertex v, then vertex v must not precede vertex u. Constraints (15) establish that if vertex u precedes vertex v and vertex v precedes vertex w, then vertex w cannot precede vertex u. Constraints (16) and (17) determine the graph connectivity depending on the current feasible sequence, and force all the variables y_{uv} such that the vertices u and v are not connected, i.e., $(u, v) \notin E$, to take the value of 0, respectively. Constraints (18) obtain the relative position of vertex u in the sequence by subtracting from n, the number of vertices in front of u in the sequence.

Constraints (19) and (20) force variable R_{up} to take the value of 1 if and only if $\varphi(u) > p$, i.e., $R_{up} = 1 \leftrightarrow p_u \geq p + 1$. Constraints (21) and (22) define the variable δ_{uvp} as the product: $\delta_{uvp} = y_{uv}(1 - R_{up})R_{vp}$ by linearizing it applying the cubic linearization technique reported by Chen et al. in [11]. Like in ILP$_1$, constraints (23) compute the binary value of z_{up} from the variable δ_{uvp}. Constraints (24) compute the cut-values from each position p and keep the maximum among them. Constraints (25) to (29) are only defining the variables used in ILP$_2$.

4.3 Theoretical Analysis of ILP$_1$ and ILP$_2$

At first, it seems ILP$_2$ generates a larger computational cost than ILP$_1$. However, if we analyze both models by quantifying their exact number of variables and constraints, it is possible to make a fair comparison without ambiguity. Table 1 summarizes the number of variables and constraints generated by each model.

From Table 1 we can observe that when $m \cong n$, ILP$_1$ and ILP$_2$ generate both $\mathcal{O}(n^3)$ variables and constraints, which could result in a similar behavior between them. However, when $m \cong n^2$, ILP$_1$ increases the number of variables and constraints to $\mathcal{O}(n^4)$, while ILP$_2$ still generates $\mathcal{O}(n^3)$ variables and constraints, which yields to a better performance of ILP$_2$ when the graphs tend to be denser.

Fig. 1 Lexicographic grid
graph with $n = 15$ vertices
and $m = 22$ edges

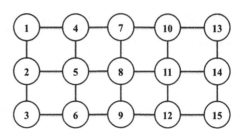

5 Algorithm to Identify Grid Graphs

In this section we propose an algorithm that verifies whether or not an arbitrary graph has a grid structure, and if so, the theorem proposed in [10] is applied to solve the problem for the given graph. The central idea of this algorithm is to compare the *main numerical properties* of any grid graph against the properties of the input graph. Figure 1 shows an example of a grid graph.

From the graph presented in Fig. 1, we can determine a series of useful numerical properties. First, we can see that any grid graph can be drawn as a mesh with r rows and c columns, in our example $r = 3$ and $c = 5$. The number of vertices and edges are $n = c \times r$ and $m = c \times (r - 1) + r \times (c - 1)$, respectively.

Furthermore, we can distinguish among certain classes of vertices based on their adjacency degree. The vertices appearing in the *corners* ($\{1, 3, 13, 15\}$ in our example) present always an adjacency degree of 2. The vertices in the *perimeter*, excluding the corner nodes, have an adjacency degree of 3. In our example those vertices are $\{2, 4, 6, 7, 9, 10, 12, 14\}$. Finally, the rest of the vertices, which are placed in the *middle* of the grid, have an adjacency degree of 4. Vertices $\{5, 8, 11\}$ are an example of these nodes.

We can quantify the number of vertices of each class of nodes, according to the number of rows and columns in the grid. Corner nodes (denoted by C) are always 4. The number of perimeter nodes (P) can be obtained by noticing that in each row and column in the perimeter (2 rows and 2 columns) there are $c - 2$ and $r - 2$ perimeter nodes, respectively. So we have $|P| = 2(c - 2) + 2(r - 2)$, whose algebraic reduction give us $|P| = 2(r + c - 4)$. The number of middle nodes (M) is obtained easily by subtracting from the total number of vertices, the number of corner and perimeter nodes, i.e., $|M| = cr - 2(r + c - 4) - 4$. Additionally, we can notice that certain classes of vertices will never be joined by an edge. For example, a *corner* node will not be adjacent to a *middle* node in any grid graph.

So far we have obtained the main numerical properties of a grid graph based on the values of r and c. However, for any input graph we cannot know these values. Therefore, we have to obtain them by using only the input data.

To calculate the values of r and c, we use the number of vertices and edges from the input graph. Specifically, we use the equation to compute the number of edges given r and c, i.e., $m = c(r - 1) + r(c - 1)$.

From the previous formula, we have $m = cr - c + cr - r = 2cr - c - r$, but $n = cr$, so $m = 2n - c - r$, which leads to $2n - m = c + r$. Besides, $n = cr$ implies $c = n/r$, so we have $2n - m = n/r + r = (n + r^2)/r$. Therefore, we have $r(2n - m) = n + r^2$, which finally yields to the following quadratic equation:

$$r^2 - (2n - m)r + n = 0. \tag{30}$$

To obtain the roots of Eq. (30) we used the well-known quadratic formula: $x = \left(-b \pm \sqrt{b^2 - 4ac}\right)/2a$, to solve quadratic equations of type $ax^2 + bx + c = 0$. In our case $x = r$, $a = 1$, $b = -(2n - m)$, and $c = n$. Substituting and reducing similar terms give us Eqs. (31) and (32):

$$r_1 = \left((2n - m) + \sqrt{(2n - m)^2 - 4n}\right)/2 \tag{31}$$

$$r_2 = \left((2n - m) - \sqrt{(2n - m)^2 - 4n}\right)/2. \tag{32}$$

It is important to point out that the resulting values r_1 and r_2 are actually the correct values for c and r (if the input graph is a grid). Figure 2 presents the algorithm that identifies if an arbitrary graph has a grid structure. Lines 1 to 6 present the basic definitions for our algorithm, which starts by computing the possible number of rows and columns (r and c) of the input graph. If the discriminant [see Eqs. (31) and (32)] is negative, or any of the values of c and r is not a positive integer, then the input graph is not a grid. These conditions are verified at the beginning because it is more likely that an arbitrary graph does not satisfy them. However, if the input graph satisfies the previous conditions, it means that the graph has the expected number of vertices and edges, but it is not necessarily a grid graph, since its edges might not connect the correct vertices.

To verify the correct connectivity of the graph, the algorithm counts the number of vertices whose adjacency degree is 2, 3 and 4, and the number of edges with a specific adjacency degree for both endpoints. If the number of vertices with a determined adjacency degree or the number of different types of edges does not match with the expected ones (lines 14 to 34), it means that the graph is not a grid. On the other hand, if the given graph satisfies all the conditions, then the input graph has a grid structure, and the algorithm returns the minimum value between r and c, which is the optimal separation value [3, 10] (lines 35 and 36).

The worst case is presented when the input graph has a grid structure, then our algorithm must verify the adjacency degree of each vertex and the graph connectivity in the edge list. Therefore, the computational complexity of this algorithm is $\mathcal{O}(n + m)$. On the other hand, for any other kind of graph, the algorithm computes only two values: r and c, in this case its complexity is $\theta(2)$.

$IsGRID\big(G = (V,E)\big)$

1. Let $n = |V|$, $m = |E|$ be the number of vertices and edges, respectively.
2. Let c and r be the number of columns and rows, respectively.
3. Let vs^* be the optimal vertex separation value for the grid graph.
4. Let $N(u) = \{v \in V : (u,v) \in E\}$ be the set of vertices adjacent to vertex u, such that $|N(u)|$ represents the adjacency degree of vertex u.
5. Let C, P, M be the subsets of *corner*, *perimeter*, and *middle* nodes, respectively; such that $C, P, M \subset V$, $C \cup P \cup M \equiv V$, and $C \cap P \cap M \equiv \{\emptyset\}$.
6. Let $E_{CC}, E_{PP}, E_{MM}, E_{CP}, E_{CM}, E_{PM}$ be the subsets of edges joining *corner-corner, perimeter-perimeter, middle-middle, corner-perimeter, corner-middle,* and *perimeter-middle* vertices, respectively; such that $E_{CC}, E_{PP}, E_{MM} E_{CP}, E_{CM}, E_{PM} \subseteq E$, $E_{CC} \cup E_{PP} \cup E_{MM} \cup E_{CP} \cup E_{CM} \cup E_{PM} \equiv E$, and $E_{CC} \cap E_{PP} \cap E_{MM} \cap E_{CP} \cap E_{CM} \cap E_{PM} \equiv \{\emptyset\}$.
7. $\Delta = (2n - m)^2 - 4n$ /* Discriminant */
8. if $(\Delta < 0)$
9.   return false
10. $c = \big((2n - m) + \sqrt{\Delta}\big)/2$
11. $r = \big((2n - m) - \sqrt{\Delta}\big)/2$
12. if $(c \notin \mathbb{Z}^+ \vee r \notin \mathbb{Z}^+)$
13.   return false
14. $C = \{v \in V : |N(v)| = 2\}$
15. $P = \{v \in V : |N(v)| = 3\}$
16. $M = \{v \in V : |N(v)| = 4\}$
17. if $(|C| \neq 4 \vee |P| \neq 2(r + c - 4) \vee |M| \neq cr - 2(r + c - 4) - 4)$
18.   return false
19. $E_{CC} = \{(u,v) \in E | u, v \in C\}$
20. $E_{PP} = \{(u,v) \in E | u, v \in P\}$
21. $E_{MM} = \{(u,v) \in E | u, v \in M\}$
22. $E_{CP} = \{(u,v) \in E | u \in C \wedge v \in P\}$
23. $E_{CM} = \{(u,v) \in E | u \in C \wedge v \in M\}$
24. $E_{PM} = \{(u,v) \in E | u \in P \wedge v \in M\}$
25. if $(r = c = 2)$
26.   if $(|E_{CC}| \neq 4 \vee |E_{PP}| \neq 0 \vee |E_{MM}| \neq 0 \vee |E_{CP}| \neq 0 \vee |E_{CM}| \neq 0 \vee |E_{PM}| \neq 0)$
27.   return false
28. if $((r = 2 \wedge c \geq 3) \vee (r \geq 3 \wedge c = 2))$
29. $r = \max\ \{c,r\}$, $c = \min\ \{c,r\}$
30.   if $(|E_{CC}| \neq 2 \vee |E_{PP}| \neq 3r - 8 \vee |E_{MM}| \neq 0 \vee |E_{CP}| \neq 4 \vee |E_{CM}| \neq 0 \vee |E_{PM}| \neq 0)$
31.   return false
32. if $(r \geq 3 \wedge c \geq 3)$
33.   if $(\ |E_{CC}| \neq 0 \vee |E_{PP}| \neq 2(r + c) - 12 \vee |E_{MM}| \neq 2cr - 5(r + c) + 12 \vee |E_{CP}| \neq 8 \vee |E_{CM}| \neq 0 \vee |E_{PM}| \neq |P|)$
34.   return false
35. $vs^* = \min\ \{c,r\}$
36. return (true, vs^*)

Fig. 2 Algorithm that verifies whether or not a given graph has a grid structure

6 Benchmark of VSP Instances

The set of *standard instances* for VSP, VSPLIB_2012 (available at http://www.optsicom.es/vsp/) was proposed in [2]. It basically includes three main datasets, two of them contain structured graphs: *Trees* and *Grids,* and one dataset of hard instances: *Harwell-Boeing.* Additionally, we used the same set of *Small* instances used by Pantrigo et al. in [12].

6.1 Proposed Benchmark of Scale-Free Instances

The use of real-world instances (or a close enough model) is an important issue in an experimental research. In this section a benchmark of similar to (*scale-free*) Internet network instances is proposed. The main characteristic of *scale-free* networks is their *power law* degree distribution, that is, a small group of vertices have a large adjacency degree, and a large number of vertices have a small adjacency degree [13]. The model to generate this kind of networks is presented in Eq. (33).

$$P(k) \sim k^{-\gamma} \tag{33}$$

where $P(k)$ is the probability that a vertex has an adjacency degree of k, and γ is a parameter that decreases from ∞ to 0 and describes how fast disappears the frequency of degree k in the network, in such a way that the average degree of the network increases when γ decreases [14].

In this work the algorithm of Eppstein et al. was used to generate our proposed *scale-free* instances [15]. This algorithm is implemented in the *scale-free* network generator of JUNG API (version 1.7.5) for Java.

We generated five dataset with instances of sizes $n = \{5, 6, 7, 8, 9\}$. For all datasets, the number of edges was fixed to $m = 2n$ and the number of iterations to $r = 1'000,000$. For each dataset, a number of 30 different instances were generated according to the following criteria.

- A specific seed (s) was set to each instance to ensure that the generated instances can be replicated exactly. The set of used seeds is $S = \{1, 2, \ldots, 30\}$. The first instance was created using seed 1, the second one using seed 2, and so on.
- We verified that the generated instance did not contain isolated vertices, that is, a vertex whose adjacency degree is 0. In case of having isolated vertices, they were deleted from the graph, which generated instances with a different number of vertices in the graph n' than the number of vertices originally requested n.
- Finally, we verified the resulting network was connected, that is, for each vertex u, there must exist a path connecting it to other vertex v in the network. When there is no such path from u to v, it implies the existence of more than one cluster in the graph, so the instance is discarded. In this case, we try to generate other instance by using a different seed ($s \geq 31$) until the graph is completely connected.

The name of the instances was codified as: $N\_n\_n'\_s$' where N comes from the word *Network*, n represents the number of requested vertices, n' is the real number of vertices in the instance ($n = n'$ if there were not isolated vertices), and s codifies the seed used to generate the instance.

Table 2 Experimental results of ILP$_1$ and ILP$_2$ with the *standard* and *scale-free* instances

Inst.	Mod.	O.V.	Time	# V	# C	E.N.	S.E.
grid_3	ILP$_1$	3	7.36	2062	5930	*1129*	153.23
	ILP$_2$	3	*3.16*	*874*	*2058*	1130	*356.57*
$n = 5$	ILP$_1$	4	0.255	536	1534	*31*	121.5
	ILP$_2$	4	*0.172*	*166*	*350*	207	*1203.4*
$n = 6$	ILP$_1$	3.4	0.689	916	2639	*163.06*	236.6
	ILP$_2$	3.4	*0.322*	*277*	*606*	407.83	*1266.5*
$n = 7$	ILP$_1$	3.633	1.723	1443	4178	*321.56*	186.6
	ILP$_2$	3.633	*0.692*	*428*	*966*	520.86	*752.6*
$n = 8$	ILP$_1$	3.8	6.555	2107.5	6125.8	1077.63	164.3
	ILP$_2$	3.8	*1.710*	*611.8*	*1416.1*	*630.66*	*368.8*
$n = 9$	ILP$_1$	3.9	39.264	2906.6	8475	3079.46	78.4
	ILP$_2$	3.9	*5.454*	*824.2*	*1946.4*	*1440.93*	*264.1*
Total	ILP$_1$	21.733	55.846	9971.1	28881.8	5801.71	940.63
	ILP$_2$	21.733	*11.513*	*3181.0*	*7342.5*	*4337.28*	*4211.97*

7 Experiments and Results

All the experiments were carried out in a computer with an Intel Core 2 Duo
processor and 4 GB in RAM. The source codes were implemented in Java (JRE
1.6.0_33) on a Mac OSX 10.6.8 operating system. We used CPLEX (v12.5) API
for Java to solve both ILP models to optimality.

The first experiment consists of testing the performance of the ILP models by
solving one instance from the standard benchmark and all of the 150 instances
from our proposed benchmark of *scale-free* networks. We only solved the grid_3
instance (from the standard benchmark) because it is the only one that can be
solved by both models in a reasonable time.

In this experiment we used the default settings of CPLEX, that is, we did not fix
a time limit or a cut-off (incumbent) value. We let CPLEX to be able to use all of
the two available processors and the pre-solver was enabled.

Table 2 shows the results of this experiment. The headings of this table are:
instance identifier (*Inst.*), ILP model (*Mod.*), objective value found by the model
(*O.V.*), computing time (*Time*), number of generated variables (*# V*), number of
generated constraints (*# C*), number of explored nodes reported by CPLEX (*E.N.*),
and speed of exploration (*S.E.*) which is defined as the number of explored nodes
per second (*E.N./Time*).

From Table 2 we can observe that ILP$_2$ outperforms ILP$_1$ in all the measured
attributes, except in the objective value found, where both models achieved the
same solution quality. Specifically, ILP$_2$ *reduces* the computing time of ILP$_1$ in
79.38 %. Furthermore, according to CPLEX, ILP$_2$ produces a smaller number of
variables, constraints and explored nodes than ILP$_1$ in 68.09, 74.57 and 25.24 %,
respectively. Additionally, ILP$_2$ improves the speed of exploration of ILP$_1$ in

Table 3 Results of the algorithm that identifies grid-structured graphs for the benchmark of *standard* instances

Dataset	# Corr.	O.V.	Effect (%)	Time
Grid (52)	52	28.5	100	0.012
Tree (50)	50	–	100	0.0005
Small (84)	84	–	100	0.001
HB (73)	73	–	100	0.004
Total	*259*	–	*100*	*1.113*

347.78 %, which indicates that ILP_2 is 4.47 times faster than ILP_1, for the evaluated set of instances.

The second experiment is intended to test the algorithm to identify and solve grid graphs. For this experiment we used the entire set of 259 standard instances. Table 3 shows the experimental results. The headings of Table 3 are: name of the dataset (*Dataset*), number of the instances correctly classified (*# Corr.*), objective value obtained (if applies) by the theorem of Bollobás et al. (*O.V.*), percentage of correctly classified instances (*Effect.*), and average computing time (*Time*).

As we can see in Table 3, our proposed algorithm achieves 100 % of effectiveness, and a remarkable running time. Specifically, it correctly classified 259 instances in 1.1 CPU seconds. It is important to point out that our algorithm only solved those instances identified as grids.

8 Conclusions

In this chapter the vertex separation problem was approached. To solve this problem we proposed: (1) a new ILP model based on a precedence scheme to represent a candidate solution, (2) an algorithm to identify grid graphs which experimentally showed to be really fast, and (3) a new benchmark of small-sized *scale-free* instances, which were used to test the integer linear programming models.

Experimental results revealed that our proposed ILP model (ILP_2) outperforms the reference model taken from the literature (ILP_1) in all the measured attributes. Moreover, ILP_2 *reduces* the average computing time of ILP_1 in 79.38 %, and *increases* the speed of exploration in 347.78 % with respect to ILP_1. Our algorithm that identifies grid-structured graphs has an effectiveness of 100 % and a remarkable running time of 1.1 s to compute 259 instances with different sizes and densities.

All the methods proposed in this research work can be applied to solve other graph layout problems. Specifically, the new modeling of the candidate solution proposed here can be applied to model other linear arrangement problems, such as: cut-width, sum-cut or bandwidth, among others.

Acknowledgments We would like to thank the National Council of Science and Technology of Mexico (CONACYT), the General Direction of Higher Technological Education (DGEST) and the Ciudad Madero Institute of Technology (ITCM) for their financial support. We also thank IBM Academic Initiative for allowing us to use their optimization engine CPLEX v12.5.

References

1. Lengauer, T.: Black-white pebbles and graph separation. Acta Informatica **16**, 465–475 (1981)
2. Duarte, A., Escudero, L., Martí, R., Mladenovic, N., Pantrigo, J., Sánchez-Oro, J.: Variable neighborhood search for the vertex separation problem. Comput. Oper. Res. **39**(12), 3247–3255 (2012)
3. Díaz, J., Petit, J., Serna, M.: A survey of graph layout problems. ACM Comput. Surv. **34**(3), 313–356 (2002)
4. Leiserson, C.: Area-efficient graph layouts (for VLSI). In: Proceedings of IEEE Symposium on Foundations of Computer Science, pp. 270–281 (1980)
5. Bodlaender, H., Gustedt, J., Telle, J.: Linear time register allocation for a fixed number of registers. In: Proceedings of the Symposium on Discrete Algorithms. (1998)
6. Kornai, A.: Narrowness, path-width, and their application in natural language processing. Discrete Appl. Math. **36**, 87–92 (1997). Elsevier Science Publishers B. V. (1992)
7. Lopes, I., de Carvalho, J.: Minimization of open orders using interval graphs. IAENG Int. J. Appl. Math. **40**(4), 297–306 (2010)
8. Ellis, J., Sudborough, I., Turner, J.: The vertex separation and search number of a graph. Inf. Comput. **113**, 50–79 (1994)
9. Skodinis, K.: Computing optimal linear layouts of trees in linear time. In: Paterson, M. (ed.) Proceedings of 8th Annual European Symposium on Algorithms. LNCS, vol. 1879, pp. 403–414. Springer, London (2000)
10. Bollobás, B., Leader, I.: Edge-Isoperimetric inequalities in the grid. Combinatorica **11**, 299–314 (1991)
11. Chen, D., Batson, R., Dang, Y.: Applied Integer Programming: Modeling and Solution. Wiley publisher. ISBN: 978-1-118-21002-4. (2010)
12. Pantrigo, J., Martí, R., Duarte, A., Pardo, E.: Scatter search for the Cutwidth minimization problem. Ann. Oper. Res. **199**(1), 285–304 (2012)
13. Barabasí, A.: Emergence of Scaling in Complex Networks. Handbook of graphs and networks: from the Genome of the Internet, pp. 69–84. (2005)
14. López, T.: Complejidad Computacional Estructural en Redes Complejas. PhD Thesis. Universidad Autónoma de Nuevo León, México (2012)
15. Eppstein, D., Wang, J.: A steady state model for graphs power laws. arXiv preprint cs/0204001. (2002)

Preference Incorporation into Evolutionary Multiobjective Optimization Using a Multi-Criteria Evaluation Method

Laura Cruz-Reyes, Eduardo Fernandez, Claudia Gomez and Patricia Sanchez

Abstract Most approaches in the evolutionary multiobjective optimization literature concentrate mainly on generating an approximation of the Pareto front. However, this does not completely solve the problem since the Decision Maker (DM) still has to choose the best compromise solution out of that set. This task becomes difficult when the number of criteria increases. In this chapter, we introduce a new way to incorporate and update the DM's preferences into a Multiobjective Evolutionary Algorithm, expressed in a set of solutions assigned to ordered categories. We propose a variant of the well-known Non-dominated Sorting Genetic Algorithm II (NSGA-II), called Hybrid-MultiCriteria Sorting Genetic Algorithm (H-MCSGA). In this algorithm, we strengthen the selective pressure based on dominance adding selective pressure based on assignments to categories. Particularly, we make selective pressure towards non-dominated solutions that belong to the best category. In instances with 9 objectives on the project portfolio problem, H-MCSGA outperforms NSGA-II obtaining non-dominated solutions that belong to the most preferred category.

L. Cruz-Reyes (✉) · C. Gomez · P. Sanchez
Instituto Tecnologico de Ciudad Madero, Tamaulipas, Mexico
e-mail: lauracruzreyes@itcm.edu.mx

C. Gomez
e-mail: cggs71@hotmail.com

P. Sanchez
e-mail: jpatricia.sanchez@gmail.com

E. Fernandez
Universidad Autonoma de Sinaloa, Sinaloa, Mexico
e-mail: eddyf@uas.edu.mx

O. Castillo et al. (eds.), *Recent Advances on Hybrid Approaches for Designing* 533
Intelligent Systems, Studies in Computational Intelligence 547,
DOI: 10.1007/978-3-319-05170-3_37, © Springer International Publishing Switzerland 2014

1 Introduction

A wide variety of problems in the real world often involves multiple objectives to be minimized or maximized simultaneously [1, 2]. These objectives present some degree of conflict among them; therefore, there is no single solution that is optimal with respect to all the objectives. These problems are called Multiobjective Optimization Problems (MOPs).

In single-objective optimization, it is possible to determine a single optimal solution. However, in the resolution of a MOP there does not exist a method to determine if a solution is better than other. These problems give rise to a set of optimal solutions, that is, a set of compromise solutions with different trade-offs among the objectives. The set of compromise solutions is popularly known as Pareto optimal solutions or non-dominated solutions. Although there are multiple Pareto optimal solutions, only one solution will be selected and implemented. As was stated by Fernandez et al. [3], to solve a MOP means to find the best compromise solution according to the decision maker's (DM's) preferences. Therefore, according to Lopez et al. [4], in the multiobjective optimization process two main tasks can be identified: (1) find a set of Pareto optimal solutions, and (2) choose the most preferred solution out of this set. Since all the compromise solutions are mathematically equivalent, the DM should provide some additional information for choosing the most preferred one.

According to Lopez et al. [4], there are two main approaches to solve multi-objective optimization problems. The first is the Multi-Criteria Decision Making (MCDM) approach characterized by the use of mathematical programming techniques and a decision making method. In most of the MCDM's methods, the DM provides information to build a preference model. This model is exploited by a mathematical programming method with the aim to find solutions that better match the DM's preferences [5]. The second approach is the Evolutionary Multiobjective Optimization (EMO). Since evolutionary algorithms use a population based approach, they are able to find an approximation of the whole Pareto front as output on a single run. Nevertheless, finding this set does not completely solve the problem. The DM still has to choose the best compromise solution out of that set. Hence, in recent years several works have addressed the incorporation of preference in evolutionary algorithms, giving rise to the Multi-Objective Evolutionary Algorithms (MOEAs). The goal of MOEAs is to assist the DM to select the final solution that best matches her/his preferences. This is not a difficult task when dealing with problems having two or three objectives. However, according to Fernandez et al. [6], when the number of criteria increases the algorithm's capacity to find the Pareto frontier quickly degrades (e.g. [2, 7]) and the size of a representative portion of the known Pareto frontier may be too large. It becomes harder for the DM to establish valid judgments in order to compare many solutions with several conflicting criteria.

In order to make the decision making phase easier, the DM would agree with incorporating his/her multicriteria preferences into the search process. This preference information is used to guide the search towards the Region of Interest (ROI) [8], the privileged zone of the Pareto frontier that best matches the DM's preferences.

The DM's preference information can be expressed in different ways. According to Bechikh [9], the following are the explicit expressions most commonly used: *weights, solution ranking, objective ranking, reference point, reservation point, trade-off between objectives, outranking thresholds and desirability thresholds.*

Preferences may be also implicitly given. Let us suppose that a set of solutions has been sorted by the DM on a set of ordered categories like {Very Good, Good, Acceptable, Bad}. These assignments express the DM's assignment policy. Then, the set of assignments can be considered preference knowledge. Using this knowledge in the framework of a multicriteria sorting method, each new solution generated by the search process can be assigned to one of those categories. To some extent, such preference information surrogates the DM in judging new solutions.

In this chapter, we describe a solution evaluation method to make selective pressure towards solutions that are non-dominated and that belong to the best category according to the DM preferences. This method is incorporated into the classical NSGA-II, and it is used for solving the project portfolio problem.

2 Background

2.1 The Project Portfolio Problem

A project is a temporary, unique and unrepeatable process which pursues a set of aims [10]. Moreover, a portfolio is a set of projects that can be done on the same period [10]. For this reason, projects that belong to the same portfolio share the organization's available resources. Therefore, it is not sufficient to compare the projects individually, but they must be compared by project groups, by identifying which portfolio gives more contribution to the objectives of the organization.

Proper selection of the projects that will integrate the portfolio to receive resources of the organization is one of the most important decision issues for any institution [11, 12]. The main economic and mathematical models that address the problem of portfolio assume that there exists a set of N projects, where each project is characterized by costs and revenues. The DM is responsible for the selection of the group of projects (portfolio) that the institution will implement [13].

Formalizing these concepts, let us consider a set of N projects, which are competing to be funded by a total budget denoted by B. On the other hand, a portfolio is a subset of these projects which is usually modeled as a binary vector $x = \langle x_1, x_2, \ldots, x_N \rangle$. In this vector, x_i is a binary variable that is equal to '1' if the

ith project will be supported and '0' in the other case. Each project i has an associated cost that is identified as c_i. A portfolio is feasible when it satisfies the constraint of the total budget B:

$$\left(\sum_{i=1}^{N} x_i c_i\right) \leq B \tag{1}$$

Each project i corresponds to an area (health, education, etc.) denoted by a_i. Each area has budgetary limits defined by the DM or any other competent authority. Let us consider L_j and U_j as lower and upper limits respectively for each area j. Based on this, a feasible portfolio must satisfy the following constraint for each area j:

$$L_j \leq \sum_{i=1}^{N} x_i g_{i,j} \leq U_j \tag{2}$$

where $g_{i,j}$ can be defined as:

$$g_i(j) = \begin{cases} 0 & if\ a_i = j, \\ 1 & otherwise \end{cases} \tag{3}$$

In the same way, each project i corresponds to a geographical region to which it will benefit. In the same way as areas, each region has lower and upper limits as another constraint that must be fulfilled by a feasible portfolio.

Each project i is modeled as a p-dimensional vector $f(i) = \langle f_1(i),$ $f_2(i), f_3(i), \ldots, f_p(i)\rangle$, where each $f_j(i)$ indicates the benefit of project i to objective j into a problem with p objectives. The quality of a portfolio x is determined by the union of the benefits of each of the projects that compose it. This can be expressed as:

$$z(x) = \langle z_1(x), z_2(x), z_3(x), \ldots, z_p(x)\rangle \tag{4}$$

where $z_j(x)$ in its simplest form, is defined as:

$$z_j(x) = \sum_{i=1}^{N} x_i f_j(i) \tag{5}$$

If we denote by R_F the region of feasible portfolios, the problem of project portfolio is to identify one or more portfolios that solve:

$$\max_{x \in R_F} \{z(x)\} \tag{6}$$

subject to the constraints in (1–2) and of the geographic region.

2.2 The THESEUS Method

Proposed by Fernandez and Navarro [14], the THESEUS multicriteria evaluation method is based on comparing a new object, to be assigned, with reference objects through models of preference and indifference relations. The assignment is the result of comparisons with other objects whose assignments are known. In the following, $C(x)$ denotes a potential assignment of object x and $C(b)$ is the current assignment of b. According to THESEUS, $C(x)$ should satisfy:

$$\forall x \in U, \quad \forall b \in T$$
$$xP(\lambda)b \Rightarrow C(x) \succeq C(b) \tag{7.a}$$
$$bP(\lambda)x \Rightarrow C(b) \succeq C(x)$$

$$xQ(\lambda)b(x) \succeq C(b)$$
$$bQ(\lambda)x \Rightarrow C(b) \succeq C(x) \tag{7.b}$$

$$xI(\lambda)b(x) \succeq C(b) \wedge C(b) \succeq C(x) \Leftrightarrow C(x) = C(b) \tag{7.c}$$

As was stated in [14], $C(x)$ is a variable whose domain is the set of ordered categories. Equations (7.a–7.c) express the necessary consistency amongst the preference model, the reference set and the appropriate assignments of x. The assignment $C(x)$ should be as compatible as possible with the current knowledge about the assignment policy [14].

THESEUS uses the inconsistencies with Eqs. (7.a–7.c) to compare the possible assignments of x. More specifically:

- The set of $P(\lambda)$-inconsistencies for x and $C(x)$ is defined as $D_P = \{(x, b), (b, x), b \in T$ such that (7.a) is FALSE$\}$.
- The set of $Q(\lambda)$-inconsistencies for x and $C(x)$ is defined as $D_Q = \{(x, b), (b, x)\, b \in T$ such that (7.b) is FALSE$\}$.
- The set of $I(\lambda)$-inconsistencies for x and $C(x)$ is defined as $D_I = \{(x, b), b \in T$ such that (7.c) is FALSE$\}$.

Some $I(\lambda)$-inconsistencies might be explained by 'discontinuity' of the description based on the set of categories. The cases in which $xI(\lambda)b \wedge |k - j| = 1$ will be called second-order $I(\lambda)$-inconsistencies and grouped in the set D_{2I}. The set $D_{1I} = D_I - D_{2I}$ contains the so-called first-order $I(\lambda)$-inconsistencies, which are not consequences of the described discontinuity effect. Let n_P, n_Q, n_{1I}, n_{2I} denote the cardinality of the above-defined inconsistency sets, and $N_1 = n_P + n_Q + n_{1I}$, $N_2 = n_{2I}$.

THESEUS suggests an assignment that minimizes the above inconsistencies with lexicographic priority favoring N_1, which is the most important criterion (cf. [14]). The basic assignment rule is:

1. Assign the minimum credibility level $\lambda > 0.5$.
2. Starting with $k = 1$ $(k = 1,...,M)$ and considering each $b_{k,h} \in T$, calculate $N_1(C_k)$.
3. Identify the set $\{C_j\}$ whose elements hold $C_j = argminN_1 (C_k)$.
4. Select $C_{k*} = argminN_2 (C_i)$;

   $\{C_j\}$

5. If C_{k*} is a single solution, assign x_j to C_{k*}; other situations are approached below.

The suggestion may be a single category or a sequence of categories. The first case is called a well-defined assignment; otherwise, the obtained solution highlights the highest category (C_H) and the lowest category (C_L) which is appropriate for assigning the object, but fails in determining the most appropriate.

3 The Proposed Method

In a MOP, the ROI is characterized by solutions that remain two features: (1) to be non-dominated; (2) to be assigned to the most preferred category C_M. The idea behind this new proposal consists in making selective pressure towards solutions that hold both properties.

In [15], we proposed to use a variant of the popular NSGA-II, which is called the MultiCriteria Sorting Genetic Algorithm (MCSGA). It Works like NSGA-II, with the following differences:

– Each solution of the first front of NSGA-II (the non-dominated front) is assigned by THESEUS to one category of the set $Ct = \{C_1,..., C_M\}$, which is a set of ordered categories defined by the DM;
– The first front of NSGA-II is divided into $M' \leq M$ sub-fronts; the first ranked sub-front contains the solutions that were assigned to the most preferred category;
– The remaining fronts are re-ordered by considering each sub-front of the original non-dominated front as a new front;
– The same operations of NSGA-II are applied, but considering the new fronts; now the new first front is composed of non-dominated solutions belonging to the most preferred category.

Remarks In order to form the reference set T, the DM should evaluate several existing solutions. In the beginning of the search process, the DM does not know even a rough approximation to his/her ROI. Then, he/she can hardly identify feasible solutions belonging to the most preferred categories. Hence, the selective pressure towards the ROI is weak.   □

To overcome this drawback, we propose to hybridize the MCSGA (see Cruz-Reyes et al. [15]). This will be combined with another multiobjective metaheuristic approach, which provides an approximation to the Pareto frontier. The DM's preferences will be expressed on a subset of that approximation, and then T will be constructed. The hybrid procedure is described below:

PROCEDURE H-MCSGA (*L, Number_of_generations*)
1. Run a multiobjective optimization method and obtain an approximation to the Pareto frontier *PF*
2.   Initialize reference set T by using a subset of *PF*
3.   Set σ-parameters agreeing with T
4. Initialize population P
5. Generate random population with size L
6. Evaluate objective values
7. Generate non-dominated fronts on P
8. Assign to these fronts a rank (level)
9. Calculate σ on $R_1 \times R_1$
10. For each $x \in R_1$, assign x to one preference category
11. Form M' sub-fronts of R_1
12. Assign to sub-fronts a rank and update the remaining fronts
13. Generate child population Q with size L
14.   Perform binary tournament selection
15.   Perform recombination and mutation
16. FOR $I = 1$ to *Number_of_generations* DO
17.   Assign $P' = P \cup Q$
18.   Generate non-dominated fronts on P'
19.   Assign to these fronts a rank (level)
20.   Calculate σ on $R_1 \times R_1$
21.   For each $x \in R_1$, assign x to one preference category
22.   Form M' sub-fronts of R_1
23.   Assign to sub-fronts a rank and update the remaining fronts
24.   FOR each parent and child in P' DO
25.   Calculate crowding distance
26.   Loop (inside) by adding solutions to the next generation until L individuals have been found
27.   End FOR
28.   Replace P by the L individuals found
29.   Generate child population Q with size L
30.   Perform binary tournament selection
31.   Perform recombination and mutation
32.   Update the reference set T
33. End FOR
34. End PROCEDURE

Remarks The differences between the procedure presented here and the procedure in [15] are the following:

- The reference set T is updated with better solutions found during the search process.
- The update is carried out a certain number of times and in a pre-defined iteration.
- The new reference set is constructed from the solutions of the current iteration. The solutions are categorized, simulating a DM, using the outranking model described in [6].
- If the new reference set is consistent, it replaces the current reference set; otherwise the reference set is not updated.

4 Experimentation

Let us consider as an illustration the same situation addressed by Cruz-Reyes et al. [15] to the application of the new proposal of the H-MCSGA. The problem is distributing 2.5 billion dollars among a set of applicant projects. This set is composed of 100 proposals and nine objectives each, all of which deserve financing. Information about some projects (values of objectives, costs, area, and region) is shown in Table 1. In the first phase of H-MCSGA, the Ant Colony Optimization Algorithm (ACO) by Rivera et al. [16] was used. This algorithm incorporates preferences by using the outranking model from [6]. The ACO parameters were the same as those reported in the article, but we did not use local search or consider synergy. The solutions were categorized by simulating a DM whose preferences are compatible with the outranking model from [6]. There are four categories being the fourth the best. The classified solutions were processed in order to construct a consistent reference set that satisfies Eqs. (7.a–7.c).

We experimented with five random instances, creating for each of them one reference set. The parameters of the evolutionary search in the second phase of the H-MCSGA were: crossover probability = 1; mutation probability = 0.05; population size = 100, number of generations = 500. The update of the reference set was carried out every 150 generations. The second phase of the H-MCSGA was run, and the obtained solutions were compared with the solutions from the standard NSGA-II. The results are illustrated in Table 2.

We can see that, our proposal satisfies the ROI conditions since always finds non-dominated solutions and of the most preferred category. On the other hand, some solutions from NSGA-II are dominated and moreover do not belong to the best category.

Table 1 Information about some projects

Project	N_1	N_2	...	N_8	N_9	Cost	Area	Region
1	36,955	34,670	...	50,175	15,720	8,130	1	2
2	33,265	28,710	...	39,605	21,575	8,620	1	2
3	17,340	32,630	...	48,975	24,520	7,620	2	2
4	39,490	18,000	...	48,135	12,410	9,380	1	1
5	46,565	13,760	...	55,120	64,200	5,410	3	1

Table 2 Comparative results between NSGA-II and H-MCSGA

Instance	Instance description	Algorithm	Size of the solution set	Solutions that remain non-dominated in A ∪ B	Highest category
1	9 objectives	NSGA-II	116	79	3
	100 projects	H-MCSGA	10	10	4
2	9 objectives	NSGA-II	119	59	3
	100 projects	H-MCSGA	8	8	4
3	9 objectives	NSGA-II	117	88	3
	100 projects	H-MCSGA	8	8	4
4	9 objectives	NSGA-II	120	103	3
	100 projects	H-MCSGA	18	18	4
5	9 objectives	NSGA-II	104	95	3
	100 projects	H-MCSGA	9	9	4

Note A is the set of solutions obtained by NSGA-II; B is the set obtained by H-MCSGA

5 Conclusions

An original idea to incorporate and update the DM's preferences into multi-objective evolutionary optimization has been presented.

The DM's preferences are captured in a reference set of solutions assigned to ordered categories. This set is updated during the search process; consequently, the DM's preferences are updated.

Our proposal is a derivation from NSGA-II, where we strengthen the selective pressure based on dominance with selective pressure based on evaluation.

Unlike Pareto-based evolutionary algorithms, the selective pressure is not degraded with the number of objective functions.

In examples with 9 objectives, this proposal outperforms solutions obtained by NSGA-II. Experimentation in the small-scale is required to reach definitive conclusions.

Acknowledgments This work was partially financed by CONACYT, PROMEP and DGEST.

References

1. Deb, K.: Multi-Objective Optimization using Evolutionary Algorithms, pp. 13–46. Wiley, Chichester (2001). (Weinheim-Brisbane-Singapore-Toronto)
2. Coello, C.A., Lamont, G.B., Van Veldhuizen, D.A.: Evolutionary Algorithms for Solving MultiObjective Problems, 2nd edn. Springer, New York (2007)
3. Fernandez, E., Lopez, E., Bernal, S., Coello, C.A., Navarro, J.: Evolutionary multiobjective optimization using an outranking-based dominance generalization. Comput. Oper. Res. **37**(2), 390–395 (2010)
4. Jaimes, A.L., Martinez, S.Z., Coello, C.A.: An introduction to multiobjective optimization techniques. Optim. Polym. Process. 29–57 (2009)
5. Miettinen, K.: Introduction to multiobjective optimization: noninteractive approaches. In: Branke, J., Deb, K., Miettinen, K., Slowinski, R. (eds.) Multiobjective Optimization: Interactive and Evolutionary Approaches, pp. 1–26. Springer, Berlin (2008)
6. Fernandez, E., Lopez, E., Lopez, F., Coello, C.: Increasing selective pressure toward the best compromise in evolutionary multiobjective optimization: the extended NOSGA method. Inf. Sci. **181**, 44–56 (2011)
7. Wang, Y., Yang, Y.: Particle swarm optimization with preference order ranking for multi-objective optimization. Inf. Sci. **179**(12), 1944–1959 (2009)
8. Deb, K., Chaudhuri S., Miettinen, K.: Towards estimating nadir objective vector using evolutionary approaches. In: Proceedings of the 8th Genetic and Evolutionary Computation COnference (GECCO'O6), pp. 643–650. (2006)
9. Bechikh, S.: Incorporating decision maker's preference information in evolutionary multi-objective optimization. Dissertation Ph.D. thesis, High Institute of Management of Tunis, University of Tunis, Tunisia. http://delta.cs.cinvestav.mx/~ccoello/EMOO/thesis-bechikh.pdf.gz (2013)
10. Carazo, A., Gomez, T., Molina, J., Hernandez-Diaz, A., Guerrero, F., Caballero, R.: Solving a comprehensive model for multiobjective project portfolio selection. Comput. Oper. Res. **37**(4), 630–639 (2010)
11. Castro, M.: Development and implementation of a framework for I&D in public organizations. Master's thesis, Universidad Autonoma de Nuevo León (2007)
12. Garcia R.: Hyper-Heuristicforsolving social portfolio problem. Master's thesis, Instituto Tecnológico de Cd., Madero (2010)
13. Fernandez, E., Navarro, J.: A genetic search for exploiting a fuzzy preference model of portfolio problems with public projects. Ann. OR **117**(191–213), 191–213 (2002)
14. Fernandez, E., Navarro, J.: A new approach to multicriteria sorting problems based on fuzzy outranking relations: the THESEUS method. Eur. J. Oper. Res. **213**, 405–413 (2011)
15. Cruz-Reyes, L., Fernandez, E., Olmedo, R., Sanchez, P., Navarro, J.: Preference Incorporation into evolutionary multiobjective optimization using preference information implicit in a set of assignment examples. In: Proceedings of Fourth International Workshop on Knowledge Discovery, Knowledge Management and Decision Support. Atlantis Press (2013)
16. Rivera, G., Gomez, C., Fernandez, E., Cruz, L., Castillo, O., Bastiani, S.: Handling of synergy into an algorithm for project portfolio selection. In: Castillo, O., Melin, P., Kacprzyk, J. (eds.) Recent Advances on Hybrid Intelligent Systems, pp. 417–430. Springer, Berlin (2013)

A Loading Procedure
for the Containership Stowage Problem

**Laura Cruz-Reyes, Paula Hernández Hernández, Patricia Melin,
Héctor Joaquín Fraire Huacuja, Julio Mar-Ortiz, Héctor José Puga
Soberanes and Juan Javier González Barbosa**

Abstract This chapter deals with the containership stowage problem. It is an NP-hard combinatorial optimization whose goal is to find optimal plans for stowing containers into a containership with low operational costs, subject to a set of structural and operational constraints. In order to optimize a stowage planning, like in the literature, we have developed an approach that decomposes the problem hierarchically. This approach divides the problem into two phases: the first one consists of generating a relaxed initial solution, and the second phase is intended to make this solution feasible. In this chapter, we focus on the first phase of this approach, and a new loading procedure to generate an initial solution is proposed. This procedure produces solutions in short running time, so that, it could be applied to solve real instances.

L. Cruz-Reyes (✉) · P. Hernández Hernández · H. J. Fraire Huacuja · J. J. González Barbosa
Instituto Tecnológico de Ciudad Madero, Ciudad Madero, México
e-mail: lauracruzreyes@itcm.edu.mx; lcruzreyes@prodigy.net.mx

P. Hernández Hernández
e-mail: paulahdz314@hotmail.com

H. J. Fraire Huacuja
e-mail: hfraire@prodigy.net.mx

J. J. González Barbosa
e-mail: jjgonzalezbarbosa@hotmail.com

P. Melin
Tijuana Institute of Technology, Tijuana, México
e-mail: pmelin@tectijuana.mx

J. Mar-Ortiz
Universidad Autónoma de Tamaulipas, Tampico, México
e-mail: jmar@uat.edu.mx

H. J. Puga Soberanes
Instituto Tecnológico de León, León, México
e-mail: pugahector@yahoo.com

O. Castillo et al. (eds.), *Recent Advances on Hybrid Approaches for Designing
Intelligent Systems*, Studies in Computational Intelligence 547,
DOI: 10.1007/978-3-319-05170-3_38, © Springer International Publishing Switzerland 2014

1 Introduction

The efficiency of a maritime container terminal mainly depends on the process of handling containers, especially during the ships loading process. A good stowage planning facilitates these processes.

This chapter deals with the containership stowage problem, which is referred to as the Master Bay Plan Problem (MBPP) [1]. It is an NP-hard combinatorial optimization whose goal is to find optimal plans for stowing containers into a containership with low operational costs, subject to a set of structural and operational constraints.

In order to optimize a stowage planning, we decompose the problem hierarchically like in recent approaches [2, 3]. The problem is divided into two phases: the first one consists of generating a relaxed solution, that is, we remove the constraints of stability, and the second phase is intended to make this solution feasible [4].

This chapter is organized into five parts. Section 2 describes MBPP. Section 3 presents our proposed heuristic to generate a relaxed solution for MBPP. Section 4 shows the experimental results. Finally, Sect. 5 shows the conclusions and future work.

2 Description of Master Bay Plan Problem

The problem of stowage a set of containers into a containership can be defined as follows [5]:

Given a set C of n containers of different types to be loaded on the ship and a set S of m available locations on the containership, we have to determine the assignment of each container to a location of the ship, in such a way, to satisfy all the given structural and operational constraints related to ship and containers, and to minimize the total stowage time.

In MBPP, each container $c \in C$ must be stowed in a location $l \in S$ of the ship. The l-th location is actually addressed by the indices i, j, k representing, respectively: the bay (i), the row (j), and the tier (k). We denote by I, J and K, respectively, the set of bays, rows and tiers of the ship, and by b, r and s their corresponding cardinality.

The objective function is expressed in terms of the sum of the time t_{lc} required for loading a container c, $\forall c \in C$, in location l, $\forall l \in S$, such that $L = \sum_{lc} t_{lc}$. However, when two or more quay cranes are used for the loading operations the objective function is given by the maximum over the minimum loading time (L_q) for handling all containers in the corresponding ship partition by each quay crane q, that is, $L = \max_{QC}\{L_q\}$, where QC, is the set of available quay cranes.

The main constraints that must be considered for the stowage planning process for an individual port are related to the structure of the ship and focused on the size, type, weight, destination and distribution of the containers to be loaded. The description of each of them will be found below.

Size of containers. Usually, set C of containers is considered as the union of two subsets, T and F, consisting, respectively, of 20 and 40 ft containers, such that $T \cap F = \varnothing$ and $T \cup F = C$. Containers of 40 ft require two contiguous locations of 20 ft. Note that according to a practice adopted by the majority of maritime companies, bays with even number are used for stowing $40'$ containers and correspond to two contiguous odd bays that are used for the stowage of $20'$ containers. Consequently, if a $40'$ container is stowed in an even bay (for instance bay 02) the locations of the same row and tier corresponding to two contiguous odd bays (e.g. bay 01 and bay 03) are not anymore available for stowing containers of $20'$.

Type of containers. Different types of containers can usually be stowed in a containership, such as standard, carriageable, reefer, out of gauge and hazardous. The location of reefer containers is defined by the ship coordinator (who has a global vision of the trip), so that we know their exact position. This is generally near power points in order to maintain the required temperature during transportation. Hazardous containers are also assigned by the harbour-master's office, which authorizes their loading. In particular, hazardous containers cannot be stowed either in the upper deck or in adjacent locations. They are considered in the same way as $40'$ containers. Note that, for the definition of the stowage plan we consider only dry and dry high cube containers, having exterior dimensions conforming to ISO standards of 20 and 40 ft long, either 8 ft 6″ or 9 ft 6″ high and 8 ft depth.

Weight of containers. The standard weight of an empty container ranges from 2 to 3.5 t, while the maximum weight of a full container to be stowed in a containership ranges from 20–32 and 30–48 t for $20'$ and $40'$ containers, respectively. The weight constraints force the weight of a stack of containers to be less than a given tolerance value. In particular the weight of a stack of 3 containers of $20'$ and $40'$ cannot be greater than an a priori established value, say MT and MF, respectively. Moreover, the weight of a container located in a tier cannot be greater than the weight of the container located in a lower tier having the same row and bay and, as it is always required for security reasons; both $20'$ and $40'$ containers cannot be located over empty locations.

Destination of containers A good general stowing rule suggests to load first (i.e., in the lower tiers) those containers having as destination the final stop of the ship and load last those containers that have to be unloaded first.

Distribution of containers. Such constraints, also denoted stability constraints, are related to a proper weight distribution in the ship. In particular, we have to verify different kinds of equilibrium, namely:

- Horizontal equilibrium: the weight on the right side of the ship, including the odd rows of the hold and upper deck, must be equal (within a given tolerance, say Q_1) to the weight on the left side of the ship.
- Cross equilibrium: the weight on the stern must be equal (within a given tolerance, say Q_2) to the weight on the bow.

- Vertical equilibrium: the weight on each tier must be greater or equal than the weight on the tier immediately over it. Let us denote by L and R, respectively, the set of rows belonging to the left/right side of the ship and by A and P, respectively, the sets of anterior and posterior bays of the ship.

The MBPP is NP-hard combinatorial optimization problem. In Avriel et al. [6] the demonstration of the complexity is presented. In the next section, the 0/1 Linear Programming Model proposed by Ambrosino et al. [1] is presented. They assume that all containers are ready to be loaded on the quay without considering their stock position in the yard.

2.1 Binary Linear Programming Model for MBPP

Ambrosino et al. [1, 5] assume x_{lc}, $l = 1, \ldots, m$, $c = 1, \ldots, n$, are decision variables of the problem, with the following specification:

$$x_{lc} = \begin{cases} 1 & \text{if a container } c \text{ is stowed in location } l \\ 0 & \text{otherwise} \end{cases}$$

The lth location is identified by indices i, j, k representing, respectively, its bay, row and tier address, while c identifies the number (or code) of the cth stowed container. This means that variable $x_{lc} = x_{ijkc}$, directly gives the location where container c is stowed, when it is set to 1. Therefore, at the optimal solution we have the exact position of each container in the ship.

The definition of variable x_{lc}, $\forall l \in S$, $\forall c \in C$, enables an easy formulation of the underlying model for MBPP, reported below.

$$\text{Min} L = \sum_l \sum_c t_{lc} x_{lc} \tag{1}$$

$$\sum_l \sum_c x_{lc} = n \tag{2}$$

$$\sum_l x_{lc} \leq 1 \quad \forall c \tag{3}$$

$$\sum_c x_{lc} \leq 1 \quad \forall l \tag{4}$$

$$\sum_l \sum_c w_c x_{lc} \leq Q \tag{5}$$

$$\sum_{c \in T} x_{ijkc} = 0 \quad \forall i \in E, j \in J, k \in K \tag{6}$$

$$\sum_{c \in F} x_{ijkc} = 0 \quad \forall i \in O, j \in J, k \in K \tag{7}$$

$$\sum_{c \in T} x_{i+1jkc} + \sum_{c \in F} x_{ijkc} \leq 1 \quad \forall i \in E, j \in J, k \in K \tag{8}$$

$$\sum_{c \in T} x_{i-1jkc} + \sum_{c \in F} x_{ijkc} \leq 1 \quad \forall i \in E, j \in J, k \in K \tag{9}$$

$$\sum_{c \in T} x_{i+1jk+1c} + \sum_{c \in F} x_{ijkc} \leq 1 \quad \forall i \in E, j \in J, k = 1, \ldots, |K| - 1 \tag{10}$$

$$\sum_{c \in T} x_{i-1jk+1c} + \sum_{c \in F} x_{ijkc} \leq 1 \quad \forall i \in E, j \in J, k = 1, \ldots, |K| - 1 \tag{11}$$

$$\sum_{c \in T} w_c x_{ijkc} + \sum_{c \in T} w_c x_{ijk+1c} + \sum_{c \in T} w_c x_{ijk+2c} \leq MT \quad \forall i \in I, j \in J, k = 1, \ldots, |K| - 2 \tag{12}$$

$$\sum_{c \in F} w_c x_{ijkc} + \sum_{c \in F} w_c x_{ijk+1c} + \sum_{c \in F} w_c x_{ijk+2c} \leq MF \quad \forall i \in I, j \in J, k = 1, \ldots, |K| - 2 \tag{13}$$

$$\sum_c w_c x_{ijkc} - \sum_c w_c x_{ijk+1c} \geq 0 \quad \forall i \in I, j \in J, k = 1, \ldots, |K| - 1 \tag{14}$$

$$\sum_c d_c x_{ijkc} - \sum_c d_c x_{ijk+1c} \geq 0 \quad \forall i \in I, j \in J, k = 1, \ldots, |K| - 1 \tag{15}$$

$$-Q_2 \leq \sum_{i \in A, j, k} \sum_c w_c x_{ijkc} - \sum_{i \in P, j, k} \sum_c w_c x_{ijkc} \leq Q_2 \tag{16}$$

$$-Q_1 \leq \sum_{i, j \in L, k} \sum_c w_c x_{ijkc} - \sum_{i, j \in R, k} \sum_c w_c x_{ijkc} \leq Q_1 \tag{17}$$

$$x_{lc} \in \{0, 1\} \quad \forall l, c \tag{18}$$

Equation (1) is the objective function that minimizes the total stowage time L, expressed in terms of the sum of time t_{lc} required for loading a container c, $\forall c \in C$, in location l, $\forall l \in S$. Note that t_{lc} are the loading times expressed in $1/100$ of minute, and depend of the row and tier address in the ship in which the container will be loaded.

Constraint (2) defines the number of locations to select for stowing the given containers. Constraints (3) and (4) are the well-known assignment constraints forcing each container to be stowed only in one ship location and each location to have at most one container.

The capacity constraint (5) establishes that the total weight of all containers cannot exceed the maximum weight capacity Q of the containership.

Constraints (6)–(11) are the size conditions. In particular, (6) and (7) force, respectively, $40'$ containers to be stowed in even bays and $20'$ container to be stowed in odd bays, while (8) and (9) make unfeasible the stowage of $20'$ containers in those odd bays that are contiguous to even locations already chosen for stowing $40'$ containers, and inversely; (10) and (11) prevent $20'$ containers being positioned over $40'$ ones.

Weight constraints (12) and (13) say that a stack of at most three containers of either $20'$ or $40'$ cannot exceed values MT and MF, respectively, that usually correspond to 45 and 66 t; note that such constraints verify the corresponding tolerance value in all occupied tiers in the same row and bay for any possible stack of three containers, as it is required by the weight constraints.

Constraints (14) force heavier containers not to be put over lighter ones. It is worth noting that constraints (14) also avoid the stowing of both $20'$ and $40'$ containers over empty locations. The destination constraints (15) avoid positioning containers that have to be unloaded first below those containers that have a later destination port.

Constraints (16) and (17) are the horizontal and cross equilibrium conditions, stating that the difference in weight between the anterior and the posterior bays and between the left and right side must be at most Q_2 and Q_1 tons respectively. Finally, in (18) the binary decision variables of the problem are defined.

In this formulation of the problem is assumed that the ship starts its journey in one port, and successively visits a given number of other ports where only unloading operations are allowed.

Moreover, it is assumed that the number of containers to load on board is not greater than the number of available locations; this means that we are not concerned with the problem of selecting some containers to be loaded among all and that the capacity constraint (5) is only related to the maximum weight available for all containers.

3 A Loading Procedure

In this section, we describe a new procedure to generate a relaxed initial solution for MBPP. This initial solution is obtained by a loading heuristic (LH), which is described in Algorithm 1. This proposal consider only the constraints of size, weight and destination of the problem, that is, the constraints 2–15 of the 0/1 Linear Programming (LP) model proposed in Ambrosino et al. [1], which was described in Sect. 2.1.

1. Let $C' = \{C_1, C_2, ..., C_h, ..., C_D\}$ be a partition of C, where C_h denotes the set of containers having as the destination port h. Split C_h according to the type of containers, that is 20' and 40', thus obtaining sets C_{hT} (Twenty) and C_{hF} (Forty), respectively.
2. Split the total locations in the containership in four quadrants.
3. For each destination h, starting from the last one (D) back to 1, assign first containers belonging to C_{hT} as follows: {the cycle of the Lines 3 to 15 is repeated for containers belonging to C_{hF}}
4.   Sort C_{hX} in decreasing order of the weights, where $X \in \{T, F\}$.
5.   For $cont=1$ to $|C_{hX}|$
6.   Select a quadrant according to stability conditions.
7.   For $k=1$ to K
8.   For $j=1$ to J
9.   For $i=1$ to O {O for C_{hT} and E for C_{hF}}
10.   If location (i,j,k) is available, assign container $c \in C_{hX}$ to location (i,j,k), then set $C_{hX} = C_{hX} \setminus \{c\}$, $cont=cont-1$, $k=K$, $j=J$ and $i=O/E$ {According to C_{hX}}.
11.   End For
12.   End For
13.   End For
14.   If the container was not assign go to step 6, and would try to assign it in other quadrant.

15.   End For

The operation of the Algorithm 1 can be divided into two phases. In the initial phase (Line 1 and 2), the set of containers C is partitioned in $2 \times D$ subsets (Line 1), where D is the total number of destination ports. Each subset is identified by C_{hX}, where $h = 1, ..., D$ and $X \in \{T, F\}$.

Line 2 shows that the total locations of the containership are partitioned in four quadrants. Each quadrant contains all locations whose address is formed by:

- First quadrant: a number of bay between 1 and $\lfloor I/2 \rfloor$, an odd row and any tier.
- Second quadrant: a number of bay between 1 and $\lfloor I/2 \rfloor$, an even row and any tier.
- Third quadrant: a number of bay between $\lfloor I/2 \rfloor$ and I, an odd row and any tier.
- Fourth quadrant: a number of bay between $\lfloor I/2 \rfloor$ and I, an even row and any tier.

The last phase (Lines 4–15), the algorithm tries to load containers into the containership. First, containers belonging to C_{hT} are assigned, and subsequently, this procedure is repeated again to load containers belonging to C_{hF}. In this procedure, first the containers of each subset is sorted in decreasing order of the weights (Line 4). Then, the largest number of containers to locations available is trying to assign (Lines 5–15), so that the heaviest are loaded first.

The selection of a quadrant is according to stability conditions, and they follow the next order: first, fourth, second and third quadrant (Line 6).

If the container was not assigned, then go to Line 6, and try to assign it in other quadrant.

4 Experimental Results

In this section, the performance of the LH algorithm is tested. In the following subsections, we describe the test instances, experimental environment and the performance analysis.

4.1 Test Cases

The test cases are composed by four characteristic sets:

1. Container set
2. Containership characteristics
3. Tolerances
4. Loading times

In order to validate our heuristic approach, we generate two different types of container sets. Table 1 shows the first one, called pseudo-random, which was obtained by fixing some parameters; and Table 2 shows the second type, named random, which was generated without fixing any parameters. Both of them were constructed according to the format and conditions defined in [7].

Tables 1 and 2 report the characteristics of the mentioned set of containers, which show the total number of containers, expressed in TEU and absolute number (n); the numbers of containers whose type is $20'$ (T) and $40'$ (F), respectively; the number of containers for three classes of weight (L: low, M: medium, H: high); and the partition of containers by destination. Three classes of weight are considered, namely low (from 5 to 15 t), medium (from 16 to 25 t) and high containers (from 26 to 32 t).

The set of pseudo-random containers was generated according to the following specifications. Each row in the Table 1 corresponds to a specific instance. In each instance, the first $|T|$ containers of set C are of $20'$ and the following $|F|$ containers are of $40'$. In relation to the characteristic of *Weight*, the first $|L|$ containers of the set C are light-weight, the next $|M|$ are medium-weight and the last $|H|$ are high-weight. Finally, the characteristic of *Destination* of the containers of the set C, are established in the same sequence that *Weight* characteristic.

These instances concern a small size containership, with a maximum capacity of 240 TEU, composed of 12 odd bays, 6 even bays, 4 rows and 5 tiers (3 in the hold and 2 in the upper deck, respectively). Table 3 shows the loading times for a small containership.

The tolerances for the set of pseudo-random instances (see Table 1) are the following. The maximum horizontal weight tolerance (Q_1) was fixed to 18 % of the total weight of all the containers to be loaded. While the maximum cross weight tolerance (Q_2) was fixed to 9 % of the total weight, expressed in tons.

Table 1 Containers for the set of small-sized instances (pseudo-random)

Instance	TEU	n	Type (n)		Weight (n)			Destination (n)		
			T	F	L	M	H	D_1	D_2	D_3
1	69	50	31	19	23	25	2	23	27	0
2	83	60	37	23	26	32	2	27	33	0
3	85	65	45	20	30	33	2	31	34	0
4	88	65	42	23	29	34	2	31	34	0
5	90	70	50	20	31	37	2	30	40	0
6	90	75	60	15	35	38	2	32	43	0
7	93	65	37	28	30	33	2	31	34	0
8	93	70	47	23	29	39	2	32	38	0
9	93	70	47	23	31	36	3	25	20	25
10	94	74	54	20	34	38	2	25	25	24

Table 2 Containers for the set of small-sized instances (random)

Instance	TEU	n	Type (n)		Weight (n)			Destination (n)		
			T	F	L	M	H	1	2	3
1	138	100	62	38	46	50	4	47	53	0
2	165	120	75	45	52	64	4	55	65	0
3	170	130	90	40	60	66	4	62	68	0
4	175	130	85	45	58	68	4	62	68	0
5	180	140	100	40	62	74	4	61	79	0
6	180	150	120	30	70	76	4	65	85	0
7	185	130	75	55	60	66	4	62	68	0
8	185	140	95	45	58	78	4	65	75	0
9	185	140	95	45	62	73	5	50	40	50
10	188	148	108	40	68	76	4	50	50	48

Table 3 Loading times for the set of small-sized instances, the times are expressed in 1/100 of minute, taken from [3]

Tier	Row			
	3	1	2	4
2	240	250	260	270
4	230	240	250	260
6	220	230	240	250
82	210	220	230	240
84	200	210	220	230

Regarding MT, which is the maximum weight tolerance of three containers stack of 20′ was fixed to 45 t and; MF (the maximum weight tolerance of three containers stack of 40′) was fixed to 66 t. It is important to point out that MT and MF were used in both set of containers (pseudo-random and random).

Table 4 LH performance for the pseudo-random instances set

Inst.	Relaxed 0/1 LP model		LH	
	Obj_1	Time	Obj_2	Time
1	11,970	4.081	12,180	0.001
2	14,590	23.975	14,430	0.001
3	15,610	14.723	15,510	0.001
4	15,650	19.874	15,610	0.002
5	16,790	29.659	17,060	0.004
6	18,000	31.245	18,200	0.051
7	15,200	8.505	15,840	0.005
8	16,760	25.651	16,800	0
9	16,820	36.064	16,920	0
10	17,980	40.735	17,780	0.001
Avg	15,937	23.451	16,033	0.0066

4.2 Infrastructure

The following configuration corresponds to the experimental conditions:

- *Software*: Operating system Microsoft Windows 7 Home Premium; Java programming language, Java Platform, JDK 1.6; and integrated development, NetBeans 7.2. Solver Gurobi 5.0.1.
- *Hardware*: Computer equipment with processor Intel (R) Core (TM) i5 CPU M430 2.27 GHz and RAM memory of 4 GB.'

4.3 Performance Analysis

Tables 4 and 5 show the comparison of the results obtained by two solution methods for the set of small-sized instances.

In Table 4, we can observe the loading heuristic performance for the pseudo-random set of containers. This approach is able to load 100 % of the total number of containers to be loaded. The objective values are given in 1/100 of minute (Obj), and CPU time, expressed in seconds (Time). The results are divided into two relevant columns according to solutions found by relaxed 0/1 LP model and LH algorithm. The computational execution of relaxed 0/1 LP model were stopped when the first feasible solution is reached by the commercial software Gurobi.

As we can see, in the last row of the Table 4, our proposed procedure loses a little bit solution quality with respect to the relaxed 0/1 LP model. Nevertheless, the running time for our heuristic is really small, much less than one second in average for the entire set of tested instances.

Table 5 shows the results for the random set of containers. The *C* column indicates the total number of containers to be loaded, the *IC* column reports the containers loaded by the LH, and the last column indicates the total loading time reported by LH. The objective values are given in 1/100 of minute (Obj).

Table 5 LH performance for the random instances set

Instance	C	IC	Obj
1	100	95	23,070
2	120	104	25,420
3	130	113	27,830
4	130	106	25,990
5	140	98	24,570
6	150	110	27,300
7	130	96	23,790
8	140	113	27,820
9	140	121	29,280
10	148	129	30,860
Percent	100 %	81.7 %	–

We are not comparing our approach against the initial solution obtained by the LP solver used because it takes a lot of time to get a relaxed initial solution. However, we can see that our proposed approach is able to load 81.7 % of the total number of containers to be loaded.

5 Conclusions

In this work, we developed a heuristic algorithm to generate an initial solution. In order to build an initial solution, the constraints of size, weight and destination are considered.

The experimental results reveal that the heuristic achieves to load 100 % of the containers in pseudo-random instances and 81.7 % in random instances.

The loading heuristic produces feasible solutions in a short running time, so that, it could be applied to solve real-world instances.

As a future work, it is considered to incorporate the two-phases of a hierarchical approach. In addition, we could consider using other intelligent computing techniques, like the ones in [8–13].

Acknowledgments This work was partially financed by CONACYT, PROMEP and DGEST. We also thank Gurobi for allowing us to use their optimization engine.

References

1. Ambrosino, D., Sciomachen, A., Tanfani, E.: Stowing a containership: the master bay plan problem. Transp. Res. Part A: Policy Pract. **38**, 81–99 (2004)
2. Delgado, A., Jensen, R.M., Janstrup, K., Rose, T.H., Andersen, K.H.: A constraint programming model for fast optimal stowage of container vessel bays. Eur. J. Oper. Res. **220**, 251–261 (2012)

3. Ambrosino, D., Anghinolfi, D., Paolucci, M., Sciomachen, A.: A new three-stepheuristic for the master bay plan problem. Marit. Econ. Logistics **11**, 98–120 (2009)
4. Hernández, P.H., Cruz-Reyes, L., Melin, P., Mar-Ortiz, J., Huacuja, H.J.F., Soberanes, H.J.P., Barbosa, J.J.G.: An ant colony algorithm for improving ship stability in the containership stowage problem. In: Castro, F., Gelbukh, A., González, M. (eds.) Advances in Soft Computing and Its Applications. Lecture Notes in Computer Science, vol. 8266, pp. 93–104. Springer, Berlin (2013)
5. Ambrosino, D., Sciomachen, A., Tanfani, E.: A decomposition heuristics for the container ship stowage problem. J. Heuristics **12**, 211–233 (2006)
6. Avriel, M., Penn, M., Shpirer, N.: Container ship Stowage problem: complexity and connection to the coloring of circle graphs. Discrete Appl. Math. **103**(1), 271–279 (2000)
7. Cruz-Reyes, L., Hernández, P., Melin, P., et al.: Constructive algorithm for a benchmark in ship stowage planning. In: Castillo, O., Melin, P., Kacprzyk, J. (eds.) Recent Advances on Hybrid Intelligent Systems, pp. 393–408. Springer, Berlin (2013)
8. Melin, P., Olivas, F., Castillo, O., Valdez, F., Soria, J., Valdez, M.: Optimal design of fuzzy classification systems using PSO with dynamic parameter adaptation through fuzzy logic. Expert Syst. Appl. **40**(8), 3196–3206 (2013)
9. Montiel, O., Camacho, J., Sepúlveda, R., Castillo, O.: Fuzzy system to control the movement of a wheeled mobile robot. Soft Comput. Intell. Control Mobile Robot. **318**, 445–463 (2011)
10. Montiel, O., Sepulveda, R., Melin, P., Castillo, O., Porta, M. A., Meza, I.M.: Performance of a simple tuned fuzzy controller and a PID controller on a DC motor. In: FOCI 2007, pp. 531–537 (2007)
11. Sombra A., Valdez F., Melin P., Castillo O.: A new gravitational search algorithm using fuzzy logic to parameter adaptation. In: IEEE Congress on Evolutionary Computation 2013, pp. 1068–1074 (2013)
12. Valdez F., Melin P., Castillo O.: Evolutionary method combining particle swarm optimization and genetic algorithms using fuzzy logic for decision making. In: Proceedings of the IEEE International Conference on Fuzzy Systems, 2009, pp. 2114–2119 (2009)
13. Valdez F., Melin P., Castillo O.: Fuzzy logic for parameter tuning in evolutionary computation and bio-inspired methods. In: MICAI (2) 2010, pp. 465–474 (2010)

Quality-Assessment Model for Portfolios of Projects Expressed by a Priority Ranking

S. Samantha Bastiani, Laura Cruz-Reyes, Eduardo Fernandez, Claudia Gómez and Gilberto Rivera

Abstract Organizations need to make decisions about how to invest and manage the resources to get more benefits, but, commonly the organization's resources are not enough to support all project proposals. Thus, the decision maker (DM) wants to select the portfolio with the highest contribution to the organizational objectives. But in many practical cases, to know exactly the benefits associated to implement each proposal is too difficult, therefore it is questionable the issue of evaluating portfolio quality in these conditions In order to face these uncertainty situations, the DM usually ranks the applicant projects according to his/her preferences about an estimated impact of each portfolio. However, a correct modeling of the quality of the portfolio is indispensable to develop a model of coherent optimization to the ranking given by the DM. In the literature, this type of problems has been scantily approached in spite of being present in many practical situations of assignment of resources. In this Chapter we propose a quality model of portfolio and an algorithm that solves it. The experimental results show that the algorithm that includes our model offers benefits to the decision maker, and his advantages highlighted with respect to the related works reported in the state of the art.

S. Samantha Bastiani · L. Cruz-Reyes · C. Gómez (✉) · G. Rivera
Instituto Tecnológico de Ciudad Madero, 1ro. de Mayo Y Sor Juana I. de La Cruz S/N CP,
89440 Tamaulipas, México
e-mail: cggs71@hotmail.com

S. Samantha Bastiani
e-mail: b_shulamith@hotmail.com

L. Cruz-Reyes
e-mail: lcruzreyes@prodigy.net.mx

G. Rivera
e-mail: riveragil@gmail.com

E. Fernandez
Universidad Autónoma de Sinaloa, Facultad de Ingeniería Culiacán, Josefa Ortiz de
Domínguez S/N Ciudad Universitaria, Culiacán, Sinaloa, Mexico
e-mail: eddyf@uas.edu.mx

O. Castillo et al. (eds.), *Recent Advances on Hybrid Approaches for Designing
Intelligent Systems*, Studies in Computational Intelligence 547,
DOI: 10.1007/978-3-319-05170-3_39, © Springer International Publishing Switzerland 2014

1 Introduction

Usually, methods for scoring, ranking or evaluating projects contain some way of aggregating multicriteria descriptions of projects (e.g. [1]). The validity of these methods depends on how accurately ranking and scores reflect DM preferences over portfolios. In fact, the portfolio's score should be a value function on the portfolio set, but this requires a proper elicitation of DM preferences inside the portfolio's space.

Ranking is also used in problems where a "Participatory budgeting" is involved. "Participatory budgeting" can be defined as a public space in which government and society agree on how to adapt priorities of citizenship to public policy agenda. The utility of these participatory exercises is that the government obtains information about priorities of the participating social sectors, and might identify programs with a consensual benefit.

Ranking of public actions given by the participants is an expression of their preferences on projects, not on portfolios. Let us assume that a method of integrating the individual ranking on a collective order is applied, as the Borda score or a procedure based on the exploitation of collective fuzzy preference relations (e.g. [2–4]). With the obtained group order, the DM has more information about social preferences regarding the different actions to budget. The DM should use that information to find the best portfolio.

Project portfolio selection is one of the most difficult, yet most important decision-making problems faced by many organizations in government and business sectors. To carry out the project selection, the DM usually starts with limited information about projects and portfolios. His/her time is often the most critical scarce resource. In multiple situations the DM feels more comfortable employing simple decision procedures, because of lack of available information, lack of time, aversion to more elaborated decision methods, and even because of his/her fondness for established organizational practices. Cooper et al. [5] argues about popularity of scoring and ranking methods in R&D project management in most American enterprises. Methods of scoring and ranking are used by most of the government organizations that fund R&D projects.

This chapter proposes a genetic algorithm called ESPRI (Evolutionary algorithm for solving the public portfolio problem from ranking information), which includes: (1) a local search which is a process of improving a solution and (2) a new model for project portfolio selection, which makes use of the ranking of a set of projects according to the preferences of a decision maker. This chapter is structured as follows: the background is briefly described in the second section. It is also presented the algorithm, which includes a diversification strategy that is a multiobjective local search, which leads to optimize the proposed impact model. Section 3 presents such model, followed by a description of the solution algorithm (Sect. 4). Finally in Sect. 5 we give empirical evidence that supports our results and some conclusions are given in Sect. 6.

2 Background

A project is a temporary process, singular and unrepeatable, pursuing a specific set of objectives [6]. In this chapter, it is not considered that the projects can be divided into smaller units, such as assignments or activities.

A portfolio consists of a set of projects that can be performed in the same period of time [6]. Due to that, projects in the same portfolio can share the available resources of the funding organization. Therefore, it is not sufficient to compare the projects individually; the DM should compare project groups to identify which portfolio makes a major contribution to the objectives of the organization.

Selecting portfolios integrated by properly selected projects is one of the most important decision problems for public and private institutions [7, 8]. The DM is in charge for selecting a set of projects to be supported [9–11].

2.1 Related Works

Gabriel et al. [12] proposed an additive function as a portfolio's score. This function aggregates the rank of projects. The simplest model is to assign project priorities in correspondence to the project rank (the highest priority is assigned to the best ranked project and so on). The portfolio's score is the sum of priorities associated with its projects. 0–1 mathematical programming is used to maximize the score.

Mavrotas et al. [13] proposed an additive function depending on a project's augmented score. This augmented score is built according to the project's specific rank. The augmented score of a project A holds that no combination of projects with worse ranking positions and a lower total cost can have a score bigger than A. The augmented score is obtained by solving a knapsack problem for each project. The portfolio's score is the sum of its projects' augmented scores.

In order to illustrate limitations of those methods, consider the following example: Let us suppose a 20-project strict ranking; priority 20 is assigned to the best project; 19 is assigned to the second one; 1 is assigned to the worst ranked project. Considering a score given by the sum of priorities, the portfolio containing the best and the worst projects (score $= 21$) should be indifferent to the portfolio containing the second best project and the second-to-last one (score $= 21$). The DM could hardly agree with such a proposition.

It is necessary to compare impact of possible portfolios in order to find the best one. The information provided by the simple project ranking is very poor for portfolio optimization purposes. Hence, some proxy impact measures should be defined. This problem was approached by [13] under the assumption that "the portfolio impact on a DM mind is determined by the number of supported projects and their particular rank". If project A is clearly better ranked than B, then A is admitted to have "more social impact" than B.

The DM should consider this information from the ranking. The appropriateness of a portfolio is not only defined by the quality of the included projects, but also by the amount of supported projects. The purpose is to build a good portfolio by increasing the number of supported projects and controlling the possible disagreements regarding DM preferences, which are assumed as incorporated in input ranking. For Fernandez and Olmedo [14], a discrepancy is the fact that given a pair of projects (A, B) (being B worse ranked than A), B belongs to the portfolio and A does not. Different categories of discrepancy are defined according to the relative rank of the concerning projects. Some discrepancies might be acceptable between the information provided by the ranking and the decisions concerning the approval (hence supporting) of projects, whenever this fact increases the number of projects in the portfolio. However, this inclusion should be controlled because the admission of unnecessary discrepancies is equivalent to underestimating the ranking information. A multi-objective optimization problem is solved by using NSGA-II (Non-dominated Sorting Genetic Algorithm), in which the objective functions are the number of supported projects and the number of discrepancies (separately in several functions, in regard to the importance of each kind of discrepancy) [14].

Main drawback: a portfolio quality measure model based solely on discrepancies and the number of supported projects is highly questionable; more information is required about project impacts. If DM thinks in terms of priority and relatively important projects, their numbers and ranks should be considered.

2.2 NSGA-II (Non-Dominated Sorting Genetic Algorithm-II)

The research problem is not traditional and its solution implicates the use of techniques of multi-objective optimization, which were implemented by previously identified multi-objective evolutionary algorithms (MOEAs). Several works have reported successful results with this kind of algorithms [15–17].

One of the most used algorithms for solving multi-objective problems is the algorithm NSGA-II, which is shown in Fig. 1.

The procedure "Fast non-dominated-sort" (shown in Fig. 2) optimizes the algorithm NSGA-II. Finally this algorithm has a diversity indicator whose evaluation is shown in Fig. 3. This indicator favors solutions in less populated regions of the search space; these solutions will be advantaged by the selection mechanism [15].

2.3 Multi-Objective Local Search

A multi-objective local search is a strategy that improves an initial solution C. Which have demonstrated good performance. One of the local searches used to multi-objective problems is reported in [18, 19].

Fig. 1 Structure of the
algorithm NSGA-II

```
1    Repeat
2        R_t = P_t ∪ Q_t
3        F = fast-nondominated-sort (R_t)
4    until |P_{t+1}| < N
5        crowding-distance-assignment (F)
6        P_{t+1} = P_{t+1} ∪ F_i
7    Sort (P_{t+1}, ≥n)
8        P_{t+1} = P_{t+1} [0:N]
9        Q_{t+1} = make-new-pop(P_{t+1})
10       t = t+1
11   Until (stop condition)
```

Fig. 2 Structure of the
algorithm fast non-
dominated-sort

```
1    for each p ∈ P
2        for each q ∈ P
3            if(p ≺ q) then
4                S_p = S_p ∪ {q}
5            else if(q ≺ p) then
6                n_p = n_p + 1
7        if n_p = 0 then
8            F_1 = F_1 ∪ {p}
9        i = 1
10       while F_i ≠ 0
11           H = 0
12           for each p ∈ F_i
13               for each q ∈ S_p
14                   n_q = n_q - 1
15                   if n_q = 0 then H = H{q}
16           i = i+1
17           F_i = H
```

Fig. 3 Algorithm structure
of the crowding-distance-
assignment

```
1    l = |I|
2    for each i, set I [i]_distance = 0
3    for each objective m
4            I = sort (I, m)
5                for i = 2 to(l-1)
6    I[i]_distance = I[i]_distance + (I[i+1]).m − I[i-1].m)
```

This search explores regions near to the best known solutions by a simple scheme consisting of randomly selecting ls projects, and generating all possible combinations of them for each solution in the non-strictly-outranked frontier. Small values for ls provoke behavior that is too greedy, whereas large values produce intolerable computation times. In our experiments we obtained a good balance between these by using $ls = \lceil \ln N \rceil$.

3 Proposed Model

The new model overcomes the idea proposed by Fernandez [14, 20]. In this model the optimization is performed over indicators, which positively provide indirect information on the impact of the portfolio.

This model handles three categories: priority, satisfiability and acceptability, besides incorporating a ranking in descending order, based on the original ranking ascending.

The description of the model is present below:

$$I_1(\overrightarrow{x}) = \sum_{i=1}^{n} x_i F(i, 1) \quad F(i, 1) = \begin{cases} 1 & \text{if } i \in G_1, \\ 0 & \text{otherwise,} \end{cases} \tag{1}$$

where the binary variable x_i indicates whether the i-th project belongs to the portfolio or does not. That is, $x_i = 1$ if the i-th project belongs to portfolio; otherwise $x_i = 0$. Note that function I_1 counts how many projects belonging to the priority category (Group 1) are contained in the portfolio.

$$I_2(\overrightarrow{x}) = \sum_{i=1}^{n} x_i(n - i) F(i, 1), \tag{2}$$

The better the position of projects, the greater value of I_2. This objective is a proxy variable that measures how good are the projects according to their category.

$$I_3(\overrightarrow{x}) = \sum_{i=1}^{n} x_i F(i, 2) \quad F(i, 2) = \begin{cases} 1 & \text{if } i \in G_2, \\ 0 & \text{otherwise,} \end{cases} \tag{3}$$

where the binary variable x_i indicates, if the i-th project belongs to the portfolio or does not. That is, $x_i = 1$ if the i-th project belongs to the portfolio; otherwise $x_i = 0$. Note that function I_3 counts how many projects belonging to the satisfactory category (Group 2) are contained in the portfolio.

$$I_4(\overrightarrow{x}) = \sum_{i=1}^{n} x_i(n - i) F(i, 2), \tag{4}$$

Equation (4) measures in proxy way, how good the supported satisfactory projects are.

Similarly we define

$$I_5(\overrightarrow{x}) = \sum_{i=1}^{n} x_i F(i,3) \quad F(i,3) = \begin{cases} 1 & \text{if } i \in G_3, \\ 0 & \text{otherwise,} \end{cases} \tag{5}$$

Function I_5 counts how many projects belonging to the acceptable category (Group 3) are contained in the portfolio.

Finally

$$I_6 = \sum_{i=1}^{n} x_i. \tag{6}$$

represents the portfolio cardinality.

We assume that the DM "feels" the potential impact of the portfolio in terms of the numbers of projects for category and the positions they occupy.

The best portfolio should be the best solution of the multi-objective problem:

$$Max\,(I_1,\ I_2,\ I_3,\ I_4,I_5,\ I_6) \\ C \in R_F \tag{7}$$

where R_F is the feasible region determined by budgetary constraints.

In this case the DM, based on his/her preferences, should select the best portfolio. DM "feels" the potential impact of the portfolio in terms of the number of supported projects in each category and their position.

4 Proposed Genetic Algorithm

The algorithm developed in this research work is called ESPRI. It is inspired by the NSGA-II algorithm developed by Deb et al. [15], which successfully manages exponential complexity [13]. ESPRI searches the portfolios that optimizes Eq. (7).

For illustrating ESPRI algorithm process, a set of n projects is taken as example, with its respective total budget as well as the budget for each project. Previously, such projects were ranked according to DM preferences. Heuristically, the projects were separated in three categories: priority, satisfiability and acceptability. Once this process is complete, the algorithm generates random portfolios, which form the NSGA-II initial population.

Later, the following procedures are applied: fast non-dominated sort, crowding distance and the remaining genetic operators. Finally, the algorithm shows the estimated non-dominated solutions. Figure 4 present ESPRI algorithm.

Fig. 4 Structure of the
ESPRI algorithm

```
1    Repeat
2        R_t = P_t∪Q_t
3        quality assessment: Impact Indicators Model
4        Multi-objetive Local Search (R_t)
5        F= fast non-dominated sort (R_t)
6        while: |P_t+1|< N
7            crowding-distance-assignment (F)
8        P_t+1=P_t+1∪ F
9        Sort (P_t+1, ≥n)
10       P_t+1=P_t+1[0:N]
11       Q_t+1=Do-new-pop(P_t+1)
12       t=t+1
13   Until (stop condition)
```

5 Computational Experiments

This section describes the experiments with the proposed evolutionary algorithm ESPRI. The aim of this experiment is to study the capacity of the proposed indicator model, as well as to compare ESPRI solutions against the state of the art.

5.1 Experimental Environment

The following configuration corresponds to the experimental conditions for tests described in this chapter:

1. Software: Operating System, Mac OS X Lion 10.7.5 (11G63b), Java Programming Language, Compiler NetBeans 7.2.1.
2. Hardware: Intel Core i7 2.8 GHz CPU and 4 GB of RAM.
3. Instances: was taken from state of the art, reported in Fernandez [14, 20].
4. Performance measure: In this case the performance is measured through the aforementioned objectives.

5.2 An Illustrative Example

Within the public portfolio problem, an instance for ranking strategy consists of four attributes: Id, total amount to be distributed, project cost and ranking. The test ranking denoted by P example is taken from the state of the art [14], which works with an instance that consists of 100 projects. The projects are separated into three categories: priority, satisfiability and acceptability, approximately uniform. ESPRI

Table 1 Instance of 100 projects

P	Budget	P	Budget	P	Budget	P	Budget
1	84.00	26	31.25	51	27.50	76	46.50
2	124.50	27	26.50	52	41.25	77	44.00
3	129.75	28	36.25	53	29.50	78	25.75
4	147.75	29	50.00	54	25.25	79	38.25
5	126.00	30	34.75	55	40.00	80	40.75
6	137.25	31	48.25	56	30.75	81	42.75
7	96.00	32	46.00	57	39.00	82	43.00
8	84.75	33	36.75	58	44.50	83	32.25
9	93.00	34	34.00	59	47.50	84	37.75
10	121.50	35	26.00	60	36.00	85	44.75
11	102.75	36	31.75	61	28.50	86	27.00
12	141.75	37	29.75	62	29.00	87	39.50
13	105.75	38	37.25	63	30.25	88	30.00
14	98.25	39	26.75	64	49.50	89	37.50
15	101.25	40	43.75	65	33.00	90	49.00
16	83.25	41	27.25	66	38.50	91	41.75
17	109.50	42	47.00	67	33.50	92	39.25
18	107.25	43	41.00	68	48.50	93	34.50
19	135.00	44	30.50	69	35.00	94	49.75
20	97.50	45	45.25	70	28.75	95	48.00
21	127.50	46	26.25	71	25.50	96	29.25
22	114.00	47	45.50	72	40.25	97	47.75
23	106.50	48	44.25	73	38.75	98	42.25
24	94.50	49	48.75	74	46.75	99	46.25
25	43.50	50	33.25	75	37.00	100	39.75
						Total	5,542.00

algorithm was run 20 times; in each run 200 iterations were performed. The algorithm solved an instance of 100 projects with a total amount of 2.5 billion to be distributed; this instance can be seen in Table 1.

ESPRI algorithm was set as: one-point crossover, strategy of mutation is swap with mutation probability of 0.5, population size $= 200$ and number of generations $= 100$. Note that if the resources were distributed strictly following the ranking order, the resulting portfolio would have 22 projects, all belonging to the priority category.

Table 2 shows a representative sample of the approximation to the Pareto frontier, which our proposal might reach. The data marked represent a set of solutions preferred by a DM interested in increasing the number of priority projects that are supported, as well as the total number of projects, but with emphasis on those considered satisfactory (category 2).

Table 2 Experimental results obtained with ESPRI algorithm and multi-objective local search

Objectives					
I_1	I_2	I_3	I_4	I_5	I_6
26	2168	8	466	3	37
21	1764	5	283	12	38
25	2077	13	733	0	38
18	1484	10	547	19	47
24	**2008**	**14**	**839**	**1**	**39**
22	1846	6	332	11	39
23	1936	6	360	9	38
22	1796	19	1075	0	41
24	**1963**	**12**	**696**	**5**	**41**
24	**2003**	**10**	**552**	**7**	**41**
20	1651	8	442	15	43
24	**1955**	**14**	**821**	**2**	**40**

Table 3 Experimental results obtained with ESPRI algorithm and multi-objective local search

Paper	Max $(I_1, I_2, I_3, I_4, I_5, I_6)$
[9]	24, 1995, 0, 0, 0, 24
Our proposal	24, 2003, 10, 552, 7, 41
Our proposal	24, 1963, 12, 696, 7, 41

Table 3 shows the results on the instance that was used in [9]. As can be seen, the estimated non-dominated solutions seem to be satisfactory for the DM. The solutions obtained by our proposal should be more preferred than the best solution in [9] because the last one contains less projects and less priority projects.

6 Conclusions

The proposed model of impact indicators of the portfolio can explore the solution space and generate potential best portfolios. This model reasonably represents DM preferences on portfolios under limited information about projects. The portfolios by ESPRI are more satisfactory than those obtained by the state-of-the-art procedures.

Above experiments were performed with the basic algorithm ESPRI provided solutions whose performance was similar to those reported in the literature [14, 20]. To improve the quality of these solutions, we integrated a multiobjective local search. The ESPRI algorithm in conjunction with this strategy improved the quality of the solutions, which indicates that the algorithm converges in areas where the previous version of the algorithm ESPRI had not touched. The solutions provided to the DM with this version of the algorithm ESPRI, could find a better compromise solution.

References

1. Henriksen, A.D., Traynor, A.J.: A practical R&D project selection scoring tool. IEEE Trans. Eng. Manage. **46**(2), 158–170 (1999)
2. Fernández, E., López, E., Bernal, S., Coello Coello, C.A., Navarro, J.: Evolutionary multiobjective optimization using an outranking-based dominance generalization. Comput. Oper. Res. **37**(2), 390–395 (2010)
3. Leyva, J.C., Fernandez, E.: A new method for group decision support based on ELECTRE-III methodology. Eur. J. Oper. Res. **148**(1), 14–27 (2003)
4. Macharis, C., Brans, J.P., Mareschal, B.: The GDSS PROMETHEE procedure. A PROMETHEE-GAIA based procedure for group decision support. J. Decis. Sys. **7**, 283–307 (1998)
5. Cooper, R., Edgett, S., Kleinschmidt, E.: Portfolio management for new product development: results of an industry practices study. R&D Manage. **31**(4), 361–380 (2001)
6. Carazo, A.F., Gómez, T., Molina, J., Hernández-Díaz, A.G., Guerreo, F.M., Caballero, R.: Solving a comprehensive model for multi-objective project portfolio selection. Comput. Oper. Res. **37**(4), 630–639 (2010)
7. Castro, M.: Development and implementation of a framework for I&D in public organizations. Master's thesis, Universidad Autónoma de Nuevo León (2007)
8. M, Caballero, R.: Solving a comprehensive model for multi-objective project portfolio selection. Comput. Oper. Res. **37**(4), 630–639 (2010)
9. García, R., Gilberto, Rivera, Claudia, Gómez, Laura, Cruz: Solution to the social portfolio problem by evolutionary algorithms. Int. J. Comb. Optim. Probl. Inf. **3**(2), 21–30 (2012)
10. Gabriel, S., Kumar, S., Ordoñez, J., Nasserian, A.: A multi-objective optimization model for project selection with probabilistic consideration. Socio Econ. Plann. Sci. **40**(4), 297–313 (2006)
11. Mavrotas, G., Diakoulaki, D., Koutentsis, A.: Selection among ranked projects under segmentation, policy and logical constraints. Euro. J. Oper. Res. **187**(1), 177–192 (2008). 2009
12. Fernández, E., Olmedo, R.: Public project portfolio optimization under a participatory paradigm. Appl. Computat. Intell. Soft Comput. Arc. **2013**, 4 (2013)
13. Deb, K.: Multi-objective optimization using evolutionary algorithms. Wiley, Chichester-New York-Weinheim-Brisbane-Singapore- Toronto (2001)
14. Cruz, L., Fernandez, E.R., Gomez, C.G., Rivera, G.: Multicriteria optimization of interdependent project portfolios with 'a priori' incorporation of decision maker preferences. Eureka-2013. Fourth International Workshop Proceedings (2013)
15. Pineda, A.A.S, Estrategias de Búsqueda Local para el problema de programación de tareas en sistemas de procesamiento paralelo. Thesis (2013)
16. Fernández, E., Luz Flerida, Félix, Mazcorro, Gustavo: Multi-objective optimization of an out- ranking model for public resources allocation on competing projects. Int. J Oper. Res. **5**(2), 190–210 (2009)
17. Coello Coello, C.A., Lamont, G.B., Van Veldhuizen, D.A.: Evolutionary algorithms for solving multi-objective problems. Genetic and Evolutionary Computation, 2nd edn. Springer, Berlin (2007)
18. Fernández, E., Navarro, J.: A genetic search for exploiting a fuzzy preference model of portfolio problems with public projects. Ann. OR **117**, 191–213 (2002)
19. Fernández Eduardo, R., Navarro Jorge, A., Olmedo Rafael, A.: Modelos y herramientas computacionales para el análisis de proyectos y la formación de carteras de I&D. Revista Iberoamericana de Sistemas, Cibernética e Informática **1**(1), 59–64 (2004)
20. Fernández, E., López, E., López, F., Coello Coello, C.A.: Increasing selective pressure towards the best compromise in evolutionary multiobjective optimization: the extended NOSGA method. Inf. Sci. **181**(1), 44–56 (2010)

Exact Methods for the Vertex Bisection Problem

Héctor Fraire, J. David Terán-Villanueva, Norberto Castillo García,
Juan Javier Gonzalez Barbosa, Eduardo Rodríguez del Angel
and Yazmín Gómez Rojas

Abstract In this chapter we approach the vertex bisection problem (VB), which is relevant in the context of communication networks. A literature review shows that the reported exact methods are restricted to solve particular graph cases. As a first step to solve the problem for general graphs using soft computing techniques, we propose two new integer-linear programming models and a new branch and bound algorithm (B&B). For the first time, the optimal solutions for an extensive set of standard instances were obtained.

1 Introduction

In the context of network communications, a network can be represented by a graph $G = (V, E)$, where V represents the nodes on the network and E represents the communication channels between nodes. Each vertex represents a device that

H. Fraire · J. David Terán-Villanueva (✉) · N. C. García · J. J. G. Barbosa · E. R. d. Angel ·
Y. G. Rojas
Instituto Tecnológico de Ciudad Madero, Ciudad Madero, Mexico
e-mail: david_teran00@yahoo.com.mx

H. Fraire
e-mail: automatas2002@yahoo.com.mx

N. C. García
e-mail: norberto_castillo15@hotmail.com

J. J. G. Barbosa
e-mail: jjgonzalezbarbosa@hotmail.com

E. R. d. Angel
e-mail: rodela22@hotmail.com

Y. G. Rojas
e-mail: yaz.gomezr@gmail.com

O. Castillo et al. (eds.), *Recent Advances on Hybrid Approaches for Designing*   567
Intelligent Systems, Studies in Computational Intelligence 547,
DOI: 10.1007/978-3-319-05170-3_40, © Springer International Publishing Switzerland 2014

stores and distributes information. Several tasks are carried out on a communication network. However, the most significant ones are: broadcast, accumulation and gossip [1].

- The *broadcast* task consists in divulge a message from one node to every node on the network.
- The *accumulation* task consists in storage information coming from other nodes.
- The *gossip* task is similar to the *broadcast* task; it divulges the information to all the nodes. However, instead of distributing the information from one node to all the nodes; the graph is divided into two sets of the same size (L and R) and the nodes that have a connection from set L to set R are identified ($\delta \in L | \forall u \in \delta \,\exists (u, v) \in E$, where $v \in R$). So the information is accumulated in δ and then divulged to set R. The elements in L and R are selected with the intention of minimizing the number of nodes that have a connection from L to R. This problem is an NP-Hard problem [2] known as the vertex bisection problem.

2 Problem Description

To define formally the vertex bisection problem (VB), we use the variables described in [3]:

$G = (V, E)$	A cyclic non-directed graph.
V	Vertices of the graph.
E	Edges of the graph.
φ	A labeling for the vertices of the graph.
Φ	Set of every possible labeling for the graph.
n	Number of vertex of the graph. $n = |V|$.
$L\left(\left\lfloor\frac{n}{2}\right\rfloor, \varphi, G\right)$	Set of vertex on the left side of the permutation (L).
$R\left(\left\lfloor\frac{n}{2}\right\rfloor, \varphi, G\right)$	Set of vertex on the right side of the permutation (R).
$\delta\left(\left\lfloor\frac{n}{2}\right\rfloor, \varphi, G\right)$	Set of vertices on L that have a connection with at least one vertex on R.

Given a non-directed graph $G = (V, E)$ and a labeling $\varphi \in \Phi$, the vertex bisection of the graph is defined as [4]:

$$VB(\varphi, G) = \left| \delta\left(\left\lfloor\frac{n}{2}\right\rfloor, \varphi, G\right) \right|.$$

To calculate the vertex bisection, we need to rearrange the graph on a linear way. Where the position of each vertex defines its label; and the label of all the vertices defines a permutation. This problem associates each permutation with an objective function value. Once a permutation is produced, a cut between the two elements in the middle of the permutation is carried out.

Every vertex that is on the left side of the cut belongs to set L, and every vertex that is on the right side of the cut belongs to set R. Those elements from L that have an edge that connects them with at least one element in R will be included in the set δ. Thus, the objective function value for that particular permutation will be the cardinality of the set δ.

2.1 Vertex Bisection Problem

Given a non-directed graph $G = (V, E)$, the vertex bisection problem consists in finding the permutation $\varphi' \in \Phi$ that minimizes the vertex bisection for G. This problem is defined as follows:

$$VB(\varphi', G) = \min_{\varphi \in \Phi}\{VB(\varphi, G)\}$$

2.2 Calculus of the Objective Function Value

The objective function value can be calculated by using a binary variable that indicates if any given vertex i belongs to the set L and has a connection with at least one vertex of the set R. As it was said, the connection between the two vertices (v_i, v_j) must exist in the set of edges E, where the values of i and j represents the position of the vertices in the permutation.

$$e_i = \begin{cases} 1 & \text{if} (v_i, v_j) \in E \quad \text{for } 1 \leq i \leq \lfloor \frac{n}{2} \rfloor, \lfloor \frac{n}{2} \rfloor < j \leq n \\ 0 & \text{otherwise} \end{cases}$$

Once set the binary variables for the elements in L of the current permutation, the vertex bisection value for that permutation $VB(\varphi, G)$ is calculated by adding all the binary variables.

$$VB(\varphi, G) = \sum_{i=1}^{\lfloor \frac{n}{2} \rfloor} e_i$$

3 Exact Models

3.1 Binary Variables

For the linear programming models proposed to solve the vertex bisection problem, three binary variables were used: x_u^p, $y_{u,v}^{p,q}$ and z_p.

The variable x_u^p is used to represent the current permutation and is defined as:

$$x_u^p = \begin{cases} 1, & p = \varphi(u) \\ 0, & \text{otherwise} \end{cases}.$$

The variable $y_{u,v}^{p,q}$ is used to model the connectivity of the graph and is defined as:

$$y_{u,v}^{p,q} = x_u^p \wedge x_v^q$$

I.e., if there exists an edge that connects the vertex u with the vertex v. We can also define this variable as follows:

$$y_{u,v}^{p,q} = \begin{cases} 1, & x_u^p = 1 \wedge x_v^q = 1 \\ 0, & \text{otherwise} \end{cases}$$

The last variable z_p represents the connections among the elements in the left side of the cut ($c = \lfloor n/2 \rfloor$) with the elements on the right side of the same cut, and is defined as:

$$z_p = \begin{cases} 1, & y_{u,v}^{p,q} = 1 \wedge p \leq \lfloor n/2 \rfloor \wedge q > \lfloor n/2 \rfloor \\ 0, & \text{otherwise} \end{cases}$$

3.2 Quadratic programming model

The quadratic programming model tries to minimize the objective function (1) and is defined using the previously defined variables x_u^p, $y_{u,v}^{p,q}$ y z_p.

$$\min \sum_{p=1}^{\lfloor n/2 \rfloor} z_p \tag{1}$$

This function is subject to the next constraints:

$$\sum_{p \in \{1,\dots,n\}} x_u^p = 1 \qquad \forall u = 1,\dots,n \tag{2}$$

$$\sum_{u \in \{1,\dots,n\}} x_u^p = 1 \qquad \forall p = 1,\dots,n \tag{3}$$

$$y_{u,v}^{p,q} = x_u^p x_v^q \qquad \forall (u, v) \in E \vee (v, u) \in E, p, q = 1, \ldots, n \qquad (4)$$

$$z_p \leq \sum_{q=\lfloor n/2 \rfloor+1}^{n} \sum_{u=1}^{n} \sum_{v=1}^{n} y_{u,v}^{p,q} \leq (\lfloor n/2 \rfloor) z_p \qquad \forall p = 1, \ldots, \lfloor n/2 \rfloor \qquad (5)$$

Constraints (2) and (3) verify that every label has a corresponding vertex and every vertex has a corresponding label. Thus, these constraints ensure that a feasible permutation is provided.

Constraint (4) verifies the connectivity (edges) between vertices $u, v \in E$ in function of their relative positions.

Constraint (5) identifies the set of vertices that belongs to L, that have at least one connection to a vertex in set R.

This value is the vertex bisection value, minimized by the objective function (1), of G for a particular permutation φ. This model functions as the basis for two linear integer programming models resulting from the linearization of the constraint (4).

3.3 Integer Linear Programming Model

The linear integer programming model (ILP1) minimizes the objective function shown in (6). This model is defined using the previously defined variables x_u^p, $y_{u,v}^{p,q}$ y z_p and the traditional linearization technique.

$$\min \sum_{p=1}^{\lfloor n/2 \rfloor} z_p \qquad (6)$$

This function is subject to the following constraints:

$$\sum_{p \in \{1,\ldots,n\}} x_u^p = 1 \qquad \forall u = 1, \ldots, n \qquad (7)$$

$$\sum_{u \in \{1,\ldots,n\}} x_u^p = 1 \qquad \forall p = 1, \ldots, n \qquad (8)$$

$$y_{u,v}^{p,q} \leq x_u^p \qquad \forall (u, v) \in E \vee (v, u) \in E, p, q = 1, \ldots, n \qquad (9)$$

$$y_{u,v}^{p,q} \leq x_v^q \qquad \forall (u, v) \in E \vee (v, u) \in E, p, q = 1, \ldots, n \qquad (10)$$

$$x_u^p + x_v^q \leq y_{u,v}^{p,q} + 1 \qquad \forall (u, v) \in E \vee (v, u) \in E, p, q = 1, \ldots, n \qquad (11)$$

$$z_p \leq \sum_{q=\lfloor n/2 \rfloor+1}^{n} \sum_{u=1}^{n} \sum_{v=1}^{n} y_{u,v}^{p,q} \leq \left(\left\lfloor \frac{n}{2} \right\rfloor\right) z_p \qquad \forall p = 1, \ldots, \lfloor n/2 \rfloor \qquad (12)$$

Constraints (7) and (8) are the same to constraints (2) and (3) of the quadratic model. These constraints ensure that a feasible permutation is provided.

Constraints (9), (10) and (11) are the result of the traditional linearization of the constraint (4) of the quadratic model, and have the purpose of identifying if u and v have an edge that communicates them.

Constraint (12) calculates the number of vertex in L that has a connection with at least one vertex in R.

3.4 Compacted Integer Linear Programming Model

The compacted integer linear programming model (ILP2), minimizes the objective function (13). It also uses the variables x_u^p, $y_{u,v}^{p,q}$ y z_p and the linearization proposed by Leo Liberti.

$$\min \sum_{p=1}^{\lfloor n/2 \rfloor} z_p \tag{13}$$

This function is subject to the following constraints:

$$\sum_{p \in \{1,\ldots,n\}} x_u^p = 1 \qquad \forall u = 1,\ldots,n \tag{14}$$

$$\sum_{u \in \{1,\ldots,n\}} x_u^p = 1 \qquad \forall p = 1,\ldots,n \tag{15}$$

$$\sum_{p=1}^{n} y_{u,v}^{p,q} = x_v^q \qquad \forall (u,v) \in E \vee (v,u) \in E, p,q = 1,\ldots,n \tag{16}$$

$$y_{u,v}^{p,q} = y_{v,u}^{q,p} \qquad \forall (u,v) \in E \vee (v,u) \in E, p,q = 1,\ldots,n \tag{17}$$

$$z_p \leq \sum_{q=\lfloor n/2 \rfloor+1}^{n} \sum_{u=1}^{n} \sum_{v=1}^{n} y_{u,v}^{p,q} \leq (\lfloor n/2 \rfloor) z_p \qquad \forall p = 1,\ldots,\lfloor n/2 \rfloor \tag{18}$$

Constraints (14), (15) and (18) are the same as (2), (3) and (5) in the quadratic model. However, constraint (16) and (17) are a compacted linearization for constraint (4) on the quadratic model.

3.5 Branch and Bound Algorithm

This method generates the permutation element by element. A partial evaluation is calculated until the permutation has reached half its size, i.e., because we need the complete set of elements in one side of the permutation in order to calculate the elements in the other side of it. Suppose that we have $n/4$ elements in the current permutation, the calculus of the objective function value requires to know which elements in L are connected to at least one element in R. Thus, we need to know which elements are in R, but R has not been constructed. However, if we know all the elements in L then $R = V \backslash L$, and so at least we need to know which elements are in L. Hence, at least half the permutation needs to be constructed in order to know L.

Algorithm 1 shows the recursive branch and bound algorithm, where line 3 has a cycle that repeats the whole algorithm depending on the depth of the tree (i) and the amount of time left to use (Lines 3 and 4). The algorithm produces half the permutation, and at that point (Lines 6–12) it evaluates the vertex bisection (Line 7). If a better solution is found (Line 8), the current partial solution and its best value are stored in *bestPerm* and *bestVal* respectively (Lines 9 and 10). If the depth of the tree (i) is lower than half the size of the permutation (Line 13), then the algorithm is executed again for $i + 1$ (line 14). Lines 5 and 16 are the ones that manage the queue, and that structure is the one that produces the permutations.

Algorithm 1. B&B algorithm

```
1    bestVal = maxValue;
2    GeneratePermutations(i){
3        for (j = 0; j<n − j + 1; j++){
4            if (time <maxTime){
5                perm [i] = TakeFromQueue();
6                if (i == ⌊n / 2⌋){
7                    currentVal = PartialEvaluation(perm);
8                    if (currentVal< bestVal){
9                        bestPerm = perm;
10                       bestVal = currentVal;
11                   }
12               }
13               if ( i<⌊n / 2⌋){
14                   GeneratePermutations(i + 1);
15               }
16               InsertInQueue(perm[i]);
17           }
18       }
19   }
20   finalSolution = GenerateCompleteSolution(bestPerm);
```

3.5.1 Permutation generation example

In order to generate the permutations a queue is used; at the beginning of the program, this queue is filled with the elements on the permutation. Figure 1 shows, in green circles, the number of the iteration where an element is taken from the queue. And in red circles the iteration when an element is inserted back into the queue. Figure 1 also contains a table for the iteration and the current elements in the queue for that iteration.

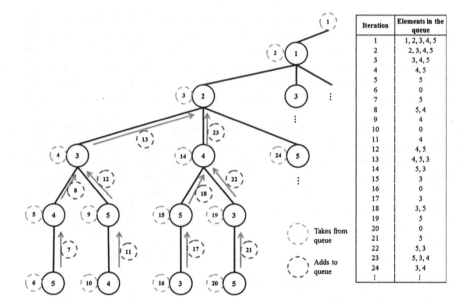

Fig. 1 Tree construction

Once the queue is depleted, then the same elements are inserted back into the queue but in the reverse order. Thus, a new branch can be explored.

4 Experimentation

All the experimentation was carried out on a laptop with the following characteristics:

- Operative system: Windows 7, service pack 1, 64 bit system.
- Intel core i7 at 2.9 GHz with 8 GB of RAM.
- Programming Language C#.
- Standard set of instances, Grid, Small, Tree, and Harwell-Boeing.

The linear integer programming models were implemented on Cplex version 10.9. The maximum amount of time for the Grid, Small and Tree sets of instances was set to 5 min. While, for the Harwell-Boeing, a time limit of 1 h was established. Table 1 shows the names of the instances used for the experimentation [5].

Table 1 Instances used in the experimentation

Small	p17_16_24_n.txt	p45_19_25_n.txt	p73_22_29_n.txt
	p18_16_21_n.txt	p46_19_20_n.txt	p74_22_30_n.txt
	p19_16_19_n.txt	p47_19_21_n.txt	p75_22_25_n.txt
	p20_16_18_n.txt	p48_19_21_n.txt	p76_22_30_n.txt
	p21_17_20_n.txt	p49_19_22_n.txt	p77_22_37_n.txt
	p22_17_19_n.txt	p50_19_25_n.txt	p78_22_31_n.txt
	p23_17_23_n.txt	p51_20_28_n.txt	p79_22_29_n.txt
	p24_17_29_n.txt	p52_20_27_n.txt	p80_22_30_n.txt
	p25_17_20_n.txt	p53_20_22_n.txt	p81_23_46_n.txt
	p26_17_19_n.txt	p54_20_28_n.txt	p82_23_24_n.txt
	p27_17_19_n.txt	p55_20_24_n.txt	p83_23_24_n.txt
	p28_17_18_n.txt	p56_20_23_n.txt	p84_23_26_n.txt
	p29_17_18_n.txt	p57_20_24_n.txt	p85_23_26_n.txt
	p30_17_19_n.txt	p58_20_21_n.txt	p86_23_24_n.txt
	p31_18_21_n.txt	p59_20_23_n.txt	p87_23_30_n.txt
	p32_18_20_n.txt	p60_20_22_n.txt	p88_23_26_n.txt
	p33_18_21_n.txt	p61_21_22_n.txt	p89_23_27_n.txt
	p34_18_21_n.txt	p62_21_30_n.txt	p90_23_35_n.txt
	p35_18_19_n.txt	p63_21_42_n.txt	p91_24_33_n.txt
	p36_18_20_n.txt	p64_21_22_n.txt	p92_24_26_n.txt
	p37_18_20_n.txt	p65_21_24_n.txt	p93_24_27_n.txt
	p38_18_19_n.txt	p66_21_28_n.txt	p94_24_31_n.txt
	p39_18_19_n.txt	p67_21_22_n.txt	p95_24_27_n.txt
	p40_18_32_n.txt	p68_21_27_n.txt	p96_24_27_n.txt
	p41_19_20_n.txt	p69_21_23_n.txt	p97_24_26_n.txt
	p42_19_24_n.txt	p70_21_25_n.txt	p98_24_29_n.txt
	p43_19_22_n.txt	p71_22_29_n.txt	p99_24_27_n.txt
	p44_19_25_n.txt	p72_22_49_n.txt	p100_24_34_n.txt
Tree	Tree_22_3_rot1.mtx	Tree_22_3_rot1.mtx	Tree_22_3_rot1.mtx
	Tree_22_3_rot2.mtx	Tree_22_3_rot2.mtx	Tree_22_3_rot2.mtx
	Tree_22_3_rot3.mtx	Tree_22_3_rot3.mtx	Tree_22_3_rot3.mtx
	Tree_22_3_rot4.mtx	Tree_22_3_rot4.mtx	Tree_22_3_rot4.mtx
	Tree_22_3_rot5.mtx	Tree_22_3_rot5.mtx	Tree_22_3_rot5.mtx
Grid	Grid3x3.mtx.rnd	grid_5.mtx.rnd	grid_7.mtx.rnd
	Grid4x4.mtx.rnd	grid_6.mtx.rnd	
Harwell–Boeing	bcspwr01.mtx.rnd	bcsstk01.mtx.rnd	
	bcspwr02.mtx.rnd	can___24.mtx.rnd	

Table 2 ILP1 results

ILP1					
Instances	# Instances	CPU Sec.	Avg. Err.	# Optimal	Percentage of optimal (%)
Grid	5	1203.664	6.75	1	20
Small	84	20660.723	3.2738	17	20.2381
Tree	15	4504.692	2	0	0
H-B	4	14403.138	8.75	0	0
Total	108	40772.217	3.5926	18	16.6667

Table 3 ILP2 results

ILP2

Instances	# Instances	CPU Sec.	Avg. Err.	# Optimal	Percentage of optimal (%)
Grid	5	907.021	5.8	2	40
Small	84	1598.902	2.4286	69	82.1429
Tree	15	73.198	2	15	100
H–B	4	14402.205	7.75	0	0
Total	108	16981.326	2.7222	86	79.6296

Table 4 Branch and bound results

B & B

Instances	# Instances	CPU Sec.	Avg. Err.	# Optimal	Percentage of optimal (%)
Grid	5	1203.664	6.75	1	20
Small	84	29802.476	2.25	0	0
Tree	15	4504.692	2	0	0
H-B	4	14403.353	7.5	0	0
Total	108	49914.185	2.6181	1	0.9259

5 Results

Tables 2 and 3 show the outperformance of ILP2 over ILP1 by finding the optimal solution of 63 % more of the instances than the previous model, also it improves the solving time by a 58.4 % and the average error is reduced by a 24.2 %.

The B&B proposed, when compared with ILP2 does not improve neither the time nor the percentage of optimal, but it improves the average error given by 3.8 % (see Table 4); due the time limit, some results given by the B&B cannot be taken as optimal. Because in that time the algorithm could not finish exploring the solution tree, so it does not guarantees that the result found in that time was an optimal.

6 Conclusions

As it is shown in the result tables, there is a large difference in quality and performance between the two linear integer programming models. This is the impact of the use of a compact linearization for the constraint (4) of the quadratic model. However, the branch and bound algorithm, despite its large time consumption and low percentage of optimal values found, presents a lower average error than the other two integer linear programming. At this point, it is important to explain that even if the branch and bound were to find the optimal value of an instance, we did not add it to the number of optimal solutions found if the algorithm did not finish the exploration of the whole tree of permutations.

Acknowledgments The authors thank Consejo Nacional de Ciencia y Tecnología (CONACYT), Dirección General de Educación Superior Tecnológica (DGEST) and Instituto Tecnológico de Ciudad Madero (ITCM) for their financial support to this project. Also we thank to IBM Academic Initiative for the software support (CPLEX 12.5).

References

1. Böckenhauer, H.J.: Two open problems in communication in edge-disjoint paths modes. Acta Mathematica et Informatica Universitatis Ostraviensis. University of Ostrava. **7**(1), 109–117 (1999)
2. Fleischer, B.U.D.: Vertex bisection is hard, too. J. Graph Algorithms Appl. **13**(2), 119–131 (2009)
3. Díaz, J., Petit, J., Serna, M.:A Survey on graph layout problems. ACM Comput. Surv. (CSUR). ACM. **34**(3), 313–356 (2002)
4. Petit J.: Addenda to the Survey of Layout Problems. Bulletin of EATCS **3** (105), (2013)
5. Duarte, A., Martí, R., Mladenovic, N., Pantrigo, J.J., Sánchez-Oro, J.: Variable neighborhood search for the vertex separation problem. Comput. Oper. Res. **39**(12), 3247–3255 (2012). Elsevier

Part V
Evolutionary and Intelligent Methods

Using a Graph Based Database to Support Collaborative Interactive Evolutionary Systems

J. C. Romero and M. García-Valdez

Abstract Web based, collaborative, interactive evolutionary computational systems, can generate a high amount of information. There is information regarding users and their collaborations, the interaction and subjective evaluation of individuals of the population. There is also information about the actual evolutionary process: relationships between individuals and evolutionary operators, used. In this work we propose the use of graph-based databases as back end storage of the evolutionary process of collaborative interactive evolutionary systems due to the expressiveness and flexibility provided by the model and how relationships found on the system can be easily mapped to graphs. The flexibility provided enables the design of user models, social network modules that can enhance the system. As a proof of concept, a comparative implementation against a relational database is presented.

1 Introduction

Interactive evolutionary computation (IEC) is a branch of evolutionary computation where users become a part of the evolutionary process by replacing the fitness function; evaluating individuals of a population based on their personal preferences [1]. These evaluations are subjective according to the user point of view and are based on their perceptions, interests and desires.

Normally such systems require users to evaluate large amounts of individuals iteratively, causing them to lose interest in their participation by fatigue that is

J. C. Romero · M. García-Valdez (✉)
Tijuana Institute of Technology, Tijuana, Mexico
e-mail: mariosky@gmail.com; mario@tectijuana.mx

J. C. Romero
e-mail: jcromerohdz@gmail.com

O. Castillo et al. (eds.), *Recent Advances on Hybrid Approaches for Designing Intelligent Systems*, Studies in Computational Intelligence 547,
DOI: 10.1007/978-3-319-05170-3_41, © Springer International Publishing Switzerland 2014

generated [1]. Nowadays some of these systems are migrating to Web technologies capting volunteers users who collaborate in the evaluations, distributing the load, to hopefully lower the associated fatigue. Having Web-based interactive evolutionary systems opens the possibility of linking to social platforms in order to involve the largest number possible of users to assist in the evaluation of individuals produced by these systems applications. There are some works that are based on the above. For example the work of Secretan et al. [2, 3] uses a Web-based application called Picbreeder that allows users to evolve images in a collaborative way; EvoSpace interactive is a framework for developing collaborative distributed interactive Evolutionary Algorithms for Artistic Design proposed by Garcia-Valdez et al. [4, 5]. Finally the work of Clune and Lipson [6] which is a web application called EndLessForms that explores the design of 3D objects selected by the user likes to evolve objects into the next generation. These systems need to store the evaluations given by users, these can be as ratings giving to an individual, the selections of individuals based on their preferences, and so on. These systems need to store the related information in a database management system. In the sense for storing the ratings given by users, individual evolution, selections based on user preferences (likes) and so on. Also developers of such systems must design complex queries for the database in order to enhance the application, for instance selecting similar users, similar individuals, popular individuals and so on. In this work, we argue that the relationship between the users and individuals can be represented naturally in a graph. We propose the use of graph-based databases due to the expressiveness and flexibility provided. This chapter is organized as follows: The related work is presented in Sects. 2 and 3 discus how we can support collaborative interactive evolutionary systems with graph-based databases. Section 4 presents experimental results and in Sect. 5, the conclusions of this work are presented.

2 Related Work

In this section the related work of collaborative interactive evolutionary systems is presented.

2.1 Picbreeder

Picbreeder is a Web-based application that allows users to evolve images in a collaborative way maintaining a large catalog of user-created content allowing users collaboration by searching through extensive design spaces [2]. Picbreeder provides to users of all experience levels to enjoy all the creative contributions produced by other users. In this way users experience a new form of recreation called creative social recreation through collaborative exploration. In this sense

Fig. 1 Picbreeder browse
searching categories [2]

Browse Categories

<u>All</u>>

<u>Alien</u> (230)   <u>Animal</u> (66)   <u>bird</u> (122)   <u>Bug</u> (51)
<u>butterfly</u> (69) <u>cool pattern</u> (94) <u>creature</u> (103) <u>Dog</u> (88)
<u>eye</u> (263)   <u>eyeball</u> (72)   <u>Face</u> (370)   <u>Fish</u> (129)
<u>ghost</u> (77)   <u>happy</u> (64)   <u>head</u> (56)   <u>insect</u> (56)
<u>mask</u> (64)   <u>monster</u> (51)   <u>rainbow</u> (87)   <u>Spaceship</u> (61)

Fig. 2 Picbreeder
architecture [2]

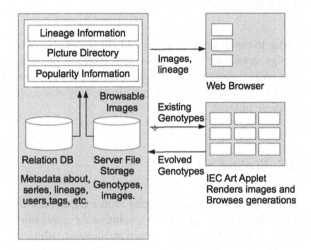

these systems helps their users to find interesting images through tagging,
browsing and searching. Picbreeder provides a search-engine style interface for
tags search that are automatically grouped into browsed categories and subcate-
gories. This representation is shown in Fig. 1.

Figure 2 represents the architecture of Picbreeder, which is a Web-based
application that uses a relational database. This database stores meta-data about
general information of the images, for example their tags, ratings, lineage and
authorship, also users basic information for the user account. The database helps to
generated search by browse categories.

2.2 EndlessForms

EndlessForms is a Web application that explores object designs by choosing those
the users like most. These selected objects become the parents of the next generation of objects [6]. EndLessForms proposes a new way to evolve 3D objects
inspired by biological morphologies using generative encoding. One of the
experiments proposed in this chapter was to use interactive evolutionary systems
to determine the potential for generating complex and interesting 3D objects. They

alien   alien   alien   alien   alien

Fig. 3 Alien objects from EndlessForms [6]

chose the interactive evolution, because that allows open-ended exploration of the design space of objects that can produce by their method. Additionally, the interactive evolution avoids the greedy nature of evolution by objectives, which potentially allows to access more interesting objects [6]. In Fig. 3 we can see some of these objects.

2.3 EvoSpace-I

EvoSpace-Interactive is an open source framework focused on Web environments for collaborative interactive evolutionary applications. This framework defines three main components for each application, which are:

- Individual.
- Processing Script.
- Worker Script.

The individual is represented internally as a data dictionary stored in Redis [7]; the individual contains three main attributes: id, chromosome, mom, dad, and views. This attributes represents the key information of the individual as the individual offspring, the number of times that the individual has been selected, etcetera, as we can see in Fig. 4 [4].

As we can observe on Fig. 5, this work uses database management systems to implement collaborative interactive evolutionary applications. One of the reasons that this framework is using Redis [8] is because it provides a hash-based implementation of sets and queues, which are natural data structures for the EvoSpace model. On the other hand this framework uses a relational database to save basic information about the user extracted from the social platform (Facebook) through open graph API and OAuth2 authentication.

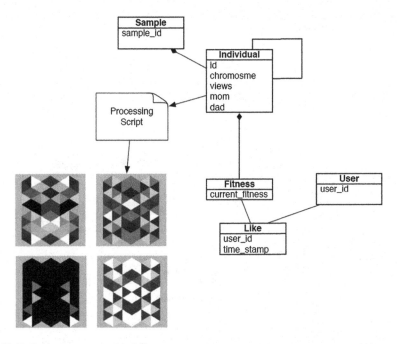

Fig. 4 Individual representation [4]

3 Supporting Collaborative Interactive Evolutionary Systems with a Graph Database Management System

In this section we discuss why we can use a graph-based database management system in order to support the collaborative interactive evolutionary systems.

3.1 Graph Database

A graph is a set of nodes and relationships that are connected to each other. Formally a graph is a collection of edges and vertices [9]. Where vertices represent the nodes and edges represents the relations between the nodes. Using graph structures allows us to model a variety of scenarios, for example road systems, medical history, etcetera. A graph database is based on the graph theory and is a database management system with ability to use methods like Create, Read, Update and Delete [8, 10, 11]. These databases use a property graph to save the information of the nodes and relationships as we can see in Fig. 4. They have a flexible way for traversing through node and relationships on the information generated by graph model. Also the expressiveness for modeling data (whiteboard-friendly), it means we can start drawing nodes with their respective relationships and then start to implement the data model of a specific domain (Fig. 6).

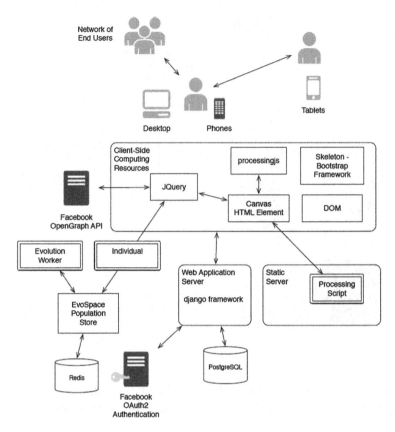

Fig. 5 EvoSpace-I framework [4]

3.2 Supporting Collaborative Interactive Evolutionary Systems with a Graph-Based Database

The advantage of using a graph gives us the ability to represent any kind of data in an accessible and flexible form to be implemented on a graph-based database. Also the graph-based database offers storage for several billions of nodes, relationships and properties [11]. However, the data presented in almost all collaborative interactive evolutionary systems is related to the individual and users. We consider that this relationship could be represented naturally in graph due to flexibility of expressing for design data models as we can see in Fig. 7.

In Fig. 7 we present a graph model design to represent elements of collaborative interactive evolutionary systems. With this representation we can access the information through the paths formed by the graphs, either to obtain a set of information or any pattern that is formed by the data, all depending on the questions we ask to perform the desired queries in the information networks that graphs generate. This is the reason why we propose to use a graph-based database

Fig. 6 Small social network

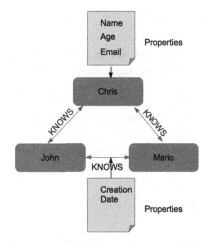

Fig. 7 Graph representation
of user-individual

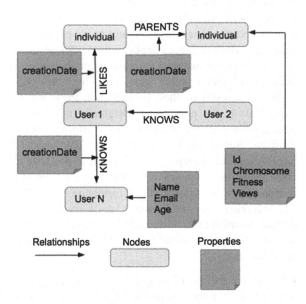

to support such systems. In other approaches could be complicated to query the data when it starts to be more complex. Sometimes is necessary to use libraries, using some programming language in order to elaborate these queries. Having the data in a graph-based database simplify the way to explore all the complex data generated with them. Some examples of complex queries are: Get the descendants of individuals in the population; another could be how many users they know each other, and how many users have more participation. Here is where we can exploit the full potential of having the data model in a graph-based database in these systems. For the purposes of answering the questions mentioned, we just traverse the paths of the graph.

3.3 Neo4j

In this work we decided to use Neo4j due to is an open source graph-based database management system database. Neo4j can store large quantities of nodes, relationships and properties [8]. It has the characteristics of not designing data schemas; with this we have the ability to design flexible and expressive models that they called whiteboard-friendly. It also provides a human-readable query language called cypher. With cypher we can traverse the paths of information networks formed by the graphs and get the information that we desire [8, 11].

4 Results

In this chapter we consider to implement a subsystem to design a data modeling that represents the relationship between the users and individuals using a graph-based database management system as we explained in Sect. 3. We used the data generated with the Shapes application [4]. These data were created during a period of time of approximately 3 months; while the application was online, where more than 60 volunteers active users providing their information through Facebook login. As a result of their participation these users generate more than 12,000 individuals in the population. Originally this data was store in Redis database management system and the user basic information was store in Postgres relational database. Subsequently a graph-based data model was designed, and all the information was migrated to a graph-base database management system (Neo4j). Once the information was stored in this way, we proceeded to exploit the information, that was formed in the information network, related to the user-individual and individual-user relationships. As a result we answered questions that were difficult to implement in the original model. The questions that we were able to answer based on the benefit and flexibility provided by this graph-based model is the following:

- Which are the most likely individuals of the population?
- Which are the most seen individuals of the population?
- Who are the most participants users?
- Which is the offspring of the individual?

The results of the answers of the questions above can be displayed in the following tables. This was possible thanks to the technology we used in this work (Neo4j) because it provides its own query language for traversing the paths formed by the information networks in the database.

The Table 1 presents the top ten results of individuals who have more likes in the population.

The Table 2 presents the top ten results of individuals who have more views in the population.

Table 1 Top 10 most likely individual of the population

Id_node	Individual	Ind_likes
5906	"pop:individual:9199"	18
12194	"pop:individual:9078"	17
9633	"pop:individual:8771"	16
11957	"pop:individual:9027"	16
6406	"pop:individual:9014"	15
11992	"pop:individual:9158"	15
10571	"pop:individual:11347"	15
11136	"pop:individual:8955"	15
5128	"pop:individual:9119"	14
5360	"pop:individual:10559"	13

Table 2 Top ten most viewed individuals of the population

Id_node	Individual	Views
3215	"pop:individual:110"	77
5814	"pop:individual:9078"	64
189	"pop:individual:3055"	61
845	"pop:individual:7652"	55
1698	"pop:individual:1143"	53
5894	"pop:individual:4063"	51
3331	"pop:individual:9158"	50
497	"pop:individual:9027"	49
7546	"pop:individual:1639"	49
8488	"pop:individual:9200"	47

Table 3 Individual pop:individual:9078 ancestries

Id_node	Individual	User	Ind_likes
5906	"pop:individual:9199"	"Anonymous"	18
12194	"pop:individual:9078"	"Anonymous"	17
9633	"pop:individual:8771"	"Anonymous"	16
11957	"pop:individual:9027"	"Anonymous"	16
6406	"pop:individual:9014"	"Anonymous"	15
11992	"pop:individual:9158"	"anonymous"	15
10571	"pop:individual:11347"	"Frank Arce"	15
11136	"pop:individual:8955"	"Anonymous"	15
5128	"pop:individual:9119"	"Anonymous"	14
5360	"pop:individual:10559"	"Frank Arce"	13

The Table 3 presents the top ten results of users who have more participations based on their likes in the population.

The Table 4 presents the ancestries of the individual "pop: individual: 9078".

Table 4 Individual
pop:individual:9078
ancestries

Id_node	Individual	Views
8367	"pop:individual:9038"	"14"
11776	"pop:individual:9298"	"17"
11719	"pop:individual:9313"	"3"
7021	"pop:individual:10777"	"3"
6768	"pop:individual:9096"	"11"
5128	"pop:individual:9119"	"46"
4645	"pop:individual:10136"	"10"
2520	"pop:individual:10927"	"5"
1688	"pop:individual:10778"	"13"
1646	"pop:individual:9314"	"21"
1538	"pop:individual:10631"	"2"
1345	"pop:individual:9312"	"7"
849	"pop:individual:9152"	"14"
553	"pop:individual:9317"	"4"
488	"pop:individual:9223"	"8"
406	"pop:individual:10135"	"4"

5 Conclusions

In conclusion we can say that graph-based databases provide benefits to collaborative interactive evolution systems designing flexible data models and is easy to exploit information networks generated by the graphs. They gives us the ability to traverse the paths that we wish to explore in the networks formed by these, thus obtaining all the knowledge of user interests with respect to individuals and patterns formed in these relationships. This knowledge is very useful because the more we know about the behavior of the users according to their interests, we could improve the user models in order to adapt better their participation on collaborative interactive evolution systems. As future work fuzzy logic could also be used to improve the model [12–16]. In addition, other evolutionary approaches [17–19] or bio-inspired algorithms [20–24] could be tried in this problem.

References

1. Takagi, H.: Interactive evolutionary computation: Fusion of the capacities of EC optimization and human evaluation. Proc. IEEE **89**(9), 1275–1296 (2001)
2. Secretan, J., Beato, N., D'Ambrosio, D.B., Rodriguez, A., Campbell, A., Stanley, K.O.: School of electrical engineering and computer science university of central Florida, Orlando, FL 32816-2362 {jsecreta,nbeato, ddambro, adeleinr, acampbel,kstanley}@eecs.ucf.edu Picbreeder: Evolving pictures collaboratively online
3. Secretan, J., Beato, N., D'Ambrosio, D.B., Rodriguez, A., Campbell, A., Folsom-Kovarik, J.T., Stanley, K.O.: Picbreeder: A case study in collaborative evolutionary exploration of design space. Evol. Comput. **19**(3), 373–403 (2011)

4. Garcia, M., Trujillo, L., Fernández-de-Vega, F., Merelo-Guervós, J.J., Olague, G.: EvoSpace-interactive: A framework to develop distributed collaborative-interactive evolutionary algorithms for artistic design. In: Proceedings of the 2nd International Conference on Evolutionary and Biologically Inspired Music, Sound, Art and Design, EvoMUSART (2013)
5. García-Valdez, M., Trujillo, L., de Vega, F.F., Guervós, J.J.M., Olague, G.: EvoSpace: A distributed evolutionary platform based on the tuple space model. In: Proceedings of the 16th European Conference, EvoApplications 2013, pp. 499–508. Vienna, Austria, 3–5 Apr 2013
6. Clune, J., Lipson, H.: Evolving three-dimensional objects with a generative encoding inspired by developmental biology. In: Proceedings of the European Conference on Artificial Life (2011)
7. Redis. http://redis.io/ (2013)
8. Neo4j. http://www.neo4j.org/ (2013)
9. Biggs, N., Lloyd, E., Wilson, R.: Graph Theory, 1736–1936. Oxford University Press, Oxford (1986)
10. Marko, A.: Rodriguez and neubauer, Peter the graph traversal pattern. Graph data management: Techniques and applications (2011)
11. Robinson, I., Webber, J., Eifrém, E.: Graph databases. O'Reilly media (2013)
12. Hidalgo, D., Castillo, O., Melin, P.: Type-1 and type-2 fuzzy inference systems as integration methods in modular neural networks for multimodal biometry and its optimization with genetic algorithms. Inf. Sci. 179(13), 2123–2145 (2009)
13. Leal-Ramirez, C., Castillo, O., Melin, P., Rodriguez-Diaz, A.: Simulation of the bird age-structured population growth based on an interval type-2 fuzzy cellular structure. Inf. Sci. 181, 519–535 (2011)
14. Castillo, O., Huesca, G., Valdez, F.: Evolutionary computing for topology optimization of type-2 fuzzy controllers. Stud. Fuzziness Soft Comput. 208, 163–178 (2008)
15. Mendoza, O., Melin, P., Castillo, O.: Interval type-2 fuzzy logic and modular neural networks for face recognition applications. Appl. Soft Comput. J. 9, 1377–1387 (2009)
16. Mendoza, O., Melin, P., Licea, G.: Interval type-2 fuzzy logic for edges detection in digital images. Int. J. Intell. Syst. 24, 1115–1133 (2009)
17. Valdez, F., Melin, P., Castillo, O.: Evolutionary method combining particle swarm optimization and genetic algorithms using fuzzy logic for decision making. In: Proceedings of the IEEE International Conference on Fuzzy Systems, pp. 2114–2119 (2009)
18. Castillo, O., Melin, P.: A review on the design and optimization of interval type-2 fuzzy controllers. Appl. Soft Comput. 12(4), 1267–1278 (2012)
19. Cervantes, L., Castillo, O., Melin, P.: Intelligent control of nonlinear dynamic plants using a hierarchical modular approach and type-1 fuzzy logic. MICAI, pp. 1–12 (2011)
20. Sombra, A., Valdez, F., Melin, P., Castillo, O.: A new gravitational search algorithm using fuzzy logic to parameter adaptation. In: IEEE Congress on Evolutionary Computation, pp. 1068–1074 (2013)
21. Valdez, F., Melin, P., Castillo, O.: Parallel particle swarm optimization with parameters adaptation using fuzzy logic. MICAI 2, 374–385 (2012)
22. Valdez, F., Melin, P., Castillo, O.: Bio-inspired optimization methods on graphic processing unit for minimization of complex mathematical functions. In: Recent Advances on Hybrid Intelligent Systems, pp. 313–322 (2013)
23. Melendez, A., Castillo, O.: Optimization of type-2 fuzzy reactive controllers for an autonomous mobile robot. NaBIC 2012, pp. 207-211
24. Castillo, O., Melin, P.: Optimization of type-2 fuzzy systems based on bio-inspired methods: A concise review. Inf. Sci. 205, 1–19 (2012)

Fuzzy Labeling of Users in an Educational Intelligent Environment Using an Activity Stream

Francisco Arce and Mario García-Valdez

Abstract This chapter presents a method for labeling users in an intelligent environment according to activities drawn from an activity stream. The activity stream is composed by all the activities that are registered in a certain window of time. Using a fuzzy inference engine labels are assigned to students, in order to aggregate the generated data. To evaluate the scalability of the approach several simulations where executed, results show the method is viable tool.

1 Introduction

Intelligent environments are spaces with embedded systems, sensors, information and communication technologies that are becoming invisible to the user as they are being integrated into physical objects, infrastructure, the environment in which we live, work and many other environments. Fuzzy logic in contrast to a classical set, which has a crisp boundary, the boundary of a fuzzy set is blurred [1–3]. This smooth transition is characterized by membership functions which give fuzzy sets flexibility in modeling linguistic expressions [4, 5]. Because our language is fuzzy, they share the uncertainty and impreciseness [6, 7]. This idea provides a good way of bridging the gap between human users and computing systems, and this motivates related research into Computing with Word helps to give a meaning [8, 9]. This chapter presents a fuzzy labeling of users in an intelligent environment [10] by their activities. Activity Stream is a collection of activities. An activity consists of an actor, a verb, an object, and a target, the goal of this specification is to provide

F. Arce · M. García-Valdez (✉)
Tijuana Institute of Technology, Tijuana, Mexico
e-mail: mariosky@gmail.com; mario@tectijuana.edu.mx

F. Arce
e-mail: francisco.j.arce@gmail.com

O. Castillo et al. (eds.), *Recent Advances on Hybrid Approaches for Designing*   593
Intelligent Systems, Studies in Computational Intelligence 547,
DOI: 10.1007/978-3-319-05170-3_42, © Springer International Publishing Switzerland 2014

sufficient metadata about an activity such that a consumer of the data can present it to a user in a rich human-friendly format. This work aims to detect the individual level of participation of the user and label it, to determine the level of participation the Fuzzy labeler needs the list of the activities of the user. Which can perform 3 different activities "Login", "Answer" and "Approach". With this information the system can now label the level of the user participation in "Low" participation, "Medium" participation and "High" participation and adapt the content the next time this user use the interactive environment. We propose to label users from a list of activities (Activity stream) randomly generated with certain parameters that will be explained later, these events will be consistent with a group of users would have on the environment as a user who has not entered the environment cannot use or approximated to a device. This fuzzy labeling will tell us the level of user participation and with it the system can take an action either change it to another group that has a lower level of participation, perhaps because his group was more participatory users which engulfed activities or change a user that have more participation to a group with the same level of participation, this would generate a greater number of activities in the session of the group with the highest level of participation. This chapter is organized as follows: In the Sect. 2 the related work is discussed. In Sect. 3 the proposed method is presented. Section 4 presents experimental results and in Sect. 5 the conclusions of this work are presented.

2 Related Work

In this section the related work of this research work are presented.

2.1 Activity Stream

JSON Activity Streams 2.0 is a standard. This specification details a model for representing potential and completed activities using the JSON format. In JAS activity stream is a list of activities performed by an individual; this list consists of an actor, a verb, an object, and a target as we can see in the Fig. 1. An activity tells the story of a person performing an action on or with an object for instance activities in a social network, could be "Geraldine posted a photo to her album" or "John shared a video"; in most cases these components will be explicit, but they may also be implied. It is a goal of this specification to provide sufficient metadata about an activity such that a consumer of the data can present it to a user in a human-friendly format. Within this specification, an object is a thing, real or imaginary, which participates in an activity. It may be the entity performing the activity, or the entity on which the activity was performed.

An object consists of properties defined in the following sections. Certain object types may further refine the meaning of these properties, or they may define

```
{
  "published": "2011-02-10T15:04:55Z",
  "actor": {
    "url": "http://example.org/martin",
    "objectType" : "person",
    "id": "tag:example.org,2011:martin",
    "image": {
      "url": "http://example.org/martin/image",
      "width": 250,
      "height": 250
    },
    "displayName": "Martin Smith"
  },
  "verb": "post",
  "object" : {
    "url": "http://example.org/blog/2011/02/entry",
    "id": "tag:example.org,2011:abc123/xyz"
  },
  "target" : {
    "url": "http://example.org/blog/",
    "objectType": "blog",
    "id": "tag:example.org,2011:abc123",
    "displayName": "Martin's Blog"
  }
}
```

Fig. 1 Example of JAS 2.0

additional properties. Some types of objects may have an alternative visual representation in the form of an image, video or embedded HTML fragments. An Activity Stream is a collection one or more individual activities. The relationship between the activities within the collection is undefined by this specification [11]. Most of the current work with activity streams are oriented to the web like [12] in this work the authors analyze the activity stream of the social media usage per employee extracting different characteristics of the stream and compare the way they search on the web [13]. In this other work they study the stream to personalize the stream of relevant news for this user. Also there's a way to modeling an activity stream [14]. In our case we use JAS2 format for the stream of activities but exclude some of the features of the standard as the Target or subsections of the Actor then the stream is analyzed with the fuzzy labeler.

2.2 Fuzzy Labels

The concept of Fuzzy Logic was conceived by Lotfi Zadeh, a professor at the University of California at Berkley, and presented not as a control methodology, but as a way of processing data by allowing partial set membership rather than crisp set membership or non-membership. Fuzzy logic is based on the observed relation as a differential position. Best suited to the real world in which we live, and can even understand and work with our expressions, like "it's too hot", "he's not too tall". Fuzzy Logic incorporates a simple, rule-based IF X AND Y THEN Z approach to a solving control problem rather than attempting to model a system mathematically. Fuzzy Logic is also a technique with application in different fields and is rated one of the applications of interest by the ease with which assigns percentages belonged. Almost all the labels we give to groups of objects are fuzzy. For example, friends, tall trees etc. An object may belong to the set of objects with a certain label, with a certain membership value. In traditional set theory, this membership value only has two possible values, 1 and 0, representing the case where the object belongs to or does not belong to the set, respectively. We use a fuzzy term such as 'big' to label a particular group, because they share the property of objects within this group (i.e., they are big). The objects within this group will have different membership values varying from 0 to 1 qualifying the degree to which they satisfy. Zengchang Qin work with fuzzy labeling for modeling uncertainty [15, 16] also this can be applied to label objects like [17–19]. Our work will label the individual based on the activity on the environment giving a meaning to the individual for the environment.

3 Proposed Method

The architecture of the proposed labeler method can be observed in Fig. 2.

This environment is composed of various devices and is used by user groups which can at the same time sign to the environment their devices the condition for these devices can enter the environment (with sign we mean the environment will recognize them and send content) is to be able to run a web browser like their Smartphone's, tablet's, laptop's etc., users can have four activities for now "Login", "Logout", "Answered" and "Approached" where the "Login" is access the environment, "Logout" is to leave the environment, "Answered" is to answer a questionnaire on their tablet or Smartphone and "approached" is approach one of the devices of the environment as a projection of a projector or a monitor. The activities are generated in the JAS2 format as seen in Fig. 2 but only considering the actor who gives us information on who performed that activity either a user or a device name, in the verb we have an action, the date and time that took place (Fig. 3).

The fuzzy system is configured as follows we handle three input variables "Logins", "Answered" and "Approached" where "Logins" has 3 membership functions Low, Medium and High with ranges of [−4 0 4], [1 5 9] y [6 10 14] respectively, "Answered" and "Approached" with two membership functions Low and High with ranges of [−4 0 8] y [2 10 14] respectively. The output (Participation) has three membership functions Low, Medium and High with

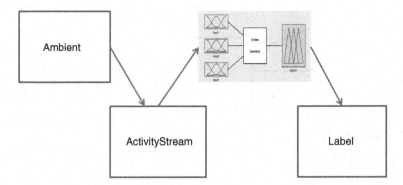

Fig. 2 Architecture

Fig. 3 Example of our activity

```
{    "actor":
        {"objectType":"User",
         "displayName":"Francisco Arce"},
     "verb":"Login",
     "Date":"11/12/13",
     "Time:" 1384231639"
}
```

ranges of [−0.4 0 0.4], [0.1 0.5 0.9] y [0.6 1 1.4] respectively this can be seen graphically in Fig. 4.

In Fig. 5 we can observe the rules used.

4 Experimental Results

In this section the results obtained in this work are presented. We use Activity streams generated randomly and stochastically, the users are not real and activities are consistent with normal use of the environment as a user who has not paid into the environment cannot perform activities by the fact of not being present in the environment in Fig. 6 we can see a sample Activity stream. To generate an activity there is a 50 % chance of it being a user and the other 50 % to be a device, if the "Actor" is a user we take randomly a user from a list of 30 where as explained above can perform the following activities "Login", "Logout", "Answered" and "Approached".

If the "Actor" is a device, we will choose from a list of 5 devices (Monitor, projector, tablet, Smartphone, speaker) and which can do one of the following activities "Start", "Shutdown", "Playing", "Asking", "Showing" and "Waiting".

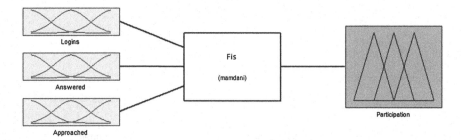

Fig. 4 Fis structure used

1. If (Logins is Low) and (Answers is Low) and (Approach is Low) then (output1 is Low) (1)
2. If (Logins is Low) and (Answers is Low) and (Approach is High) then (output1 is Low) (1)
3. If (Logins is Low) and (Answers is High) and (Approach is Low) then (output1 is Low) (1)
4. If (Logins is Low) and (Answers is High) and (Approach is High) then (output1 is Low) (1)
5. If (Logins is Med) and (Answers is Low) and (Approach is Low) then (output1 is Med) (1)
6. If (Logins is Med) and (Answers is Low) and (Approach is High) then (output1 is Med) (1)
7. If (Logins is Med) and (Answers is High) and (Approach is Low) then (output1 is Med) (1)
8. If (Logins is Med) and (Answers is High) and (Approach is High) then (output1 is Med) (1)
9. If (Logins is High) and (Answers is Low) and (Approach is Low) then (output1 is High) (1)
10. If (Logins is High) and (Answers is Low) and (Approach is High) then (output1 is High) (1)
11. If (Logins is High) and (Answers is High) and (Approach is Low) then (output1 is High) (1)
12. If (Logins is High) and (Answers is High) and (Approach is High) then (output1 is High) (1)

Fig. 5 Rules

```
{'Date': '2013-11-12', 'verb': 'Logout', 'actor': {'displayName': 'Kristi Sabala', 'objectType': 'User'}, 'Time': '1384321001'}
{'Date': '2013-11-12', 'verb': 'Logout', 'actor': {'displayName': 'Brandie Coursey', 'objectType': 'User'}, 'Time': '1384321080'}
{'Date': '2013-11-12', 'verb': 'Login', 'actor': {'displayName': 'Lashay Struck', 'objectType': 'User'}, 'Time': '1384321180'}
{'Date': '2013-11-12', 'verb': 'Logout', 'actor': {'displayName': 'Lashay Struck', 'objectType': 'User'}, 'Time': '1384321261'}
{'Date': '2013-11-12', 'verb': 'Logout', 'actor': {'displayName': 'Soraya Glorioso', 'objectType': 'User'}, 'Time': '1384321471'}
{'Date': '2013-11-12', 'verb': 'Login', 'actor': {'displayName': 'Soraya Glorioso', 'objectType': 'User'}, 'Time': '1384321511'}
{'Date': '2013-11-12', 'verb': 'Login', 'actor': {'displayName': 'Zachary Elliff', 'objectType': 'User'}, 'Time': '1384321539'}
{'Date': '2013-11-12', 'verb': 'Login', 'actor': {'displayName': 'Kristian Bonenfant', 'objectType': 'User'}, 'Time': '1384321650'}
{'Date': '2013-11-12', 'verb': 'Login', 'actor': {'displayName': 'Reanna Fausnaught', 'objectType': 'User'}, 'Time': '1384321838'}
{'Date': '2013-11-12', 'verb': 'Login', 'actor': {'displayName': 'Enola Wakeland', 'objectType': 'User'}, 'Time': '1384321842'}
{'Date': '2013-11-12', 'verb': 'Approached', 'actor': {'displayName': 'Soraya Glorioso', 'objectType': 'User'}, 'Time': '1384321849'}
{'Date': '2013-11-12', 'verb': 'Logout', 'actor': {'displayName': 'Temeka Armand', 'objectType': 'User'}, 'Time': '1384322015'}
{'Date': '2013-11-12', 'verb': 'Answer', 'actor': {'displayName': 'Soraya Glorioso', 'objectType': 'User'}, 'Time': '1384322135'}
{'Date': '2013-11-12', 'verb': 'Logout', 'actor': {'displayName': 'Lashay Struck', 'objectType': 'User'}, 'Time': '1384322161'}
{'Date': '2013-11-12', 'verb': 'Login', 'actor': {'displayName': 'Barbar Plate', 'objectType': 'User'}, 'Time': '1384322291'}
{'Date': '2013-11-12', 'verb': 'Approached', 'actor': {'displayName': 'Zachary Elliff', 'objectType': 'User'}, 'Time': '1384322350'}
{'Date': '2013-11-12', 'verb': 'Login', 'actor': {'displayName': 'Lashay Struck', 'objectType': 'User'}, 'Time': '1384322442'}
{'Date': '2013-11-12', 'verb': 'Logout', 'actor': {'displayName': 'Celsa Gullette', 'objectType': 'User'}, 'Time': '1384322593'}
{'Date': '2013-11-12', 'verb': 'Approached', 'actor': {'displayName': 'Reanna Fausnaught', 'objectType': 'User'}, 'Time': '1384322697'}
{'Date': '2013-11-12', 'verb': 'Logout', 'actor': {'displayName': 'America Peloquin', 'objectType': 'User'}, 'Time': '1384322745'}
{'Date': '2013-11-12', 'verb': 'Approached', 'actor': {'displayName': 'Barbar Plate', 'objectType': 'User'}, 'Time': '1384322793'}
{'Date': '2013-11-12', 'verb': 'Logout', 'actor': {'displayName': 'Azalee Callaway', 'objectType': 'User'}, 'Time': '1384322833'}
{'Date': '2013-11-12', 'verb': 'Logout', 'actor': {'displayName': 'Yahaira Brissette', 'objectType': 'User'}, 'Time': '1384323061'}
{'Date': '2013-11-12', 'verb': 'Approached', 'actor': {'displayName': 'Enola Wakeland', 'objectType': 'User'}, 'Time': '1384323108'}
{'Date': '2013-11-12', 'verb': 'Logout', 'actor': {'displayName': 'Jeremy Rarick', 'objectType': 'User'}, 'Time': '1384323132'}
{'Date': '2013-11-12', 'verb': 'Logout', 'actor': {'displayName': 'Josh Balzer', 'objectType': 'User'}, 'Time': '1384323173'}
{'Date': '2013-11-12', 'verb': 'Login', 'actor': {'displayName': 'Angele Mcnab', 'objectType': 'User'}, 'Time': '1384323490'}
{'Date': '2013-11-12', 'verb': 'Answer', 'actor': {'displayName': 'Angele Mcnab', 'objectType': 'User'}, 'Time': '1384323797'}
{'Date': '2013-11-12', 'verb': 'Answer', 'actor': {'displayName': 'Angele Mcnab', 'objectType': 'User'}, 'Time': '1384323803'}
```

Fig. 6 Sample of activities

Table 1 The results for each user with random

User	Logins	Answers	Approaches	Participation
1	6	4	7	0.5
2	3	2	5	0.44
3	6	6	4	0.5
4	6	3	5	0.5
5	6	4	4	0.5
6	8	6	6	0.637
7	4	8	2	0.5
8	10	7	0	0.849
9	4	4	2	0.5
10	6	6	5	0.5
11	5	4	3	0.5
12	4	9	4	0.5
13	6	5	2	0.5
14	10	5	5	0.83
15	8	3	4	0.637
16	8	5	6	0.61
17	8	2	5	0.61
18	4	2	2	0.5
19	7	3	4	0.544
20	8	6	5	0.61
21	6	8	7	0.5
22	9	5	3	0.83
23	3	4	0	0.45
24	8	2	6	0.637
25	10	6	6	0.84
26	7	6	5	0.553
27	4	3	3	0.5
28	5	6	5	0.5
29	10	3	3	0.849
30	5	5	10	0.5

4.1 Activity Generating Random Streams

In this experiment 1,000 activities were generated among which was a 50 % chance of being user activity and the other 50 % of a device, hence the probability of choosing a person or device is the same for everyone (Table 1).

4.2 Generation Activity Streams with Low Weight in Logins

In this experiment 1,000 activities were generated among which there was a 50 % to be activity of a user and the other 50 % of a device, hence the probability of choosing a person or device is the same for everyone, in the previous case the

Table 2 The results for each
user with random and lower
weight on "Logins"

User	Logins	Answers	Approaches	Participation
1	4	8	5	0.5
2	3	11	7	0.45
3	5	10	8	0.5
4	4	4	12	0.5
5	3	3	4	0.45
6	4	6	8	0.5
7	5	15	8	0.558
8	2	7	13	0.421
9	2	3	9	0.352
10	3	10	7	0.45
11	5	7	12	0.5
12	4	7	9	0.5
13	3	9	9	0.45
14	4	8	6	0.5
15	4	7	4	0.5
16	5	5	7	0.5
17	3	7	11	0.45
18	2	7	7	0.352
19	2	7	6	0.352
20	4	7	11	0.5
21	4	10	7	0.5
22	3	8	4	0.45
23	4	10	5	0.5
24	7	5	10	0.553
25	5	8	7	0.5
26	5	4	5	0.5
27	4	6	12	0.5
28	3	5	5	0.44
29	5	8	9	0.5
30	4	11	13	0.5

activity "Login" was predominant so we decided to give a low weight to activity
"Login" to reduce the likelihood of the occurrence (Table 2).

4.3 Activity Streams Generating with Random Triangular Distribution in the Selection of Users

In this experiment 1,000 activities were generated among which there was a 50 %
to be activity of a user and the other 50 % of a device, now we use a triangular
probability distribution [20] where the users that are at the center of the list are
more likely to be chosen that users who are in the extremes as seen in Fig. 7.
Similarly the "Logins" have a lower weight (Table 3).

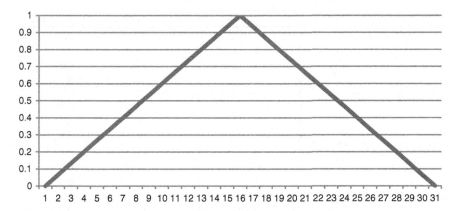

Fig. 7 Triangular probability distribution to be selected

	User	Logins	Answers	Approaches	Participation
Table 3 The results for each user with random triangular distribution	1	5	4	6	0.5
	2	4	8	12	0.5
	3	4	5	5	0.5
	4	6	7	7	0.5
	5	2	0	0	0.352
	6	7	10	6	0.549
	7	0	0	0	0.128
	8	6	7	11	0.5
	9	6	12	12	0.5
	10	6	5	20	0.5
	11	2	2	3	0.352
	12	6	12	8	0.5
	13	3	5	7	0.44
	14	2	6	6	0.352
	15	3	3	2	0.45
	16	2	0	0	0.352
	17	4	13	12	0.5
	18	0	0	0	0.128
	19	5	9	11	0.5
	20	5	4	4	0.5
	21	5	10	7	0.5
	22	9	8	10	0.855
	23	3	6	2	0.45
	24	5	17	12	0.5
	25	0	0	0	0.128
	26	6	9	7	0.5
	27	2	8	2	0.352
	28	7	14	10	0.549
	29	3	12	17	0.451
	30	3	3	3	0.45

Table 4 The results for each user with random

User	Logins	Answers	Approaches	Participation
1	3	9	4	0.45
2	4	6	7	0.5
3	5	5	3	0.5
4	2	4	6	0.352
5	3	2	8	0.45
6	5	9	13	0.5
7	5	6	8	0.5
8	4	4	8	0.5
9	6	11	13	0.5
10	6	3	9	0.5
11	3	3	8	0.45
12	4	5	15	0.5
13	3	14	14	0.451
14	3	2	3	0.45
15	3	12	8	0.45
16	3	9	6	0.45
17	3	11	9	0.45
18	3	0	2	0.45
19	9	11	13	0.863
20	5	3	0	0.5
21	2	8	7	0.352
22	3	15	8	0.451
23	6	7	7	0.5
24	8	8	18	0.646
25	3	4	2	0.45
26	3	6	12	0.45
27	2	5	5	0.381
28	4	6	7	0.5
29	7	11	11	0.549
30	3	6	4	0.45

4.4 Activity Streams Generating with Random Triangular Distribution in the Selection of Users and a Shuffle in the List of Users at 500 Activities

In this experiment 1,000 activities were generated among which there was a 50 % to be activity of a user and the other 50 % of a device, now we use a triangular probability distribution where the users that are at the center of the list are more likely to be chosen that users who are in the extremes as seen. Similarly the "Logins" have a lower weight; where to reach 500 events performed a shuffle in the list of users to alter the odds (Table 4).

Fig. 8 Comparison of the level of participation of each user for each experiment

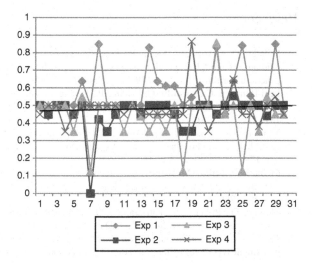

5 Conclusions

This chapter presents a fuzzy labeling of users in an intelligent environment where we use virtual users with different probabilities of being selected to observe the behavior of the labeler before using it with real users, the level of participation we obtained for 4 different experiments where we observe in Fig. 8 that experiment 1 where the probability of selection for each user is the same level of participation is a little high to have a greater number of user Login's, Experiment 2 was oriented more to a medium level of participation, but we found that the user 7 was very little participation may therefore be a candidate to be removed from the group, experiment 3 tube users less participatory and experiment 4 participation improved and there was a very active user. These results prove that this classification identifies users highly participative and less participative allowing us to make appropriate action with the environment and with the user group.

As future work is planned to do dynamic ranges of the membership functions, that by the fact that in this case the activity stream was fixed in 1,000 but in a real case this number will vary therefore tend to change values. Furthermore, this stream of activities will be processed by a event handler [11, 19, 21] activity per activity in which these activities will have a meaning for the interactive environment and in turn generate other events. As future work evolutionary or bio-inspired methods could be used to improve the proposed model, as in [22–25].

References

1. Hidalgo, D., Castillo, O., Melin, P.: Type-1 and type-2 fuzzy inference systems as integration methods in modular neural networks for multimodal biometry and its optimization with genetic algorithms. Inf. Sci. **179**(13), 2123–2145 (2009)
2. Leal-Ramirez, C., Castillo, O., Melin, P., Rodriguez-Diaz, A.: Simulation of the bird age-structured population growth based on an interval type-2 fuzzy cellular structure. Inf. Sci. **181**, 519–535 (2011)
3. Martinez, R., Castillo, O., Aguilar, L.T.: Optimization of interval type-2 fuzzy logic controllers for a perturbed autonomous wheeled mobile robot using genetic algorithms. Inf. Sci. **179**(13), 2158–2174 (2009)
4. Melin, P., Castillo, O.: Adaptive intelligent control of aircraft systems with a hybrid approach combining neural networks, fuzzy logic and fractal theory. J. Appl. Soft Comput. **3**(4), 353–362 (2003)
5. Montiel, O., Roberto, S., Patricia, M., Oscar, C., Miguel, P., Iliana, M.: Performance of a simple tuned fuzzy controller and a PID controller on a DC motor. FOCI, pp. 531–537 (2007)
6. Sepulveda, R., Castillo, O., Melin, P., Montiel, O.: An efficient computational method to implement type-2 fuzzy logic in control applications. Adv. Soft Comput. **41**, 45–52 (2007)
7. Sepulveda, R., Castillo, O., Melin, P., Rodriguez-Diaz, A., Montiel, O.: Experimental study of intelligent controllers under uncertainty using type-1 and type-2 fuzzy logic. Inf. Sci. **177**(10), 2023–2048 (2007)
8. Sepulveda, R., Montiel, O., Lizarraga, G., Castillo, O.: Modeling and simulation of the defuzzification stage of a type-2 fuzzy controller using the Xilinx system generator and Simulink. Stud. Comput. Intell. **257**, 309–325 (2009)
9. Sepulveda, R., Montiel, O., Castillo, O., Melin, P.: Optimizing the MFs in type-2 fuzzy logic controllers, using the human evolutionary model. Int. Rev. Autom. Control **3**(1), 1–10 (2011)
10. Arce, F., García, M.: Personalization of learning object sequencing and deployment in intelligent learning environments. In: Recent Advances on Hybrid Intelligent Systems, pp. 559–567 (2013)
11. Ming, L., Mo, L., Luping, D., Elke, R., Murali, M.: Event stream processing with out-of-order data arrival department of computer science, Worcester Polytechnic Institute Worcester, Massachusetts 01609, USA (2007)
12. Ido, G., Tal, S., Maya, B., Inbal, R., Tal, D.: Swimming against the Streamz: Search and analytics over the enterprise activity stream. In: CIKM '12 Proceedings of the 21st ACM International Conference on Information and Knowledge Management, Israel, pp. 1587–1591 (2012)
13. Ido, G., Ariel, R., Inbal, R.: Personalized activity streams: sifting through the "river of news" RecSys '11. In: Proceedings of the Fifth ACM Conference on Recommender Systems, Israel, pp. 181–188 (2011)
14. Daniel, A., Don, C., Ugur, Ç., Mitch, C., Christian, C., Sangdon, L., Michael, S., Nesime, T., Stan, Z.: Aurora: A new model and architecture for data stream management. VLDB J. Int. J. Very Large Data Bases **12**(2), 120–139 (2003)
15. Wang, F., Liu, S., Liu, P., Bai, Y.: Bridging physical and virtual worlds: complex event processing for RFID data streams. In: Proceedings of the International Conference on Extending Database Technology, pp. 588–607 (2006)
16. Zengchang, Q.: Learning with fuzzy labels, submitted for the degree of doctor of philosophy department of engineering mathematics. University of Bristol, Bristol (2005)
17. Frits, T., Duin, R.: Dealing with a priori knowledge by fuzzy labels, pattern recognition group. In: Department of Applied Physics, Delft University of Technology, The Netherlands Received for Publication 22 Dec 1980
18. Gutiérrez, J., Hervé, J.: Using the theory of fuzzy logic in the classification of satellite images with mixed coverages: the urban case in Mérida, Venezuela. Interciencia **30**(5), 261–266 (2005)

19. Johannes, A., Magne, S., Janos, A.: Learning fuzzy classification rules from labeled data, department of process engineering, university of Veszprem, P.O. Box 158, H-8201 Veszprem, Hungary Received 5 Aug 2000; Accepted 7 Apr 2000
20. Hesse, R.: Triangle distribution: Mathematica link for excel, Graziadia Graduate School of Business, Pepperdine University U.S.A. (2000)
21. Snell, J., Atkins, M., Norris, W., Messina, C., Wilkinson, M., Dolin, R.: JSON activity streams 2.0 (2013)
22. Sombra, A., Valdez, F., Melin, P., Castillo, O.: A new gravitational search algorithm using fuzzy logic to parameter adaptation. In: IEEE Congress on Evolutionary Computation 2013, pp. 1068–1074 (2013)
23. Valdez, F., Melin, P., Castillo, O.: Evolutionary method combining particle swarm optimization and genetic algorithms using fuzzy logic for decision making. In: Proceedings of the IEEE International Conference on Fuzzy Systems, pp. 2114–2119 (2009)
24. Valdez, F., Melin, P., Castillo, O.: Parallel particle swarm optimization with parameters adaptation using fuzzy logic. MICAI **2**, 374–385 (2012)
25. Valdez, F., Melin, P., Castillo, O.: Bio-inspired optimization methods on graphic processing unit for minimization of complex mathematical functions. In: Recent Advances on Hybrid Intelligent Systems, pp. 313–322 (2013)

Automatic Estimation of Flow in Intelligent Tutoring Systems Using Neural Networks

Amaury Hernandez, Mario Garcia and Alejandra Mancilla

Abstract Flow is a mental state where a person is fully focused on an activity, and is enjoying performing it. Mihaly Csikszentmihalyi, who coined the concept, defines flow in terms of the skill and challenge levels of the activity as perceived by the person performing such activity. In this chapter, we propose the use of neural networks to predict if a student, after completing a computer-programming problem, is in a state of flow or not. To do so, we performed an experiment where we apply a very basic computer-programming tutorial to 21 students. We registered in a database how much time it took the students to finish the test, how many keystrokes they needed to press before achieving the goals of each exercise, how much time it took the student to start trying to solve the problem, the time between each keystroke, and how many attempts the student needed before successfully completing each exercise. Using these variables, we built a neural network that was capable of predicting if a student was in flow or not after the completion of each problem in the tutorial.

1 Introduction

A student interacting with a computer tutoring system will experience different mental states while performing the learning activities presented by the system [9]. There are several mental states the student can experience, such as apathy, if the student perceives the learning activity to be too easy, but yet, he perceives his

A. Hernandez · M. Garcia (✉) · A. Mancilla
Tijuana Institute of Technology, Calzada Tecnologico s/n, Tijuana, Mexico
e-mail: mariosky@gmail.com

A. Hernandez
e-mail: amherag@gmail.com

A. Mancilla
e-mail: alejandra.mancilla@gmail.com

O. Castillo et al. (eds.), *Recent Advances on Hybrid Approaches for Designing Intelligent Systems*, Studies in Computational Intelligence 547,
DOI: 10.1007/978-3-319-05170-3_43, © Springer International Publishing Switzerland 2014

skills not to be high enough while performing the activity; anxiety, if the difficulty is too high and his skill level is low; relaxation, if the difficulty is low, but the student's skills are high; and finally, the student should experience flow if the difficulty of an activity is high, yet, the student perceives himself as being capable of solving the problem. Mihaly Csikszentmihalyi established this relationship between difficulty and skill, and different mental states [3]. It is important to note that the difficulty and skill levels used for defining a student's flow must be the levels as perceived by the student himself, and not an observer, for example. Ensuring a high level of flow would certainly help in the learning process of a student. However, the first problem we encounter when trying to achieve this is the actual calculation of flow. How can we estimate a person's flow during the course of a learning activity? That's the question we try to answer in this chapter, and we believe the answer lies on the use of classification algorithms [22].

2 Basic Concepts

To fully understand the problem we are trying to solve, and the methodology used to solve it, we first need to know what is an interactive tutoring system, and what is flow.

2.1 Interactive Tutoring Systems

Interactive tutoring systems, also known as intelligent tutoring systems (ITS), are computer systems designed to help their users learn, by providing user-customized instruction. This customization of the user experience can be achieved by using intelligent computing techniques [13], such as recommender systems, fuzzy inference systems, and clustering techniques. These methods are usually used to optimize certain parts of the system, like the navigation and the structure of the learning activity, what learning objects must be presented to the learner to improve the user experience, the learning objects themselves and the capture of the user model.

Intelligent tutoring systems are often divided into four components (see Fig. 1): the domain model, the student model, the tutoring model, and the user interface model [14].

2.1.1 The Domain Model

This model, also known as the cognitive model, specifies all the requirements that must be met for each learning activity to be considered complete. Then, for example, we can define in this model the standards to be used to evaluate the

Fig. 1 Classical architecture
of an ITS

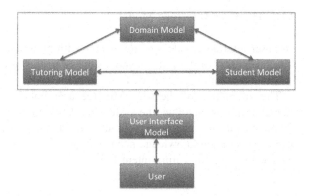

student's performance, and use this level of performance to determine if the system can continue with the next activity. It is noteworthy that this model does not determine what activity comes next, as the tutor model is the one that determines this. The domain model also has the content that is going to be shown to the learner.

2.1.2 The Student Model

Just as the domain model defines the concepts to be learned, the strategies to be followed by the student, the conditions to enable the continuation of the activity, i.e., the definition of the content of the intelligent tutoring system, the student model defines information of the student. From a more general perspective, this model would be called the user model.

The automatic estimation of the student's flow would be determined and stored in this model.

2.1.3 The Tutoring Model

The domain model defines the rules to be met before a student can continue with the next learning activity, but the tutoring model determines what activity follows. To do so, the tutoring model receives data from the domain and student models, and estimates a probability that the student has learned certain production rule. If the probability estimate reaches certain percentage, the system now determines which production rule should be next on trying to be satisfied by the student, and what activity should be used as the medium to estimate such probability. This probability is often 95 % [5].

2.1.4 The User Interface Model

The user interface model communicates with the other three models and decides what elements must be displayed to the user and how, in order to enable and facilitate a learning experience to the student.

The interaction of the user with the user interface can be thought of as a dialogue. In this dialogue, the system must be capable of understanding the input of the student and generating output understandable by the student. The system also needs domain knowledge needed for communicating content, and knowledge needed for communicating intent [15].

2.2 Flow

Sometimes, a person, while performing an activity, will experience a mental state where he feels focused, involved and is enjoying the course of the activity. This mental state is known as flow, a term proposed by Mihaly Csikszentmihalyi [4]. Flow is closely related to motivation, the main difference being that motivation is the purpose or psychological cause of any action [16]. Thus, a person can be very motivated to execute and continue the execution of certain action, but not necessarily be in a state of flow.

Often, people trying to focus his attention on a task will find they can not achieve a full focus, i.e., they are still aware of their surroundings. When a person is in a state of flow, this awareness disappears and all of the attention is given on the task at hand.

2.2.1 Conditions for Flow

In Fig. 2 we can see illustrated the relationship between an individual's skill level and the challenge or difficulty level of an activity. This is a model proposed by Csikszentmihalyi in 1997. As has been noted before, the skill and difficulty levels must be the levels as perceived by the person while performing the task.

As represented by the graph, to achieve a state of flow a person must perceive a high challenge level on the task at hand, but they need to feel comfortable performing such task. We can also see how other mental states can be achieved by varying the skill and challenge levels.

This flow model gives us a general idea of how an individual can attain flow. However, several researchers have found flaws in the model. One of the most recent models is the one proposed by Schaffer. Schaffer states that an individual must meet the following 7 conditions before one can achieve flow:

Fig. 2 Relationship between perceived skill level and perceived challenge level, and mental state [3]

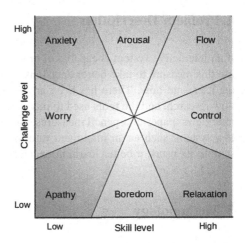

1. Knowing what to do
2. Knowing how to do it
3. Knowing how well you are doing
4. Knowing where to go
5. High perceived challenges
6. High perceived skills
7. Freedom from distractions [20].

It is noteworthy that the fourth condition is only needed when some sort of navigation is involved in the task being performed.

3 Related Work

Automatic calculation of flow has been covered in several studies. In [6], flow is automatically measured using image processing on the facial expressions of people. In [20], a system that monitors conversational cues, gross body language, and facial features to detect boredom, engagement, confusion, and frustration is presented. Most studies employ manual means to measure the flow of an individual after or during the course of an activity [11]. There are, nevertheless, studies where the objective is to automatically estimate an individual's performance [7], or other emotional states, such as interest [17]. We can find plenty of research on the relationship between several emotions and the interaction of a user and an intelligent tutoring system, such as motivation [1, 12], and anxiety [8].

4 Problem Description

If an interactive system is capable of predicting if a user is in flow, the system can use this measure to take better decisions for determining what content is shown to the user, and how. There has been an increasing interest in the methods used in flow research, and the tools used to measure flow are still manual methods, such as interviews, paper-and-pencil measures, and the Experience Sampling Method [11]. If a system, such as an intelligent tutoring system, could ensure a constant flow state to its users, a better learning experience could be achieved [5].

5 Methodology

We created a web application where a user interacts with a computer-programming tutorial. The tutorial is made of six questions about the programming language new LISP, with an increasing complexity and difficulty. To know if the user was experiencing flow, or not, he needs to respond two questions at the end of each exercise he has responded successfully: "How difficult did you find this exercise?" and "How do you feel after finishing this exercise?" The possible answers for the first questions were: "very easy", "easy", "normal", "hard", and "very hard". For the second question, the possible answers were (translated from Spanish): "I want to do more exercises" (flow), "attentive" (arousal), "desperate" (anxiety), "I'm not interested" (apathy), "relaxed" (relaxation), "everything's under control" (control), "I'm having difficulties" (worry), and "bored" (boredom). We used other answers that represented the mental states, as we discovered that students had difficulty understanding what the mental states really meant. In addition to the capture of this information, while the user is trying to solve each problem, the software records each keystroke the user is typing, along with how much time passed before the next keystroke was made; how much time he took to complete each question; how long it took him to start typing a possible solution; and how many attempts it took him to reach the correct solution. 21 students interacted with the system, generating 108 records that represent the student trying to solve one exercise. Not every student finished all the six questions, but this doesn't affect the experiment.

Our hypothesis is that the number of keystrokes and the total time it took the user to solve each problem could represent the user's perceived skill, and the time before beginning to write a possible solution and the time between each keystroke (we actually average these times, and then we average all the times that are greater than the latter calculated average; this represents an approximation of how much time the individual needed to think on how he could continue with his solution) could represent the user's perceived difficulty of the problem; then, using a classification algorithm, we can use these calculations as input variables to a model that determines if a person is experiencing flow at the end of an exercise. Although this is subjective guidance, as there is no scientific evidence proving these measures could represent skill and difficulty, the system is very limited on what information it can extract from the user. Also, we are not explicitly proposing a model where these measures represent skill and difficulty; a machine learning classification algorithm will be used to deduce a model from the data. Two models were built using these input variables: one that estimates if a user is in flow or not, and one that determines how difficult a user perceives a task. The models were built using the software RapidMiner 5.3 and the data sets we generated by the users of our tutorial.

For the validation of the models, k-fold cross validation and Cohen's kappa coefficient are used. In k-fold cross validation, the dataset used to train the model is split into k mutually exclusive subsets (the folds) of approximately equal size, and the model is trained k times. The cross validation estimate of accuracy is the overall number of correct classifications, divided by the number of instances in the dataset [10]. The kappa coefficient measures pairwise agreement among a set of reviewers making category judgments, correcting for expected chance of agreement [2].

5.1 Flow Model

For the estimation of flow, we created a classification neural network [21]. We filtered the data and regarded all the mental states but flow, to be "non-flow." This way, the model only needs to be able to classify a student as being on flow or not.

5.2 Difficulty Model

As in the flow model, we decided to use a classification neural network. We also filtered the data to generalize "very hard" and "hard" as just "hard", and "very easy" and "easy" as just "easy." The answer "normal" isn't transformed. Thus, our model is a polynomial classification neural network. We decided to create this additional model in order to observe how our input variables would behave while being used to predict another measure other than flow.

6 Results

The following results were gathered from neural networks taking as input variables the measures explained in our methodology. Genetic algorithms were used to optimize how many and what attributes to be used as input variables, and to optimize the learning rate and momentum parameters of the neural networks. Although we used 10-fold cross-validation to estimate the accuracy of our

Table 1 Accuracy and Cohen's kappa of the neural network for each exercise

Exercise #	Accuracy	Cohen's kappa
1	83.00 % ± 2.45 %	0.574 ± 0.056
2	70.00 % ± 8.94 %	0.155 ± 0.161
3	90.00 % ± 2.22 %	0.680 ± 0.078
4	61.18 % ± 18.45 %	0.078 ± 0.077
5	93.75 % ± 0.00 %	0.765 ± 0.000
6	100.00 % ± 0.00 %	1.000 ± 0.000

Table 2 Accuracy and Cohen's kappa of the neural network for each exercise

Exercise #	Accuracy	Cohen's kappa
1	83.00 % ± 9.80 %	0.442 ± 0.102
2	84.21 % ± 0.00 %	0.186 ± 0.000
3	82.22 % ± 10.77 %	0.248 ± 0.238
4	61.18 % ± 4.71 %	0.335 ± 0.078
5	83.75 % ± 5.00 %	0.518 ± 0.108
6	55.29 % ± 9.56 %	0.200 ± 0.114

predictive models, we also looped the process 5 times and averaged the accuracies of the neural network. At the end of the learning process of each neural network, we calculate Cohen's kappa coefficient to assess the true quality of the generated model. It is important to note that, as we're working with human beings, not machines, the results must be taken with skepticism, meaning that if the Cohen's kappa coefficient for an exercise was low, this doesn't necessarily mean the model isn't reliable. What we are trying to achieve is a sufficiently high accuracy and Cohen's kappa in a number of exercises, such that we can be certain that a model that can estimate one's flow and perceived difficulty, using the input variables described, should exist.

6.1 Flow Model

As we can see in Table 1, exercises 1, 3, 5 and 6 achieved very satisfactory results.

6.2 Difficulty Model

The results shown in Table 2 weren't very satisfactory, mainly because of the low Cohen's kappa coefficients in most of the exercises. However, the results of exercises 1 and 5 are potential evidence that difficulty could also be calculated using the proposed input variables.

7 Conclusion

We are certain that a model that estimates if an individual is in a state of flow or not can be effectively generated by using as input variables the number of keystrokes required to complete a task, the number of attempts before arriving to a correct solution, the time required to complete the task, the time it took an individual to start trying to solve the problem at hand, and the average time the user needed for thinking on how to continue with his solution. We are currently working on a recommender system that recommends learning objects to a user of an interactive intelligent system based on the user's flow. We'll use the models proposed in this work to increase the effectiveness and reliability of the system. Fuzzy logic could also be used to improve the model as in [18, 19].

References

1. Astleitner, H., Koller, M.: An aptitude-treatment-interaction-approach on motivation and student's self-regulated multimedia-based learning. Interact. Edu. Multimedia **13**, 11–23 (2006)
2. Carletta, J.: Assessing agreement on classification tasks: the kappa statistic. Comput. Linguist. **22**(2), 249–254 (1996)
3. Csikszentmihalyi, M.: Finding Flow: The Psychology of Engagement with Everyday Life. Basic Books, New York (1997)
4. Csikszentmihalyi, M.: Flow: The Psychology of Optimal Experience. LidovéNoviny, Praha (1990)
5. Corbett, A., John, A.: Student modeling and mastery learning in a computer-based programming tutor. In: Intelligent Tutoring Systems. Springer, Berlin (1992)
6. D'Mello, S., et al.: AutoTutor detects and responds to learners affective and cognitive states. In: Workshop on Emotional and Cognitive Issues at the International Conference on Intelligent Tutoring Systems (2008)
7. Endler, A., Günter, R., Martin, B.: Towards motivation-based adaptation of difficulty in e-learning programs. Australas. J. Educ. Technol. **28**(7), 1119–1135 (2012)
8. Keeley, J., Zayac, R., Correia, C.: Curvilinear relationships between statistics anxiety and performance among undergraduate students: evidence for optimal anxiety. Stat. Educ. Res. J. **7**(1), 4–15 (2008)
9. Kim, J., Frick, W.: Changes in student motivation during online learning. J. Educ. Comput. Res. **44**(1), 1–23 (2011)
10. Kohavi, R.: A study of cross-validation and bootstrap for accuracy estimation and model selection. IJCAI **14**(2), 1137 (1995)
11. Lopez, J., Charles, S. (eds.): The Oxford Handbook of Positive Psychology. Oxford University Press, Oxford (2011)
12. Martens, L., Gulikers, J., Bastiaens, T.: The impact of intrinsic motivation on e-learning in authentic computer tasks. J. Comput. Assist. Learn. **20**, 368–376 (2004)
13. Mödritscher, F., Garcia, M., Gütl, C.: The past, the present and the future of adaptive e-Learning. In: Proceedings of the International Conference on Interactive Computer Aided Learning (ICL2004) (2004)
14. Nkambou, R., Riichiro, M., Jacqueline, B. (eds.): Advances in intelligent tutoring systems, vol. 308. Springer, Berlin (2010)

15. Padayachee, I.: Intelligent Tutoring Systems: Architecture and Characteristics. University of Natal, Information Systems & Technology, School of Accounting & Finance, Durban (2002)
16. Schacter, L., Daniel, G., Daniel, W.: Psychology (Loose Leaf). Macmillan Higher Education, NY (2010)
17. Schiefele, U., Krapp, A., Winteler, A.: Interest as a predictor of academic achievement: a meta-analysis of research. In: Renninger, K.A., Hidi, S., Krapp, A. (eds.), The Role of Interest in Learning and Development. (pp. 183–212). Erlbaum, Hillsdale, NJ (1992)
18. Sepulveda, R., Castillo, O., Melin, P., Montiel, O.: An efficient computational method to implement type-2 fuzzy logic in control applications. Adv. Soft Comput. **41**, 45–52 (2007)
19. Sepulveda, R., Castillo, O., Melin, P., Rodriguez-Diaz, A., Montiel, O.: Experimental study of intelligent controllers under uncertainty using type-1 and type-2 fuzzy logic. Inf. Sci. **177**(10), 2023–2048 (2007)
20. Shaffer, O.: Crafting Fun User Experiences: A Method to Facilitate Flow. Human Factors International. Online White paper. http://www.humanfactors.com/FunExperiences.asp (2013)
21. Urias, J., Hidalgo, D., Melin, P., Castillo, O.: A method for response integration in modular neural networks with type-2 fuzzy logic for biometric systems. Adv. Soft Comput. **41**, 5–15 (2007)
22. Woolf, B., Burleson, W., Arroyo, I.: Emotional intelligence for computer tutors. In: Workshop on Modeling and Scaffolding Affective Experiences to Impact Learning at 13th International Conference on Artificial Intelligence in Education (2007)

Features and Pitfalls that Users Should Seek in Natural Language Interfaces to Databases

Rodolfo A. Pazos Rangel, Marco A. Aguirre, Juan J. González
and Juan Martín Carpio

Abstract Natural Language Interfaces to Databases (NLIDBs) are tools that can be useful in making decisions, allowing different types of users to get information they need using natural language communication. Despite their important features and that for more than 50 years NLIDBs have been developed, their acceptance by end users is very low due to extremely complex problems inherent to natural language, their customization and internal operation, which has produced poor performance regarding queries correctly translated. This chapter presents a study on the main desirable features that NLIDBs should have as well as their pitfalls, describing some study cases that occur in some interfaces to illustrate the flaws of their approach.

1 Introduction

Increasingly, a vast number of people in different areas use databases (DBs) for storing important information. Obtaining unusual or complex information from a database demands a broad knowledge on computing from users, such as a language for querying databases, and to be acquainted with the schema of the database to be queried.

R. A. Pazos Rangel (✉) · M. A. Aguirre · J. J. González
Instituto Tecnológico de Ciudad Madero, Ciudad Madero, Mexico
e-mail: r_pazos_r@yahoo.com.mx

M. A. Aguirre
e-mail: marco.itcm@gmail.com

J. J. González
e-mail: jjgonzalezbarbosa@hotmail.com

J. M. Carpio
Instituto Tecnológico de León, León, Mexico
e-mail: jmcarpio61@hotmail.com

O. Castillo et al. (eds.), *Recent Advances on Hybrid Approaches for Designing
Intelligent Systems*, Studies in Computational Intelligence 547,
DOI: 10.1007/978-3-319-05170-3_44, © Springer International Publishing Switzerland 2014

Currently, a growing need for using systems that allows common users (different from computer professionals) to obtain important information from databases is growing. This capability can be extremely important for making business decisions.

Natural language interfaces to databases (NLIDBs) emerged as an attractive solution that serves as a bridge between common users and the information they need for decision making.

Since the early 1960s, researchers have undertaken the development of NLIDBs capable of satisfying the needs of final users.

However, the complexity of the problems involved in the development of effective interfaces has prevented their widespread acceptance.

A comprehensive review of the state of the art and some of the most important problems in NLIDBs are described in [1].

This study of the state of the art on NLIDBs is intended mainly for DB administrators that are considering installing a NLIDB for querying a database. This chapter defines the desirable features that DB administrators should look for in a system of this type. Finally, since there exist different types of problems that prevent NLIDBs to be attractive to users (regarding portability, functionality and performance), this chapter describes the pitfalls that some interfaces have.

2 Desirable Features in a NLIDB

NLIDBs were created for retrieving information from DBs by means of NL communication. What is expected to obtain from this interaction is the correct answer to the query input by a user. However, besides this characteristic, there exist others that are considered important for a NLIDB for being considered attractive enough to meet the expectations of end users, mainly those from businesses. Next we describe desirable characteristics that a NLIDB should have.

2.1 Ease of Customization

One of the main characteristics that a NLIDB must have is easiness of customization: the interface should be easily and rapidly customized, requiring a minimal intervention by the DB administrator.

There exist interfaces such as C-Phrase that have an authoring tool, which facilitates the customization task.

2.2 Operability

Like all Graphical User Interfaces (GUIs), NLIDBs are expected to have a friendly user interface that enables the system to be easily operated.

Earlier NLIDBs had a command-line-based user interface, where users formulated NL queries by typing them in a simple text line; this was so because of the hardware and software limitations at that time. Simplicity and fastness were the main advantages of the usage of those NLIDBs; however, the very same simplicity limited their operation because of the lack of tools for user-interface interaction. An example of this type of interfaces is NLPQC.

As technology improved and the boom of graphical interfaces grew larger, many different types of GUIs were available, such as: menu-based, dialogue/discourse-based and multimodal.

Preferably, users should be guided by NLIDBs along query formulation.

2.3 Authoring

Customizing a NLIDB for a new domain is a complex task that could be time consuming. For facilitating the customization process, it is necessary that interfaces have a tool that permits modifying the knowledge base or the data dictionary.

An authoring tool should permit associating elements (DB tables, columns, relations between tables, etc.) of the DB schema to the linguistic elements (words or phrases) that can occur in NL queries, as well as including other elements that the interface might need, without compromising friendliness.

One of the few important NLIDBs that have an authoring tool is C-Phrase.

2.4 Habitability

The habitability is the ability of a NLIDB of knowing user expectations, without surpassing the linguistic capabilities of the system (limited grammar and linguistic coverage).

Obviously users expect the NLIDB to answer correctly all their queries; however, it can not be expected from the interface to be able of interpreting the queries as a human formulates them (including some jargon, words whose meaning is not in the data dictionary, etc.). For example, EasyAsk keeps terminology for business applications, but this would not be adequate if the interface were ported to an application of a different domain.

Some interfaces intend to increase their linguistic capacity by including domain-independent dictionaries, which are complemented with all the semantic information that is expected to be needed by users when querying a database for some specific domain.

2.5 Transparency

This characteristic establishes that the capabilities and limitations of a NLIDB must be evident to users. Though this information should be clear, usually the interfaces only describe their operation without making clear their capabilities and limitations. When something unexpected occurs (such as an erroneous translation or misinterpreted query, something that can not be customized, etc.), users might be disappointed because they do not know what has happened.

2.6 Robustness

A NLIDB must be capable of answering all the queries formulated by users. The system must cope with the problems that may arise when processing a query, as well as dealing with queries that involve aggregate functions, temporal queries, deductive queries, nested queries, and queries that involve negation.

2.7 Efficiency

A reason why a user would like to use a NLIDB is because he/she expects the interface to answer quickly. In the experiments reported on NLIDBs, the response time is not evaluated. Usually, the databases used for testing are small, and therefore response times are small. However, some NLIDBs (for example, ELF) have data dictionaries whose volume of information is proportional to the size of the database to be queried; therefore, the time for translating a query (from NL to SQL) increases with the size of the database, which would prevent using this type of NLIDBs for querying very large databases.

2.8 Accuracy

NLIDB users expect all the queries they formulate are correctly answered. This is considered the most important characteristic; however, many factors influence the results that can be obtained from an interface. These factors are related to natural language issues, interface customization, and the inner workings of the interface.

The average recalls achieved by state-of-the-art NLIDBs vary from 60 to 90 %, which are unsatisfactory for considering using them in businesses for decision making.

2.9 Intelligence

NLIDBs must be capable of answering temporal and deductive queries. Temporal queries are issued to a database that permits storing the history of values that a piece of data may adopt over time; while deductive queries are those that are based on deductions by using inference. Additionally, NLIDBs should be able to improve their performance (percentage of correctly answered queries) by feedback received from user-interface interaction.

Though intelligence is a desirable feature in an interface, the customization of this type of systems requires deep knowledge of their inner workings, which might make difficult their customization.

Interfaces such as DaNaLIX permit feedbacking the system by user-interface interaction.

2.10 Multimodality

Normally, most NLIDBs permit formulating queries through a keyboard; however, this might not be the most adequate in some situations. Current technology permits users to interact using mobile devices by means of speech, menus, graphical objects, touch screens, etc., which is expected to be adopted by NLIDBs for increasing their functionality.

There exist applications that use these types of interactions that operate as NLIDBs; however, they are designed for operating with just one domain, and consequently they have very good performance. Unfortunately, porting them to another domain would imply modifying their inner workings so they could answer correctly for the new domain.

Interfaces such as EasyAsk include voice recognition, and others like EDITE [2] use graphical objects for facilitating user-interface interaction.

2.11 Independence

There exist four types of independence in NLIDBs: domain, database management system, natural language, and hardware and software independence.

2.11.1 Domain Independence

Domain independence is one of the most important characteristics, since a NLIDB is expected to be portable to any database, so that it may answer queries related to the domain of interest.

An interface must have an architecture, and together with its data dictionary should permit customizing the interface for a new or different domain without requiring a lengthy intervention by the DB administrator.

An example of the information that must contain a data dictionary model can be seen in [3].

As previously mentioned, commercial interfaces such as ELF and EasyAsk have tools for semiautomatically customizing them for a specific domain, which usually is not good enough for correctly answering all the queries formulated by users. Additionally, EasyAsk provides utilities for customizing the interface to any domain by carrying out software design tasks, such as defining user requirements, design and implementation of the application and support, which can make more expensive using the interface.

2.11.2 DBMS Independence

A NLIDB must be able to retrieve information from a database independently of the database management system (DBMS). Some examples of DBMSs used are Oracle, Sybase SQL Server, PostgrestSQL, Microsoft SQL Server, Access, DB2, Informix, and MySQL.

To get an idea on the DBMSs that can be queried by NLIDBs, let us consider ELF which supports only Microsoft Access versions 2007 and 2010. Interfaces like DaNaLIX interact with XML databases; however, this type of databases does not require a typical DBMS, they simply use a framework that offers functionalities similar to those of a DBMS: query processing, rule insertion, etc.

DBMS independence is related to domain independence since it is useful for adapting the interface to any database regardless of its format (relational, ontology oriented, object oriented, etc.). Unfortunately, developing a NLIDB that includes this feature requires many implementation details.

2.11.3 Natural Language Independence

Most of the NLIDBs implemented support only English for formulating queries. Some, such as the ones described in [4–6], support a few other languages.

Some efforts have been devoted to developing NLIDBs capable of supporting multiple languages. To this end, it has been attempted to separate the syntactic parsing from the semantic analysis; however, the translation effectiveness varies widely depending on the syntactic and semantic complexity of every language, and only languages with similar syntax have been supported by multilingual interfaces. Some of the interfaces that have attempted to provide NL independence are those described in [7–11].

2.11.4 Hardware and Software Independence

A NLIDB must be able to be ported to any type of computer, and nowadays, to any wireless device (smart phone, tablet, notebook, PDA, etc.) regardless of the hardware and software that it uses.

The architecture of modern computers permits using many applications; however, more than hardware options, what might affect the use of a NLIDB in different devices is the software that supports it. This difficulty arises from the diversity of operating systems in the market together with the framework needed for supporting the interface.

2.12 Handling of Linguistic Phenomena

A NLIDB must consider the linguistic problems that may affect the meaning of a query.

Some linguistic problems are extremely complex, which can be observed when people communicate with each other, they might not follow syntactic or semantic rules, and their expressions usually involve syntactic and semantic ellipsis. Some of the most important linguistic problems are: anaphora, ellipsis and ambiguity. There exist several works that deal with these problems from the computational linguistic area; however, these problems have been overlooked in most NLIDBs.

3 Pitfalls in NLIDBs

Users expect NLIDBs to be intelligent and robust enough to answer correctly all their queries, as it has been observed in investigations based on the Wizard of Oz experiment. However, results are usually disappointing because of the extremely difficult problems that NLIDBs have to cope with.

Despite NL processing is one of the main fields of artificial intelligence, progress has been frustratingly slow, because researchers have underestimated the problems involved in NL processing.

Many interfaces have used different types of approaches and architectures for addressing existing problems, but they have not been satisfactory enough to date so as to achieve a performance (i.e., recall) close to 100 %. Furthermore, some NLIDB have design flaws that render them inefficient for large and complex databases (as illustrated in Subsects. 3.4 and 3.5).

In this section we describe the most important pitfalls that prospective users of NLIDBs should be aware of, grouped in six main categories: type of graphical user interface, domain indpendence, customization, escalability, translation process, and performance evaluation.

3.1 Type of Graphical User Interface

The purpose of using different types of GUI design techniques is facilitating user-NLIDB communication. As it has been mentioned, the easiest interface to use is the one that permits introducing a query in the command line; for this type of interfaces, their main limitation is that they do not permit a sophisticated user-interface interaction.

There exist many menu-based interfaces like the one described in [12]. Though it is desirable that an interface guides the user along query formulation, this type of interfaces does not permit formulating queries in free-text NL. This technique may require users to go through several steps for formulating a query, which might be lengthy.

On the other hand, the introduction of queries by voice (like the one used in EasyAsk) requires voice recognition that obtains a performance close to 100 %, which remains being a challenge.

The use of different techniques for developing GUIs provides more functionality to NLIDBs; however, adding more functionality to these systems involves the use of more computer resources and usually requires more training for operating them, which may result in making their use more difficult.

3.2 Domain Independence

Achieving domain independence in NLIDBs requires customizing them for a new or different domain. Additionally, it is very difficult for domain-independent interfaces to obtain recalls above 90 %, and achieving this performance usually involves a lengthy and complex customization process.

For making shure that a NLIDB is domain-indpendent, it must be tested with different databases, and these should have a complexity similar to databases found in large businesses. The complexity of a database can be measured by the number of tables, the overall number of columns, the number of columns with the same or similar information (for example, the number of columns for storing people names), and the structure of relations among tables (number of relations and cycles in the graph that represents the DB schema).

Despite the large number of NLIDBs developed, there only exist a small number of databases that have been used for testing them. Most have been evaluated with the Tang and Mooney databases [13], which involve three different domains (geography, restaurants and jobs). These databases were implemented in Prolog, do not have a relational structure, and their structure is simple since they have from 1 to 8 tables.

One of the most complex databases used for evaluation is ATIS [14] which is a relational DB that stores information on airline flights. Its complexity resides in its structure, since it involves 27 tables and 123 columns. Additionally, the graph that represents the DB schema has many relations among tables and cycles, and 85 % of the query corpus involves semantic ellipsis.

In conclusion, potential users of a NLIDB should find out if it has been tested with several databases of different domains and how good is its performance with complex databases.

3.3 Customization

The customization process consists of populating the data dictionary of a NLIDB with the information necessary for correctly interpreting queries formulated by users. This information consists of the association of elements of the DB schema (tables, columns and relations) with words/phrases that occur in NL queries.

Some interfaces have tools for carrying out a semiautomatic customization (such as EasyAsk, ELF, EnglishQuery, etc.), some others use learning techniques (like those described in [15–19]). However, these types of customization are not good enough for answering all the queries formulated by users; therefore, it is usually necessary to fine-tune the customization for improving interface performance.

Other interfaces (like C-Phrase) have an authoring tool that facilitates the customization process; however, a factor that greatly affects the quality of the customization is the ability of the customizer. Therefore, a desirable feature in a NLIDB is that it should have a tool that permits an easy and quick customization, and that it renders a customization quality that is independent of the customizer ability.

NLIDBs that use supervised learning techniques for query translation need to be trained with a set of queries, which can be considered as a customization process. Unfortunately, the performance of this type of NLIDBs depends on the corpus of training queries, which leads to two problems: selecting an adequate training corpus and fine-tuning the learned knowledge when the interface fails in answering some queries.

It is expected that users with a bachelor's degree in engineering in computer science be able to customize a NLIDB. However, a desirable feature of a NLIDB, concerning customization, is that it does not require from the customizer to learn neither specialized concepts (for example, linguistic terminology, logic programming terms, etc.) nor the inner workings of the interface.

3.4 Scalability

Some NLIDBs use approaches that are not adequate for very large databases; i.e., those with tables that have over one million rows. For example, some interfaces look into the database for search values present in queries in order to identify the column name needed for generating the SLQ expression when translating from NL to SQL. This approach, though effective for small databases, is not efficient for large ones.

Most of the NLIDBs have been tested with small databases that contain tables with less than 100,000 rows. In a test carried out with ELF, a database was used that included verbs, conjugations and lexical tags from Freeling [10], which contained approximately 497,000 rows in a table with three columns. In this test the compilation of the dictionary took around 5 min, and the result was that ELF could not answer any query, because the translation approach is not adequate for dealing with large databases.

3.5 Translation Process

The translation process of a NLIDB must be able to deal with all the possible problems that may occur when translating a NL query. Unfortunately, the complexity of NL query processing has been underestimated; therefore, most NLIDBs do not effectively deal with all the problems, rendering an unsatisfactory performance, especially for decision making in businesses.

In some NLIDBs statistical and machine learning approaches are used for translating queries; however, these techniques involve some degree of uncertainty, which has caused that their recalls reach only 60–85 %.

Other NLIDBs use templates or pattern matching. These types of approaches limit their performance, since they can only permit translating those queries that are predefined in the templates or those that match the pre-established patterns.

There exist some processing flaws that may affect query translation; for example, NLIDBs whose translation approach requires looking into the database for search values present in queries. In this translation approach, each value stored in the database is saved in the data dictionary together with the id of the column where the value is stored. This approach has a major drawback: when a new value is stored in the database, then the NLIDB will not be able to translate a query that involves this value because it will not be found in the data dictionary. Therefore, to mitigate this situation it is necessary to frequently update the data dictionary for databases that have many insertions and updates.

For exemplifying this situation, ELF was tested using the Northwind database. In this experiment a new row was inserted in table *Products* that contained the value "Cheddar" in column *ProductName*, and next the following query was formulated:

How much does Cheddar cost?

In this case the interface could not identify the column *ProductName* needed for generating the SQL expression because "Cheddar" was not in the data dictionary.

Another type of error occurs when a value is present in two or more DB columns. For example, in this experiment table *Products* had a row that contained the value "Chang" in column *ProductName*, next a row was inserted into table *Employees* that had the same value "Chang" in column *LastName*, then upon formulating the following query:

Show the data of the employee Chang

The interface got confused because it found the value "Chang" in tables *Products* and *Employees* and generated the following SQL expression (which involves information from tables *Employees, Products, OrderDetails* and *Orders*): *SELECT DISTINCT Employees.EmployeeID, OrderDetails.OrderID, Orders.ShipName, Employees.LastName FROM Orders, Employees, OrderDetails, Products, Orders INNER JOIN Employees ON Orders.EmployeeID = Employees.EmployeeID, Orders INNER JOIN OrderDetails ON Orders.OrderID = OrderDetails.OrderID, OrderDetails INNER JOIN Products ON OrderDetails.ProductID = Products.ProductID WHERE (Products.ProductName = "Chang").*

Finally, prospective users of NLIDBs should be aware that, in general, the more complex a database is the more likely it is that an interface will fail when translating a query from NL to SQL.

3.6 Evaluation

One of the problems that has prevented the progress of NLIDB technology is the lack of evaluation standards and well established benchmarks.

Another situation that usually occurs in NLIDB evaluation is that most interfaces are customized and tested by the authors, which does not guarantees the impartiality of the results.

Potential users of NLIDBs should be aware that there exist several metrics for measuring the performance of interfaces: accuracy, recall and F1. The most widely used for evaluating performance is accuracy, which is defined as the percentage of correctly answered queries with respect to the number of translated queries. However, this metric is not the one that end users should be more interested in but recall, which is defined as the number of correctly, answered queries with respect to all the queries input to the interface. It is important to point out that the value of recall is less than or equal to accuracy, and it is usually the case that some NLIDBs report very good accuracies but their recalls are not so good.

Concerning performance measurement, it is difficult to compare the performance reported for different NLIDBs, since there is no uniformity on what a *correct answer* means. The evaluations of some interfaces consider an answer *correct* even if it includes information additional to the one requested, while other evaluations consider an answer correct only if it includes just the expected information. This distinction might be extremely important for a business that intends using a NLIDB for critical decision making.

An example of this situation is shown in a test carried out using ELF and the Northwind database, where the following query was formulated:

Show the units in stock of product Chang

Though the phrase "units in stock" explicitly refers to column *Products.UnitsInStock*, the interface considers the word "units" as referring also to column *OrderDetails.Quantity*, and it also adds to the output the name of the product and the order number; which were not requested in the query, as the resulting SQL expression shows: *SELECT DISTINCT OrderDetails.Quantity, Products.UnitsInStock, OrderDetails.OrderID, Products.ProductName FROM OrderDetails, Products, OrderDetails INNER JOIN Products ON OrderDetails.ProductID = Products.ProductID WHERE (Products.ProductName = "Chang").*

Similarly to the previous example, when the following query was formulated:

How much does Chai cost?

One would expect that the result of the query only included the price of product "Chai"; however, the interface returned all the information of the row that contains "Chai", as the following resulting SQL expression shows: *SELECT DISTINCTROW Products.* FROM Products WHERE (Products.ProductName = "Chai").*

Strictly, there should only exist two types of evaluation for a query translation: correct and incorrect, which is what it really matters for end users. However, there exist performance evaluations, such as those reported in [20], where the interface generates more than one semantic interpretation of a query, in which the answer is evaluated with some percentage ρ of recall (where $0 < \rho < 100$ %). Considering partially-correct answers is not adequate, since end users are usually interested in fully correct answers, not partially correct answers. Another detail that potential users of NLIDBs should be aware of is that the evaluation of some interfaces report recall values, which are actually accuracy figures (see definition at the beginning of this subsection).

4 Conclusion

A little over 30 years ago, considerable progress in NLIDB technology was made. However, NLIDBs have not been widely accepted as expected because of the complexity of NL processing, customization and internal operation, which has resulted in unsatisfactory performance (low percentage of correctly answered queries).

Currently, the information contained in databases has achieved great importance in many areas (especially businesses), which has fostered the development of NLIDBs for facilitating inexperienced users accessing information.

In this chapter, the features that a DB administrator should seek when considering the use of a NLIDB are presented, which are: operability, authoring, habitability, transparency, robustness, efficiency, accuracy, intelligence, multimodality, and independence. Though these characteristics are very important, it is difficult that an interface includes all of them. Even with existing technology, developing a robust enough interface requires solving most of the problems, which is a complex task.

One of the main flaws in the design of NLIDBs is the approach used for dealing with the translation process, which has caused the stagnation of this research area: recall of most NLIDBs lies in the range 60–85 %, and it has been extremely difficult to increase it. Unfortunately, many approaches for NLIDB design have overlooked the complexity of the problems involved. Therefore, we think that an approach for dealing with very complex problems has to be used; for example, an approach that analyzes all the problems that may occur in NL to SQL translation and affective techniques for dealing with each problem.

References

1. Pazos, R., González, J., Aguirre, M., Martínez, J., Fraire, H.: Natural language interfaces to databases: an analysis of the state of the art. Recent Adv. Hybrid Intell. Syst. Stud. Comput. Intell. **451**, 463–480 (2013)
2. Reis, P., Mamede, N., Matias, J.: Edite: a natural language interface to databases: a new dimension for an old approach. In: Proceedings of the 4th International Conference on Information and Communication Technology in Tourism (1997)
3. Pazos, R., González, J., Aguirre, M.: Semantic model for improving the performance of natural language interfaces. In: Proceedings of the MICAI 2011 Mexican International Conference on Advances in Artificial Intelligence, pp. 277–290 (2011)
4. Jain, H.: Hindi language interface to databases. Master's thesis, Thapar University (2011)
5. Kovacs, L.: SQL generation for natural language interface. J. Comput. Sci. Control Syst. **2**(18), 19–22 (2009)
6. Meng, X., Wang, S.: NChiql: the Chinese natural language interface to databases. Lecture Notes in Computer Science 2113, pp. 145–154 (2001)
7. Boldasov, M., Sokolova, E., Malkovsky, M.: User query understanding by the InBASE system as a source for a multilingual NL generation module. Lect. Notes Comput. Sci. **2448**, 33–40 (2002)
8. Jung, H., Geunbae, G.: Multilingual question answering with high portability on relational databases. In: Proceedings Conference on Multilingual Summarization and Question Answering 19, pp. 1–8 (2002)
9. Kwiatkowski, T., Zettlemoyer, L., Goldwater, S., Steedman, M.: Inducing probabilistic CCG grammars from logical form with higher-order unification. In: Proceedings of the Conference on Empirical Methods in Natural Language Processing, pp. 1223–1233 (2010)
10. Padró Ll., Stanilovsky, E.: FreeLing 3.0: towards wider multilinguality. In: Proceedings of the Language Resources and Evaluation Conference (2012)
11. Pakray, P.: Keyword based multilingual restricted domain question answering. Master's thesis, Jadavpur University (2007)
12. Hallet, C., Scott, D., Power, R.: Composing questions through conceptual authoring. Comput. Linguist. **33**, 105–133 (2007)
13. Tang, L., Mooney, R.: Using multiple clause constructors in inductive logic programming for semantic parsing. In: Proceedings of 12th European Conference on Machine Learning, pp. 466–477 (2001)
14. Price, P.: Evaluation of spoken language systems: the ATIS domain. In: Proceedings of the DARPA Speech and Natural Language Workshop, pp. 91–95

15. Clarke, J., Goldwasser, D., Chang, M.W., Roth, D.: Driving semantic parsing from the world's response. In: Proceedings 14th Conference on Computational Natural Language Learning, pp. 18–27 (2010)
16. Giordani, A., Moschitti, A.: Translating questions to SQL queries with generative parsers discriminatively reranked. In: Proceedings of the Conference on Computational Linguistics (Posters), 401–410 (2012)
17. Kate, R., Mooney, R.: Using string-kernels for learning semantic parsers. In Proceedings of 21st ICCL and 44th Annual Meeting of the Association for Computational Linguistics, pp. 913–920 (2006)
18. Liang, P., Jordan, M., Klein, D.: Learning dependency-based compositional semantics. In: Proceedings of 49th Annual Meeting of the Association for Computational Linguistics, pp. 590–599 (2011)
19. Lu, W., Tou, H.N., Lee, W.S., Zettlemoyer, L.: A generative model for parsing natural language to meaning representations. In Proceedings of Conference on Empirical Methods in Natural Language Processing, pp. 783–792 (2008)
20. Kaufman, E., Bernstein, A., Fischer, L.: NLP-Reduce: A "naïve" but domain-independent natural language interface for querying ontologies. In: Proceedings of the 4th European Semantic Web Conference, pp. 1–2 (2007)

Step Length Estimation and Activity Detection in a PDR System Based on a Fuzzy Model with Inertial Sensors

Mariana Natalia Ibarra-Bonilla, Ponciano Jorge Escamilla-Ambrosio, Juan Manuel Ramirez-Cortes, Jose Rangel-Magdaleno and Pilar Gomez-Gil

Abstract This chapter presents an approach on pedestrian dead reckoning (PDR) which incorporates activity classification over a fuzzy inference system (FIS) for step length estimation. In the proposed algorithm, the pedestrian is equipped with an inertial measurement unit attached to the waist, which provides three-axis accelerometer and gyroscope signals. The main goal is to integrate the activity classification and step-length estimation algorithms into a PDR system. In order to improve the step-length estimation, several types of activities are classified using a multi-layer perceptron (MLP) neural network with feature extraction based on statistical parameters from wavelet decomposition. This work focuses on classifying activities that a pedestrian performs routinely in his daily life, such as walking, walking fast, jogging and running. The step-length is dynamically estimated using a multiple-input–single-output (MISO) fuzzy inference system. Results provide an average classification rate of 87.49 % with an accuracy on step-length estimation about 92.57 % in average.

M. N. Ibarra-Bonilla · J. M. Ramirez-Cortes (✉) · J. Rangel-Magdaleno
Department of Electronics, Instituto Nacional de Astrofísica,
Óptica y Electrónica, Tonantzintla, Puebla, México
e-mail: jmram@inaoep.mx

P. J. Escamilla-Ambrosio
Centro de Investigacion en Computacion, Instituto Politécnico Nacional,
Mexico City, Mexico

P. Gomez-Gil
Computer Science Department, Instituto Nacional de Astrofísica,
Óptica y Electrónica, Tonantzintla, Puebla, México
e-mail: pgomezexttu@gmail.com

O. Castillo et al. (eds.), *Recent Advances on Hybrid Approaches for Designing*
Intelligent Systems, Studies in Computational Intelligence 547,
DOI: 10.1007/978-3-319-05170-3_45, © Springer International Publishing Switzerland 2014

1 Introduction

Pedestrian dead reckoning is a navigation technique that provides and maintains the geographical position for a person travelling on foot by using self-contained sensors. The current advancements of personal mobile devices with low cost MEMS sensors such as accelerometers, gyroscopes and magnetometers, have made PDR a relevant approach for the development of applications called location based services (LBS), with potential use in GPS-denied environments. A broad range of LBS applications are continuously emerging in the following market segments: emergency location, personal child security, information directory service, push/pull advertising, and many others. A fundamental component common to all LBS is the use of positioning technologies to track the movement of mobile users and to deliver information services to them on the move at the right time and right location [1].

In PDR techniques, estimation of each new position is based on the previous one derived from the last step, taking advantage of the sequential nature of pedestrian motion. Furthermore, displacements are calculated based on the estimation of a step length. The step length is a time-varying process which is strongly correlated to the velocity and the step frequency of the pedestrian [2]. For that reason some works have incorporated pedestrian activity classification in order to improve the accuracy of the distance traveled calculation, and consequently the localization accuracy [2–6]. In [2] a probabilistic neural network (PNN) is employed to classify acceleration samples with information about walking, irregular motions and standing still. The acceleration samples are obtained from an Inertial Measurement Unit (IMU) attached to the user's waist. In Ref. [3] five IMUs, containing a triaxial accelerometer, triaxial gyroscope and a triaxial magnetomer, are placed on five different position on the user's body, performing localization simultaneously with activity recognition. They present experimental results in 2D and 3D for classifying three activities: walking, standing and turning activities. In order to recognize those activities they use sensors located in right and left legs. In the 2D case they use a rule-based classifier and in 3D case they analyze stairs activity using a K-nearest neighbors (k-nn) classifier.

Reference [4] presents activity and environment classification using a foot mounted device with IMUs and a GPS receiver. They propose an algorithm that classifies the following activities: stationary, crawling, walking, running, biking, moving in vehicle, level up or down elevator and up or down stairs. Multiple probability density functions that map each feature (i.e. metric) to an activity are provided, and a naive Bayesian probabilistic model is used to determine the probability of each activity. The principal disadvantage of the systems described in references [3, 4] is the use of many sensors attached on the user's body or on the foot, which is not appropriated for some LBS applications, for instance, a situation involving information directory services. References [5, 6] use Hidden Markov Models (HMM) for classification of the following activities: standing, walking, going up and down stairs, jogging and running. Both approaches combine accelerometer and gyroscope measurements, and [5] incorporates the use of a magnetometer.

In this work, we present a fuzzy model aiming to perform step length estimation and activity detection in the context of a pedestrian dead reckoning project, using inertial sensors. The project uses a single inertial measurement unit with triaxial accelerometer and gyroscope, which have characteristics adequate to be used in human activity analysis [7, 8].

Effective algorithms are required to interpret the accelerometer data in the context of different activities. Some common approaches to automatic classification of human activity are based on machine learning techniques, especially Hidden Markov Models (HMMs) [5, 6, 9]. The use of HMM is attractive, although they present some difficulties on parameter estimation. These techniques typically operate through a two-stage process: first, features are obtained from sliding windows of accelerometer data and then a classifier is used to identify the activity [10]. A range of different approaches has been used to obtain features from accelerometer data; some works obtain features directly from the time-varying acceleration signal [11, 12] and others from frequency analysis [2, 13]. More recently, wavelet analysis has been used to derive time–frequency features [10, 14, 15].

In this work a wavelet analysis is applied to triaxial accelerometer data in order to identify points in the signal where a pedestrian changes from one activity to another. In search of incorporating the classification algorithm into a PDR system we propose a dynamical method for estimating the step length using a Fuzzy Inference System (FIS), which uses the activity pedestrian as an additional input. In this work we proposes the use of a single device attached to the pedestrian waist rather than multiple devices distributed across the body or on the user's foot. In that way, the developed techniques can be easily incorporated into personal smartphones with IMU sensors, which nowadays are becoming of popular use. With that purpose we are considering activities that a pedestrian performs in his daily life, such as: walking, walking fast, jogging and running.

The organization of this chapter is as follows: Sect. 2 describes the proposed methodology on step-length estimation incorporating pedestrian activity detection. Section 3 presents some experimental results. Conclusions and future work are discussed in Sect. 4.

2 Fuzzy Model for Step Length Estimation

The task of determining the activity a pedestrian is executing is inherently a classification problem. Some researchers, particularly in the field of biomechanics, have determined that gait trajectory signals have nonlinear and non-stationary characteristics [16]. In nonlinear or complex classification problems, neural networks, which have gained prominence in the area of pattern recognition, have several properties that make them attractive. Figure 1 presents the general block diagram of the proposed algorithm. Step length is estimated using a fuzzy inference system, with information obtained from positive-going and negative-going zero crossing direction of normalized signals obtained from the three-axis

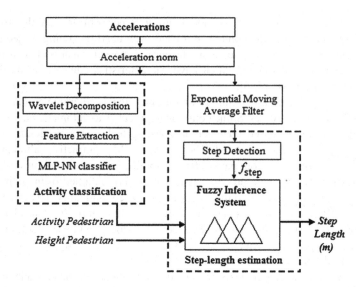

Fig. 1 Block diagram of the proposed architecture for activity classification and step-length estimation

accelerometer. The system incorporates information corresponding to the subject activity, which is simultaneously obtained from the IMU signals. A feature vector constructed with statistical information obtained from wavelet coefficients, is fed into a multilayer perceptron neural network, which provides information about the subject activity.

2.1 Pedestrian Activity Classification Algorithm

Classification of pedestrian activity is achieved by using the measurements from the three axial accelerometer clusters of the STEVAL-MKI062V2 [17]. The STEVAL-MKI062V2 is an inertial measurements unit (IMU) which includes accelerometers, gyroscopes and magnetometers, as well as pressure and temperature sensors to provide 3-axis sensing of linear, angular and magnetic motion, complemented with temperature and barometer/altitude readings. Therefore, this unit constitutes a platform with 10 degrees of freedom (DOF). The STEVAL-MKI062V2 includes the LSM303DLH, which is a system-in-package featuring a 3D digital linear acceleration sensor. The LSM303DLH has a linear acceleration full-scale of ± 2 g/± 4 g/± 8 g, which can be selected by the user as needed. Figure 2 shows the wearable sensor module and the way it is attached to the test subject.

Four types of activity patterns were collected from 26 subjects. The height of these subjects is located in a range between 1.50 and 1.78 m. The four activity

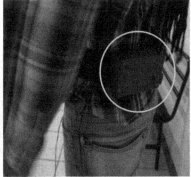

Fig. 2 Wearable sensor module and the way it is attached to the test subject

patterns were specifically walking, walking fast, jogging and running. The data were collected from the triaxial accelerometer every 20 ms, which corresponds to 50 samples per second.

2.1.1 Feature Extraction

Features were extracted from the norm of the acceleration raw signals corresponding to the three axis Ax, Ay, Az, at the time k. The norm vector is obtained as:

$$A_k = \sqrt{A_{x_k}^2 + A_{y_k}^2 + A_{z_k}^2},$$ (1)

The mean of the norm signal is obtained through averaging over a sliding window of N samples, with $N = 100$ in this experiment.

$$\bar{A}_k = \frac{1}{N} \sum_{k-N-1}^{k} A_k,$$ (2)

which in this case is equivalent to 2 s of data at a sampling frequency of 50 Hz. Subtracting (2) from the norm acceleration signal it results in:

$$Acc_k = A_k - \bar{A}_k,$$ (3)

where Acc_k is the normalized acceleration signal at time k, with the effect of gravity removed.

In this work, a feature extraction process is implemented using wavelet decomposition. The original signal is decomposed into a series of approximation and detail coefficients containing spectral and temporal information. Statistical information of these coefficients is used to form the feature vector for classification purposes in a dynamical way [10].

The Discrete Wavelet Transform (DWT) is a transformation that can be used to analyze the temporal and spectral properties of non-stationary signals. The DWT is defined by the following Eq. [18]:

$$W(j,k) = \sum_j \sum_k f(x) 2^{-j/2} \psi(2^{-j}x - k) \tag{4}$$

The set of functions $\psi_{j,k}(n)$ is referred to as the family of wavelets derived from $\psi(n)$, which is a time function with finite energy and fast decay called the mother wavelet. The basis of the wavelet space corresponds then, to the orthonormal functions obtained from the mother wavelet after scale and translation operations. The definition indicates the projection of the input signal into the wavelet space through the inner product, then, the function $f(x)$ can be represented in the form:

$$f(x) = \sum_{j,k} d_j(k) \psi_{j,k}, \tag{5}$$

where $d_j(k)$ are the wavelet coefficients at level j. The coefficients at different levels can be obtained through the projection of the signal into the wavelets family as expressed in Eqs. (6) and (7).

$$\langle f, \psi_{j,k} \rangle = \sum_l d_l \langle f, \varphi_{j,k+l} \rangle \tag{6}$$

$$\langle f, \phi_{j,k} \rangle = \frac{1}{\sqrt{2}} \sum_l c_l \langle f, \varphi_{j-1,2k+l} \rangle \tag{7}$$

The DWT analysis can be performed using a fast, pyramidal algorithm described in terms of multi-rate filter banks. The DWT can be viewed as a filter bank with octave spacing between filters. Each sub-band contains half the samples of the neighboring higher frequency sub-band. In the pyramidal algorithm the signal is analyzed at different frequency bands with different resolution by decomposing the signal into a coarse approximation and detail information. The coarse approximation is then further decomposed using the same wavelet decomposition step. This is achieved by successive high-pass and low-pass filtering of the time signal and a down-sampling by two [19], as defined by the following Eqs. (8) and (9):

$$a_j(k) = \sum_m h(m - 2k) a_{j+1}(m) \tag{8}$$

$$d_j(k) = \sum_m g(m - 2k) a_{j+1}(m) \tag{9}$$

Figure 1 shows a three-level filter bank. Sets cA and cD are known as approximation and detail coefficients, respectively. The wavelet decomposition is obtained through a sliding window of 100 samples, using the wavelet toolbox included in Matlab, with a three-level wavelet decomposition applied to the norm of the acceleration signal and a Daubechies 5/7 mother wavelet. Figure 3 shows in

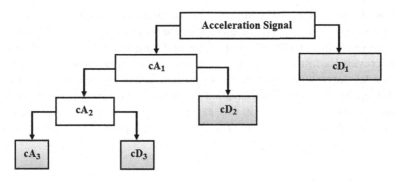

Fig. 3 Wavelet decomposition

shaded boxes the target frequency band signals selected for feature extraction. Two statistical features, mean and standard deviation, were extracted from each coefficient, giving a total of eight attributes in order to discriminate the four activities; walking, walking fast, jogging and running.

2.1.2 Classification Using a Multi-layer Perceptron Neural Network

Activity classification was performed using a feed forward multilayer perceptron neural network. The perceptron used in this work was trained using the Levenberg–Marquardt back propagation algorithm [20]. The backpropagation algorithm used in the training of multilayer perceptrons, is formulated as a non-linear least-squares problem. Essentially, the Levenberg–Marquardt algorithm is a least-squares estimation method based on the maximum neighborhood idea. Let $E(w)$ be an objective error function made up of m individual error terms $e_i^2(w)$ as follows:

$$E(w) = \sum_{i=1}^{m} e_i^2(w) = \|f(w)\|^2, \tag{10}$$

where

$$e_i^2(w) = (y_{di} - y_i)^2, \tag{11}$$

y_{di} is the desired value of output neuron i, and y_i is the actual output of that neuron. The aim of the Levenberg–Marquardt algorithm is to compute the weight vector $w$such as $E(w)$ is minimized. In each iteration the weight vector is updated according to Eq. (12):

$$w_{k+1} = w_k + \delta w_k, \tag{12}$$

where

$$\delta w_k = -\left(J_k^T f(w_k)\right) \left(J_k^T J_k + \lambda I\right)^{-1}. \tag{13}$$

J_k is the Jacobian of f evaluated at w_k, λ is the Marquardt parameter, and I is the identity matrix. The number of epochs in the training phase differs from one example to another, however, the Levenberg–Marquardt back propagation algorithm provided a fast convergence. Structure of the neural network consisted of two hidden layer with eight neurons and one output layer with four neurons corresponding to the four classification conditions. The neural networks toolbox included in Matlab was used in this experiment. Two-fold cross validation was employed in the validation stage. The procedure is explained in detail in Sect. 3.

2.2 Step Length Estimation Algorithm

Pedestrian dead reckoning (PDR) is a navigation technique based on position estimation of a person travelling on foot. A PDR algorithm is expected to provide information about a pedestrian position during a natural walk. A PDR algorithm can be divided into three principal parts: step detection, step length estimation and heading determination [21]. In this work, a proposal for step detection and step length estimation to be used in a PDR algorithm is presented. Next subsections describe in detail the proposed procedures.

2.2.1 Step Detection

Step detecting stage is fundamental in a PDR algorithm; if the step detection is inaccurate is not possible to determine the distance traveled by the subject, and consequently an estimation of the current position is not feasible [21]. There are two basic methods to detect pedestrian steps based on the measurements provided by the accelerometers: peak detection [22] and zero-crossing detections [23]. These methods consist of comparing the acceleration values with the predefined thresholds and taking the minimum step period into account. In waist attached devices, peak detection is not appropriate because peaks of acceleration also occur in irregular motions, such as turning when avoiding obstacles on a crowded road. It is difficult to distinguish between these peaks and those measured during regular walking, thus step misdetections are prone to occur. Furthermore, methods based on comparing the acceleration values with the predefined thresholds are not appropriate when the pedestrian activity is not constant because the time and amplitude characteristics of the acceleration measures change during the walk. So, in order to make the step count more reliable and to improve the accuracy of the distance traveled calculation we propose an algorithm to identify valid steps. Detection is performed by the positive-going and negative-going zero crossing direction of the normalized signal obtained from the three-axis accelerometer according to Eq. (3). In order to improve the performance of the step detection, an initial smoothing of the acceleration signal is performed. Raw normalized

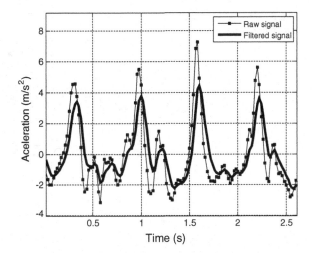

Fig. 4 Normalized raw and filtered signals obtained during the walk

acceleration signals are passed through an exponential weighted moving average filter defined in Eq. 4, as follows:

$$\hat{Acc}_k = \alpha \, Acc_k + (1 - \alpha) \, \hat{Acc}_{k-1}, \tag{14}$$

where $\hat{Acc}$ indicates the estimated filtered acceleration signal, Acc is the raw acceleration signal, $0 < \alpha < 1$ is a filter parameter, and k is the discrete time variable. In this application $\alpha = 0.3$ and the initial condition is assumed as the initial measured value of the acceleration signal. An example of the raw and filtered signals is shown in Fig. 4.

Then, positive-going and negative-going zero crossing direction of the signal $\hat{Acc}$ are detected in order to determine the interval where a step as occurred. Once the algorithm determines the zero crossing positions it proceeds to detect the maximum peak position which is present between two consecutive zero crossings. An example of this procedure is shown in Fig. 5. Squares indicate the maximum peaks and circles indicate the zero crossings detected. The next procedure is to determine which of those peaks represent valid steps. A valid step occurs when the foot touches the ground and this is recorded by the accelerometer as a peak of large magnitude with short duration. As shown in Fig. 5, the valid steps correspond with the largest peaks. It can be noticed the presence of peaks with smaller magnitudes which do not correspond to valid steps.

In some reported works, valid steps are detected by comparing the acceleration signals with some predefined threshold [23, 24]. The proposed algorithm identifies the valid steps by measuring the width of each detected peak in the time domain, which corresponds to the range comprised between two consecutive zero crossing with alternate directions. Length of the measured width is stored in a variable called *range* (R_{ng}) as show in Fig 6. The algorithm compares the value of R_{ng} against a sliding window average (R_{mean}). This comparison is performed by computing the ratio between R_{ng} and R_{mean}. If the ratio is greater than some value

Fig. 5 Zero crossing and
maximum peaks detection

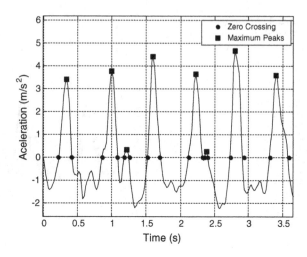

Fig. 6 Scheme of peaks
discrimination

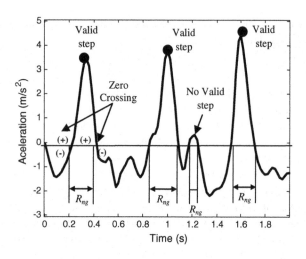

previously specified the event is considered a valid step and its position is stored.
Once the pedestrian steps are detected, the step frequency is calculated based on
the step period, which is obtained by measuring the time between consecutive
valid steps.

2.2.2 Step Length Estimation

The step length assessment is fundamental to the calculation of distance traveled.
In this work the step length is dynamically estimated using a fuzzy inference
system (FIS). In the proposed FIS, three inputs linguistic variables are defined:
step frequency, with partition in three fuzzy sets labeled as slow, regular and fast,

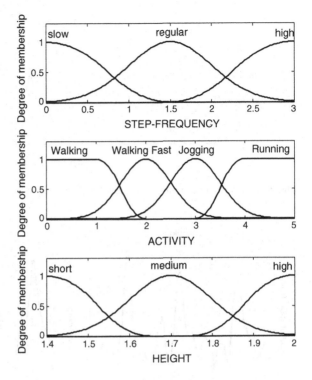

Fig. 7 Membership functions of the input linguistic variables

pedestrian activity, defined by four fuzzy sets with labels walking, walking fast, jogging and running, and *pedestrian height*, which is defined by three membership functions with labels short, medium and high. The membership functions of the input linguistic variables are shown in Fig. 7. The output variable is defined as *step length*, with linguistic values defined by eight Gaussian fuzzy sets, as shown in Fig. 8. Therefore, 36 rules complete the rule base defined in the FIS system. An example of one rule is:

IF (step-freq is slow) AND (activity is walk) AND (height is short) THEN (step-length is very small)

The fuzzy operators min, max-product and centroid method are used as the fuzzy operators of intersection compositional rule of inference and defuzzification, respectively.

3 Experimental Results

Several experiments were carried out with 26 subjects wearing the IMU STEVAL-MKI062V2 attached to the waist. The subjects were asked to walk in a straight line a distance of 100 m in each activity pattern. Data processing for activity

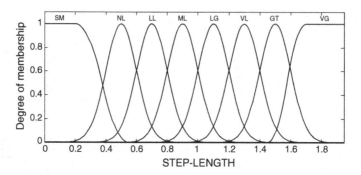

Fig. 8 Membership functions of the output linguistic variable

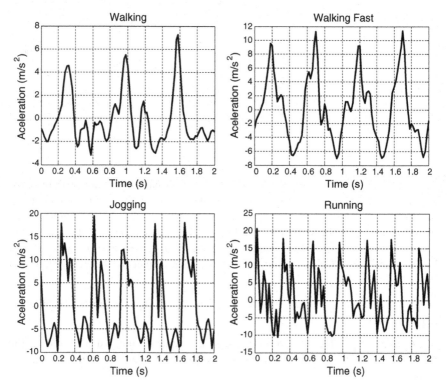

Fig. 9 Raw and filtered normalized acceleration signal acquired in walking

classification and step-length estimation was carried out off-line using the described algorithm. A typical example of 2 s raw acceleration signal is shown in Fig. 9. The first experiment corresponds to neural network-based activity classification. The available data is divided in two groups in order to perform two-fold cross validation. The first set is used for training and the second set is used for

Table 1 Classification accuracy (%) confussion matrix for the four walking patterns by applying the training process

	Walking	Walking fast	Jogging	Running
Walking	92.3	7.7	0	0
Walking fast	7.7	92.3	0	0
Jogging	0	0	88.45	11.55
Running	0	0	7.7	92.3
Overall accuracy	91.33			

Table 2 Classification accuracy (%) confussion matrix for the four walking patterns by applying the testing process

	Walking	Walking fast	Jogging	Running
Walking	84.61	15.39	0	0
Walking fast	3.85	88.45	7.7	0
Jogging	0	7.7	84.6	7.7
Running	0	0	7.7	92.3
Overall accuracy	87.49			

Table 3 Accuracy results of step length estimation

	Walking	Walking fast	Jogging	Running
Accuracy (%)	94.88	93.91	91.3	90.19
Overall accuracy	92.57 %			

generalization; further, both sets are interchanged and the process is repeated. Average classification rate is presented in the form of confusion matrices. Table 1 presents the confusion matrix obtained from the classification rate of the four activity patterns during the training process. Table 2 show similar results corresponding to the generalization process.

The activity detection procedure was then incorporated in step detection as described in the block diagram of Fig. 1. In general the step detection algorithm detected successfully all pedestrian steps, even when a change in the activity was presented, with a detection rate of 100 %. Final results corresponding to step-length estimation, and its effect in the computation of total travelled distance presented an accuracy of 92.57 % in average. Table 3 presents the accuracy of the step-length estimation algorithm when a pedestrian travels a distance of 100\,m.

4 Conclusions

In this work, an approach on the use of IMUs for pedestrian activity classification and step-length estimation has been presented. The results obtained show that wavelet decomposition and neural networks represents a good choice of feature extraction and classification system, respectively. The step length is dynamically estimated using a fuzzy inference system (FIS) which incorporates classification activity during the estimation process. This operation allows the system to adjust computation of the total distance travelled by a pedestrian according to the ongoing activity. The described system has been developed in the context of a PDR system aiming to provide information about location and movements of a person travelling in a GPS-denied environment. Future work includes integration of the presented system with height and attitude estimation algorithms, and extensive testing in walking paths including stairs and ramps.

Acknowledgments The first author acknowledges the financial support from the Mexican National Council for Science and Technology (CONACYT), scholarship No. 237756.

References

1. Gartner, G., Ortag, F.: Advances in Location-based Services, Lecture Notes in Geoinformation and Cartography. Springer-Verlag, Berlin (2012)
Sun, Z., Mao, X., Tian, W., Zhang, X.: Activity classification and dead reckoning for pedestrian navigation with wearable sensors. Measur. Sci. Technol. **20**(1), 1–10 (2009)
3. Altun, K., Barshan, B.: Pedestrian dead reckoning employing simultaneous activity recognition cues. Measur. Sci. Technol. **23**(2), 1–20 (2012)
4. Bancroft, J.B., Garrett, D., Lachapelle, G.: Activity and environment classification using foot mounted navigation sensors. In: Proceedings of International Conference on Indoor Positioning and Indoor Navigation, pp. 13–15. Sydney, NSW (2012)
5. Chen, X., Hu, S., Shao, Z., Tan, J.: Pedestrian positioning with physical activity classification for indoors. In: Proceedings of IEEE International Conference on Robotics and Automation, pp. 1311–1316. Shanghai, China May, 2011
6. Panahandeh, G., Mohammadiha, N., Leijon, A., Handel, P.: Continuous hidden markov model for pedestrian activity classification and gait analysis. IEEE Trans. Instrum. Meas. **62**(5), 1073–1083 (2013)
7. Mathie, M.J., Celler, B.G., Lovell, N.H., Coster, A.C.F.: Classification of basic daily movements using a triaxial accelerometer. Med. Biol. Eng. Comput. **42**(5), 679–687 (2004)
8. Yang, C.C., Hsu, Y.L.: A review of accelerometry-based wearable motion detectors for physical activity monitoring. Sensors **10**(8), 7772–7788 (2010)
9. Mannini, A., Sabatini, A.M.: Accelerometry-based classification of human activities using markov modeling. Comput. Intell. Neurosci. **2011**, 2–9 (2011)
10. Preece, S.J., Goulermas, J.Y., Kenney, L.P.J., Howard, D.: A comparison of feature extraction methods for the classification of dynamic activities from accelerometer data. IEEE Trans. Biomed. Eng. **56**(3), 871–879 (2009)
11. Ravi, N., Dandekar, N., Mysore, P., Littman, M.L.: Activity recognition from accelerometer data. In: Proceedings of 17th Conference on Innovative Applications of Artificial Intelligence, vol. 5, pp. 1541–1546. Pittsburgh, Pennsylvania (2005)
12. Kwapisz, J.R., Weiss, G.M., Moore, S.A.: Activity recognition using cell phone accelerometers. ACM SIGKDD Explor. Newslett. **12**(2), 74–82 (2011)
13. Karantonis, D.M., Narayanan, M.R., Mathie, M., Lovell, N.H., Celler, B.G.: Implementation of a real-time human movement classifier using a triaxial accelerometer for ambulatory monitoring. IEEE Trans. Inf. Technol. Biomed. **10**(1), 156–167 (2006)
14. Ning, W., Ambikairajah, E., Lovell, N.H., Celler, B.G.: Accelerometry based classification of walking patterns using time-frequency analysis. In: Proceedings of 29th IEEE Annual

International Conference Engineering in Medicine and Biology Society, pp. 4899–4902. Lyon, France, 22–26 Aug 2007

15. Yunqian, M.A., Hesch, J.A.: Gait classification using wavelet descriptors in pedestrian navigation. In: Proceedings of 24th Institute of Navigation GNSS Conference, pp. 1328–1337. Portland, Oregon (2011)

16. West, B.J., Scafetta, N.: Nonlinear dynamical model of human gait. Phys. Rev. E. **67**(5), 1–10 (2003)

17. ST Microelectronics: UM0937 User manual. http://www.st.com/st-web-ui/static/active/en/resource/technical/document/user_manual/CD00271225.pdf. Visited 05 Dec 2013

18. Priestley, M.B.: Wavelets and time-dependent spectral analysis. J. Time Ser. Anal. **17**(1), 85–103 (2008)

19. Pinsky, M.A.: Introduction to Fourier Analysis and Wavelets. Graduate Studies in Mathematics, (102), American Mathematical Society (2009)

20. Demuth, H., Beale, M.: Neural Network Toolbox for Use with MATLAB, pp. 5-28–5-30. The Mathworks Inc, Natick (2001)

21. Kim, Y.K., Park, J.H., Kim, H.W., Hwang, S.Y., Lee, J.M.: Step estimation in accordance with wear position using the 3-axis accelerometer. In: Proceedings of 3rd SPENALO International Symposium. Bexco, Busan, Korea, Sep 2011

22. Nam, Y.: Map-based indoor people localization using an inertial measurement unit. J. Inf. Sci. Eng. **27**(4), 1233–1248 (2011)

23. Ibarra-Bonilla, M.N., Escamilla-Ambrosio, P.J., Ramírez-Cortes, J.M.: Pedestrian dead reckoning towards indoor location based applications. In: Proceedings of 8th International Conference on Electrical Engineering Computing Science and Automatic Control, Yucatán, México, Oct 2011

24. Park, S.K., Suh, Y.S.: A zero velocity detection algorithm using inertial sensors for pedestrian navigation systems. Sensors **10**(10), 9163–9178 (2010)

Geo-Navigation for a Mobile Robot and Obstacle Avoidance Using Fuzzy Controllers

Oscar Montiel, Roberto Sepúlveda, Ignacio Murcio
and Ulises Orozco-Rosas

Abstract This chapter presents the design of a system of fuzzy controllers for a differential mobile robot that was developed to navigate in outdoors environments over a predetermined route from point A to point B without human intervention. The mobile robot has the main features of geo-navigation to obtain its current position during the navigation, obstacles detection and the avoidance of these obstacles in an autonomous form. In this work to achieve the autonomous navigation in real-time, it was necessary to design a system based on fuzzy controllers. The system performs the detection and the analysis of the surrounding environment of the mobile robot to take actions that allow achieving the target point in a safe way. The position and orientation of the mobile robot is achieved with the use of geographical coordinates, through a GPS and the use of a magnetic compass which determines the steering angle. The detection of the environment is through ultrasonic sensors mounted on the mobile robot. All the inputs are taken by the system to compute through fuzzy rules the motion control of the mobile robot, to estimate the position and orientation accurately and to control the speed of the two DC motors to drive the wheels. In this work, the experiments were performed in dynamic outdoors environments, where the mobile robot performed successfully the navigation and the obstacles avoidance. In all the experiments, the mobile

O. Montiel (✉) · R. Sepúlveda · I. Murcio · U. Orozco-Rosas
Instituto Politécnico Nacional, Centro de Investigación y Desarrollo de Tecnología Digital
(CITEDI-IPN), Av. Del Parque No. 1310, Mesa de Otay, Tijuana 22510, BC, México
e-mail: oross@ipn.mx; oross@citedi.mx

R. Sepúlveda
e-mail: rsepulvedac@ipn.mx

I. Murcio
e-mail: imurcio@citedi.mx

U. Orozco-Rosas
e-mail: ulises.or@gmail.com

O. Castillo et al. (eds.), *Recent Advances on Hybrid Approaches for Designing*
Intelligent Systems, Studies in Computational Intelligence 547,
DOI: 10.1007/978-3-319-05170-3_46, © Springer International Publishing Switzerland 2014

robot achieved its mission to reach the target position without human intervention; the results show the validity of the developed system. The experimental framework, experiments and results are explained in terms of performance and accuracy.

1 Introduction

The development of mobile robots responds to the need to extend the application field of robotics. At the beginning, the robotics was restricted to the extent of a mechanical structure anchored at one end. Nowadays, the autonomy is increasing, and human intervention is limited [1].

For a mobile robot a locomotion system is required that allows the motion freely in its work space. That is the reason why we have robots that can walk, jump, run, slide, skate, swim, fly and roll [2, 3].

For a long time, the wheels have been the most popular locomotion system on mobile robots. Hence the wheels have been the simplest and efficient solution for mobility on hard terrains and free of obstacles; this condition allows obtaining relatively high speeds on mobile robots [1]. In order to engage in unstructured environments with obstacles that can be dangerous to humans, the mobile robots present advantages. An example of application of fuzzy logic in mobile robotics is presented in [4] with the work named "Intelligent Mobile Robot Motion Control in Unstructured Environments" in which a fuzzy control is used for robot movement with remote control sensor motion using Sun SPOT technology in order to engage in unstructured environments with obstacles that can be hazardous to human beings.

In this work, a differential system was used for the motion system of the mobile robot, since it is the simplest and inexpensive system. The mobile robot adapts easily to its environment due to its size, it is easy to access to almost anywhere to be required.

We designed and developed an autonomous mobile robot, which it follows a previously generated path that is free of obstacles, but when the mobile robot is working in on-line mode most of the time the environment is not known perfectly, then the mobile robot must have the ability to react to unexpected situations and the changes in its environment, this work will be achieved through the perception of the environment by using the sensors mounted on the mobile robot.

One of the main problems in this type of robots is the necessity for algorithms capable of real-time decision making, where the mobile robot has to perceive simultaneously and analyze the surrounding environment, locate obstacles and generate the activities to evade obstacles and follow the desired path, to allow a safe navigation in dynamic environments and achieve successfully its mission.

Another problem to be considered during the development of this work is to estimate the position and orientation of the mobile robot with high precision. To solve this problem and minimizing the error we use the geographical coordinates for navigation. With this method, we avoid to use other methods such as odometry and inertial navigation which introduce errors that increase with the distance traveled.

To solve all the problems mentioned above, in this work, we propose the design of fuzzy controllers for the position, orientation and obstacle avoidance. Due to the versatility and efficiency of the fuzzy controllers, they have been demonstrated in different applications to solve real-world problems of mobile robotics where is difficult to estimate and solve with classic mathematical models.

In this work, a hybrid architecture is used in the same way that modern mobile robots combine reactive control and deliberative architecture in layers models [5]. The robotic system will have an unobstructed path previously established that meets deliberative architecture. The system will have to avoid unforeseen obstacles when the navigation is performed, in this case, the system will have to make use of the reactive architecture.

The design is composed of two servomotors to control and regulate the pose of a non-holonomic mobile robot with frontal differential driving wheels, and a swivel castor wheel in rear. We use the power control pulse width modulation (PWM) for controlling the rotational speed of the servomotors in individually form. The fuzzy controllers in this mobile robot are dedicated to orientation, object detection and position of the mobile robot in the geographic space. All the code for the designed algorithms was developed in the programming language C. The proposed controllers are based on Mamdani Type Fuzzy Logic, with a layer of obstacle avoidance control, which is reactive by the use of the signals from the ultrasonic sensors. The system has another deliberative layer for a medium term planning navigation. By the use of the Global Positioning System (GPS) and a magnetic compass, the system computes the estimate position and geographic orientation of the mobile robot, the result information provides the current location and the destination direction of the route.

2 Global Positioning System

The Global Positioning System (GPS) is a set of 24 satellites circling the Earth, it sends radio signals to the surface that determines the position anywhere on the globe. Each satellite sends to earth a radio wave carrying sequences of numbers called code. GPS satellites send two sequences of numbers: one for precision (P) and other one for common acquisition (CA), each satellite has only one P code and one CA code, so that the receiver can make the difference between the signals from the different satellites. The P code is a long sequence that is repeated once every 7 days, while the CA code is much shorter, it repeats every microsecond. The P code value use a higher modulation, it provides a greater degree of precision for military receivers that is unable to understand in civilian receivers [6].

To achieve a sufficient level of confidence in the position, the information must be updated every second. However, some receivers have a mode that allows you to update the position at a slower pace (every 2–5 s), in order to save on battery consumption of the receiver. This mode can be used when changes in course and speed are of little importance.

Generally, the GPS receptors show altitude information. However, this information is of little use if a differential GPS receiver is not used [7]. An example of the use of a differential GPS in a mobile robot is presented in Heesung et al. [8], this work named "Autonomous Navigation of Mobile Robot based on DGPS/INS Senso Fusion by EKF in Semi-outdoor Structured Environment", it uses a Differential Global Positioning System (DGPS) because the technology was required with high accuracy in navigation, and represent a solution in localization in the outdoor environment [9].

The National Marine Electronic Association (NMEA) is an institution dedicated to establish a standard for communication of marine peripherals. Some decades ago, each brand worked with its own way of communication (data protocol). In order to make easier communication between various devices (both at the protocol data and electrical interfaces), the institution was created where manufacturers, distributors, educational institutions and other stakeholders in marine peripheral equipment are participating. This institution has no profit, the protocol 0183 is the latest version of the NMEA, see in Table 1 the NMEA commands.

The protocol 0183 is characterized by transmitting statements, each of the sentences beginning with '$' and ends with (CR: Carriage Return, LF: Line Feed), the two characters preceding '$' are those that identify your equipment (for the GPS "GP"), the next three characters indicate the sentence that is being sent, there are three types of statements that are query (query sentences) of origin of the equipment (proprietary sentences) and shipping (talker sentences), Table 2 shows the $ GPGGA sentence.

3 Implementation

This section describes the methodology used for the development of fuzzy systems for a mobile robot, to control the position, orientation and obstacle avoidance. Figure 1 shows the mobile robot used in this work, which is a typical example of a non-holonomic mechanical system, which uses a system of differential-type locomotion. A holonomic robot is a robot that has zero kinematic constraints, expressed as an explicit function of position. Conversely, a non-holonomic robot is a robot with one or more kinematic constraints [10].

3.1 Fuzzy Control System

To achieve the goal of orientation and position control, a hybrid architecture was determined. A-priori knowledge about the environment is considered, the path

Table 1 NMEA commands

NMEA register	Description
GGA	Global positioning system fixed data
GLL	Geographic position—latitude/longitude
GSA	Global navigation satellite system GNSS active satellites
GSV	Global navigation satellite system GNSS sight
RMC	Minimum data recommended specific GNSS
VTG	Course over ground and ground speed

Table 2 $ GPGGA sentence (fix data)

Field	Example	Comment
ID sentence	$ GPGGA	
UTC hour	092204.999	hhmmss.sss
Latitude	4250.5589	ddmm.mmmm
N/S indicator	S	N = North, S = South
Longitude	14718.5084	dddmm.mmmm
E/W indicator	E	E = Est, W = West
Fix position	1	0 = not valid, 1 = valid, 2 = valid DGPS, 3 = valid PPS
Used satellites	04	The satellites are used (0–12)
HDOP	24,4	Horizontal dilution of precision
Altitude	19,7	Altitude in meters according to WGS-84 ellipsoid
Altitude unities	M	M = Meters
Geoid separation		Geoid separation in meters according to WGS84 ellipsoid
Separation unities	In blank	M = Meters
DGPS age	In blank	DGPS data age in seconds
GDPS station ID	0000	
Checksum	*1F	
Terminator	CR/LF	

* indicates the end of the argument, and it is followed by a small-size datum (checksum)

Fig. 1 Differential mobile robot designed for this work

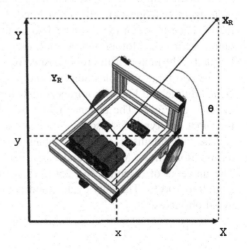

where the mobile robot will navigate is a free obstacle path. This path must be followed using a fuzzy control system, with geographic coordinates as inputs.

The idea is to perform a sequence of actions to reduce the error generated in the mobile robot movement direction. In addition, the system is oriented to providing flexibility, adaptability and the ability to react to unforeseen circumstances. In case of obstacles on the path, the system uses local navigation, which it will come into action when encountering an obstacle. The fuzzy obstacle avoidance system uses presence sensors to obtain a perception of the world. The robot has an efficient communication between sensors and actuators using feedback to react according to the environment conditions, the Fig. 2 shows the obstacle avoidance, orientation and position system for the mobile robot.

The orientation control is based on the navigation system with GPS and magnetic compass, these two systems provide the current geographical position and orientation of the mobile robot. The path tracking is performed in a dynamically form considering the current position of the mobile robot and the way points in the path planned, this points had been chosen to be free of obstacles.

For position and obstacle avoidance, the system uses proximity devices that detect events that are not considered, giving to the robot the perception of the world around it and the ability to navigate in a safe form.

3.1.1 Fuzzy System for Obstacle Avoidance

To solve the problem of unexpected obstacles in mobile robot navigation, a fuzzy controller obstacle avoidance using ultrasonic sensors is used. These sensors are free of mechanical friction and detect the presence of objects as well as the distance found with respect to the robot.

The distance obtained by the ultrasonic sensors is processed and evaluated using a fuzzy system, which makes the movement of the mobile robot and obstacle avoidance. The distance from the obstacle to the mobile robot is the instantaneous radius of curvature "R". A Mamdani type fuzzy system is used with two linguistic input variables, Sensor1 (left sensor) and Sensor2 (right sensor), which represent the distance between the robot and some unexpected object in its path. The two output variables, Motor1 and Motor2, they correspond to the speed control of the left and right motor of the robot respectively, see Fig. 3.

The two input variables Sensor1 and Sensor2, they have three linguistic terms: Near, Middle and Far with a trapezoidal shape. The universe of discourse [0, 1.5] meters, see Fig. 4. The distance range was determined by considering the size of the robot and type of sensors.

The two linguistic output variables Motor2 and Motor1, with the linguistic terms: Stop, Slow, Medium and Fast, they have a trapezoidal shape, see Fig. 5. The universe of discourse [0, 250] where in each variable, the duty cycle goes from 0 to 100 %. The duty cycle controls the speed of the motors of the robot, to avoid obstacles.

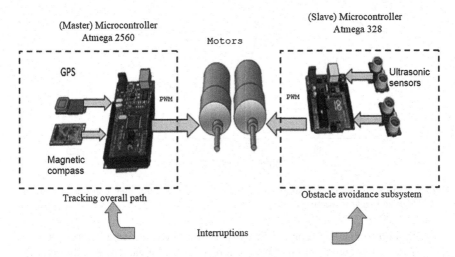

Fig. 2 Obstacle avoidance, orientation and position system for the mobile robot

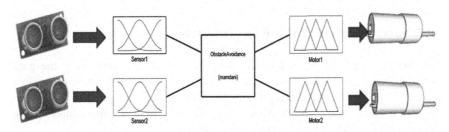

Fig. 3 Fuzzy system with two linguistic input variables and two output

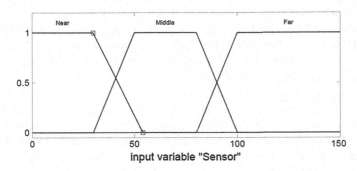

Fig. 4 Distribution of the membership functions with the linguistic input variables Sensor1 and Sensor2

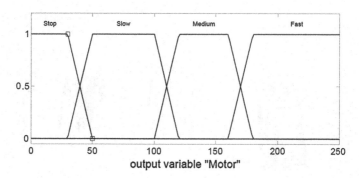

Fig. 5 Distribution of membership functions Motor1 and Motor2, they correspond to motor control *left* and *right* respectively

From knowledge based on existing experience, the rules were determined as in [17]. These rules consist of nine fuzzy inference rules for each motor, the rules are expressed in Tables 3 and 4. The rules indicate what action to take in each motor of the mobile robot according to the linguistic output variables.

3.1.2 Fuzzy System for Position and Orientation

The fuzzy system for orientation is used for control and monitoring the path tracking. The system uses a Mamdani type fuzzy controller with an input variable called Theta (θ), this variable represents the orientation error, and also the system has two outputs, which are the Motor1 and Motor2, these output variables make the correction of the orientation error, see Fig. 6.

The Fig. 7 shows the linguistic input variable Theta with their respective linguistic terms, which are represented in nine trapezoidal shape membership functions, under the universe of discourse [−359, 359] degrees. The linguistic terms associated with the Theta variable refer to the orientation error, which is: Very Large Negative Error (EMGN), Large Negative Error (EGN), Medium Negative Error (EMN), Small Negative Error (EPN), Minimum Error (EM), Very Large Positive Error (EMGP), Large Positive Error (EGP), Medium Positive Error (EMP), and Small Positive Error (EPP).

The fuzzy system for the orientation is considered with two identical output linguistic variables. The name of the output variables are Motor1 and Motor2 representing the speed of each motor to make corrections to the orientation and position. The universe of discourse of each variable is [0, 255] PWM. The linguistic terms associated with these variables are: Stop, Slow, Medium, Fast, and Very Fast, see Fig. 8.

The orientation and position control is the difference between the current position of the robot $q_a = (x_a, y_a, \theta_a)$ and desired $q_d = (x_d, y_d, \theta_d)$, this difference represents the orientation error which should tend to zero as shown in the formula (1), by the correction of orientation made by the fuzzy controller.

Table 3 Fuzzy rules for obstacle avoidance with the linguistic variables of speed for the right motor

		Sensor2		
		Near	Middle	Far
Sensor1	Near	Stop	Slow	Slow
	Middle	Stop	Medium	Fast
	Far	Stop	Medium	Fast

Table 4 Fuzzy rules for obstacle avoidance with the linguistic variables of speed for the left motor

		Sensor2		
		Near	Middle	Far
Sensor1	Near	Stop	Stop	Stop
	Middle	Medium	Medium	Slow
	Far	Slow	Fast	Fast

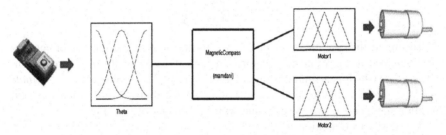

Fig. 6 Fuzzy system for orientation with one linguistic input and two linguistic output variables

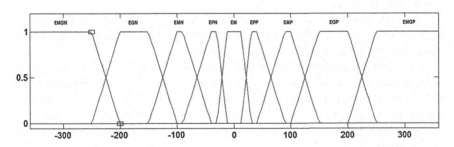

Fig. 7 Distribution of the membership functions of linguistic input variable Theta

$$\lim_{t \to \infty} (q_d - q(t)) = 0 \tag{1}$$

$$d = \sqrt{(X_d - X_a)^2 - (Y_d - Y_a)^2} \tag{2}$$

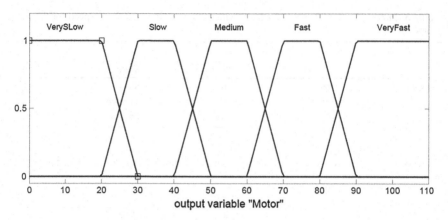

Fig. 8 Distribution of the membership functions of linguistic output variables Motor1 and Motor2

$$\theta_d = A \tan\left(\frac{Y_d - Y_a}{X_d - X_a}\right) \tag{3}$$

To obtain the orientation error (θ), we must compare the current orientation (θ_a) obtained using a magnetic compass (LSM303DLM) with the desired orientation (θ_d), which is acquired using a GPS (MTK3329), we obtain $x_a = longitude$ and $y_a = latitude$ of the point representing the current position, then the next position is given (x_d, y_d), to continue with the navigation, see Fig. 9.

The desired orientation of the robot (θ_d) and the distance (d) to the target, require two points of geographic coordinates, as shown in (2) and (3). The desired orientation is compared with the current orientation (θ_a) to calculate (θ), which should tend to zero.

The path tracking can be performed dynamically considering the current position of the mobile robot and the way points in the path planned, the points along the path were chosen to be free of obstacles.

Three data are taken from the GPS and processed using the statistical median in order to improve the location of a point due to a lack of precision, which has the GPS in their reading. If we take more data from GPS, the precision is improved, but increases latency in the algorithm considerably. Therefore, it was decided to use only three data because it has an improvement in the accuracy, and latency is not excessive.

From the linguistic terms associated with the input variable, we have nine fuzzy inference rules, which are expressed in Table 5.

Fig. 9 Correction of position

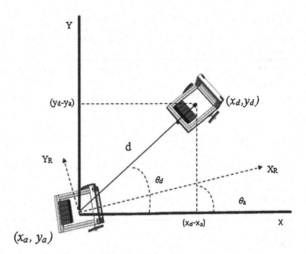

Table 5 Fuzzy inference rules for the control of Motor1 and Motor2

	EMGN	EGN	EMN	EPN	EM	EPP	EMP	EGP	EMGP
Motor1	Very Fast	Fast	Medium	Slow	Stop	Stop	Stop	Stop	Stop
Motor2	Stop	Stop	Stop	Stop	Stop	Slow	Medium	Fast	Very fast

4 Experiments and Results

This section shows the results from the experiments. In the first experiment, the correction of the orientation error Theta is analyzed. In the second experiment, the obstacle avoidance is analyzed. The third experiment consists in moving the robot from point A to point B in one direction. The fourth experiment is the same displacement from point A to point B and return to point A. The fifth experiment is conducted to follow-up on a path which consists of an initial point A and a target point B, where this path has several intermediate points where the mobile robot must pass.

The design of the fuzzy inference system (FIS) of each controller was done in the fuzzy logic toolbox of Matlab. The FIS for obstacle avoidance use two linguistic input variables Sensor1 and Sensor2 and two output Motor1 and Motor2. The parameters used for obstacle avoidance controller are shown in Tables 6 and 7.

For the orientation and position fuzzy system we have a linguistic input variable called Theta and two output variables Motor1 and Motor2 representing orientation error correction, the orientation controller parameters are shown in Tables 8 and 9.

The fuzzy controllers system was implemented in two Arduino Mega microcontrollers, one for orientation and position and the second one for obstacle avoidance. The fuzzy controller system algorithm was programmed using the C language.

It was verified that the fuzzy inference system (FIS) for obstacle avoidance has an expected performance in the fuzzy logic Toolbox of Matlab to meet the

Table 6 Specs for linguistic input variables Sensor1 and Sensor2 to get the obstacles distance

Linguistic variable	Universe of discourse	Linguistic terms	Membership function shape	Parameters
Sensor1 Sensor2	[0 150]	Near	Trapezoidal	[−54 −6 30 54]
		Middle	Trapezoidal	[30 50 80 100]
		Far	Trapezoidal	[80 100 156 204]

Table 7 Specs for linguistic output variables Motor1 and Motor2 to obtain the speed of motors to avoid obstacles

Linguistic variable	Universe of discourse	Linguistic terms	Membership function shape	Parameters
Motor1 Motor2	[0 250]	Stop	Trapezoidal	[−90 −10 30 50]
		Slow	Trapezoidal	[30 50 100 120]
		Medium	Trapezoidal	[100 120 160 180]
		Fast	Trapezoidal	[160 180 250 300]

Table 8 Specs for linguistic input variable Theta for orientation error

Linguistic variable	Universe of discourse	Linguistic terms	Membership function shape	Parameters
Theta	[−360 360]	EMN	Trapezoidal	[−400 −370 −250 −200]
		EGN	Trapezoidal	[−250 −200 −150 −100]
		EMN	Trapezoidal	[−150 −100 −90 −40]
		EPN	Trapezoidal	[−90 −40 −30 −12]
		EM	Trapezoidal	[−30 −12 12 30]
		EPP	Trapezoidal	[12 30 40 90]
		EMP	Trapezoidal	[40 90 100 150]
		EGP	Trapezoidal	[100 150 200 250]
		EMGP	Trapezoidal	[200 250 360 400]

requirements of the mobile robot. The Fig. 10 shows the rules to simulate the behavior of the robot if an unexpected object is detected by the Sensor1 or Sensor2.

In the same way, the orientation controller was designed using Matlab Toolbox where the input Theta is simulated as shown in Fig. 11.

Table 9 Specs for linguistic output variables Motor1 and Motor2 to control the orientation

Linguistic variable	Universe of discourse	Linguistic terms	Membership function shape	Parameters
Motor1	[0 220]	Stop	Trapezoidal	[−10 0 10 30]
Motor2		Slow	Trapezoidal	[10 30 50 70]
		Medium	Trapezoidal	[50 70 100 120]
		Fast	Trapezoidal	[100 120 150 170]
		Very fast	Trapezoidal	[150 170 220 300]

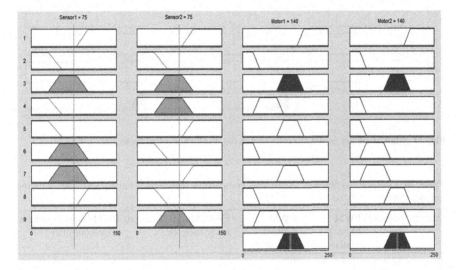

Fig. 10 Matlab viewer of the fuzzy rules for the obstacle avoidance controller

4.1 Comparison of the Results from Obstacle Avoidance and Orientation Controllers with Matlab

This section shows the results from the comparison between the controllers implemented in Matlab using the fuzzy toolbox and the implemented in the microcontroller Arduino programmed in C. The distances from the linguistic inputs variables Sensor1 and Sensor2 were simulated in Matlab in order to compare with the results from the fuzzy algorithms implemented in the microcontroller, see Table 10. The linguistic input variable Theta were simulated in Matlab in order to compare with the results from the implementation in the microcontroller, see Table 11. The Tables 10 and 11 clearly show the good performance of the fuzzy controllers.

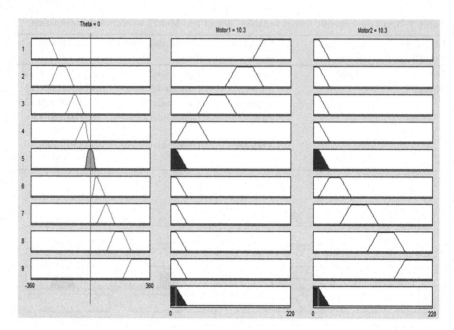

Fig. 11 Matlab viewer of the fuzzy rules for the orientation error controller

Table 10 Obstacle avoidance controller comparison between Matlab and the Arduino microcontroller

Sensor1	Sensor2	PWM Motor1 Matlab	PWM Motor2 Matlab	PWM Motor1 Arduino microcontroller	PWM Motor2 Arduino microcontroller
90	24.9	50	0	50	0
90	90	120	120	120	120
90	153	50	120	50	120
64	118	112	91	105	84

Table 11 Orientation controller comparison between Matlab and the Arduino

GPS data	PWM Motor1 Matlab	PWM Motor2 Matlab	PWM Motor1 Arduino microcontroller	PWM Motor2 Arduino microcontroller
22	13.4	24.6	12.5	26
164	12.4	75	12.5	75
−244	100	12.5	97.6	12.5

To get a better insight into the functioning of the fuzzy controllers made in this work, the Fig. 12 shows the control surface of the fuzzy obstacle avoidance system between the linguistic inputs Sensor1 and Sensor2 with the linguistic outputs

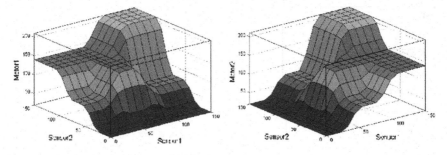

Fig. 12 Surface from the obstacle avoidance fuzzy controller

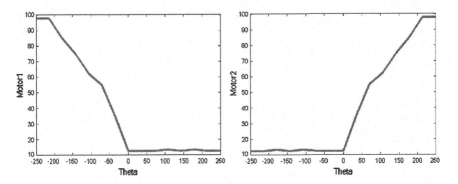

Fig. 13 Graph from the orientation fuzzy controller

Motor1 and Motor2. The Fig. 13 shows the graph generated between the input Theta and output Motor1 and Motor2. These surfaces and graphs were obtained from the toolbox of Matlab.

4.2 Set-up for the Input Variable Theta for the Orientation and Position Fuzzy System

As mentioned, the mobile robot used in this work uses a GPS for location. Due to the physical size of the robot, the error introduced by the GPS is very important. Considering the acquisition of 40 data from a geographic set point 32°32.3105′ latitude and 116°56.5755′ longitude, only 45 % of the latitude data and 65 % of the longitude data are within the desired class, and the rest has a high dispersal as shown in Figs. 14a and 15a. Classes are determined with a range of 20 units equivalent to 3.5 m, each second is 30 m, and the GPS used in this work divides every minute in ten thousand units.

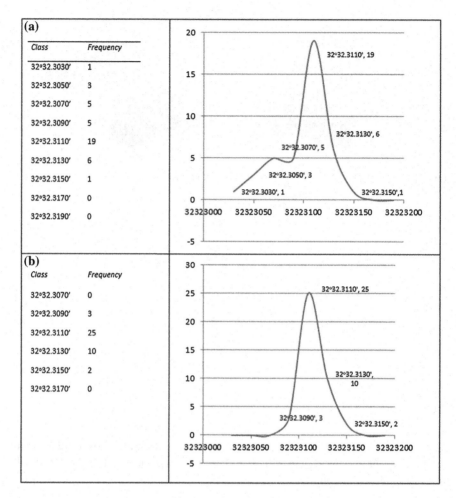

Fig. 14 a Distribution of the latitude of a geographic point without the use of a statistical method. **b** Distribution of the latitude of a geographic point with the use of the statistical method median

To decrease the error rate was necessary to make use of statistical method of median with three data reading. The Figs. 14b, and 15b show the improved data latitude and longitude respectively, in which the 62.5 % latitude data and 62.5 % longitude data are within the desired class, and the rest with a minimal dispersal.

The geographical points along the route of the mobile robot were selected using the statistical median, taking 40 data for each geographic location desired. Each data is made up of 15 readings, and using the statistical method of the median was improved by 75 % in the latitude and 80 % in longitude, but each reading has a cost of 2 s, which means 30 s per data acquired.

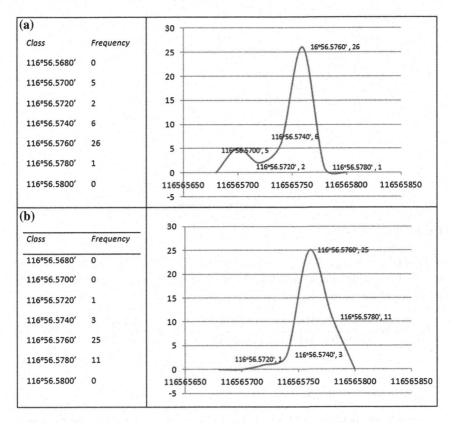

Fig. 15 a Distribution of the latitude of a geographic point without the use of a statistical method. **b** Distribution of the latitude of a geographic point with the use of the statistical method median

4.3 Experiments

The next set of navigation experiments were designed to demonstrate the fuzzy systems implemented in this work.

4.3.1 Experiment 1

This experiment aims to show that the mobile robot is able to correct the orientation error Theta, making a comparison between the current position and a desired one. The current and desired geographic coordinates were simulated to verify the correct operation of the orientation controller. In an experimentally form we take the coordinates (32°32.3474′, 116°56.5355′) as the current, and (32°32.3516′, 116°56.5344′) as the desired coordinates. The steering angle is obtained as shown

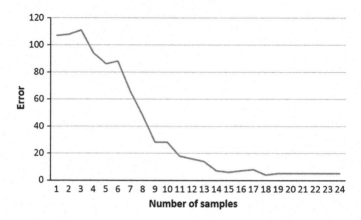

Fig. 16 Orientation error

in Eq. (4), which it gives an angle of $-75°$. By using of (5) the result is converted to a positive angle to be compared with the current direction, which it is acquired by the magnetic compass.

$$\theta_d = A \tan\left(\frac{32°32.3516' - 32°32.3474'}{116°56.5344' - 116°56.5355'}\right) = A \tan\left(\frac{0.0042'}{-0.0011'}\right) = -75° \quad (4)$$

$$\theta_d = 360 - 75 = 285° \quad (5)$$

$$\text{Error} = \theta_d - \theta_a = 285° - 182° = 103° \quad (6)$$

The current coordinate ($32°32.3474'$, $116°56.5355'$) was given to the robot with the current orientation an angle of $182°$, which it is compared with the desired orientation using the Eq. (6) until the error tends to zero, as shown in Fig. 16.

The Fig. 17 shows the corrections made by the differential robot motors, which it shows that the left motor (Motor 2) made the corrections, and the right motor (Motor 1) did not make any correction.

4.3.2 Experiment 2

In this experiment, the behavior of the mobile robot to avoid an obstacle was analyzed. An obstacle was placed opposite to the mobile robot at a distance of 84 cm from the sensor right and 75 cm from the left sensor.

The controller shows the results of evasion on the left side, where the robot detected a greater distance to avoid the obstacle. The action taken by the controller is to reduce the motors speed because the object is too close to the robot, but with a difference in speed as shown in Fig. 18. Where it can be observed that the speed of Motor1 (left side) was greater than Motor2 (right side) until the sample number 15 where the obstacle is avoided. Next to the sample number 15 in Fig. 18, it can be

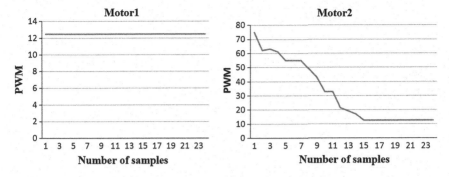

Fig. 17 Correction of the orientation controller for the *right* and *left* motor

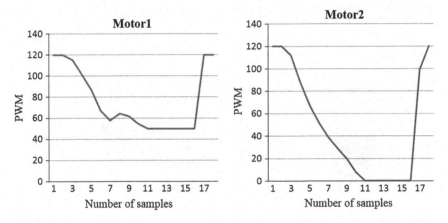

Fig. 18 Correction of the obstacle avoidance controller for the speed of Motor1 and Motor2

observed that the speed of motors is increased because the obstacle already was avoided.

Figure 19 shows that the mobile robot moved closer to the right side to achieve evade the object. The figure shows after the obstacle was avoided in the sample number 15, the distance from the Sensor1 increases, on the Sensor2 the change was less abrupt due to less proximity to the obstacle.

4.3.3 Experiment 3

This experiment consists of moving the robot from one point A to point B in one direction, see Fig. 20. Where the robot is located at the geographical coordinates 32°32.3550′ latitude and 116°56.5420′ longitude, this information for the point A and for the point B the coordinates are 32°32.3532′ latitude and 116°56.5373′ longitude. This test was performed with less than three meters radius error. In this

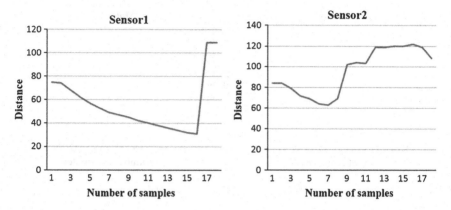

Fig. 19 Distance from Sensor1 and Sensor2 with respect to the obstacle

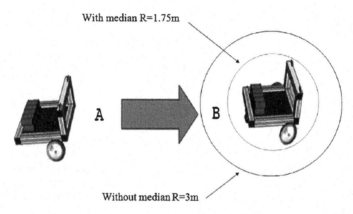

Fig. 20 Experiment 3, mobile robot navigation in one direction from point *A* to point *B*

work, the acceptable range of error was set in three meters. At the end of the test the mobile robot reach the target point B in the geographic location 32°32.3544′ latitude and 116°56.5392′ longitude, this result is due to the characteristics of GPS resolution. The result was achieved, and it is within the allowed error.

4.3.4 Experiment 4

The following experiment was performed to demonstrate that the mobile robot is able to go from point A to point B and back to point A which is the beginning of the path. The geographic coordinates of point A are 32°32.3550′ latitude and 116°56.5420′ longitude, and the coordinates of point B are 32°32.3500′ latitude and 116°56.5385′ longitude.

With median R=1.75m

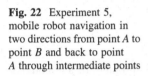

A

B

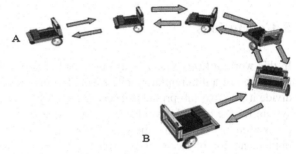

Without median R=3m

Fig. 21 Experiment 4, mobile robot navigation in two directions from point *A* to point *B* and back to point *A*

Fig. 22 Experiment 5, mobile robot navigation in two directions from point *A* to point *B* and back to point *A* through intermediate points

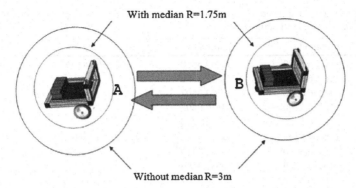

A

B

The test showed a result of the location of the robot at the point B with the coordinates 32°32.3480′ latitude and 116°56.5400′ longitude, this result is within the allowed error, and back to point A the mobile robot obtained the target at coordinates 32°32.3558′ latitude and 116°56.5435′ longitude, also within the three meters error allowed, see Fig. 21.

4.3.5 Experiment 5

This test is very similar to the previous one, the difference is that the robot will have to go through intermediate points to reach the point B and then return to the starting point through the same points as seen in Fig. 22. Where the geographic coordinates of point A are 32°32.3550′ latitude and 116°56.3422′ longitude and the coordinates of point B are 32°32.3530′ latitude and 116°56.5344′ longitude.

The test was performed successfully, the geographic location reached for the mobile robot at point B was 32°32.3544′ latitude and 116°56.5360′ longitude. The back to point A the mobile robot reached the coordinates 32°32.3542′ latitude and 116°56.5410′ longitude, as seen in Table 12.

Table 12 Geographical locations along the path in two directions from point A to point B and back to point A through intermediate points

	Longitude	Latitude	Longitude	Latitude
A	116°56.5420′	32°32.3550′	116°56.5420′	32°32.3550′
	116°56.5370′	32°32.3530′		
	116°56.5300′	32°32.3507′		
	116°56.5200′	32°32.3496′		
B	116°56.5320′	32°32.3410′		
	116°56.5344′	32°32.3422′	116°56.5360′	32°32.3544′
	116°56.5320′	32°32.3410′		
A	116°56.5200′	32°32.3496′		
	116°56.5300′	32°32.3507′		
	116°56.5370′	32°32.3530′		
	116°56.5420′	32°32.3550′	116°56.5410′	32°32.3542′

5 Conclusions

In this work, a fuzzy system was developed to solve the problem of autonomous navigation of a differential mobile robot. To solve this problem, it is necessary to divide it into two sub-problems because the proposal of this work is to use a fuzzy system for the obstacle avoidance and a second fuzzy system for the position and orientation, in order to satisfy the proposed control. Through the design and implementation of the fuzzy controls the following points of interest were found. The use of geographical coordinates for navigation is important to estimate the position and the magnetic compass for the orientation of the mobile robot as accurate as possible, thus reducing the accumulated error that increases with the distance if an odometer system is used [11]. To reduce the inherent error in GPS, the tests were done using the statistical method of the median. It was determined that the measurements made with the GPS using the median generates radius errors below 1.75 m compared with measurements without the statistical method where the results go up to 3 m of error radius or more. The calculus of the median generates considerable data latency; this is the main reason why it is not recommended for a robot that requires high speed reaction. Another consideration to reduce the error caused by the use of a conventional GPS is the use of a DGPS, due to the given accuracy in centimeters by the DGPS. One of the problems using this type of technology (GPS) is the dependency of the good weather conditions, and the dependency of a line of sight between the satellites and the GPS device. In future works other approaches will be considered [12–16].

References

1. Baturone, A.O.: Robótica, Manipuladores y Robots Móviles, Barcelona. Marcombo, Spain (2007)
2. Siegwart, R., Nourbakhsh, I.R., Scaramuzza, D.: Introduction to Autonomous Mobile Robots, 2nd edn. The MIT Press, London (2011)
3. Torres, F., Pomares, J., Gil, P., Puente, S.T., Rafael, R.: Robots y Sistemas Sensoriales, Madrid. Prentice Hall, Spain (2002)
4. Mester, G.: Intelligent mobile robot motion control in unstructured environments. Acta Polytech. Hung. 7(4), 153–165 (2010)
5. Hjortland, E., Nyberg, K., Salisbury, J.V., Chen, L., Hasan, A.: Navigation of and Outdoor robot using a fuzzy logic controller. In: Canadian Conference on Electrical and Computer Engineering, pp. 1037–1040 (2005)
6. Letham, L.: GPS Facil, Barcelona. Paidotribo, Spain (2001)
7. Correia, P.: Guía práctica del GPS. Marcombo, Barcelona (2002)
8. Christiand, C.H., Sunglok, C., Wonpil, Y.: Autonomous navigation of mobile robot based on DGPS/INS senso fusion by EKF in semi-outdoor structured environment. In: International Conference on Intelligent Robots and Systems, Taipei (2010)
9. Kim, S.H., Roh, C.W., Kang, S.C., Park, M.Y.: Outdoor navigation of a mobile robot using differential GPS and curb detection. In: International Conference on Robotics and Automation, Roma (2007)
10. Nourbakhsh, S.R., Siegwart, R.: Autonomous Mobile Robots. The MIT Press, Massachusett (2004)
11. Montiel, O., Sepúlveda, R., Castillo, O., Melin, P.: Ant colony test center for planning autonomous mobile robot navigation. Comput. Appl. Eng. Educ. 21(2), 214–229 (2013)
12. Aguilar, L., Melin, P., Castillo, O.: Intelligent control of a stepping motor drive using a hybrid neuro-fuzzy ANFIS approach. J. Appl. Soft Comput. 3(3), 209–219 (2003)
13. Melin, P., Castillo, O.: Adaptive intelligent control of aircraft systems with a hybrid approach combining neural networks, fuzzy logic and fractal theory. Appl. Soft Comput. 3(4), 353–362 (2003)
14. Montiel, O., Sepúlveda, R., Melin, P., Castillo, O., Porta, M., Meza, I.: Performance of a Simple Tuned Fuzzy Controller and a PID Controller on a DC Motor. In: FOCI 2007, pp. 531–537
15. Gonzalez, J.L., Castillo, O., Aguilar, L.T.: FPGA as a tool for implementing non-fixed structure fuzzy logic controllers. In: FOCI 2007, pp. 523–530
16. Melendez, M., Castillo, O.: Hierarchical genetic optimization of the fuzzy integrator for navigation of a mobile robot. In: Melin, P., Castillo, O. (eds.) Soft Computing Applications in Optimization, Control, and Recognition, pp. 77–96. Springer, Heidelberg (2013)
17. Leal-Ramirez, C., Castillo, O., Melin, P., Rodriguez-Diaz, A.: Simulation of the bird age-structured population growth based on an interval type-2 fuzzy cellular structure. Inf. Sci. 181, 519–535 (2011)

Ad Text Optimization Using Interactive Evolutionary Computation Techniques

Quetzali Madera, Mario García-Valdez and Alejandra Mancilla

Abstract The description of a product or an ad's text can be rewritten in many ways if other text fragments similar in meaning substitute different words or phrases. A good selection of words or phrases, composing an ad, is very important for the creation of an advertisement text, as the meaning of the text depends on this and it affects in a positive or a negative way the interest of the possible consumers towards the advertised product. In this chapter we present a method for the optimization of advertisement texts through the use of interactive evolutionary computing techniques. The EvoSpace platform is used to perform the evolution of a text, resulting in an optimized text, which should have a better impact on its readers in terms of persuasion.

1 Introduction

Text content is plays a very important role in e-commerce applications, as this is one of the most common ways of giving information about a commercial product to the consumers [1]. When the author of an advertisement text is an expert ad writer, the product should have a better chance of receiving a positive response from the consumers. The combination of words or phrases (blocks of text) that the experts decide to use when writing the text of an ad is important, because this particular combination could be the one that persuades the consumer into buying

Q. Madera · M. García-Valdez · A. Mancilla (✉)
Tijuana Institute of Technology, Calzada Tecnologico s/n, Tijuana, Mexico
e-mail: alejandra.mancilla@gmail.com

Q. Madera
e-mail: quetzalimadera@tectijuana.edu.mx

M. García-Valdez
e-mail: mario@tectijuana.edu.mx

O. Castillo et al. (eds.), *Recent Advances on Hybrid Approaches for Designing Intelligent Systems*, Studies in Computational Intelligence 547,
DOI: 10.1007/978-3-319-05170-3_47, © Springer International Publishing Switzerland 2014

the product. If an inexperienced writer decides to write an advertisement text, it would be very difficult for him to choose a correct combination of the blocks of text that is successful into persuading the majority of the consumers. In this work, we propose that a writer of any level of experience can create an ad with different interchangeable blocks of text carrying the same meaning, and a third party could optimize it.

Evolutionary algorithms are commonly used to solve optimization problems [2] and that's why we decided to use this kind of techniques to optimize the advertisement texts. We believe that if a group of people can evaluate, in terms of persuasion, different combinations of the same ad, after many generations of evolution we can find an optimal ad, which will have a better impact on the majority of the consumers.

2 Basic Concepts

In this section we present some basic concepts for understanding this work, and the essential parts that compose this project. One of the most important components is EvoSpace, which turned to be a crucial tool for the implementation of the genetic algorithm used in this work.

2.1 Evolutionary Algorithms

Evolutionary algorithms are a subfield of artificial intelligence. They are mainly used in optimization problems where the search space is very large and aren't lineal. These algorithms search for solutions based on the theory of the Darwinian evolution.

The methods of this kind generate a set of individuals that represent possible solutions [3]. These solutions are usually generated randomly at the beginning of the evolution process. After each generation, the best solutions share part of their information to create other possible, better, solutions. All of the individuals compete to be the more fit solutions; the better solutions are conserved, while the worse are destroyed, according to a fitness function that evaluates their performance [4].

2.2 Genetic Algorithms

Genetic algorithms (see Fig. 1) are inspired in biological evolution; they evolve a population of individuals by performing genetic recombination and mutation [5–9]. A selection of the best solutions is made by the use of certain criterion and a fitness

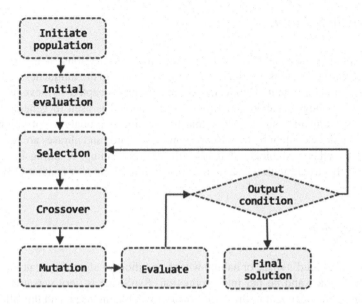

Fig. 1 Genetic algorithm diagram

function, and based on their performances, the more fit individuals survive and the less fit are discarded. Optimization based on genetic algorithms is a search method based on probability [10, 11].

This is an elitist algorithm as it always conserves the best individual of the population unchanged. As the number of generations or iterations increases, the probability of finding the optimum solution tends to increase.

2.3 Interactive Evolutionary Computation

Interactive evolutionary computing is a variation of evolutionary computing where the fitness of an individual is determined through the subjective evaluation performed by a human being. In traditional evolutionary computing, a human being requires a computational process to solve a problem. To do this, a person gives a problem's description as input to a solution model, and this model returns a result that has to be interpreted by a human being. But in interactive evolutionary computing the roles are inverted: there's an algorithm that asks a human being or a group of human beings to solve a problem, and then it gathers this information to interpret it later [10].

2.4 Article Spinning

Article spinning is a method used to create multiple versions of a text article without creating versions considered as plagiarism, due to the uniqueness achieved of the generated content. Duplicated content is not accepted by several search engines like Google, Yahoo and Bing, so this method is used to generate many different versions of a single article that have a higher probability of being considered as unique content by these search engines. Words and phrases are randomly changed by other text blocks that have the same meaning, resulting in another version of the article with the same meaning, but different text content [12].

2.5 EvoSpace

EvoSpace is a cloud's space or habitat where evolutionary algorithms can be stored, developed, tested and be put into production. EvoSpace is very versatile, as the population is independent to the evolutionary model being used, and this allows us to make modifications to the evolutionary algorithm at any time. The client processes, called EvoWorkers, interact dynamically and asynchronously, and they can be displayed in remote clients like in the platform storing the server [13].

2.6 Online Advertisement

Online advertisement is performed based on the content of a website [14]. For the creation of this type of advertisement, which has the objective of giving information about a product on Internet, it must contain different media elements such as text, links, images, videos, animations, etc. There are companies like Google that have created systems for the creation of online advertisement campaigns, like AdSense and AdWords [15]. AdSense positions ads in websites related to the textual content being displayed in the web pages. Users owning these ads pay a certain amount of money for each click on their ads.

3 Related Work

In this section we will explain the platform EvoSpace-Interactive, which is the platform that we used to build the graphical interfaced used by the users to choose what ads they considered were more persuasive.

3.1 EvoSpace-Interactive

EvoSpace-interactive [16] was initially tested by implementing an interactive evolutionary computation program called Shapes. This software evolved images formed by equilateral triangles that could have one of twelve possible colors. Currently, EvoSpace has made modifications to the images displayed by changing the shape to more attractive animations that last for a short period of time. For this work, we modified EvoSpace to be capable of displaying text ads instead of images [17].

4 Problem Description

If an inexperienced article writer decided to create many different versions of a text ad, where he changed some words and phrases to other text blocks of similar meaning, and he showed them to some friends and asked them to vote for the version they think is the best one, he would be performing a small optimization to his text. This is because his text isn't based only on his opinion, but on the opinion of all of his friends and his own, and now the winning text could be considered more attractive to a higher percentage of readers.

Based on this example, we believe that a more effective optimization of the text for an ad should take into consideration thousands of possible versions, and dozens of people should give their opinion on what blocks of text are better to increase the persuasion of the text.

5 Methodology

5.1 Article Format

The text (see Fig. 2) is contains different sections enclosed by curly braces, which contain different blocks of text (the blocks of text can be of any size, from a single word to whole sentences and paragraphs). These blocks of text are separated by bars. The text blocks that are outside of the curly braces (represented by grey text in Fig. 2) won't have any modifications when the texts evolve.

A vector generates each version of the text, where each number in it represents the corresponding option. In this example (see Fig. 3), if we want to know what number is the word "Fabricamos", we have to look in the options to find out it's the third option. So, in the first position of the vector we'll have the number 3. The rest of the vector represents the rest of the options that should be printed in the generated text.

{Diseñamos | Creamos | Fabricamos | Construimos | Desarrollamos} {este auto | este carro | esta pieza de arte | este impresionante transporte | este auto único | este carro único}, {para ser | para convertirse en | para que sea | con el fin de ser} la mejor forma {de viajar | de transportarte | para ir de un lugar a otro | de moverte | de trasladarte} {con tu familia | con tus amigos | a donde quieras | románticamente | cómodamente | de forma divertida | silenciosamente | de forma segura}. Por eso está {equipado | preparado | construido | fabricado | diseñado | creado | desarrollado} con {barras laterales | protecciones a los costados | puertas protegidas | protecciones laterales} contra {impacto | golpes | accidentes | choques}. Puede incluir {frenos ABS | un sistema antibloqueo de ruedas | frenos antibloqueo | frenos reforzados | frenos inteligentes | frenos antiderrapantes} y bolsas de aire {frontales | al frente | delanteras | grandes | suaves | seguras}. {Y por si fuera poco, es maniobrable, estable y eficiente | Cuenta con dirección hidráulica en todas sus versiones | Tiene un motor de 4 cilindros de 1.6L | Tiene transmisión manual de 5 velocidades | Tiene transmisión automática de 4 velocidades | Tiene un espacio interior y cajuela amplia | Por eso este transporte es tu mejor opción | Compra seguro, compra inteligentemente}.

Fig. 2 Text format

{Diseñamos | Creamos | Fabricamos | Construimos | Desarrollamos}

| 3 | 6 | 2 | 4 | 7 | 7 | 1 | 4 | 5 | 6 | 6 |

Fig. 3 Representation of the vector

5.2 Implementation of EvoSpace-Interactive

EvoSpace was modified to be able to evolve advertisement texts. The text, using the format explained before, must be analyzed by the program to determine how many text segments will be changing, and how many and what are their options. This way a vector can be generated, which would represent the chromosome of an individual. EvoSpace creates 100 random individuals when it initializes its population.

Changes were also made to the graphical user interface (see Fig. 4). The amount of likes an individual currently has was removed so it wouldn't act as a bias during the decision process of the user. By default, in EvoSpace users can create a collection of their favorite individuals, and this feature was also removed as we consider this isn't necessary for our experiments. The general design was modified so the texts stand out from the rest of the elements of the interface.

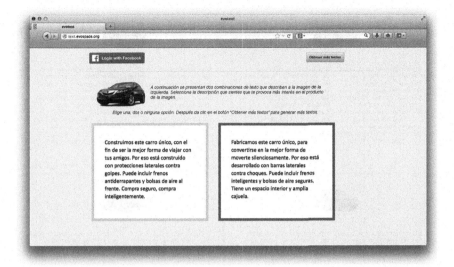

Fig. 4 User interface

5.3 System Configuration

The population is initialized with 100 randomly generated individuals. For the evaluation of the individuals, the users are presented with two texts. A user can choose what text (or texts) he considers would be more persuasive. If no text is considered as a good choice, the user can also decide not to choose any of the options. When a selection is made, the selected texts are sent to EvoSpace and stored in its database, and the user is immediately presented with two more options, so he can continue with the selection process.

6 Results

The system started operation at http://text.evospace.org since 4th December 2013, and a total of 75 users have participated in the selection process, generating more than 180 samples.

The best chromosome was extracted from the database and it was compared against the worst generated chromosome. This best chromosome was also compared against an actual ad that advertises a Chevrolet car created by an expert. This ad, similar in size to the evolved text, was taken from Chevrolet's website [18].

A total of 30 people were surveyed, showing them two paper sheets containing two texts (see Fig. 5). On the first paper sheet, the text representing our best individual and the text created by the expert were shown. The second paper sheet

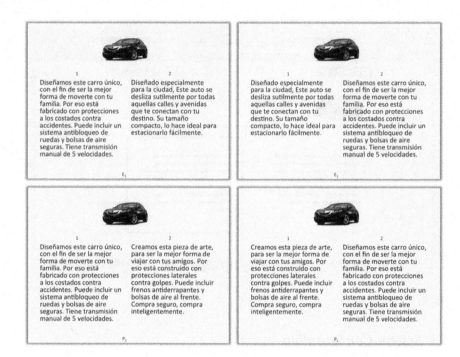

Fig. 5 Survey texts

showed the texts generated by our best and worst individuals. The texts were shown in different positions to discard the possibility of a person choosing an option due to its position (for example, if a person likes more a text because it's on the right side).

In our first case, 60 % of the people chose the text generated by our genetic algorithm, and 40 % chose the text generated by the expert. In the second case, 63.3 % of the people surveyed chose the text represented by the best individual, while 36.7 % chose the text represented by the worst individual.

7 Conclusions

The evolution of advertisement texts written by an inexperienced person in the field of marketing, through the use of interactive evolutionary techniques, is a viable alternative for the creation of texts with a higher probability of persuading consumers into buying the advertised product.

8 Future Work

We are currently working on the implementation of a clustering algorithm for grouping [19] users according to their profiles, and perform interactive evolution to generate optimal advertisement texts for each cluster of people. As future work other bio-inspired optimization methods could also be considered to solve the problem at hand, like in [20, 21].

References

1. McQuarrie, M., Edward, F., David, M.: Visual rhetoric in advertising: text-interpretive, experimental, and reader-response analyses. J. Consum. Res. **26**(1), 37–54 (1999)
2. De Jong, KA.: Evolutionary Computation: A Unified Approach. MIT Press (2006)
3. Valdez, F., Melin, P., Castillo, O.: Evolutionary method combining particle swarm optimization and genetic algorithms using fuzzy logic for decision making. In: Proceedings of the IEEE International Conference on Fuzzy Systems, pp 2114–2119 (2009)
4. Bäck, T.: Evolutionary Algorithms in Theory and Practice: Evolution Strategies, Evolutionary Programming, Genetic Algorithms, vol. 996. Oxford University Press, Oxford (1996)
5. Sanchez, D., Melin, P.: Modular neural network with fuzzy integration and its optimization using genetic algorithms for human recognition based on iris, ear and voice biometrics. Studies in Computational Intelligence, vol 312, pp 85–102 (2010)
6. Sepulveda, R., Castillo, O., Melin, P., Montiel, O.: An efficient computational method to implement type-2 fuzzy logic in control applications. Adv. Soft Comput. **41**, 45–52 (2007)
7. Sepulveda, R., Castillo, O., Melin, P., Rodriguez-Diaz, A., Montiel, O.: Experimental study of intelligent controllers under uncertainty using type-1 and type-2 fuzzy logic. Inf. Sci. **177**(10), 2023–2048 (2007)
8. Sepulveda, R., Montiel, O., Lizarraga, G., Castillo, O.: Modeling and simulation of the defuzzification stage of a type-2 fuzzy controller using the Xilinx system generator and Simulink. Stud. Comput. Intell. **257**, 309–325 (2009)
9. Sepulveda, R., Montiel, O., Castillo, O., Melin, P.: Optimizing the MFs in type-2 fuzzy logic controllers, using the human evolutionary model. Int. Rev. Autom. Control **3**(1), 1–10 (2011)
10. Takagi, H.: Interactive evolutionary computation: fusion of the capabilities of EC optimization and human evaluation. Proc. IEEE **89**(9), 1275 (2001)
11. Whitley, D.: A genetic Algorithm Tutorial. Statistics and Computing. Kluwer Academic Publishers, vol. 4.2, p. 65 (1994)
12. Malcolm, J.A., Peter, L.: An approach to detecting article spinning. Proceedings of the 3rd International Conference on Plagiarism (2008)
13. García, M., et al.: EvoSpace: a distributed evolutionary platform based on the tuple space model. In: Applications of Evolutionary Computation. Springer, Berlin Heidelberg, pp. 499–508 (2013)
14. Choi, S., Nora, J.: Antecedents and consequences of web advertising credibility: a study of consumer response to banner ads. J. Interact. Advertising **3**(1), 12–24 (2002)
15. Davis, H.: Google advertising tools: cashing in with AdSense, AdWords, and the Google APIs. O'reilly (2006)
16. García, M., et al.: EvoSpace-i: a framework for interactive evolutionary algorithms. In: Proceeding of the 15th Annual Conference Companion on Genetic and Evolutionary Computation Conference Companion. ACM, pp. 1301–1308 (2013)

17. Fernández, F., et al.: EvoSpace-interactivo: una herramienta para el arte y diseño Interactivo y colaborativo. IX Congreso Español de Metaheurísticas, Algoritmos Evolutivos y Bioinspirados, pp. 220–228 (2013)
18. Chevrolet. Chevrolet Mexico Ad description [online]. Website: http://www.chevrolet.com. mx/spark-2014.html. 12 Dec 2013
19. Hartigan, J.: Clustering Algorithms. Wiley (1975)
20. Valdez, F., Melin, P., Castillo, O.: Parallel particle swarm optimization with parameters adaptation using Fuzzy Logic. MICAI (2), 374–385 (2012)
21. Valdez, F., Melin, P., Castillo, O.: Bio-inspired optimization methods on graphic processing unit for minimization of complex mathematical functions. In: Recent Advances on Hybrid Intelligent Systems, pp. 313–322 (2013)

Using Semantic Representations to Facilitate the Domain-Knowledge Portability of a Natural Language Interface to Databases

Juan J. González B, Rogelio Florencia-Juárez,
Rodolfo A. Pazos Rangel, José A. Martínez F
and María L. Morales-Rodríguez

Abstract Our research is focused on the implementation of a Natural Language Interface to Database. We propose the use of ontologies to model the knowledge required by the interface with the aim of correctly answering natural language queries and facilitate its configuration on other databases. The knowledge of our interface is composed by modeling information about the database schema, its relationship to natural language and some linguistic functions. The design of this modeling allows users to configure the interface without performing complex and tedious tasks, facilitating its portability to other databases. To evaluate the knowledge-domain portability, we configured our interface and the commercial interface ELF in the Northwind database. The results obtained of the experimentation show that the knowledge modeled in our interface allowed it to achieve a good performance.

J. J. González B · R. Florencia-Juárez (✉) · R. A. Pazos Rangel · J. A. Martínez F ·
M. L. Morales-Rodríguez
Instituto Tecnológico de Ciudad Madero, Ciudad Madero, México
e-mail: rogelio.florencia@live.com.mx

J. J. González B
e-mail: jjgonzalezbarbosa@hotmail.com

R. A. Pazos Rangel
e-mail: r_pazos_r@yahoo.com.mx

J. A. Martínez F
e-mail: jose.mtz@gmail.com

M. L. Morales-Rodríguez
e-mail: lmoralesrdz@gmail.com

O. Castillo et al. (eds.), *Recent Advances on Hybrid Approaches for Designing Intelligent Systems*, Studies in Computational Intelligence 547,
DOI: 10.1007/978-3-319-05170-3_48, © Springer International Publishing Switzerland 2014

1 Introduction

One of the major sources of information is databases, which are collections of related information, stored in a systematic way for modeling a part of the world [1]. To access this information, sometimes it is necessary to formulate a query in a language that the computer can understand, such as database query languages like SQL (Structured Query Language). To use database query languages people require expertise, which most of them do not have. For example, let us consider the fragment of the entity-relationship diagram of the *Northwind* database shown in Fig. 1.

The first table, *employees* table, shows the employee id, first name, last name, etc. The second table, *customers* table, shows the customer id, company name, contact name, etc. The third table, *orders* table, shows the order id, customer id, employee id, etc. (see Fig. 2).

Based on the above information, a user may require a listing showing the employees named Laura and the customers of orders for which they are the employee. A SQL query to extract this information can be as follows:

```
SELECT e.FirstName, e.LastName, c.CompanyName,
        o.OrderId,o.OrderDate, o.ShippedDate
FROM Orders o, Employees e, Customers c
WHERE o.EmployeeId = e.EmployeeId AND
    o.CustomerId = c.CustomerId AND
        e.FirstName = 'Laura'
```

To formulate such queries, a user requires experience and knowledge about databases, knowledge that most users do not have. Therefore, it became necessary to design software applications for translating natural language (NL) queries into database queries, allowing users to easily access the information contained in a database. Such applications appeared in the 60s [2] and were called Natural Language Interface to Databases (NLIDB).

Considering the previous example, a user could request to an interface, extracting the same information using NL in a similar way to the following query: *Show the employees named Laura and the customers of orders for which they are the employee.*

The information displayed by the interface could be similar to the shown in Fig. 3.

Despite the large number of interfaces that have been developed, there exist problems that prevent them to attain an acceptable performance for users. To date, the percentage of performance of these interfaces has failed to reach the 100 % of correctly-answered queries. This is in part, when trying to *understand* the NL, the interfaces must face up problems such as ellipsis and ambiguity, among others. Furthermore, their linguistic coverage is limited since they can handle only a small subset of NL, and yet, some interfaces cannot respond to all questions within the same subset [3].

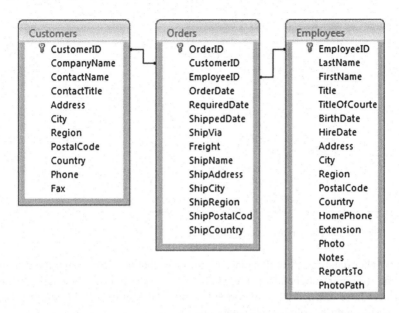

Fig. 1 Fragment of the entity-relationship diagram of the Northwind database

EmployeeID ▾	FirstName ▾	LastName ▾	Title
1	Nancy	Davolio	Sales Represe
2	Andrew	Fuller	Vice President
3	Janet	Leverling	Sales Represe
4	Margaret	Peacock	Sales Represe
5	Steven	Buchanan	Sales Manager
6	Michael	Suyama	Sales Represe
7	Robert	King	Sales Represe
8	Laura	Callahan	Inside Sales Cc
9	Anne	Dodsworth	Sales Represe

CustomerID ▾	CompanyName ▾	ContactName
ANATR	Ana Trujillo Emparedad	Ana Trujillo
ANTON	Antonio Moreno Taquei	Antonio Morenc
AROUT	Around the Horn	Thomas Hardy
BERGS	Berglunds snabbköp	Christina Berglu
BLAUS	Blauer See Delikatesser	Hanna Moos
BLONP	Blondesddsl père et fils	Frédérique Cite
BOLID	Bólido Comidas prepara	Martín Sommer
BONAP	Bon app'	Laurence Lebiha
BOTTM	Bottom-Dollar Markets	Elizabeth Lincol

OrderID ▾	CustomerID ▾	EmployeeID ◁	OrderDate
11074	SIMOB	7	06/05/19!
10262	RATTC	8	22/07/19!
10268	GROSR	8	30/07/19!
10276	TORTU	8	08/08/19!
10278	BERGS	8	12/08/19!
10279	LEHMS	8	13/08/19!
10286	QUICK	8	21/08/19!
10287	RICAR	8	22/08/19!
10290	COMMI	8	27/08/19!

Fig. 2 Employees, customers and orders tables

The first interfaces achieved high translation rates, between 69.4 and 96.2 % [4], because they were designed for a particular database. These interfaces considered both database schema and database information in the translation process (domain-dependent knowledge). This domain dependence made them difficult to

FirstName ▾	LastName ▾	CompanyName ▾	OrderId ▾	OrderDate ▾
Laura	Callahan	Queen Cozinha	11068	04/05/1998
Laura	Callahan	Old World Delicatessen	11034	20/04/1998
Laura	Callahan	Victuailles en stock	10334	21/10/1996
Laura	Callahan	Rattlesnake Canyon Grc	10262	22/07/1996
Laura	Callahan	Split Rail Beer & Ale	10756	27/11/1997
Laura	Callahan	Queen Cozinha	10786	19/12/1997
Laura	Callahan	QUICK-Stop	10845	21/01/1998
Laura	Callahan	Richter Supermarkt	11075	06/05/1998
Laura	Callahan	Piccolo und mehr	10844	21/01/1998
Laura	Callahan	Blondesddsl père et fils	10679	23/09/1997
Laura	Callahan	Hanari Carnes	10770	09/12/1997
Laura	Callahan	QUICK-Stop	10286	21/08/1996
Laura	Callahan	Pericles Comidas clásic:	10354	14/11/1996

Fig. 3 Possible information displayed by an interface

port to other databases. For this reason, interfaces that do not consider the database schema (domain-independent knowledge) were developed. Nowadays, these interfaces have reached 90 % of correctly-answered queries. In addition, they have been tested only in some small databases.

In this chapter, we describe the use of semantic representations designed to model the knowledge of an interface with the aim of increasing the number of correctly-answered queries and facilitating its knowledge-domain portability to other databases. The knowledge of the interface is composed by modeling the relationship between the database schema elements and its relation to NL, which it allows to increase the number of correctly answered queries. The design of these semantic representations, implemented in the configuration module of the interface, facilitates the configuration of the interface to other databases. To evaluate the performance of the interface using the semantic representations designed, we configured it in the *Northwind* database and we compared its performance with ELF, a commercial interface to databases. Our interface obtained a 59.26 % of recall and ELF a 55.56 %.

2 Background

There are four techniques, defined by Androutsopoulos et al. [2], used in the development of many NLIDB systems. These techniques reflect different choices of what information is to be applied and in what manner. Pazos et al. [5] describes these techniques as follows:

Pattern-matching (gNarLI, SAVVY). It employs a simple technique based on patterns or rules that are applied to the NL query.

Syntax-based systems (LUNAR, NALIX). The user's question is analyzed syntactically and the resulting parse tree is directly mapped to an expression in some database query language.

Semantic grammar systems (CoBASE, DATALOG, EASYASK, ELF, ENG-LISH QUERY, EUFID, LADDER, NCHIQL, NLPQC, PHILIQA1, PLANES, PRECISE, REL, QBI, STK, WYSWYM). Semantic grammar systems are similar to syntax-based systems. The difference is that the grammar's categories do not necessarily correspond to syntactic concepts.

Intermediate representation languages (C-PHRASE, CHAT 80, CLE, DIA-GRAM, DIALOGIC, EDITE, GINSPARG'S, IRUS, JANUS, LOQUI, MASQUE/SQL, PHILIQA, RENDEZVOUS, TAMIC, TEAM, TELI). Most current NLIDBs first transform the NL question into an intermediate logical query, expressed in some internal meaning representation language.

One of the drawbacks of the systems which employ any of these four techniques is the portability to different domains. For example, to configure a *Pattern-matching system* in a new domain is needed to have defined a set of patterns representing the structure of each possible query that users could formulate. Therefore, the performance of these systems is limited to the number of patterns defined by the user, which are usually manually designed. The *Syntax-based systems* have the problem that their syntactic structures are deeply rooted to the database information; therefore, the systems that use this technique are difficult to port to other databases. The *Semantic grammar systems* need a new semantic grammar is written whenever the system is configured for a new domain, which makes difficult to port from one database to another. Moreover, the structures of the syntactic tree can hardly be useful for other databases. The *Intermediate representation languages* permit the domain to be independent of the system modules. However the systems that use this technique use deductive databases, which are much less used than relational databases, thus, the scope of application is limited.

There are other interfaces that seek to automate the configuration of a NLIDB in a database using techniques such as Learning Semantic Parsing, process of mapping a NL sentence into a complete, formal meaning representation or logical form. Examples of these interfaces are CHILL [6], KRISP [7], SCISSOR [8] and WASP [9]. Although these interfaces demonstrated good results, they are not related to our research since these interfaces require a training set, generally designed by an expert.

The popularity of NLIDBs has been changing towards a new type of interfaces, which find answers on ontologies [10]. These interfaces to ontologies relate the meanings of the words, inherit relationships based on existing structures, treat ambiguities more effectively, allow reasoning about structured data, facilitate its portability and have the ability to combine and merge existing resources on the Web [10]. These interfaces are oriented towards the Semantic Web, where one important aspect of research is the use of ontologies as a suitable representation of knowledge available on the Web [11]. The aim is to achieve that web search engines could perform semantic searches, and they are not based anymore on coincidence of words. Among the interfaces of this type, we can mention the following: GINSENG (Guided Input Natural Language Search Engine) [12],

ORAKEL [13], NLP-REDUCE [14], PANTO (Portable Natural Language Interface to Ontologies) [15], FREyA (Feedback, Refinement and Extended Vocabulary Aggregation) [16], etc.

The main difference between interfaces to ontologies and our interface is that we do not use ontologies as a database. We use an ontology to store the needed knowledge for translating a NL query into a SQL query, by which extract information from a relational database like NLIDBs do.

3 Proposal

Our architecture is composed of two modules: the translating module and the configuration module (Authoring Tool). The translating module is responsible for converting a NL query into a SQL query, as well as running the SQL query for extracting the information requested by the user. Also, it is responsible for presenting to the user the extracted information. The configuration module is responsible for generating the information that the translating module needs for its functioning and for facilitating portability of our interface to other databases. In this chapter, we describe only the configuration module.

The configuration module was design considering two of the four aspects defined by Androutsopoulos et al. [2] to facilitate the portability of a NLIDB.

3.1 Database Management Systems

The configuration module is designed in a modular way where all access of the interface to the database is encapsulated in a *data access layer* (*CDatabaseAccess*). By this *data access layer*, the configuration module and the translation module can access the user's database to extract the requested information and to access the information from the lexicon during the queries translation process (see Fig. 4).

The modular design of our interface allows incorporate different DBMS based on the SQL standard by making some changes in the *data access layer*. Thus, the overall functioning of the interface will not be affected when porting it to a new DBMS. Currently, SQL Server 2008 is the DBMS implemented in the interface; however we will take advantage of the modular design of the interface for integrating MySQL, MS-Access, Sybase and PostgreSQL, as shown in Fig. 4.

To implement a new DBMS, we only need defining in the *data access layer* how to connect the interface to any database in the DBMS and how to extract information from the schema of any database.

This is a vital step for the functioning of the interface, since information extracted from the database schema, by the configuration module, forms the basis of knowledge of the interface. This information is part of all information that the configuration module stores in an ontology file (see Fig. 5).

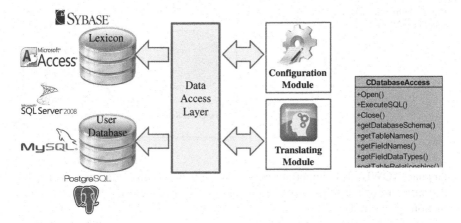

Fig. 4 Data access layer

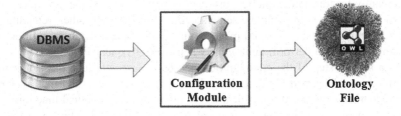

Fig. 5 Extracting and storing the database schema into an ontology file

3.2 Knowledge-Domain Portability

Another aspect considered in the portability of our interface, being perhaps the most important of all, is the knowledge-domain portability.

To achieve knowledge-domain portability, the configuration module was designed to facilitate the user, preferably a database manager (DM), the interface configuration. It is important to note that for configuring the interface, the DM does not require having a deep understanding about databases and query languages, since it is enough he/she only knows the structure of the user's database and also what tables and fields could store relevant information to users.

As a first step in porting the interface to a new knowledge-domain or database, the DM must create a *semantic representation of the database schema* (SDbS) and store it in an ontology file (see Fig. 6). In this step, the DM only have to specify on what database will be configured the interface and the ontology filename that will be generated. This process is done automatically by the configuration module by extracting information about the database schema and modeling this information using semantic representations designed for this purpose in the ontology web

Fig. 6 Semantic database
schema (SDbS)

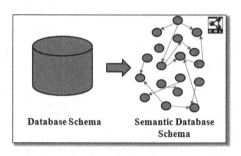

Database Schema   Semantic Database
                            Schema

language (OWL). It is important to emphasize that only information regarding to the structure of the schema of the database is modeled and not the user's information stored in the database.

The SDbS generated automatically by the configuration module is used as *domain-dependent knowledge* by the translating module. This knowledge is always loaded into memory of the interface, reducing access to the user database. It is used mainly for translating queries: (a) identifying *keywords* in the user query, (b) determining which words may be semantically related to each other, reducing on this way ambiguity, since a word may be related to different database entities, (c) determining how to perform table *joins* when the SQL query is formed.

As a second step of the configuration, the DM must introduce a set of words identified in our lexicon for describing database entities (table and field names). The lexicon contains words classified according to their grammatical function. The information of our lexicon was imported from Freeling [17], which is an open source suite of language analyzers. When a word is introduced by the DM, it must be related with those table and/or field names which describe. We named this relation, mapping (see Fig. 7).

For example, in Fig. 7, the *category* word maps to the *categories* table and to the *name* field. The *name* word maps the *categories.name* and *products.name* fields. In this way, the DM must enter words that describe the database entities.

The mappings defined by the DM are also stored in the ontology file generated by the configuration module. Thus, the database schema and the NL are linked in a single semantic representation, facilitating the translating process of the interface.

As a last step of the configuration, the DM can increase the linguistic coverage of the interface introducing words that require more than a simple mapping to be interpreted by the translating module. These words require that the interface runs a specific query on the database. As an example, we incorporated words that express a:

- *Superlatives*, which are words used to refer to the best or worst value of a set of values. For example, *What is the population of the capital of the smallest state*? To answer the above query extracted from the corpus of GeoQuery 250 [6, 18], the interface must first resolve the superlative word *smallest*, which refers to the *state* with the smallest value stored in the *area* field of the *state* table. A possible SQL query could be generated by an interface as follows:

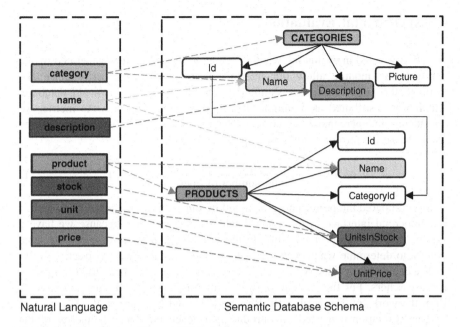

Natural Language   Semantic Database Schema

Fig. 7 Mappings between NL and SDbS elements

```
FROM CAPITAL A
WHERE A.Population = (SELECT Max(A.Population)
FROM CAPITAL A)
```

- *Ranges of values*, which are words that define an initial and final value of the information to be extracted. For example, *Are there morning flights from Pittsburgh to Dallas?*. To answer the above query extracted from the ATIS corpus [19], the interface must know that the *morning* word represents a range of hours ranging between 00:01 and 12:00 o'clock. The ranges of values are used by the user to delimit the information to be extracted from the database. A possible SQL query could be generated by an interface as follows:

```
WHERE Departure_Time BETWEEN 00001 AND 1200
```

- *Restrictions*, which are words that express a filter at the time of extracting the information from the database. For example, *Economy flights from Pittsburgh to Philadelphia*. To answer the above query extracted from the ATIS corpus [19], the interface must know that the *economy* word is a filter that must be fulfilled by the information to be extracted from the database. A possible SQL query could be generated by an interface as follows:

```
WHERE COMPOUND_CLASS.Economy = 'YES'
```

In the same way that mappings, linguistic information is added to the ontology file with the aim of increasing linguistic coverage of the interface.

4 Experimental Evaluation

We are interested in evaluating whether semantic representations generated by the configuration module allow our interface to obtain a good performance. To evaluate the performance, we configured our interface in *Northwind* database, which was distributed as a sample data base in SQL Server 7. As metric, we considered the percentage of correctly-answered queries, which we measured using the *recall* formula shown by expression (1):

$$recall = \frac{total\ number\ of\ correct\ queries}{total\ number\ of\ queries} \times 100 \qquad (1)$$

We compared the performance obtained by our interface against ELF, which is a commercial interface to relational databases. We use ELF because we have a version of this software for testing. In this way, we make sure to apply the same criteria to determine whether the answers of the interfaces were correct.

We decided to use the *Northwind* database because it has been used frequently by researchers. For the evaluation, we formed a corpus of 81 queries extracted from the works of Tan et al. [20], Bhootra [21], and Nihalani et al. [3].

The evaluation consisted on introducing to each interface all queries of the corpus. The results obtained from the experiments are shown in Table 1.

The Table 1 shows the number of correctly-answered queries, incorrectly-answered queries and the percentage of *recall* obtained for each interface evaluated.

The percentages of *recall* indicate that our interface obtained the best performance by responding correctly 59.26 % (48 queries) of the 81 queries against 55.56 % (45 queries) of ELF. The performance of our NLIDB was primarily due to:

- The database schema, modeled using the semantic representations designed, and the description of the database entities, made by the DM, allowed to our interface to identify correctly tables, fields and values in queries, which is essential for the translation process,
- The query translation algorithm of our interface identified correctly the existing semantic relationships between words in the query, discarding semantic meanings that are not consistent with the general semantics of the query,
- The tables involved in the queries were correctly joined by our interface. This was possible because the interface knows the complete schema of the database, represented semantically in the ontology.

Although the percentages of recall indicate that our interface improved the performance of ELF, its percentage scored low. The reason for this low percentage is that some queries of the corpus are beyond the scope of our interface. For example, there are queries that require some features to be answered correctly (see Table 2).

Table 1 Results obtained from the evaluation

Interface	Correctly answered	Incorrectly answered	Recall %
ELF	45	36	55.56
Our NLIDB	48	33	59.26

Table 2 Query types beyond the scope of our interface

Feature	#Queries	Example
Discourse	3	• And *their* phone numbers?
		• Tell me the age of *these* people
Aggregation	10	• What product is the *most* frequently ordered?
		• Name the *thirdmost* ordered product
Calculated Functions	6	• What is the *average* product unit price of products?
		• What is the *median* of the *age* of the employees?
Date format	5	• Give all the products and quantity ordered in *July 1996*
		• Orders that were shipped by Speedy Express in the *month* of *October*

All the above features, except discourse, can be implemented in the interface without changing the essence of our translation algorithm. We only need to design new semantic representations for indicating to the interface how it must process this kind of queries and we also need incorporate rules to identify, in this case, different date formats. To implement discourse features, we think it is necessary to design semantic representations that dynamically allow capturing entities that emerge from the interaction between the user and the interface, for solving some types of anaphora and some situations of inter-sentential ellipsis.

In general, based on the results shown in this section, we can affirm that using semantic representations for knowledge modeling allowed our interface to obtain a good performance in the corpus used, obtaining the major number of correctly-answered queries. Also, we can affirm that the configuration module, using these semantic representations, allowed us successfully configuring our interface in the *Northwind* database.

5 Conclusions

The NLIDBs are tools that allow users to access information in a database. The advantage of these tools over other query applications is that the interaction with the user is performed in a simple way through the NL.

Despite the large number of NLIDBs that have been developed, these have not yet reached 100 % of correctly-answered queries. This is because they are facing problems such as ellipsis and ambiguity. Furthermore, their linguistic coverage is limited and they are difficult to porting to other domains of knowledge. In this chapter, we describe the configuration module, which it uses the semantic representations that we designed in OWL to model the knowledge required by the interface with the aim of correctly answering NL queries and facilitate its configuration on other databases.

We evaluated whether semantic representations generated by the configuration module allow our interface to obtain a good performance. To evaluate the performance, we configured our interface in the *Northwind* database and we formed a corpus of 81 queries extracted from the works of Tan et al. [20], Bhootra [21], and Nihalani et al. [3]. To measure the performance, we consider as metric the percentage of correctly-answered queries, which we measured by the *recall*. We compared the results obtained by our interface against the commercial interface to relational databases ELF. Recall percentages indicate that our interface obtained the best performance by responding correctly to a 59.26 % of the 81 queries against a 55.56 % of ELF.

Based on the obtained results, we can affirm that using semantic representations for knowledge modeling in a NLIDB is a good alternative to increase the number of correctly-answered queries. Also, we can affirm that due to the flexibility of these semantic representations for knowledge modeling, users could introduce easily to our interface the needed knowledge to configure it in any relational database and get a good performance. As main future work, we pretend to incorporate discourse features in the interface by designing semantic representations for modeling dynamically entities mentioned in user's queries. The discourse is an issue that has not been widely addressed. Discourse allows to a NLIDB answers complex queries interactively with the user. This facilitates the information extraction since humans do not get information through isolated queries, but through queries performed interactively. It represents an advantage over query languages to databases and graphical interfaces.

As other future work, we will implement other DBMS in the interface for evaluating its portability to different databases in other DBMS and we also will evaluate the degree of complexity that represents to real users configure our interface in different databases.

References

1. Chandra, Y.: Natural Language Interfaces to Databases. University North Texas (2006)
2. Androutsopoulos, I., Ritchie, G.D., Thanish, P.: Natural language interfaces to databases-an introduction. Nat. Lang. Eng. 1(1), 29–81 (1995)
3. Nihalani, N., Motwani, M., Silaka, S.: Natural language interface to database using semantic matching. Int. J. Comput. Appl. 31, 29–34 (2011). doi:10.5120/3942-5552
4. Popescu, A.: Modern Natural Language Interfaces to Databases: Composing Statistical Parsing with Semantic Tractability. University Washington (2004)
5. Pazos, R.A., González, J.J., Aguirre, M.A., Martínez, J.A., Fraire, H.J.: Natural language interfaces to databases: an analysis of the state of the art. Recent Adv. Hybrid Intell. Syst. 463–480 (2013)

6. Zelle, J.M., Mooney, R.J.: Learning to parse database queries using inductive logic programming. In: Proceedings of the 13th National Conference on Artificial Intelligence, pp. 1050–1055 (1996)
7. Kate, R.J., Mooney, R.J.: Using string-kernels for learning semantic parsers. In: Proceedings of the 21st International Conference on Computational Linguistics and 44th Annual Meeting of the Association for Computational Linguistics (COLING/ACL-06), Sydney, Australia, pp. 913–920 (2006)
8. Ge, R., Mooney, R. J.: A statistical semantic parser that integrates syntax and semantics. In: Proceedings of the 9th Conference on Computational Natural Language Learning, pp. 9–16 (2005)
9. Wong, Y.W., Mooney, R.J.: Learning for semantic parsing with statistical machine translation. In: Proceedings of Human Language Technology Conference/North American Chapter of the Association for Computational Linguistics Annual Meeting, pp. 439–446 (2006)
10. Damljanovic, D.: Natural language interfaces to conceptual models. Ph.D. thesis, The University of Sheffield, Language Resources and Evaluation. http://etheses.whiterose.ac.uk/1630/ (2011)
11. Buraga, S.C., Cojocaru, L., Nichifor, O.C.: Survey on web ontology editing tools. Trans. Autom. Control Comput. Sci. Rom. 1–6 (2006)
12. Bernstein, A., Kaufmann, E., Kaiser, C.: Querying the semantic web with ginseng: a guided input natural language search engine. In: 15th Workshop on Information Technologies and Systems (WITS 2005), Las Vegas (2005)
13. Cimiano, P., Haase, P., Heizmann, J.: Porting natural language interfaces between domains: an experimental user study with the ORAKEL system. In: Proceedings of the 12th International Conference on Intelligent User Interfaces, Honolulu, Hawaii, USA, pp. 180–190 (2007)
14. Kaufmann, E., Bernstein, A., Fischer, L.: NLP-Reduce: A naïve but Domain-independent Natural Language Interface for Querying Ontologies. In: 4th European Semantic Web Conference (ESWC 2007), Innsbruck A (2007)
15. Wang, C., Xiong, M., Zhou, Q.: PANTO: a portable natural language interface to ontologies. LNCS **4519**, 473–487 (2007)
16. Damljanovic, D., Agatonovic, M., Cunningham, H.: Natural language interfaces to ontologies: Combining syntactic analysis and ontology-based lookup through the user interaction. In: Proceedings of the 7th Extended Semantic Web Conference (ESWC 2010), Heraklion. Springer, Berlin (2010)
17. Carreras, X., Chao, I., Padró, L., Padró M.: FreeLing: An open-source suite of language analyzers. In: Proceedings of the 4th International Conference on Language Resources and Evaluation (LREC'04) (2004)
18. Tang, L.R., Mooney, R.J.: Using multiple clause constructors in inductive logic programming for semantic parsing. In: Proceedings of the European Conference on Machine Learning (ECML) (2001)
19. Sri International: Air Travel Information Service. http://www.ai.sri.com/natural-language/projects/arpa-sls/atis.html (1990)
20. Tan, J.J., Lane, H., Rijanto, A.: Report on NLIDBs. http://www.elfsoft.com/Resources/References.htm (2001)
21. Bhootra, R.: Natural language interfaces: comparing english language front end and english query. Dissertation, Virginia Commonwealth University. http://www.elfsoft.com/Resources/References.htm (2004)

Post-Filtering for a Restaurant Context-Aware Recommender System

Xochilt Ramirez-Garcia and Mario García-Valdez

Abstract Nowadays recommender systems are successfully used in various fields. One application is the recommendation of restaurants, where even if the method of customer service is the same, the quality of service varies depending on the resources invested to improve it. Traditionally, in a restaurant a waiter takes orders from customers and then delivers the product. The motivation of this work is to make recommendations of restaurants with the aim of disseminating information about products and services offered by restaurants in the city of Tijuana through a Web based platform. The proposed recommendation algorithm is based on contextual post-filtering approach, using the output of a collaborative filtering algorithm together with contextual information of the user's current situation. The dataset used was explicitly acquired through questionnaires answered by 50 users; and the experiment was performed with a data set of 1,422 ratings of 50 users and 40 restaurants. We evaluate our approach with Mean Absolute Error (MAE) using dataset obtained of the questionnaire and the experimental results show that our approach has an acceptable accuracy for the dataset used.

1 Introduction

Humans have an excellent communication process for conveying ideas to each other and reacting appropriately. This is due to many factors: the richness of the language they share, the common understanding of how the world works and an implicit understanding of everyday situations. When humans speak using implicit

X. Ramirez-Garcia · M. García-Valdez (✉)
Instituto Tecnológico de Tijuana, Tijuana, Mexico
e-mail: mario@tectijuana.edu.mx

O. Castillo et al. (eds.), *Recent Advances on Hybrid Approaches for Designing Intelligent Systems*, Studies in Computational Intelligence 547,
DOI: 10.1007/978-3-319-05170-3_49, © Springer International Publishing Switzerland 2014

situational information (or context) they can increase the conversational band-width. Using contextual information, designers can increase the richness of communication in human–computer interaction enabling the development of better intelligent applications. This process can also be applied in recommender systems. Nowadays, recommender systems can be found in many modern applications that expose the user to a huge collection of items. Such systems typically provide the user with a list of recommended items they might prefer, or predict how much they might prefer an item. These systems help users to decide on appropriate items, and ease the task of finding preferred items in the collection.

The majority of approaches to recommender systems focus on recommending the most relevant items to individual users and do not take into consideration any contextual information, such as time, place and the company of other people (e.g., for watching movies [1] or dining out [2]). In other words, traditionally recom-mender systems deal with applications having only two types of entities, users and items, and do not put them into a context when providing recommendations.

In applications such as the recommendation of vacation packages [3, 4], per-sonalized content on a Web site [5], or a movie [6], it may not be sufficient to consider only information about users and items, it is also important to incorporate the contextual information into the recommendation process in order to recom-mend items to users in certain circumstances [7]. Recently, there has been a vast amount of research in the field of recommender systems, mostly focusing on designing new algorithms for context-aware recommendations [3, 8, 9]. This chapter proposes the use of contextual post-filtering approach that uses the tra-ditional collaborative filtering algorithm along with contextual recommendations in the domain of restaurants.

The rest of the work is presented as follows: in Sect. 2, related concepts are described; in Sect. 3, the proposed Recommender System is explained; in Sect. 4, the environment setup, data, metrics and experiments conducted to test the algo-rithm are presented; in Sect. 5, the results of the system evaluation are explained; in Sect. 6, conclusions of the study are presented; finally in Sect. 2, the most important points of Post-filtering, lessons learned and in future work are discussed.

2 Recommender Systems

In recent years, recommender systems have become part of the solution to the information overload problem faced by consumers on Internet. Collaborative fil-tering recommender systems provide users with suggestions regarding which information is most relevant to them. These systems have proved to be some of the most successful techniques to help people find contents that are most valuable to them [10].

2.1 Collaborative Filtering

A collaborative filtering algorithm [10] makes recommendations by analyzing user-to-user correlation. This method looks for similarities between users to make predictions. For example, supposing that user A is similar to user B, if user A is interested in product *a*, then it might be inferred that user B also will be interested in product *a*, and product *a* can be recommended to user B. A collaborative filtering approach needs to collect data on users' opinions of items, and identifying neighborhood according the similarity among users. This approach addresses the phenomenon that when selecting products or information, people are usually influenced by other people's experiences with the product or information. In addition, in collaborative filtering, the content analysis of items can be ignored and only user's opinions on items are considered relevant. This might be contributive to recommending those items for which content analysis is weak or impossible [11].

2.2 Context

Dey et al. [12] defines context to be the user's physical, social, emotional or informational state. Pascoe defines context to be the subset of physical and conceptual states of interest to a particular entity. These definitions are too specific. Context is all about the whole situation relevant to an application and its set of users. We cannot enumerate which aspects of all situations are important, as this will change from situation to situation. In some cases, the physical environment may be important, while in others it may be completely immaterial.

> Context is any information that can be used to characterize the situation of an entity. An entity is a person, place, or object that is considered relevant to the interaction between a user and an application, including the user and applications themselves.

This definition makes it easier for an application developer to enumerate the context for a given application scenario. If a piece of information can be used to characterize the situation of a participant in an interaction, then this information is context [13].

2.3 Obtaining Contextual Information

The contextual information can be obtained in different ways, including [14]:

- *Explicitly, i.e.*, by directly approaching relevant people and other sources of contextual information and explicitly gathering this information either by asking direct questions or eliciting this information through other means. For example, a website may obtain contextual information by asking a person to fill out a web form or to answer some specific questions before providing access to certain Web pages [4].

- *Implicitly* from the data or the environment, such as a change in location of the user detected by a smartphone application or GPS. Alternatively, temporal contextual information can be implicitly obtained from the timestamp of a transaction [4]. Nothing needs to be done in these cases in terms of user interaction or other sources of contextual information, the source of the implicit contextual information is accessed directly and the data is extracted from it [14].

2.4 Post-filtering Approach

The contextual post-filtering approach initially ignores contextual information when generating a first stage of recommendations, i.e., when generating the ranked list of all candidate items from which any number of top-N recommendations can be made, depending on specific values of N. Then, the contextual post-filtering approach adjusts the obtained recommendation list for each user using contextual information. The recommendation list adjustments can be made by [14]:

- Filtering out recommendations that are irrelevant (in a given context), or
- Adjusting the ranking of recommendations on the list (based on a given context).

In Fig. 1 the scheme of the contextual post-filtering approach is depicted. Contextual Post-Filtering analyzes the contextual preference data for a given user in a given context to find specific item usage patterns and then use these patterns to adjust the item list, resulting in more contextual recommendations. In other words, post-filtering method filters out recommended items that have low relevance in the specific context and keeps items with high relevance. A major advantage of contextual post-filtering is that it allows using any of the numerous traditional recommendation techniques previously proposed in the literature [10].

3 Proposed Restaurant Recommender System

Restaurant Recommender System (RRS) uses collaborative filtering to find restaurants for the active user [2]. The ratings of user profiles are used to determine the similarity using Pearson correlation. The similarity between the active user and top-N users is used to obtain a weighted average of ratings for each particular item; the top-N ratings are used as a list of recommended restaurants for the active user.

The output of the collaborative filtering algorithm (top-N list) is supplied to the next step of the post-filtering process. The restaurants are adjusted to the context in the next step in order to make ranking of restaurants in the current context.

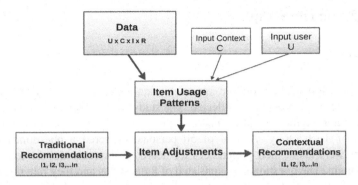

Fig. 1 Final phase of the contextual post-filtering approach: recommendation list adjustment [14]

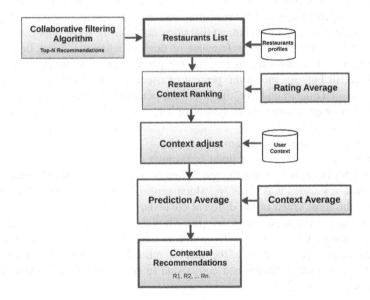

Fig. 2 The post-filtering methodology used in the domain

Post-filtering is based on the average of ratings in a specific context, so prediction is made with: (1) the average post-filtering a restaurant has in the current context (that is the mean of user ratings) and (2) the rating predicted by the collaborative filtering algorithm. The top-N list contains the restaurants with highest predictions, so each restaurant is adjusted for the user's context and listed in contextual recommendations; the process is depicted in Fig. 2.

4 Environment Setup

4.1 Dataset

In order to validate the proposed approach, data about restaurant preferences of users in different contexts was collected. The study subjects were students enrolled in a computer engineer major, a master's program and professors of the Tijuana Institute of Technology. A total of 50 persons answered a questionnaire; the questions were about their preferences for nearby restaurants and the technology used by them. The questionnaire consisted of 8 questions and they rate restaurants from a selection of 40. Each restaurant chosen was rated six times one for each context considered. At the end of the poll 1,422 ratings were stored in database. The questions are shown in the Table 1.

The user's responses of questions from 1 to 6 are shown in the Fig. 3. In Fig. 3a, the percentage of surveyed students and teachers is plotted; in Fig. 3b, shows the percentage of the element that users consider the most important to visit a restaurant; in Fig. 3c, the preferences of devices when are using Internet for restaurant recommendations; in Fig. 3d, the percentage of operating system that often used, in Fig. 3e, shows the percent of users that use the Internet to search restaurants in Tijuana; and Fig. 3f, shows the percentage of users that would like using a restaurant recommender system of Tijuana.

For questions 7 and 8 only the top-ten restaurants are shown, without and considering the contextual situation.

In Fig. 4a, the favorite restaurant is Daruma (178 votes), whereas in Fig. 4b, Daruma does not appear in the top-ten. When considering the context "midweek", the favorite restaurant was Carls Jr., which appears in both graphs; this restaurant was also the most voted in the different contexts.

Contextual recommendations of post-filtering approach depends mainly of context "midweek" or "weekend", which is the day when the restaurants were rated. Subsequently, the result of the query is refined according to the current user context; the 6 contexts mentioned correspond to combinations of contextual factors shown in Table 2.

4.2 Metrics

Recommender Systems are widely used in different domains, so the goals of each one can be different, and then, also metrics used depend on the main system goal. Sometimes prediction's accuracy is a priority and the system evaluation is focused in accuracy of values. There are several metrics to measure the accuracy: *Root Mean Square Error (RMSE)* [1, 15], *Mean Absolute Error (MAE)* [1] and *Precision and Recall* [1, 15] are some examples. By other hand, if the system tries to

Table 1 The user feedback through the explicit questionnaire was obtained

Question	Response	
1. What is your occupation?	1. Student	2. Employee
2. According your priority, order by importance the features you consider when you choose to visit a restaurant	1. Installation/decor 2. Prices 3. Service	4. Dishes 5. Atmosphere 6. Location
3. What are the devices you most frequently use?	1. Smartphone 2. Tablet	3. Laptop 4. PC
4. What are the operating systems you use?	1. Android 2. Windows 3. iOS	4. Symbian 5. BlackberryOS 6. Other
5. Have you used an application to search for restaurants in Tijuana?	1. Yes 2. No	3. Which one?
6. Would you like to use an application of recommender systems of Tijuana?	1. Yes	2. No
7. Assign rating for restaurants that you prefer without considering context situations	Restaurants list	
8. Assign rating for restaurants that you prefer considering context situations	Restaurants list	

measure subjective aspects such as user's satisfaction, recommendations quality or system utility for a cluster of users or community, others metrics are considered such as is mentioned in [16].

4.2.1 Precision and Recall

In recommender systems, from the user point of view the most useful result is to receive an ordered list of recommendations, from best to worst. In fact, in some cases the user doesn't care much about the exact ordering of the list, a set of few good recommendations is fine. Classic information retrieval metrics can be used to evaluate engines that return a list of recommended items: Precision and Recall. These metrics are widely used on information retrieving scenarios and applied to domains such as search engines, which return some set of best results for a query out of many possible results. So for example, in a search engine it should not return irrelevant results in the top results, although it should be able to return as many relevant results as possible. We could say that Precision is the proportion of top results that are relevant, considering some definition of relevant for your problem domain. In recommender systems those metrics could be adapted [15] such as:

- *Precision* is the proportion of recommendations that are good recommendations.
- *Recall* is the proportion of good recommendations that appear in top-N recommendations.

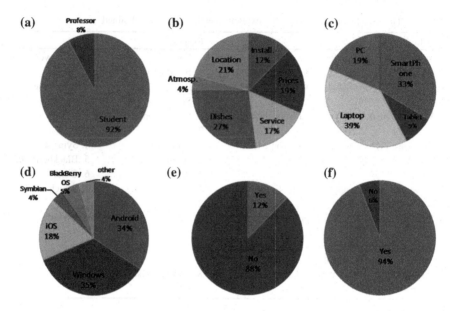

Fig. 3 The graphs showed the user's responses according their preferences

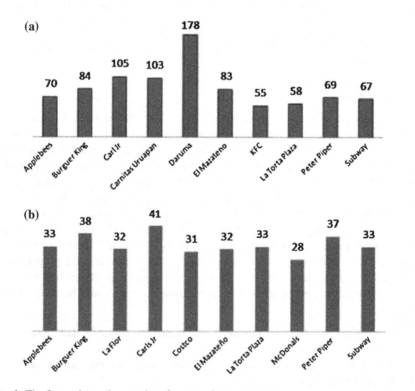

Fig. 4 The figure shows the users' preferences about restaurants

Table 2 The contextual factors considered in the approach

Contextual factor	
Day of the week	Midweek
	(*Monday, Tuesday, Wednesday, Thursday*)
	Weekend
	(*Friday, Saturday and Sunday*)
Place	School
	Home
	Work

4.2.2 Prediction Accuracy

Prediction accuracy is by far the most discussed property in the recommender system literature. At the core the vast majority of recommender systems are prediction engines. This engine may predict user opinions over items (e.g. ratings of movies) [1] or the probability of usage (e.g. purchase) [17]. A basic assumption in recommender systems is that the user will prefer a system that provides more accurate predictions. Thus, many researchers set out to find algorithms that provide better predictions. Prediction accuracy is typically independent of the user interface, and can thus be measured in an offline experiment. Measuring prediction accuracy in a user study measures the accuracy given a recommendation. This is a different concept from the prediction of user behavior without recommendations, and is closer to the true accuracy in the real system [16].

4.2.3 MAE

Mean Absolute Error (MAE) is a popular metric used in evaluating the accuracy of predicted ratings, the system generates predicted ratings $\hat{r}_{ui}$ for a test set $\mathscr{I}$ of user-item pairs (u, i) for which the true ratings r_{ui} are known. Typically, r_{ui} are known because they are hidden in an offline experiment, or because they were obtained through a user study or online experiment [16]. The MAE between the predicted and actual ratings is given by the Eq. (1):

$$\text{MAE} = \sqrt{\frac{1}{|\mathscr{I}|} \sum_{(u,i) \in \mathscr{I}} |\hat{r}_{ui} - r_{ui}|} \tag{1}$$

5 Experimental Results

The algorithm was developed in Python language and tested with a PostgreSQL database manager, this database was explicitly collected from 50 users who answered questionnaire detailed in the previous section. The following chart shows

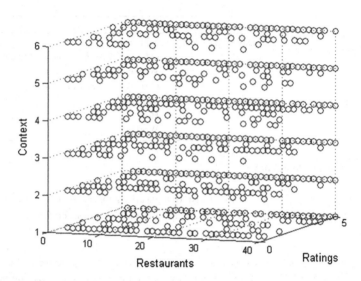

Fig. 5 The figure shows data distribution with 3-dimensions

the ratings matrix obtained from users (scale from 1 to 5) for 40 restaurants in 6 different contextual situations. The system starts with 3-dimensional data with an amount of 1,422 ratings, the data distribution in 6 contexts is in Fig. 5.

A total of 172 predictions was made for different users and the error MAE = 0.5859 when the context "midweek" for current user was considered. The rationale for this result is that using a small dataset limits the performance of the algorithm. On the other hand, having only one contextual factor does not improve the accuracy of the recommendations in this domain.

6 Conclusions

The amount of information that is supplied for the system is fundamental for accurate recommendations [11] and the scarcity problem [18] of contextual information can lead to bad recommendations. In the current experiment a small dataset was used with limited contextual information. Also a problem found when analyzing the questionnaire results, is that users didn't actually changed their ratings when asked to consider a particular context. A possible explanation of this behavior is the additional cognitive load users have when thinking a difference in preference in each scenario. Also their preference can be determined independently of the context, is not until faced with the decision problem of choosing a restaurant between various competitors in a particular context when users assign a utility rating to each option.

In order to obtain satisfactory results validation must be done with larger datasets and in other domains (for example e-learning) where contextual information is critical to recommendations. The goal is to improve the accuracy and, subsequently achieve the user satisfaction through this way. Contextual factors used by other research [3, 9, 19] are selected based on the domain of the recommender system, these factors can be included from the queries to the database and are used as a reference for ranking items [9], based on user preferences [2], or based on probability models to predict the users behavior [20].

However, in the experiment we found that users had different opinions when we use contextual assumptions and when it happens in real-life. The selection of a restaurant could be affected not only by tastes and preferences of the users but also by other factors that are not perceived in assumptions. The user no makes decisions considering contextual factors such as day of the week, the mobile device or distance. Indeed, it would be interesting to ask how the distance is important when the user is student and how the decision changes when the user is not a student, or how the influences distance in the decision when the user interact with mobile phone or laptop. Other factors (those could be subjective) are becoming more relevant: (1) the relationships (friends, family, and coworkers), (2) available money, (3) type of food the user want, or (4) the habit of going to the same place. We appreciate that there is a difference in the selection of the restaurant, the wishes of the user matter more than preferences or tastes of the user (what the user want in this moment), and then logical reasoning makes the user decide which restaurant visit.

On the other hand, contextual information obtained from users through the questionnaire was not successful at all, the reason is that most students did not show a difference between visiting a place on weekend or midweek, rating assign was the same in different contexts. In other cases, students only ignored the question leaving gaps of ratings that the system not infers whether information is null.

In this study, the users making assumptions to assign ratings it was exhausting, through the way such as is propose in [21] isn't easy to obtain relevant information in this case, probably the students' answers were not correct compared with a real-life situation where before selecting the user thinks about conditions that must be met for going to the restaurant. Then, "day of the week" is not considered such as an overriding factor for these contextual recommendations because the experiment confirmed that not often matter for students whether is weekend or midweek.

According to the observed performance of the algorithm, we are proposing in future work the following to improve the contextual recommendations:

- Changing the contextual factor "day of the week" instead of other factor that has the greatest impact on the selection of the restaurant, or combine factors to improve accuracy in the system context.
- Use other ways to get more relevant information of the context without user effort, could make collecting information implicitly such as is mentioned in [22].

- Implement other recommender system approach (pre-filtering) for compare the recommendations of post-filtering to improve the quality of the top-N recommendation list.
- Test and evaluate the algorithm in a different domain for observe the performance with contextual information.

References

1. Campochiaro, E., Cassata, R., Cremonesi, P., Turrin, R.: Do metrics make recommender algorithms? In: International Conference on Advanced Information Networking and Applications Workshops. Milano (2009)
2. Ramirez-Garcia, X., Garcia-Valdez, M.: Restaurant recommendations based on a domain model and fuzzy rules. In: International Seminary in Computer Intelligence (ISCI). Tijuana Institute of Technology, Tijuana (2012)
3. Baltrunas, L., Ludwig, B., Peer, S., Ricci, F.: Context-Aware Places of Interest Recommendations and Explanations. Free University of Bozen-Bolzano, Bolzano (2011)
4. Baltrunas, L., Ludwig, B., Peer, S., Ricci, F.: Context relevance assesment and exploitation in mobile recommender systems. In: Personal and Ubiquitous Computing. Free University of Bolzano, Bolzano (2011)
5. Martinez, L., Calles, J., Martin, E.: Ontology-based Web service to recommend spare time activities. In: International Workshop on Information Heterogeneity and Fusion in Recommender Systems (HetRec), Barcelona (2010)
6. Kim, K.R., Lee, J.H., Byeon, J.H.: Recommender system using the movie genre similarity in mobile service. In: 4th Internartional Conference on Multimedia and Ubiquitous Engineering (MUE) (2010)
7. Zimmermann, M., Lorenz, A.: Personalization and context management. User Model. User-Adap. Inter. **15**(3–4), 275–302 (2005)
8. Baltrunas L., Ricci F.: Context-based splitting of item ratings in collaborative filtering. In: The ACM Conference Series on Recommender Systems (2009)
9. Kahng, M., Lee, S., Lee, S.-G.: Ranking in Context-Aware Recommender System. School of Computer Science and Engineering. ACM, Hyderabad (2011)
10. Adomavicius, G., Tuzhilin, A.: Toward the next generation of recommender systems: a survey of the state-of-the-art and possible extensions. Trans. Knowl. Data Eng. **17**, 734–749 (2005)
11. Zheng Y., Li L., Zheng F.: Context-awareness support for content recommendation in e-learning environments. In: International Conference on Information Management, Innovation Management and Industrial Engineering (2009)
12. Dey, A.K., Abowd, G.D.: Towards a better understanding of context and context-awareness. Graphics, Visualization and Usability Center and College of Computing. Georgia Institute of Technology, Atlanta (1999)
13. Fischer, Gerhard: Context-Aware Systems—The Right Information, at the Right Time, in the Right Place, in the Right Way, to the Rigth Person. ACM University of Colorado, Boulder USA (2012)
14. Adomavicius, G., Tuzhilin, A.: Context-aware recommender systems. In: Racci, F., et al. (eds.) Recommender Systems Handbook, pp. 217–253. Springer, US (2011)
15. Caraciolo, M.: Artificial intelligence in motion. Fecha de consulta: noviembre, 2013. http://aimotion.blogspot.mx/2011/05/evaluating-recommender-systems.html (2013)
16. Shani, G., Gunawardana, A.: Evaluating Recommendation Systems, Microsoft. Springer, New York (2009)

17. Romadhony, A., Al Faraby, S., Pudjoatmodjo, B.: Online shopping recommender system using hybrid method. In: International Conference of Information and Communication Technology (ICoICT) (2013)
18. Devi, M.K.D., Samy, R.T., Kumar, S.V., Venkatesh, P.: Probabilistic neural network approach to alleviate sparsity and cold start problems in collaborative recommender systems. In: International Conference on Computational Intelligence and Computing Research (ICCIC), pp. 1–4 (2010)
19. Chu, C.H., Wu, S.H.: A Chinese restaurant recommendation system based on mobile context-aware services. In: IEEE 14th International Conference on Mobile Data Management. Taichung University, Taichung (2013)
20. Haddad, M.R., Baazaoui, H., Ziou, D., Ben Ghézala, H.: Towards a new model for context-aware recommendation. In: Department of Informatics. University of Sherbrooke, Québec (2012)
21. Baltrunas, L., Ludwig, B., Ricci, F.: Context relevance assessment for recommender systems. In: Proceedings of the International Conference on Intelligent User Interfaces, pp. 287–290 (2011)
22. Ramaswamy, L., et al.: CAESAR: a context-aware, social recommender system for low-end mobile devices. In: IEEE. International Conference on Mobile Data Management: Systems, and Services and Middleware (2009)

Design of Fuzzy Controllers for a Hexapod Robot

Roberto Sepúlveda, Oscar Montiel, Rodolfo Reyes and Josué Domínguez

Abstract The legged robots have emerged by the necessity of vehicles capable of travel and access safely on natural or unstructured terrains, in which vehicles with traditional travel systems (like the wheels) are unable to access, or if they achieve, they move on them with very low efficiency. However, despite the advantages of mobile robots with legs, there are limitations that hinder its use like the control of movement of their legs, the algorithm of locomotion, trajectory tracking and the obstacle avoidance. In our days, a very useful alternative applied to control systems is fuzzy logic; this one is capable of modeling mathematical complex systems. Therefore, fuzzy logic has been becoming popular in control systems for complex and nonlinear plants. The aim of this work is to make algorithms to control the hexapod robot body. The development of these algorithms uses fuzzy logic techniques for controlling the servomotors of the robot. Matlab algorithms are performed to establish a wireless communication using the ZigBee communication protocol, and we use the genetic algorithm toolbox from Matlab to make the control of the hexapod robot body in the "x–y" plane, this is a multi-objective optimization problem due to the stabilization of the robot body in "x" and the stabilization of the robot body in "y".

R. Sepúlveda (✉) · O. Montiel · R. Reyes · J. Domínguez
Instituto Politécnico Nacional, Centro de Investigación y Desarrollo de Tecnología Digital (CITEDI-IPN), Av. Del Parque No. 1310, Mesa de Otay, 22510 Tijuana, BC, México
e-mail: rsepulveda@ipn.mx; rsepulve@citedi.mx

O. Montiel
e-mail: oross@ipn.mx

R. Reyes
e-mail: rreyes@citedi.mx

J. Domínguez
e-mail: jdominguez@citedi.mx

O. Castillo et al. (eds.), *Recent Advances on Hybrid Approaches for Designing Intelligent Systems*, Studies in Computational Intelligence 547,
DOI: 10.1007/978-3-319-05170-3_50, © Springer International Publishing Switzerland 2014

709

1 Introduction

Mobile robots provide the ability to navigate in different terrains and have applications such as mineral exploration, planetary exploration, search and rescue missions, dangerous waste cleanup, automation process, vigilance, terrain recognition, and as mobile platforms incorporating a manipulator arm [1].

Through the years, the conventional control techniques such as PID, although have a great use in industry, the mathematics involved with the design of controllers has been a great disadvantage due to the difficulty adequately modeling many real systems [2].

The traditional approach for control systems requires an a priori model of the system. The quality of the model, it means, the loss of precision in linearization and/or uncertainties in the system parameters affect adversely the quality of the resulting control. In addition, soft computing methods like fuzzy logic possess nonlinear mapping capabilities, it does not require an analytical model and can deal with uncertainties in the system parameters.

Based on the nature of human thought, Lotfi Zadeh proposed the "fuzzy logic" in 1965. Fuzzy logic deals with imprecise or uncertainty problems. The theory of fuzzy sets based on fuzzy logic, proposes that an element belongs to a set with a certain degree. Thus, let us put knowledge in the form of rules, which is feasible to be processed by computers.

In research such as [3–5], a hexapod robot with kinematic chains and control techniques is studied; in [6, 7] a hexapod robot with the ability to walk on uneven terrain using an adaptive controller is presented.

Mănoi-Olaru and Nitulescu in [8] and [9] mention that locomotion in articulated robots is the most difficult task on these types of robots, for the control and coordination of the legs fuzzy logic, artificial neural networks, or central pattern generators, can be used. Sakr and Petriu in [10] proposes a fuzzy logic controller for a hexapod robot that can move on an irregular surface without tipping. The robot detects the surface through contact sensors located at the bottom of each of the six legs. In [11], a set of algorithms that allows a six-legged walking robot performs its movement through actions not established neither periodical using fuzzy logic techniques for making decision is shown.

In [12], is presented a set of fuzzy logic algorithms that allow a hexapod robot to achieve the mobility of their legs using the method known as free locomotion. In [13], is proposed a method to construct fuzzy rules in a fuzzy system applied to the mathematical model of a hexapod robot, the result was that the effort of trial and error can be effectively reduced to generate the rules, the establishment of these has been an important topic in design of this type of systems and their development can be delayed for a long time.

2 Robotics and Soft Computing Techniques

In this work, we present study on a mobile robot articulated with six legs; each of these consists of three rotational joints, i.e., three degrees of freedom; besides analyzing the movements of the robot, a controller was designed to stabilize it in axis "x" and "y" using fuzzy controllers.

There is no consensus on which machines can be considered robots, but there is general agreement among experts and the public on the specific tasks of robots: move around, operate a mechanical arm, sense and manipulate their environment and show intelligent behavior. Currently, it could be considered that a robot is a computer with the capability of movement, in general; it is able to develop multiple tasks according to its programming.

The mobile robots as the name suggest are robots that can move, there are different types of mobile robots (wheeled, water, air, etc.). Among the mobile robot stand out the bipeds, quadrupeds, hexapods, octopods, hybrid robots (legs and wheels) and other legged robots.

The hexapod robots mimic the structure of the limbs and the motion control of arthropod insects or animals that can walk in unstructured terrains with a high probability of success, even if one is lost. These important advantages make it reliable to use these robots for some jobs as field exploration, space exploration, disaster areas, excavations, and many other applications [14].

The hexapod robots can be classified into rectangular and hexagonal as shown in Fig. 1. The rectangular shape is inspired in hexapod insects that have six legs symmetrically distributed along its two sides. The hexagonal hexapods have six legs symmetrically distributed around the body (which may be hexagonal or circular). Typically, individual legs have from two to six degrees of freedom [15].

2.1 Fuzzy Control

Fuzzy logic is considered a generalization of the general theory of sets, which allows elements of a universe to have intermediate degrees of membership in a set by a characteristic function. This idea changed the concept of ambivalence (0 and 1 s) of Boolean logic, which happens to be a special case of fuzzy sets [16].

The theory of fuzzy logic has generated representation models of knowledge; it has also revolutionized the appliance market easily incorporating human expert knowledge in control systems with nonlinear characteristics, thanks to the simple design of fuzzy systems and the high degree of accuracy achieved in the control and decision made with this logic.

Areas of application include, but are not limited to: man–machine communications, medicine, robotics, study and estimation of natural resources, signal and image analysis, control systems, appliances and computers.

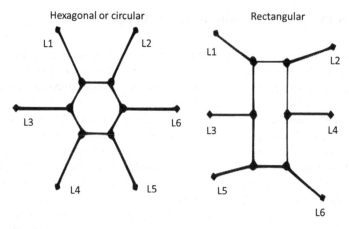

Fig. 1 Classification of the hexapod robot

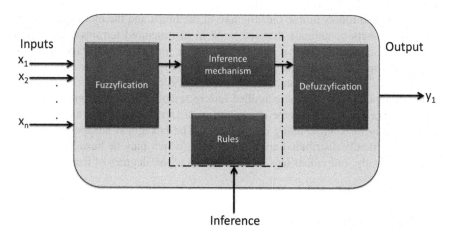

Fig. 2 A fuzzy inference system with "n" inputs and one output

Fuzzy control systems vary according to the nature of the problem to be solved; the control problems can be as complex as the typical of robotics, that require to coordinate many actions, to simple tasks such as maintaining a predefined state for a given variable. These systems are capable of using human knowledge, which is crucial in control problems where it is difficult or sometimes impossible to build accurate mathematical models, a typical fuzzy inference system is shown in Fig. 2 [17].

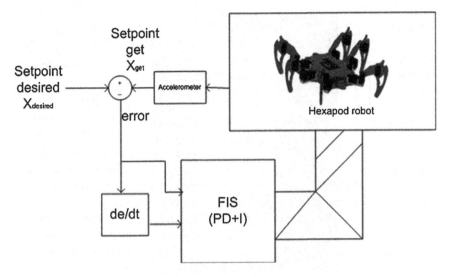

Fig. 3 Block diagram of the FIS to control de position in "x"

3 Design and Implementation of Control Algorithms

In this work, there were developed algorithms to control the position of the body of an hexapod robot with 18° of freedom in the axis "x–y" (where the problem is to solve two targets at once); so, two fuzzy controllers were proposed that are able to control the robot using an accelerometer as the sensor, allowing to know the position of the robot body on the two axis, first a fuzzy controller was designed to control the body of the hexapod robot in the "x" axis and then in "y" axis.

The control algorithms designed were not able to control the position of the robot in two axes simultaneously, so that it was required to use an evolutionary algorithm to provide a solution to this multi-objective optimization problem.

3.1 Fuzzy Control in "x"

The objective is to control the position of the hexapod robot according to a desired position value of the body; the controller must compensate the response and make the necessary changes to the output to reach the set value. To measure the position of the robot an accelerometer is used, its function is to provide feedback to the control system to determine if the controller output has been stabilized or if it is required to fit it.

In Fig. 3, the block diagram of the control system is observed, the system has the entry "$x_{desired}$" (the set point for the position of the robot), which is subtracted from the value of the variable "x_{get}" (the value measured by the accelerometer),

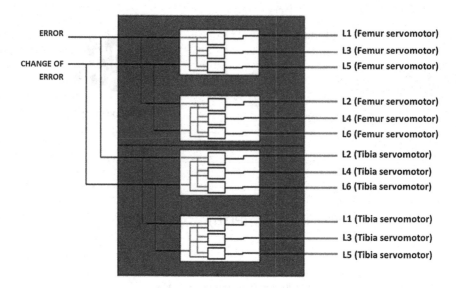

Fig. 4 Fuzzy control with two inputs and twelve outputs

the result is the value of the error, the change of error that are fed back to the FIS, and the 12 outputs of the FIS are connected by 12 servomotors of the six legs of the robot to move it.

The proposed FIS stabilizes the robot body caring always to be stable, this implies, by using the accelerometer, to determine the degree of inclination of the platform. For this, we proposed a fuzzy controller with two inputs and three outputs to control the femur servomotors, and a fuzzy controller with two inputs and three outputs for tibia servo motors to the right-side and two similar controllers to manipulate the femur and tibia servomotors on the left-side. The fuzzy controller handles two inputs with twelve independent outputs each other, as seen in Fig. 4.

The FIS uses five membership functions for controlling femur servos and three membership functions for tibia servos for both the right and to the left-side, with the right-side the legs 1, 3 and 5 are controlled, and the left-side controls the legs 2, 4, 6 of the hexapod. We used five input membership functions for the linguistic variables error, change of error and duty cycle in the controller of the femur and tibia servomotors on all legs (1–6), as it is shown in Fig. 5.

The linguistic variables for the fuzzy controller are shown in Table 1, where the meaning of the prefix used by each of these is determined. In Table 2, the fuzzy rules for the FIS that controls the femur servomotors in all legs are shown. In Table 3 the fuzzy rules for the FIS that controls tibia servomotors of all legs are shown.

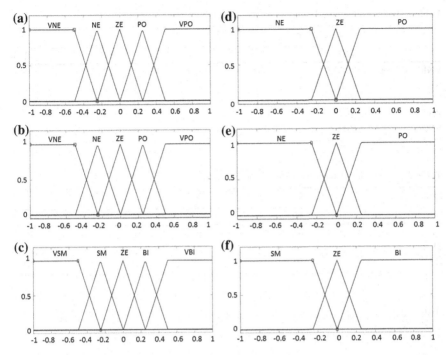

Fig. 5 **a** Position error of the femur servomotor, **b** change of error of the position of the femur servomotor, **c** duty cycle of the femur servomotor, **d** position error of the tibia servomotor, **e** change of error of the position of the tibia servomotor, **f** duty cycle of the tibia servomotor in "x"

Table 1 The linguistic variables for the fuzzy controller in "x"

Input variables		Output variables
Error	Change of error	Duty cycle (PWM)
VPO (Very positive)	VPO (Very positive)	VBI (Very big)
PO (positive)	PO (positive)	BI (big)
CE (cero)	CE (cero)	CE (cero)
NE (negative)	NE (negative)	SM (small)
VNE (very negative)	VNE (very negative)	VSM (very small)

3.2 Fuzzy Control in "y"

The controller to stabilize the platform in the "y" axis is similar to that used in "x", except that accelerometer uses different signals and the universe of discourse of the input and output variables of the FIS are different, as can be seen Fig. 6.

The linguistic values are the same as those in the FIS for "x", the fuzzy rules used in the first iteration of the FIS that control the femur servomotors in the legs 1, 3 and 5 (left-side) is shown in Table 4. Table 5 shows the femur servomotors of

Table 2 Fuzzy rule of the femur servometors in "x"

Error	Change of error				
	VNE	NE	CE	PO	VPO
VNE	VSM	SM	ZE	BI	VBI
NE	SM	SM	ZE	BI	BI
ZE	ZE	ZE	ZE	ZE	ZE
PO	BI	BI	ZE	SM	SM
VPO	VBI	BI	ZE	SM	VSM

Table 3 Fuzzy rule of the tibia servometors in "x"

Error	Change of error		
	NE	ZE	PO
NE	SM	ZE	BI
ZE	ZE	ZE	ZE
PO	BI	ZE	SM

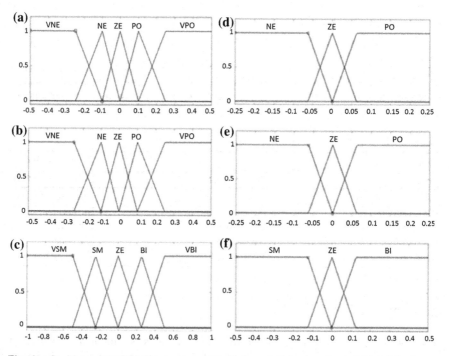

Fig. 6 **a** Position error of the femur servomotor, **b** change of error on the position of the femur servomotor, **c** duty cycle of the femur servomotor, **d** position error of the tibia servomotor, **e** change of error on the position of the tibia servomotor, **f** duty cycle of the tibia servomotor in "y"

Table 4 Fuzzy rules for the femur servomotor of the legs 1, 3 and 5 in "y"

Error	Change of error				
	VNE	NE	CE	PO	VPO
VNE	VSM	VSM	SM	SM	SM
NE	VSM	SM	SM	SM	SM
ZE	ZE	ZE	ZE	ZE	ZE
PO	VBI	BI	BI	BI	ZE
VPO	VBI	VBI	VBI	VBI	VBI

Table 5 Fuzzy rules for the tibia servomotor of the legs 1, 3 and 5 in "y"

Error	Change of error		
	NE	CE	PO
NE	SM	ZE	ZE
CE	SM	ZE	BI
PO	ZE	BI	BI

Table 6 Fuzzy rules for the femur servomotor of the legs 2, 4 and 6 in "y"

Error	Change of error				
	VNE	NE	ZE	PO	VPO
VNE	VBI	VBI	BI	BI	BI
NE	MPO	BI	BI	BI	BI
ZE	ZE	ZE	ZE	ZE	ZE
PO	VSM	SM	SM	SM	ZE
VPO	VSM	VSM	VSM	VSM	VSM

Table 7 Fuzzy rules for the tibia servomotor of the legs 2, 4 and 6 in "y"

Error	Change of error		
	NE	ZE	PO
NE	BI	ZE	ZE
ZE	BI	ZE	SM
PO	ZE	SM	SM

the legs 2, 4 and 6 (right-side), from the second iteration to the nth iteration the legs 3 and 4 are kept static. Tables 6 and 7 show the fuzzy rules for the FIS that control the tibia servomotors in the legs on the left and right-side respectively for the first iteration of the algorithm, from the second to the nth iteration the six servomotors are stationary.

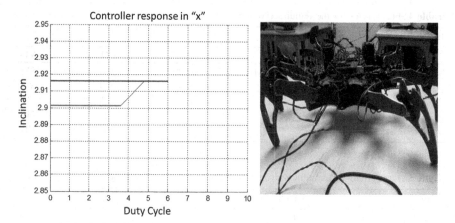

Fig. 7 FIS response in "x"

3.3 Optimization of the Hexapod in "x" and "y"

To stabilize the robot body in the "x" and "y" first the platform is stabilized in the "x" and "y" axis using the controllers described above, the values of these angles are sent wirelessly to Matlab using ZigBee communication, then the error is minimized in the "x" and "y" axis using the multi-objective optimization toolbox of Matlab (the "gamultiobj" function) and return these values wirelessly to the microcontroller to position the robot, the optimized functions are:

$$f_1(\theta_1, \theta_2, \ldots, \theta_{12}) = \sum_{i=1}^{12} (\theta_{i\alpha} - \theta_i)^2$$

$$f_2(\theta_1, \theta_2, \ldots, \theta_{12}) = \sum_{i=1}^{12} (\theta_{i\beta} - \theta_i)^2 \tag{1}$$

where:

$\theta_{i\alpha}$ = stabilization angle in x found by the FIS.
$\theta_{i\beta}$ = stabilization angle in y found by the FIS.
θ_i =
angle to optimize.

4 Results

In Fig. 7, the behavior of the controller is displayed when the robot body is stabilized in the "x" axis in the center position (this happens when the value of the accelerometer is 2.9161 cm), where the blue line represents the set point value and the red line represents the value, of the accelerometer.

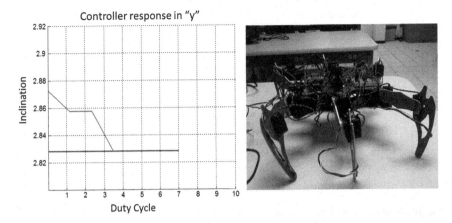

Fig. 8 FIS response in "y"

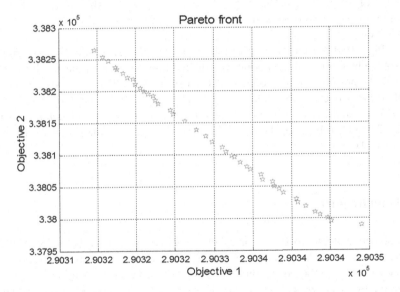

Fig. 9 Pareto front of the system

In Fig. 8, the graph of the behavior of the controller is shown when it is desired that the robot body is stabilized in "y" without this having inclination, i.e., when the accelerometer has a reading of 2.8284 cm, where the blue line represents the set point value and the red one represents the value of the accelerometer.

After obtaining the values of the servomotors when the robot body stabilizes at "x" and then in "y", we used the "gamultiobj" function to find the 12 values of the servomotors that stabilize the robot in "x–y" to a position close to the central position of both axis, during this process the Pareto front is found.

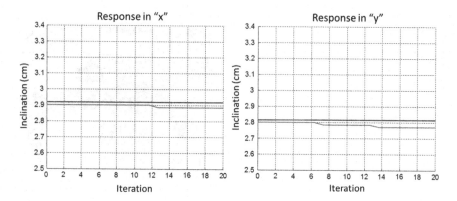

Fig. 10 Final position of the robot in "x" and "y" axis

When the instruction ends the search for better solutions (see Fig. 9), the average of all of them is made and these are sent to the Arduino to change the position of the robot body and subsequently verify the stability between the two objectives.

In Fig. 10, the graph of the position in "x" and the position in "y" are shown, where the blue line is the set point, and the red line the value of the accelerometer. The body position of the robot in "x" does not reach the set point value of 2.9161 cm and the value taken by the accelerometer at the time for "x" is 2.9013 cm, so the error is 0.0148. As in the "x" axis in "y" is not reach the desired value 2.8284 cm because the accelerometer reading is 2.7988 cm, the error produced is 0.0295.

5 Conclusions

The objective of this work is to control the body of the hexapod robot in "x–y" at the same time, to solve this problem it was necessary to divide it into three stages: first a fuzzy system is proposed to control the position of the robot body in "x", in the second a fuzzy system for controlling the robot in "y" was proposed and the last a "gamultiobj" function of Matlab was used to position the robot body in the two axis. The fuzzy controller proposed stabilizes the body of the hexapod robot in the "x" axis and takes to reach the desired value in 1.33 s, the fuzzy controller proposed for the "y" axis takes to reach the desired value in 729 ms, we can say that both controllers have a considerable response. Although fuzzy controllers stabilize the body of the hexapod robot in "x" and "y" respectively, the part of multi-objective optimization only minimizes the difference between the two fuzzy controllers, so at the end the body of the robot has a margin of error and to reduce such error would improve the proposed solution by considering the part of multi-objective optimization. As future work we could try other intelligent approaches, such as in [18–21].

References

1. Melendez, A., Castillo, O.: Hierarchical genetic optimization of the fuzzy integrator for navigation of a mobile robot. In: Soft Computing Applications in Optimization, Control, and Recognition, pp. 77–96 (2013)
2. Gonzalez, J.L., Castillo, O., Aguilar, L.T.: FPGA as a tool for implementing non-fixed structure fuzzy logic controllers. FOCI, pp. 523–530 (2007)
3. Elsevier Ltd.: Design of a hexapod robot with a servo control and a man-machine interface. Rob. Comp. Integr. Manuf. 8 (2011)
4. Kumar, V., Gardner, J.: Kinematics of redundantly actuated closed chains. IEEE Trans. Rob. Autom. 6, 6 (1990)
5. Raibert. M., Sutherland, Maquinas que caminan. Investigacion y ciencia, marzo (1983)
6. Lewinger, W.A.: A hexapod walks over irregular terrain using a controller, adapted from an insect's nervous system. In: IEEE/RSJ International Conference on Intelligent Robots and Systems, vol. 1, p. 6 (2010)
7. Mühlfriedel, T., Berns, K., Dilmanm, R.: hybrid learning concepts for a biologically inspired control of periodic movements for walking machines. In: Soft Computing in Mechatronics, Germany, pp. 38–44, 104–116 (1999)
8. Manoiu-Olaru, S., Nitulescu, M.: Hexapod robot. Mathematical support for modeling and control. IEEE Stoian Viroel, pp. 6 (2011)
9. Mǎnoi-Olaru, S., Nitulescu, M.: Hexapod robot. Virtual models for preliminary studies. IEEE, p. 6 (2011)
10. Sakr, Z., Petriu, E.M.: Hexapod robot locomotion using a fuzzy controller. In: ROSE, IEEE International Workshop on Robotic and Sensors Environments, p. 5 (2007)
11. Cruz, L., Gerrostieta, E., Ever, M.P.: Fuzzy algorithm of free locomotion for a six legged walking robot. IEEE, p. 5 (2007)
12. Yang, J.M., Kim, J.H.: Fault-tolerant locomotion of the hexapod robot. IEEE Trans. Syst. Man Cybern. (1998)
13. Chen, M.C., Lin, S.L., Lin, N.: A fuzzy controller for hexapod robot with modified fuzzy identification. Taichung, Taiwan 407, R. O. C, IEEE, p. 7 (1996)
14. Campos, R.: Hexapod Locomotion: A Nonlinear Dynamical Systems Approach. Industrial Electronics Department University of Minho, Guimaraes, Portugal, pp. 1–6
15. Ding, X.: Locomotion analysis of hexapod robot. In: Climbing and Walking Robots, Chap. 18, InTech. Beihang University Politecnico di Milano China, Italy (2010)
16. Yurkovich, S., Passino, K.M.: Fuzzy Control, Chap. 2. Addison-Wesley Longman Inc, pp. 23–50 (1998)
17. Montiel, O., Sepúlveda, R., Castillo, O., Melin, P.: Fundamentos de logica difusa, 1st edn. ILCSA, pp. 38–44, 104–116, agosto 2002
18. Montiel, O., Sepúlveda, R., Melin, P., Castillo, O., Porta, M., Meza, I.: Performance of a simple tuned fuzzy controller and a PID controller on a DC motor. FOCI, pp. 531–537 (2007)
19. Melin, P., Castillo, O.: Adaptive intelligent control of aircraft systems with a hybrid approach combining neural networks, fuzzy logic and fractal theory. Appl. Soft Comput. 3(4), 353–362 (2003)
20. Aguilar, L., Melin, P., Castillo, O.: Intelligent control of a stepping motor drive using a hybrid neuro-fuzzy ANFIS approach. J. Appl. Soft Comput. 3(3), 209–219 (2003)
21. Leal-Ramirez, C., Castillo, O., Melin, P., Rodriguez-Diaz, A.: Simulation of the bird age-structured population growth based on an interval type-2 fuzzy cellular structure. Inf. Sci. 181, 519–535 (2011)

Printed in the United States
By Bookmasters